Texts in Applied Mathematics 39

Springer
New York
Berlin
Heidelberg
Barcelona
Hong Kong
London
Milan
Paris
Singapore
Tokyo

Texts in Applied Mathematics

(continued after index)

Kendall Atkinson   Weimin Han

Theoretical Numerical Analysis

A Functional Analysis Framework

With 25 Illustrations

 Springer

Kendall Atkinson
Department of Mathematics
Department of Computer Science
University of Iowa
Iowa City, IA 52242
USA
Kendall-Atkinson@uiowa.edu

Weimin Han
Department of Mathematics
University of Iowa
Iowa City, IA 52242
USA
whan@math.uiowa.edu
and
Department of Mathematics
Zhejiang University
Hangzhou
People's Republic of China

Series Editors

J.E. Marsden
Control and Dynamical Systems, 107-81
California Institute of Technology
Pasadena, CA 91125
USA

L. Sirovich
Division of Applied Mathematics
Brown University
Providence, RI 02912
USA

M. Golubitsky
Department of Mathematics
University of Houston
Houston, TX 77204-3476
USA

Mathematics Subject Classification (2000): 65-01, 73V05, 45L05, 4601

Library of Congress Cataloging-in-Publication Data
Atkinson, Kendall E.
    Theoretical numerical analysis: a functional analysis framework
/ Kendall Atkinson, Weimin Han.
      p. cm.—(Texts in applied mathematics; 39)
    Includes bibliographical references and index.
    ISBN 0-387-95142-3 (alk. paper)
    1. Functional analysis.   I. Han, Weimin.   II. Title.   III Series.
QA320.A85 2001
515—dc21   00–061920

Printed on acid-free paper.

Production managed by Michael Koy; manufacturing supervised by Joe Quatela.
Typeset by The Bartlett Press, Inc., Marietta, GA.
Printed and bound by Maple-Vail Book Manufacturing Group, York, PA.
Printed in the United States of America.

9 8 7 6 5 4 3 2 1

ISBN 0-387-95142-3   SPIN 10780644

Springer-Verlag   New York Berlin Heidelberg
A member of BertelsmannSpringer Science+ Business Media GmbH

Dedicated to

DAISY AND CLYDE ATKINSON
HAZEL AND WRAY FLEMING

and

DAQING HAN, SUZHEN QIN
HUIDI TANG, ELIZABETH JING HAN

Series Preface

Mathematics is playing an ever more important role in the physical and biological sciences, provoking a blurring of boundaries between scientific disciplines and a resurgence of interest in the modern as well as the classical techniques of applied mathematics. This renewal of interest, both in research and teaching, has led to the establishment of the series: *Texts in Applied Mathematics* (*TAM*).

The development of new courses is a natural consequence of a high level of excitement on the research frontier as newer techniques, such as numerical and symbolic computer systems, dynamical systems, and chaos, mix with and reinforce the traditional methods of applied mathematics. Thus, the purpose of this textbook series is to meet the current and future needs of these advances and encourage the teaching of new courses.

TAM will publish textbooks suitable for use in advanced undergraduate and beginning graduate courses, and will complement the *Applied Mathematical Sciences* (*AMS*) series, which will focus on advanced textbooks and research level monographs.

California Institute of Technology   J.E. Marsden
Brown University   L. Sirovich
University of Houston   M. Golubitsky

Preface

This textbook has grown out of a course that we teach periodically at the University of Iowa. We have beginning graduate students in mathematics who wish to work in numerical analysis from a theoretical perspective, and they need a background in those "tools of the trade" that we cover in this text. Ordinarily, such students would begin with a one-year course in *real and complex analysis*, followed by a one- or two-semester course in *functional analysis* and possibly a graduate level course in *ordinary differential equations, partial differential equations,* or *integral equations*. We still expect our students to take most of these standard courses, but we also want to move them more rapidly into a research program. The course based on this book is designed to facilitate this goal.

The textbook covers basic results of functional analysis and also some additional topics that are needed in theoretical numerical analysis. Applications of this functional analysis are given by considering, at length, numerical methods for solving partial differential equations and integral equations.

The material in the text is presented in a mixed manner. Some topics are treated with complete rigor, whereas others are simply presented without proof and perhaps illustrated (e.g., the principle of uniform boundedness). We have chosen to avoid introducing a formalized framework for *Lebesgue measure and integration* and also for *distribution theory*. Instead we use standard results on the completion of normed spaces and the unique extension of densely defined bounded linear operators. This permits us to introduce the Lebesgue spaces formally and without their concrete realization using measure theory. The weak derivative can be introduced similarly using the unique extension of densely defined linear operators, avoiding the need for a formal development of distribution theory. We describe some of the standard material on measure theory and distribution theory in an intuitive manner, believing this is sufficient for much of subsequent mathematical development. In addition, we give a number of deeper results without proof, citing the existing literature. Examples of this are the *open mapping theorem*, the *Hahn-Banach theorem*, the *principle of uniform boundedness*, and a number of the results on *Sobolev spaces*.

The choice of topics has been shaped by our research program and interests at the University of Iowa. These topics are important elsewhere, and we believe this text will be useful to students at other universities as well.

The book is divided into chapters, sections, and subsections where appropriate. Mathematical relations (equalities and inequalities) are numbered by chapter, section, and their order of occurrence. For example, (1.2.3) is the third-numbered mathematical relation in Section 1.2 of Chapter 1. Definitions, examples, theorems, lemmas, propositions, corollaries, and remarks are numbered consecutively within each section, by chapter and section. For example, in Section 1.1, Definition 1.1.1 is followed by Example 1.1.2.

The first three chapters cover basic results of functional analysis and approximation theory that are needed in theoretical numerical analysis. Early on, in Chapter 4, we introduce methods of nonlinear analysis, as students should begin early to think about both linear and nonlinear problems. Chapter 5 is a short introduction to finite difference methods for solving time-dependent problems. Chapter 6 is an introduction to Sobolev spaces, giving different perspectives of them. Chapters 7 through 10 cover material related to elliptic boundary value problems and variational inequalities. Chapter 11 is a general introduction to numerical methods for solving integral equations of the second kind, and Chapter 12 gives an introduction to boundary integral equations for planar regions with a smooth boundary curve.

We give exercises at the end of most sections. The exercises are numbered consecutively by chapter and section. At the end of each chapter, we provide some short discussions of the literature, including recommendations for additional reading.

During the preparation of the book, we received helpful suggestions from numerous colleagues and friends. We particularly thank P.G. Ciarlet, William A. Kirk, Wenbin Liu, and David Stewart. We also thank the anonymous referees whose suggestions led to an improvement of the book.

Kendall Atkinson
Weimin Han

Iowa City, IA

Contents

1

Linear Spaces

Linear (or vector) spaces are the standard setting for studying and solving a large proportion of the problems in differential and integral equations, approximation theory, optimization theory, and other topics in applied mathematics. In this chapter, we gather together some concepts and results concerning various aspects of linear spaces, especially some of the more important linear spaces such as Banach spaces, Hilbert spaces, and certain other function spaces that are used frequently in this work and in applied mathematics generally.

1.1 Linear spaces

A linear space is a set of elements equipped with two binary operations, called vector addition and scalar multiplication, in such a way that the operations behave linearly.

Definition 1.1.1 *Let V be a set of objects, to be called* vectors; *and let $\mathbb{K}$ be a set of scalars, either $\mathbb{R}$ the set of real numbers, or $\mathbb{C}$ the set of complex numbers. Assume there are two operations: $(u, v) \mapsto u + v \in V$ and $(\alpha, v) \mapsto \alpha v \in V$, called* addition *and* scalar multiplication, *respectively, defined for any $u, v \in V$ and any $\alpha \in \mathbb{K}$. These operations are to satisfy the following rules.*

1. *$u + v = v + u$ for any $u, v \in V$ (commutative law);*

2. *$(u + v) + w = u + (v + w)$ for any $u, v, w \in V$ (associative law);*

3. there is an element $0 \in V$ such that $0+u = u$ for any $u \in V$ (existence of the zero element);

4. for any $u \in V$, there is an element $-u \in V$ such that $u + (-u) = 0$ (existence of negative elements);

5. $1u = u$ for any $u \in V$;

6. $\alpha(\beta u) = (\alpha\beta)u$ for any $u \in V$, any $\alpha, \beta \in \mathbb{K}$ (associative law for scalar multiplication);

7. $\alpha(u + v) = \alpha u + \alpha v$ and $(\alpha + \beta)u = \alpha u + \beta u$ for any $u, v \in V$, and any $\alpha, \beta \in \mathbb{K}$ (distributive laws).

Then V is called a linear space, or a vector space.

When $\mathbb{K}$ is the set of the real numbers, V is a real linear space; and when $\mathbb{K}$ is the set of the complex numbers, V becomes a complex linear space. In this work, most of the time we only deal with real linear spaces. So when we say V is a linear space, the reader should usually assume V is a real linear space, unless explicitly stated otherwise.

Some remarks are in order concerning the definition of a linear space. From the commutative law and the associative law, we observe that to add several elements, the order of summation does not matter, and it does not cause any ambiguity to write expressions such as $u + v + w$ or $\sum_{i=1}^{n} u_i$. By using the commutative law and the associative law, it is not difficult to verify that the zero element and the negative element $(-u)$ of a given element $u \in V$ are unique, and they can be equivalently defined through the relations $v + 0 = v$ for any $v \in V$, and $(-u) + u = 0$. Below, we write $u - v$ for $u + (-v)$.

Example 1.1.2 (a) *The set of the real numbers $\mathbb{R}$ is a real linear space when the addition and scalar multiplication are the usual addition and multiplication. Similarly, the set of complex numbers $\mathbb{C}$ is a complex linear space.*

(b) *Let d be a positive integer. The letter d is used generally in this work for the spatial dimension. The set of all vectors with d real components, with the usual vector addition and scalar multiplication, forms a linear space $\mathbb{R}^d$. A typical element in $\mathbb{R}^d$ can be expressed as $\mathbf{x} = (x_1, \dots, x_d)^T$, where $x_1, \dots, x_d \in \mathbb{R}$. Similarly, $\mathbb{C}^d$ is a complex linear space.*

(c) *Let $\Omega \subseteq \mathbb{R}^d$ be an open subset of $\mathbb{R}^d$. In this work, the symbol Ω always stands for an open subset of $\mathbb{R}^d$. The set of all the continuous functions on Ω forms a linear space $C(\Omega)$, under the usual addition and scalar multiplication of functions: For $f, g \in C(\Omega)$, the function $f + g$ defined by*

$$(f + g)(\mathbf{x}) = f(\mathbf{x}) + g(\mathbf{x}) \qquad \mathbf{x} \in \Omega,$$

belongs to $C(\Omega)$; so does the scalar multiplication function αf defined through

$$(\alpha f)(\mathbf{x}) = \alpha f(\mathbf{x}) \qquad \mathbf{x} \in \Omega.$$

Similarly, $C(\overline{\Omega})$ denotes the space of continuous functions on the closed set $\overline{\Omega}$. Clearly, $C(\overline{\Omega}) \subseteq C(\Omega)$.

(d) *A related function space is $C(D)$, containing all functions $f : D \rightarrow \mathbb{K}$ that are continuous on a general set $D \subseteq \mathbb{R}^d$. The arbitrary set D can be an open or closed set in $\mathbb{R}^d$, or perhaps neither; and it can be a lower dimensional set such as a portion of the boundary of an open set in $\mathbb{R}^d$. When D is a closed and bounded subset of $\mathbb{R}^d$, a function from the space $C(D)$ is necessarily bounded.*

(e) *For any non-negative integer m, we may define the space $C^m(\Omega)$ as the space of all the functions that together with their derivatives of orders up to m are continuous on Ω. We may also define the space $C^m(\overline{\Omega})$ to be the space of all the functions that together with their derivatives of orders up to m are continuous on $\overline{\Omega}$. These function spaces are discussed at length in Section 1.4.*

(f) *The space of continuous 2π-periodic functions is denoted by $C_p(2\pi)$. It is the set of all $f \in C(-\infty, \infty)$ for which*

$$f(x + 2\pi) = f(x) \qquad -\infty < x < \infty.$$

For an integer $k \geq 0$, the space $C_p^k(2\pi)$ denotes the set of all functions in $C_p(2\pi)$ that have k continuous derivatives on $(-\infty, \infty)$. We usually write $C_p^0(2\pi)$ as simply $C_p(2\pi)$. These spaces are used in connection with problems in which periodicity plays a major role.

Definition 1.1.3 *A subspace W of the linear space V is a subset of V that is closed under the addition and scalar multiplication operations of V, i.e., for any $u, v \in W$ and any $\alpha \in \mathbb{K}$, we have $u + v \in W$ and $\alpha v \in W$.*

It can be verified that W itself, equipped with the addition and scalar multiplication operations of V, is a linear space.

Example 1.1.4 *In the linear space $\mathbb{R}^3$,*

$$W = \{\mathbf{x} = (x_1, x_2, 0)^T \mid x_1, x_2 \in \mathbb{R}\}$$

is a subspace, consisting of all the vectors on the x_1x_2-plane. In contrast,

$$\widehat{W} = \{\mathbf{x} = (x_1, x_2, 1)^T \mid x_1, x_2 \in \mathbb{R}\}$$

is not a subspace. Nevertheless, we observe that $\widehat{W}$ is a translation of the subspace W,

$$\widehat{W} = \mathbf{x}_0 + W,$$

where $\mathbf{x}_0 = (0, 0, 1)^T$. The set $\widehat{W}$ is an example of an affine set.

Given vectors $v_1, \ldots, v_n \in V$ and scalars $\alpha_1, \ldots, \alpha_n \in \mathbb{K}$, we call

$$\sum_{i=1}^{n} \alpha_i v_i = \alpha_1 v_1 + \cdots + \alpha_n v_n$$

a *linear combination* of $v_1, \ldots, v_n$. It is meaningful to remove "redundant" vectors from the linear combination. Thus we introduce the concepts of linear dependence and independence.

Definition 1.1.5 *We say $v_1, \ldots, v_n \in V$ are* linearly dependent *if there are scalars $\alpha_i \in \mathbb{K}$, $1 \leq i \leq n$, with at least one α_i non-zero such that*

$$\sum_{i=1}^{n} \alpha_i v_i = 0. \qquad (1.1.1)$$

We say $v_1, \ldots, v_n \in V$ are linearly independent *if they are not linearly dependent, meaning that the only choice of scalars $\{\alpha_i\}$ for which (1.1.1) is valid is $\alpha_i = 0$ for $i = 1, 2, \ldots, n$.*

We observe that $v_1, \ldots, v_n$ are linearly dependent if and only if at least one of the vectors can be expressed as a linear combination of the rest of the vectors. In particular, a set of vectors containing the zero element is always linearly dependent. Similarly, $v_1, \ldots, v_n$ are linearly independent if and only if none of the vectors can be expressed as a linear combination of the rest of the vectors; in other words, none of the vectors is "redundant."

Example 1.1.6 *In $\mathbb{R}^d$, d vectors $\mathbf{x}^{(i)} = (x_1^{(i)}, \ldots, x_d^{(i)})^T$, $1 \leq i \leq d$, are linearly independent if and only if the determinant*

$$\begin{vmatrix} x_1^{(1)} & \cdots & x_1^{(d)} \\ \vdots & \ddots & \vdots \\ x_d^{(1)} & \cdots & x_d^{(d)} \end{vmatrix}$$

is non-zero. This follows from a standard result in linear algebra. The condition (1.1.1) is equivalent to a homogeneous system of linear equations, and a standard result of linear algebra says that this system has $(0, \ldots, 0)^T$ as its only solution if and only if the above determinant is non-zero.

Example 1.1.7 *Within the space $C[0, 1]$, the vectors $1, x, x^2, \ldots, x^n$ are linearly independent. This can be proven in several ways. Assuming*

$$\sum_{j=0}^{n} \alpha_j x^j = 0, \qquad 0 \leq x \leq 1,$$

we can form its first n derivatives. Setting $x = 0$ in this polynomial and its derivatives will lead to $\alpha_j = 0$ for $j = 0, 1, \ldots, n$.

Definition 1.1.8 *The span of* $v_1, \ldots, v_n \in V$ *is defined to be the set of all the linear combinations of these vectors:*

$$\text{span}\{v_1, \ldots, v_n\} = \left\{ \sum_{i=1}^{n} \alpha_i v_i \;\middle|\; \alpha_i \in \mathbb{K}, \; 1 \le i \le n \right\}.$$

Evidently, $\text{span}\{v_1, \ldots, v_n\}$ is a linear subspace of V. Most of the time, we apply this definition for the case where $v_1, \ldots, v_n$ are linearly independent.

Definition 1.1.9 *A linear space V is said to be* finite dimensional *if there exists a finite maximal set of independent vectors* $\{v_1, \ldots, v_n\}$*; i.e., the set* $\{v_1, \ldots, v_n\}$ *is linearly independent, but* $\{v_1, \ldots, v_n, v_{n+1}\}$ *is linearly dependent for any* $v_{n+1} \in V$*. The set* $\{v_1, \ldots, v_n\}$ *is called a* basis *of the space. If such a finite basis for V does not exist, then V is said to be* infinite dimensional.

We see that a basis is a set of independent vectors such that any vector in the space can be written as a linear combination of them. Obviously a basis is not unique, yet we have the following important result.

Theorem 1.1.10 *For a finite-dimensional linear space, every basis for V contains exactly the same number of vectors. This number is called the* dimension *of the space.*

A proof of this result can be found in most introductory textbooks on linear algebra; for example, see [3, Section 5.4].

Example 1.1.11 *The space $\mathbb{R}^d$ is d-dimensional. There are infinitely many possible choices for a basis of the space. A canonical basis for this space is* $\{\mathbf{e}_i = (0, \ldots, 0, 1_i, 0, \ldots, 0)^T\}_{i=1}^{d}$ *in which the single 1 is in component i.*

We introduce the concept of a linear function.

Definition 1.1.12 *Let L be a function from the linear space V to the linear space W. We say L is a* linear function *if*
(a) *for all $u, v \in V$,*

$$L(u + v) = L(u) + L(v);$$

(b) *for all $v \in V$ and all $\alpha \in \mathbb{K}$,*

$$L(\alpha v) = \alpha L(v).$$

For such a linear function, we often write

$$L(v) = Lv, \qquad v \in V.$$

This definition is extended and discussed extensively in Chapter 2. Other common notations are *linear mapping*, *linear operator*, and *linear transformation*.

Definition 1.1.13 *Two linear spaces U and V are said to be* isomorphic *if there is a linear bijective (i.e., one-to-one and onto) function $l : U \to V$.*

Many properties of a linear space U hold for any other linear space V that is isomorphic to U; and then the explicit contents of the space do not matter in the analysis of these properties. This usually proves to be convenient. One such example is that if U and V are isomorphic and are finite dimensional, then their dimensions are equal, a basis of V can be obtained from a basis of U by applying the mapping l, and a basis of U can be obtained from a basis of V by applying the inverse mapping of l.

Example 1.1.14 *The set $\mathcal{P}_k$ of all polynomials of degree less than or equal to k is a subspace of continuous function space $C[0, 1]$. An element in the space $\mathcal{P}_k$ has the form $a_0 + a_1 x + \cdots + a_k x^k$. The mapping $l : a_0 + a_1 x + \cdots + a_k x^k \mapsto (a_0, a_1, \dots, a_k)^T$ is bijective from $\mathcal{P}_k$ to $\mathbb{R}^{k+1}$. Thus, $\mathcal{P}_k$ is isomorphic to $\mathbb{R}^{k+1}$.*

Definition 1.1.15 *Let U and V be two linear spaces. The Cartesian product of the spaces, $W = U \times V$, is defined by*

$$W = \{w = (u, v) \mid u \in U, \ v \in V\}$$

endowed with componentwise addition and scalar multiplication

$$(u_1, v_1) + (u_2, v_2) = (u_1 + u_2, v_1 + v_2) \qquad \forall\, (u_1, v_1), (u_2, v_2) \in W,$$
$$\alpha\,(u, v) = (\alpha\,u, \alpha\,v) \qquad \forall\, (u, v) \in W, \ \forall\, \alpha \in \mathbb{K}.$$

It is easy to verify that W is a linear space. The definition can be extended in a straightforward way for the Cartesian product of any finite number of linear spaces.

Example 1.1.16 *The real plane can be viewed as the Cartesian product of two real lines: $\mathbb{R}^2 = \mathbb{R} \times \mathbb{R}$. In general,*

$$\mathbb{R}^d = \underbrace{\mathbb{R} \times \cdots \times \mathbb{R}}_{d \ times}.$$

Exercise 1.1.1 *Show that the set of all continuous solutions of the differential equation $u''(x) + u(x) = 0$ is a finite-dimensional linear space. Is the set of all continuous solutions of $u''(x) + u(x) = 1$ a linear space?*

Exercise 1.1.2 *When is the set $\{u \in C[0, 1] \mid u(0) = a\}$ a linear space?*

Exercise 1.1.3 *Show that in any linear space V, a set of vectors is always linearly dependent if one of the vectors is zero.*

Exercise 1.1.4 *Assume U and V are finite-dimensional linear spaces, and let $\{u_1, \dots, u_n\}$ and $\{v_1, \dots, v_m\}$ be bases for them, respectively. Using these bases, create a basis for $W = U \times V$.*

1.2   Normed spaces

In numerical analysis, we frequently need to examine the closeness of a numerical solution to the exact solution. To answer the question quantitatively, we need to have a measure on the magnitude of the difference between the numerical solution and the exact solution. A norm of a vector in a linear space provides such a measure.

Definition 1.2.1 *Given a linear space V, a norm $\|\cdot\|$ is a function from V to $\mathbb{R}$ with the following properties.*

1. *$\|v\| \geq 0$ for any $v \in V$, and $\|v\| = 0$ if and only if $v = 0$;*

2. *$\|\alpha v\| = |\alpha|\,\|v\|$ for any $v \in V$ and $\alpha \in \mathbb{K}$;*

3. *$\|u + v\| \leq \|u\| + \|v\|$ for any $u, v \in V$.*

The space V equipped with the norm $\|\cdot\|$ is called a normed linear space *or a* normed space. *We usually say V is a normed space when the definition of the norm is clear from the context.*

Some remarks are in order on the definition of a norm. The three axioms in the definition mimic the principal properties of the notion of the ordinary length of a vector in $\mathbb{R}^2$ or $\mathbb{R}^3$. The first axiom says the norm of any vector must be non-negative, and the only vector with zero norm is zero. The second axiom is usually called *positive homogeneity*. The third axiom is also called the *triangle inequality*, which is a direct extension of the triangle inequality on the plane: The length of any side of a triangle is not greater than the sum of the lengths of the other two sides. With the definition of a norm, we can use the quantity $\|u - v\|$ as a measure for the distance between u and v.

Definition 1.2.2 *Given a linear space V, a semi-norm $|\cdot|$ is a function from V to $\mathbb{R}$ with the properties of a norm except that $|v| = 0$ does not necessarily imply $v = 0$.*

One place in this work where the notion of a semi-norm plays an important role is in estimating the error of polynomial interpolation.

Example 1.2.3 (a) *For $\mathbf{x} = (x_1, \dots, x_d)^T$, the formula*

$$\|\mathbf{x}\|_2 = \left(\sum_{i=1}^{d} x_i^2 \right)^{1/2} \tag{1.2.1}$$

defines a norm in the space $\mathbb{R}^d$ (cf. Exercise 1.2.5), called the Euclidean norm, *which is the usual norm for the space $\mathbb{R}^d$. When $d = 1$, the norm coincides with the absolute value: $\|x\|_2 = |x|$ for $x \in \mathbb{R}$.*

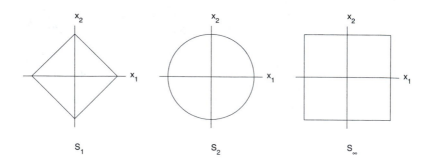

Figure 1.1. The unit ball $S_p = \left\{ x \in \mathbb{R}^2 : \|x\|_p \leq 1 \right\}$ for $p = 1, 2, \infty$

(b) *More generally, for $1 \leq p \leq \infty$, the formulas*

$$\|\mathbf{x}\|_p = \left(\sum_{i=1}^{d} |x_i|^p \right)^{1/p} \qquad \text{for } 1 \leq p < \infty, \tag{1.2.2}$$

$$\|\mathbf{x}\|_\infty = \max_{1 \leq i \leq d} |x_i| \tag{1.2.3}$$

define norms in the space $\mathbb{R}^d$ (cf. Exercise 1.2.5). The norm $\|\cdot\|_p$ is called the p-norm, and $\|\cdot\|_\infty$ is called the maximum or infinity norm. It is straightforward to show that

$$\|\mathbf{x}\|_\infty = \lim_{p \to \infty} \|\mathbf{x}\|_p$$

by using the inequality (1.2.5) given below. Again, when $d = 1$, all these norms coincide with the absolute value: $\|x\|_p = |x|$, $x \in \mathbb{R}$. Over $\mathbb{R}^d$, the most commonly used norms are $\|\cdot\|_p$, $p = 1, 2, \infty$. The unit ball in $\mathbb{R}^2$ for each of these norms is shown in Figure 1.1.

Example 1.2.4 (a) *The standard norm for $C[a, b]$ is the maximum norm*

$$\|f\|_\infty = \max_{a \leq x \leq b} |f(x)|, \qquad f \in C[a, b].$$

This is also the norm for $C_p(2\pi)$ (with $a = 0$ and $b = 2\pi$), the space of continuous 2π-periodic functions introduced in Example 1.1.2(f).

(b) *For an integer $k > 0$, the standard norm for $C^k[a, b]$ is*

$$\|f\|_{k,\infty} = \max_{0 \le j \le k} \|f^{(j)}\|_{\infty}, \qquad f \in C^k[a, b].$$

This is also the standard norm for $C_p^k(2\pi)$.

With the use of a norm for V we can introduce a topology for V, a set of *open* and *closed* sets for V.

Definition 1.2.5 *Let $(V, \|\cdot\|)$ be a normed space. Given $v_0 \in V$ and $r > 0$, the sets*

$$B(v_0, r) = \{v \in V \mid \|v - v_0\| < r\},$$
$$\overline{B}(v_0, r) = \{v \in V \mid \|v - v_0\| \le r\}$$

are called the open and closed balls centered at v_0 with radius r. When $r = 1$ and $v_0 = 0$, we have unit balls.

Definition 1.2.6 *Let $A \subseteq V$, a normed linear space. The set A is* open *if for every $v \in A$, there is a radius $r > 0$ such that $B(v, r) \subseteq A$. The set A is* closed *in V if its complement $V - A$ is open in V.*

1.2.1   Convergence

With the notion of a norm at our disposal, we can define the important concept of convergence.

Definition 1.2.7 *Let V be a normed space with the norm $\|\cdot\|$. A sequence $\{u_n\} \subseteq V$ is* convergent *to $u \in V$ if*

$$\lim_{n \to \infty} \|u_n - u\| = 0.$$

We say that u is the limit *of the sequence $\{u_n\}$, and write $u_n \to u$ as $n \to \infty$, or $\lim_{n \to \infty} u_n = u$.*

It can be verified that any sequence can have at most one limit.

Definition 1.2.8 *A function $f : V \to \mathbb{R}$ is said to be* continuous *at $u \in V$ if for any sequence $\{u_n\}$ with $u_n \to u$, we have $f(u_n) \to f(u)$ as $n \to \infty$. The function f is said to be* continuous *on V if it is continuous at every $u \in V$.*

Proposition 1.2.9 *The norm function $\|\cdot\|$ is continuous.*

Proof. We need to show that if $u_n \to u$, then $\|u_n\| \to \|u\|$. This follows from the *backward triangle inequality*

$$|\|u\| - \|v\|| \le \|u - v\| \qquad \forall\, u, v \in V \tag{1.2.4}$$

derived from the triangle inequality.   ∎

We have seen that on a linear space various norms can be defined. Different norms give different measures of size for a given vector in the space. Consequently, different norms may give rise to different forms of convergence.

Definition 1.2.10 *We say two norms* $\|\cdot\|_{(1)}$ *and* $\|\cdot\|_{(2)}$ *are equivalent if there exist positive constants* c_1, c_2 *such that*

$$c_1 \|u\|_{(1)} \leq \|u\|_{(2)} \leq c_2 \|u\|_{(1)} \qquad \forall u \in V.$$

With two such equivalent norms, a sequence $\{u_n\}$ converges in one norm if and only if it converges in the other norm:

$$\lim_{n \to \infty} \|u_n - u\|_{(1)} = 0 \quad \Longleftrightarrow \quad \lim_{n \to \infty} \|u_n - u\|_{(2)} = 0.$$

Example 1.2.11 *For the norms* (1.2.2)–(1.2.3) *on* $\mathbb{R}^d$, *it is straightforward to show*

$$\|\mathbf{x}\|_\infty \leq \|\mathbf{x}\|_p \leq d^{1/p} \|\mathbf{x}\|_\infty \qquad \forall \mathbf{x} \in \mathbb{R}^d. \tag{1.2.5}$$

Thus all the norms $\|\mathbf{x}\|_p$, $1 \leq p \leq \infty$, *on* $\mathbb{R}^d$ *are equivalent.*

More generally, we have the following well-known result. For a proof, see [11, p. 483].

Theorem 1.2.12 *Over a finite-dimensional space, any two norms are equivalent.*

Thus, on a finite-dimensional space, different norms lead to the same convergence notion. Over an infinite-dimensional space, however, such a statement is no longer valid.

Example 1.2.13 *Let* V *be the space of all bounded functions on* $[0,1]$. *For* $u \in V$, *in analogy with Example 1.2.3, we may define the following norms:*

$$\|u\|_p = \left\{ \int_0^1 |u(x)|^p dx \right\}^{1/p}, \qquad 1 \leq p < \infty, \tag{1.2.6}$$

$$\|u\|_\infty = \sup_{0 \leq x \leq 1} |u(x)|. \tag{1.2.7}$$

Now consider a sequence of functions $\{u_n\} \subseteq V$, *defined by*

$$u_n(x) = \begin{cases} 1 - nx, & 0 \leq x \leq \dfrac{1}{n}, \\ 0, & \dfrac{1}{n} < x \leq 1. \end{cases}$$

It is easy to show that

$$\|u_n\|_p = [n(p+1)]^{-1/p}, \qquad 1 \leq p < \infty.$$

Thus we see that the sequence $\{u_n\}$ *converges to* $u = 0$ *in the norm* $\|\cdot\|_p$, $1 \leq p < \infty$. *On the other hand,*

$$\|u_n\|_\infty = 1, \qquad n \geq 1,$$

so $\{u_n\}$ *does not converge to* $u = 0$ *in the norm* $\|\cdot\|_\infty$.

As we have seen in the last example, in an infinite-dimensional space, some norms are not equivalent. Convergence defined by one norm can be stronger than that by another.

Example 1.2.14 *Consider again the space of all bounded functions on* $[0, 1]$, *and the family of norms* $\| \cdot \|_p$, $1 \leq p < \infty$, *and* $\| \cdot \|_\infty$. *We have, for any* $p \in [1, \infty)$,

$$\|u\|_p \leq \|u\|_\infty, \qquad u \in V.$$

Thus, convergence in $\| \cdot \|_\infty$ *implies convergence in* $\| \cdot \|_p$, $1 \leq p < \infty$, *but not conversely. Convergence in* $\| \cdot \|_\infty$ *is usually called* uniform convergence.

1.2.2 Banach spaces

The concept of a normed space is usually too general, and special attention is given to a particular type of normed space called a *Banach space*.

Definition 1.2.15 *Let* V *be a normed space. A sequence* $\{u_n\} \subseteq V$ *is called a* Cauchy sequence *if*

$$\lim_{m,n \to \infty} \|u_m - u_n\| = 0.$$

Obviously, a convergent sequence is a Cauchy sequence. In the finite-dimensional space $\mathbb{R}^d$, any Cauchy sequence is convergent. However, in a general infinite-dimensional space, a Cauchy sequence may fail to converge; see Example 1.2.18 below.

Definition 1.2.16 *A normed space is said to be* complete *if every Cauchy sequence from the space converges to an element in the space. A complete normed space is called a* Banach space.

Example 1.2.17 *Let* $\Omega \subseteq \mathbb{R}^d$ *be a bounded open set. For* $v \in C(\overline{\Omega})$ *and* $1 \leq p < \infty$, *define the* p-norm

$$\|v\|_p = \left(\int_\Omega |v(\mathbf{x})|^p \, dx \right)^{1/p}. \qquad (1.2.8)$$

Here, $\mathbf{x} = (x_1, \ldots, x_d)^T$ *and* $dx = dx_1 dx_2 \cdots dx_d$. *In addition, define the* ∞-norm *or* maximum norm

$$\|v\|_\infty = \max_{\mathbf{x} \in \overline{\Omega}} |v(\mathbf{x})|.$$

The space $C(\overline{\Omega})$ *with* $\| \cdot \|_\infty$ *is a Banach space; i.e., the uniform limit of continuous functions is itself continuous.*

Example 1.2.18 *The space* $C(\overline{\Omega})$ *with the norm* $\| \cdot \|_p$, $1 \leq p < \infty$, *is not a Banach space. To illustrate this, we consider the space* $C[0, 1]$ *and a* *sequence in $C[0, 1]$ defined as follows:*

$$u_n(x) = \begin{cases} 0, & 0 \leq x \leq \frac{1}{2} - \frac{1}{2n}, \\ n\,x - \frac{1}{2}(n-1), & \frac{1}{2} - \frac{1}{2n} \leq x \leq \frac{1}{2} + \frac{1}{2n}, \\ 1, & \frac{1}{2} + \frac{1}{2n} \leq x \leq 1. \end{cases}$$

Let

$$u(x) = \begin{cases} 0, & 0 \leq x < \frac{1}{2}, \\ 1, & \frac{1}{2} < x \leq 1. \end{cases}$$

Then $\|u_n - u\|_p \to 0$ as $n \to \infty$; i.e., the sequence $\{u_n\}$ converges to u in the norm $\|\cdot\|_p$. But obviously no matter how we define $u(1/2)$, the limit function u is not continuous.

1.2.3 Completion of normed spaces

It is important to be able to deal with function spaces using a norm of our choice, as such a norm is often important or convenient in the formulation of a problem or in the analysis of a numerical method. The following theorem allows us to do this. A proof is discussed in [88, p. 84].

Theorem 1.2.19 *Let V be a normed space. Then there is a complete normed space W with the following properties:*
(a) There is a subspace $\widehat{V} \subseteq W$ and a bijective (one-to-one and onto) linear function $\mathcal{I} : V \to \widehat{V}$ with

$$\|\mathcal{I}v\|_W = \|v\|_V \qquad \forall\, v \in V.$$

The function $\mathcal{I}$ is called an isometric isomorphism *of the spaces V and $\widehat{V}$.*
(b) The subspace $\widehat{V}$ is dense in W; i.e., for any $w \in W$, there is a sequence $\{\widehat{v}_n\} \subseteq \widehat{V}$ such that

$$\|w - \widehat{v}_n\|_W \to 0 \qquad as \quad n \to \infty.$$

The space W is called the completion *of V, and W is unique up to an isometric isomorphism.*

The spaces V and $\widehat{V}$ are generally identified, meaning no distinction is made between them. However, we also consider cases where it is important to note the distinction. An important example of the theorem is to let V be the rational numbers and W be the real numbers $\mathbb{R}$. One way in which $\mathbb{R}$ can be defined is as a set of equivalence classes of Cauchy sequences of rational numbers, and $\widehat{V}$ can be identified with those equivalence classes of Cauchy sequences whose limit is a rational number. A proof of the above theorem can be made by mimicking this commonly used construction of the real numbers from the rational numbers.

Example 1.2.20 *Theorem 1.2.19 guarantees the existence of a unique abstract completion of an arbitrary normed vector space. However, it is often possible, and indeed desirable, to give a more concrete definition of the completion of a given normed space; much of the subject of real analysis is concerned with this topic. In particular, the subject of Lebesgue measure and integration deals with the completion of $C(\overline{\Omega})$ under the norms of (1.2.8), $\|\cdot\|_p$ for $1 \le p < \infty$. A complete development of Lebesgue measure and integration is given in any standard textbook on real analysis; for example, see Royden [141] or Rudin [142]. We do not introduce formally and rigorously the concepts of* measurable set *and* measurable function. *Rather we think of measure theory intuitively as described in the following paragraphs. Our rationale for this is that the details of Lebesgue measure and integration can often be bypassed in most of the material we present in this text.*

Measurable subsets of $\mathbb{R}$ include the standard open and closed intervals with which we are familiar. Multivariable extensions of intervals to $\mathbb{R}^d$ are also measurable, together with countable unions and intersections of them. Intuitively, the measure of a set $D \subseteq \mathbb{R}^d$ is its "length," "area," "volume," or suitable generalization; and we denote the measure *of D by* meas(D). *For a formal discussion of measurable set, see Royden [141] or Rudin [142].*

To introduce the concept of measurable function, we begin by defining a step function. *A function v on a measurable set D is a* step function *if D can be decomposed into a finite number of pairwise disjoint measurable subsets $D_1, \ldots, D_k$ with $v(\mathbf{x})$ constant over each D_j. We say a function v on D is a* measurable function *if it is the pointwise limit of a sequence of step functions. This includes, for example, all continuous functions on D. For each such measurable set D_j, we define a* characteristic function

$$\chi_j(\mathbf{x}) = \begin{cases} 1, \mathbf{x} \in D_j, \\ 0, \mathbf{x} \notin D_j. \end{cases}$$

A general step function over the decomposition $D_1, \ldots, D_k$ of D can then be written

$$v(\mathbf{x}) = \sum_{j=1}^{k} \alpha_j \chi_j(\mathbf{x}), \qquad \mathbf{x} \in D \tag{1.2.9}$$

with $\alpha_1, \ldots, \alpha_k$ scalars. For a general measurable function v over D, we write it as a limit of step functions v_k over D:

$$v(\mathbf{x}) = \lim_{k \to \infty} v_k(\mathbf{x}), \qquad \mathbf{x} \in D. \tag{1.2.10}$$

We say two measurable functions are equal almost everywhere *if the set of points on which they differ is a set of measure zero. For notation, we write*

$$v = w \ (a.e.)$$

to indicate that v and w are equal almost everywhere. Given a measurable function v on D, we introduce the concept of an equivalence class of equivalent functions:

$$[v] = \{w \mid w \text{ measurable on } D \text{ and } v = w \text{ (a.e.)}\}.$$

For most purposes, we generally consider elements of an equivalence class [v] as being a single function v.

We define the Lebesgue integral of a step function v over D, given in (1.2.9), by

$$\int_D v(\mathbf{x})\,dx = \sum_{j=1}^{k} \alpha_j \operatorname{meas}(D_j).$$

For a general measurable function, given in (1.2.10), define the Lebesgue integral of v over D by

$$\int_D v(\mathbf{x})\,dx = \lim_{k \to \infty} \int_D v_k(\mathbf{x})\,dx.$$

There are a great many properties of Lebesgue integration, and we refer the reader to any text on real analysis for further details. Note that the Lebesgue integrals of elements of an equivalence class [v] are identical.

Let Ω be an open measurable set in $\mathbb{R}^d$. Introduce

$$L^p(\Omega) = \left\{ [v] \mid v \text{ measurable on } \Omega \text{ and } \|v\|_p < \infty \right\}.$$

The norm $\|v\|_p$ is defined as in (1.2.8), although now we use Lebesgue integration rather than Riemann integration.

$$L^\infty(\Omega) = \left\{ [v] \mid v \text{ measurable on } \Omega \text{ and } \|v\|_\infty < \infty \right\}.$$

For v measurable on Ω, define

$$\|v\|_\infty = \operatorname*{ess\,sup}_{\mathbf{x} \in \Omega} |v(\mathbf{x})|$$

$$\equiv \inf_{\operatorname{meas}(\Omega')=0} \ \sup_{\mathbf{x} \in \Omega \setminus \Omega'} |v(\mathbf{x})|,$$

where "$\operatorname{meas}(\Omega') = 0$" means Ω' is a measurable set with measure zero. The spaces $L^p(\Omega)$, $1 \le p < \infty$, are Banach spaces, and they are concrete realizations of the abstract completion of $C(\overline{\Omega})$ under the norm of (1.2.8). The space $L^\infty(\Omega)$ is also a Banach space, but it is much larger than the space $C(\overline{\Omega})$ with the ∞-norm $\|\cdot\|_\infty$. Additional discussion of the spaces $L^p(\Omega)$ is given in Section 1.5.

Example 1.2.21 *More generally, let w be a positive measurable function on Ω, called a* weight function. *We can define weighted spaces $L_w^p(\Omega)$ as* *follows:*

$$L_w^p(\Omega) = \left\{ v \text{ measurable } \Big| \int_\Omega w(\mathbf{x})\,|v(\mathbf{x})|^p\,dx < \infty \right\}, \quad p \in [1,\infty),$$

$$L_w^\infty(\Omega) = \left\{ v \text{ measurable } \mid \text{ess sup}_\Omega\, w(\mathbf{x})\,|v(\mathbf{x})| < \infty \right\}.$$

These are Banach spaces with the norms

$$\|v\|_{p,w} = \left(\int_\Omega w(\mathbf{x})\,|v(\mathbf{x})|^p\,dx \right)^{1/p}, \quad p \in [1,\infty),$$

$$\|v\|_{p,\infty} = \underset{\mathbf{x}\in\Omega}{\text{ess sup}}\, w(\mathbf{x})|v(\mathbf{x})|.$$

The space $C(\overline{\Omega})$ of Example 1.1.2(c) with the norm

$$\|v\|_{C(\overline{\Omega})} = \max_{\mathbf{x}\in\overline{\Omega}} |v(\mathbf{x})|$$

is also a Banach space, and it can be considered as a proper subset of $L^\infty(\Omega)$. See Example 2.5.3 for a situation where it is necessary to distinguish between $C(\overline{\Omega})$ and the subspace of $L^\infty(\Omega)$ to which it is isometric and isomorphic.

Example 1.2.22 (a) *For any integer $m \geq 0$, the normed spaces $C^m[a,b]$ and $C_p^k(2\pi)$ of Example 1.2.4(b) are Banach spaces.*

(b) *Let $1 \leq p < \infty$. As an alternative norm on $C^m[a,b]$, introduce*

$$\|f\| = \left[\sum_{j=0}^m \|f^{(j)}\|_p^p \right]^{\frac{1}{p}}.$$

The space $C^m[a,b]$ is not complete with this norm. Its completion is denoted by $W^{m,p}(a,b)$, and it is an example of a Sobolev space. It can be shown that if $f \in W^{m,p}(a,b)$, then $f, f', \ldots, f^{(m-1)}$ are continuous, and $f^{(m)}$ exists almost everywhere and belongs to $L^p(a,b)$. This and its multivariable generalizations are discussed at length in Chapter 6.

A knowledge of the theory of Lebesgue measure and integration is very helpful in dealing with problems defined on spaces of such functions. Nonetheless, many results can be proven by referring to only the original space and its associated norm, say, $C(\overline{\Omega})$ with $\|\cdot\|_p$, from which a Banach space is obtained by a completion argument, say $L^p(\Omega)$. We return to this in Theorem 2.4.1 of Chapter 2.

Exercise 1.2.1 *Prove the backward triangle inequality of (1.2.4).*

Exercise 1.2.2 *Show that $\|\cdot\|_\infty$ is a norm on $C(\overline{\Omega})$, with Ω a bounded open set in $\mathbb{R}^d$.*

Exercise 1.2.3 *Show that $\|\cdot\|_\infty$ is a norm on $L^\infty(\Omega)$, with Ω a bounded open set in $\mathbb{R}^d$.*

Exercise 1.2.4 *Show that* $\| \cdot \|_1$ *is a norm on* $L^1(\Omega)$, *with* Ω *a bounded open set in* $\mathbb{R}^d$.

Exercise 1.2.5 *Show that for* $1 \le p \le \infty$, $\|\mathbf{x}\|_p$ *defined by* (1.2.2)–(1.2.3) *is a norm in the space* $\mathbb{R}^d$. *The main task is to verify the triangle inequality, which can be done by first proving the Hölder inequality,* $|\mathbf{x}\cdot\mathbf{y}| \le \|\mathbf{x}\|_p\|\mathbf{y}\|_{p'}$, $\mathbf{x}, \mathbf{y} \in \mathbb{R}^d$. *Here* p' *is the conjugate of* p *defined through the relation* $1/p' + 1/p = 1$; *by convention,* $p' = 1$ *if* $p = \infty$, $p' = \infty$ *if* $p = 1$.

Exercise 1.2.6 *Define* $C^\alpha[a, b]$, $0 < \alpha \le 1$, *as the set of all* $f \in C[a, b]$ *for which*

$$M_\alpha(f) \equiv \sup_{\substack{a \le x, y \le b \\ x \ne y}} \frac{|f(x) - f(y)|}{|x - y|^\alpha} < \infty.$$

Define $\|f\|_\alpha = \|f\|_\infty + M_\alpha(f)$. *Show* $C^\alpha[a, b]$ *with this norm is complete.*

Exercise 1.2.7 *Define* $C_b[0, \infty)$ *as the set of all functions* f *that are continuous on* $[0, \infty)$ *and satisfy*

$$\|f\|_\infty \equiv \sup_{x \ge 0} |f(x)| < \infty.$$

Show $C_b[0, \infty)$ *with this norm is complete.*

Exercise 1.2.8 *Does the formula* (1.2.2) *define a norm on* $\mathbb{R}^d$ *for* $0 < p < 1$?

Exercise 1.2.9 *Prove the equivalence on* $C^1[0, 1]$ *of the following norms:*

$$\|f\|_a \equiv |f(0)| + \int_0^1 |f'(x)| \, dx,$$

$$\|f\|_b \equiv \int_0^1 |f(x)| \, dx + \int_0^1 |f'(x)| \, dx.$$

Hint: You may need to use the integral mean value theorem: Given $g \in C[0, 1]$, *there is* $\xi \in [0, 1]$ *such that*

$$\int_0^1 g(x) \, dx = g(\xi).$$

Exercise 1.2.10 *Let* V_1 *and* V_2 *be normed spaces with norms* $\| \cdot \|_1$ *and* $\| \cdot \|_2$. *Recall that the product space* $V_1 \times V_2$ *is defined by*

$$V_1 \times V_2 = \{(v_1, v_2) \mid v_1 \in V_1, \, v_2 \in V_2\}.$$

Show that the quantities $\max\{\|v_1\|_1, \|v_2\|_2\}$ *and* $(\|v_1\|_1^p + \|v_2\|_2^p)^{1/p}$, $1 \le p < \infty$ *all define norms on the space* $V_1 \times V_2$.

Exercise 1.2.11 *Over the space* $C^1[0, 1]$, *determine which of the following is a norm, and which is only a semi-norm:*

(a) $\displaystyle\max_{0 \le x \le 1} |u(x)|$;

(b) $\max\limits_{0\leq x\leq 1} [|u(x)| + |u'(x)|]$;

(c) $\max\limits_{0\leq x\leq 1} |u'(x)|$;

(d) $|u(0)| + \max\limits_{0\leq x\leq 1} |u'(x)|$;

(e) $\max\limits_{0\leq x\leq 1} |u'(x)| + \int_{0.1}^{0.2} |u(x)| \, dx$.

Exercise 1.2.12 *Over a normed space $(V, \|\cdot\|)$, we define a function of two variables $d(u,v) = \|u-v\|$. Then $d(\cdot,\cdot)$ is a distance function; in other words, $d(\cdot,\cdot)$ has the following properties of an ordinary distance between two points:*

(a) *$d(u,v) \geq 0$ for any $u,v \in V$, and $d(u,v) = 0$ if and only if $u = v$;*

(b) *$d(u,v) = d(v,u)$ for any $u,v \in V$;*

(c) *(the triangle inequality) $d(u,w) \leq d(u,v) + d(v,w)$ for any $u,v,w \in V$.*

A linear space endowed with a distance function is called a metric space. Certainly a normed space can be viewed as a metric space. There are examples of metrics (distance functions) that are not generated by any norm, though.

Exercise 1.2.13 *Let V be a normed space and $\{u_n\}$ a Cauchy sequence. Suppose there is a subsequence $\{u_{n'}\} \subseteq \{u_n\}$ and some element $v \in V$ such that $u_{n'} \to u$ as $n' \to \infty$. Show that $u_n \to u$ as $n \to \infty$.*

Exercise 1.2.14 *Let V be a normed space, $V_0 \subseteq V$ a subspace. The quotient space V/V_0 is defined to be the space of all the classes $[v] = \{v + v_0 \mid v_0 \in V_0\}$. Prove that the formula*

$$\|[v]\|_{V/V_0} = \inf_{v_0 \in V_0} \|v + v_0\|_V$$

defines a norm on V/V_0. Show that if V is a Banach space and $V_0 \subseteq V$ is a closed subspace, then V/V_0 is a Banach space.

Exercise 1.2.15 *Assuming a knowledge of Lebesgue integration, show that*

$$W^{1,2}(a,b) \subseteq C[a,b].$$

Generalize this result to the space $W^{m,p}(a,b)$ with other values of m and p.
Hint: For $v \in W^{1,2}(a,b)$, use

$$v(x) - v(y) = \int_x^y v'(z) \, dz.$$

Exercise 1.2.16 *On $C^1[0,1]$, define*

$$(u,v)_* = u(0)\,v(0) + \int_0^1 u'(x)\,v'(x)\,dx$$

and

$$\|u\|_* = \sqrt{(u,u)_*}\,.$$

Show that

$$\|u\|_\infty \le c\,\|u\|_* \qquad \forall\, u \in C^1[0,1]$$

for a suitably chosen constant c.

1.3   Inner product spaces

In studying linear problems, inner product spaces are usually used. These are the spaces where a norm can be defined through the inner product and the notion of orthogonality of two elements can be introduced. The inner product in a general space is a generalization of the usual scalar product in the plane $\mathbb{R}^2$ or the space $\mathbb{R}^3$.

Definition 1.3.1 *Let V be a linear space over $\mathbb{K} = \mathbb{R}$ or $\mathbb{C}$. An inner product $(\cdot,\cdot)$ is a function from $V \times V$ to $\mathbb{K}$ with the following properties.*

1. *For any $u \in V$, $(u,u) \ge 0$ and $(u,u) = 0$ if and only if $u = 0$.*

2. *For any $u,v \in V$, $(u,v) = \overline{(v,u)}$.*

3. *For any $u,v,w \in V$, any $\alpha, \beta \in \mathbb{K}$, $(\alpha\,u + \beta\,v, w) = \alpha\,(u,w) + \beta\,(v,w)$.*

The space V together with the inner product $(\cdot,\cdot)$ is called an inner product space. When the definition of the inner product $(\cdot,\cdot)$ is clear from the context, we simply say V is an inner product space. When $\mathbb{K} = \mathbb{R}$, V is called a real inner product space, while if $\mathbb{K} = \mathbb{C}$, V is a complex inner product space.

In the case of a real inner product space, the second axiom reduces to the symmetry of the inner product:

$$(u,v) = (v,u) \qquad \forall\, u,v \in V.$$

For an inner product, there is an important property called the Schwarz inequality.

Theorem 1.3.2 (SCHWARZ INEQUALITY) *If V is an inner product space, then*

$$|(u,v)| \le \sqrt{(u,u)\,(v,v)} \qquad \forall\, u,v \in V,$$

and the equality holds if and only if u and v are linearly dependent.

Proof. We give the proof only for the real case. The result is obviously true if either $u = 0$ or $v = 0$. Now suppose $u \neq 0$, $v \neq 0$. Define

$$\phi(t) = (u + t\,v, u + t\,v) = (u, u) + 2\,(u, v)\,t + (v, v)\,t^2, \quad t \in \mathbb{R}.$$

The function ϕ is quadratic and non-negative, so its discriminant must be non-positive,

$$[2\,(u, v)]^2 - 4\,(u, u)\,(v, v) \leq 0;$$

i.e., the Schwarz inequality is valid. For $v \neq 0$, the equality holds if and only if $u = -t\,v$ for some $t \in \mathbb{R}$. ∎

An inner product $(\cdot, \cdot)$ induces a norm through the formula

$$\|u\| = \sqrt{(u, u)}, \quad u \in V.$$

In verifying the triangle inequality for the quantity thus defined, we need to use the above Schwarz inequality.

Proposition 1.3.3 *An inner product is continuous with respect to its induced norm. In other words, if $\|\cdot\|$ is the norm defined by $\|u\| = \sqrt{(u, u)}$, then $\|u_n - u\| \to 0$ and $\|v_n - v\| \to 0$ as $n \to \infty$ imply*

$$(u_n, v_n) \to (u, v) \quad \text{as } n \to \infty.$$

In particular, if $u_n \to u$, then for any v,

$$(u_n, v) \to (u, v) \quad \text{as } n \to \infty.$$

Proof. Since $\{u_n\}$ and $\{v_n\}$ are convergent, they are bounded; i.e., for some $M < \infty$, $\|u_n\| \leq M$, $\|v_n\| \leq M$ for any n. We write

$$(u_n, v_n) - (u, v) = (u_n - u, v_n) + (u, v_n - v).$$

Using the Schwarz inequality, we have

$$|(u_n, v_n) - (u, v)| \leq \|u_n - u\|\,\|v_n\| + \|u\|\,\|v_n - v\|$$
$$\leq M\,\|u_n - u\| + \|u\|\,\|v_n - v\|.$$

Hence the result holds. ∎

Commonly seen inner product spaces are usually associated with their canonical inner products. As an example, the canonical inner product for the space $\mathbb{R}^d$ is

$$(\mathbf{x}, \mathbf{y}) = \sum_{i=1}^{d} x_i y_i = \mathbf{y}^{\mathrm{T}} \mathbf{x}, \quad \forall \mathbf{x} = (x_1, \ldots, x_d)^T, \ \mathbf{y} = (y_1, \ldots, y_d)^T \in \mathbb{R}^d.$$

This inner product induces the Euclidean norm

$$\|\mathbf{x}\| = \sqrt{\sum_{i=1}^{d} |x_i|^2} = \sqrt{(\mathbf{x}, \mathbf{x})}.$$

When we talk about the space $\mathbb{R}^d$, implicitly we understand the inner product and the norm are the ones defined above, unless stated otherwise. For the complex space $\mathbb{C}^d$, the inner product and the corresponding norm are

$$(\mathbf{x}, \mathbf{y}) = \sum_{i=1}^{d} x_i \overline{y_i} = \mathbf{y}^* \mathbf{x}, \quad \forall \mathbf{x} = (x_1, \dots, x_d)^T, \ \mathbf{y} = (y_1, \dots, y_d)^T \in \mathbb{C}^d$$

and

$$\|\mathbf{x}\| = \sqrt{\sum_{i=1}^{d} |x_i|^2} = \sqrt{(\mathbf{x}, \mathbf{x})}.$$

The space $L^2(\Omega)$ is an inner product space with the canonical inner product

$$(u, v) = \int_{\Omega} u(\mathbf{x}) \, \overline{v(\mathbf{x})} \, dx.$$

This inner product induces the standard $L^2(\Omega)$-norm

$$\|u\|_2 = \sqrt{\int_{\Omega} |u(\mathbf{x})|^2 dx} = \sqrt{(u, u)}.$$

We have seen that an inner product induces a norm, which is always the norm we use on the inner product space unless stated otherwise. It is easy to show that on a complex inner product space,

$$(u, v) = \frac{1}{4}[\|u + v\|^2 - \|u - v\|^2 + i\|u + iv\|^2 - i\|u - iv\|^2],$$

and on a real inner product space,

$$(u, v) = \frac{1}{4}[\|u + v\|^2 - \|u - v\|^2]. \tag{1.3.1}$$

These relations are called the *polarization identities*. Thus in any normed linear space, there can exist at most one inner product that generates the norm.

On the other hand, not every norm can be defined through an inner product. We have the following characterization for any norm induced by an inner product.

Theorem 1.3.4 *A norm $\|\cdot\|$ on a linear space V is induced by an inner product if and only if it satisfies the parallelogram law:*

$$\|u + v\|^2 + \|u - v\|^2 = 2\|u\|^2 + 2\|v\|^2 \qquad \forall \, u, v \in V. \tag{1.3.2}$$

Note that if u and v form two adjacent sides of a parallelogram, then $\|u + v\|$ and $\|u - v\|$ represent the lengths of the diagonals of the parallelogram. This theorem can be considered to be a generalization of the theorem of Pythagoras for right triangles.

Proof. We prove the result for the case of a real space only. Assume $\|\cdot\| = \sqrt{(\cdot,\cdot)}$ for some inner product $(\cdot,\cdot)$. Then for any $u,v \in V$,

$$\|u+v\|^2 + \|u-v\|^2$$
$$= (u+v,u+v) + (u-v,u-v)$$
$$= \left[\|u\|^2 + 2(u,v) + \|v\|^2\right] + \left[\|u\|^2 - 2(u,v) + \|v\|^2\right]$$
$$= 2\|u\|^2 + 2\|v\|^2.$$

Conversely, assume the norm $\|\cdot\|$ satisfies the parallelogram law. For $u,v \in V$, let us define

$$(u,v) = \frac{1}{4}\left[\|u+v\|^2 - \|u-v\|^2\right]$$

and show that it is an inner product. First,

$$(u,u) = \frac{1}{4}\|2u\|^2 = \|u\|^2 \geq 0$$

and $(u,u) = 0$ if and only if $u = 0$. Second,

$$(u,v) = \frac{1}{4}\left[\|v+u\|^2 - \|v-u\|^2\right] = (v,u).$$

Finally, we show the linearity, which is equivalent to the following two relations:

$$(u+v,w) = (u,w) + (v,w) \quad \forall\, u,v,w \in V$$

and

$$(\alpha\, u, v) = \alpha\,(u,v) \qquad \forall\, u \in V,\ \alpha \in \mathbb{R}.$$

We have

$$(u,w) + (v,w)$$
$$= \tfrac{1}{4}\left[\|u+w\|^2 - \|u-w\|^2 + \|v+w\|^2 - \|v-w\|^2\right]$$
$$= \tfrac{1}{4}\left[(\|u+w\|^2 + \|v+w\|^2) - (\|u-w\|^2 + \|v-w\|^2)\right]$$
$$= \tfrac{1}{4}\left[\tfrac{1}{2}(\|u+v+2\,w\|^2 + \|u-v\|^2) - \tfrac{1}{2}(\|u+v-2\,w\|^2 + \|u-v\|^2)\right]$$
$$= \tfrac{1}{8}\left[\|u+v+2\,w\|^2 - \|u+v-2\,w\|^2\right]$$
$$= \tfrac{1}{8}\left[2\,(\|u+v+w\|^2 + \|w\|^2) - \|u+v\|^2\right.$$
$$\left.\qquad - 2\,(\|u+v-w\|^2 + \|w\|^2) + \|u+v\|^2\right]$$
$$= \tfrac{1}{4}\left[\|u+v+w\|^2 - \|u+v-w\|^2\right]$$
$$= (u+v,w).$$

The proof of the second relation is more involved. For fixed $u,v \in V$, let us define a function of a real variable

$$f(\alpha) = \|\alpha\,u+v\|^2 - \|\alpha\,u-v\|^2.$$

We show that $f(\alpha)$ is a linear function of α. We have

$$
\begin{aligned}
f(\alpha) &- f(\beta) \\
&= \|\alpha\,u + v\|^2 + \|\beta\,u - v\|^2 - \|\alpha\,u - v\|^2 - \|\beta\,u + v\|^2 \\
&= \tfrac{1}{2}\left[\|(\alpha + \beta)\,u\|^2 + \|(\alpha - \beta)\,u + 2\,v\|^2\right] \\
&\quad - \tfrac{1}{2}\left[\|(\alpha + \beta)\,u\|^2 + \|(\alpha - \beta)\,u - 2\,v\|^2\right] \\
&= \tfrac{1}{2}\left[\|(\alpha - \beta)\,u + 2\,v\|^2 - \|(\alpha - \beta)\,u - 2\,v\|^2\right] \\
&= 2\left[\|\tfrac{\alpha-\beta}{2}\,u + v\|^2 - \|\tfrac{\alpha-\beta}{2}\,u - v\|^2\right] \\
&= 2\,f\left(\tfrac{\alpha-\beta}{2}\right).
\end{aligned}
$$

Taking $\beta = 0$ and noticing $f(0) = 0$, we find that

$$
f(\alpha) = 2\,f\left(\frac{\alpha}{2}\right).
$$

Thus we also have the relation

$$
f(\alpha) - f(\beta) = f(\alpha - \beta).
$$

From the above relations, the continuity of f, and the value $f(0) = 0$, one concludes that (see Exercise 1.3.2)

$$
f(\alpha) = c_0\alpha = \alpha\,f(1) = \alpha\left[\|u + v\|^2 - \|u - v\|^2\right]
$$

from which, we get the second required relation. ∎

1.3.1   Hilbert spaces

Among the inner product spaces, of particular importance are the Hilbert spaces.

Definition 1.3.5 *A complete inner product space is called a* Hilbert *space.*

From the definition, we see that an inner product space V is a Hilbert space if V is a Banach space under the norm induced by the inner product.

Example 1.3.6 (SOME EXAMPLES OF HILBERT SPACES)
(a) *The Cartesian space* $\mathbb{C}^d$ *is a Hilbert space with the inner product*

$$
(x, y) = \sum_{i=1}^{d} x_i\overline{y_i}.
$$

(b) *The space* $l^2 = \{x = \{x_i\}_{i\geq 1} \mid \sum_{i=1}^{\infty} |x_i|^2 < \infty\}$ *is a linear space with*

$$
\alpha\,x + \beta\,y = \{\alpha\,x_i + \beta\,y_i\}_{i\geq 1}.
$$

It can be shown that

$$
(x, y) = \sum_{i=1}^{\infty} x_i\overline{y_i}
$$

defines an inner product on l^2. *Furthermore,* l^2 *becomes a Hilbert space under this inner product.*

(c) The space $L^2(0,1)$ *is a Hilbert space with the inner product*

$$(u,v) = \int_0^1 u(x)\,\overline{v(x)}\,dx.$$

(d) The space $L^2(\Omega)$ *is a Hilbert space with the inner product*

$$(u,v) = \int_\Omega u(\mathbf{x})\,\overline{v(\mathbf{x})}\,dx.$$

More generally, if $w(\mathbf{x})$ *is a positive function on* Ω, *then the space*

$$L_w^2(\Omega) = \left\{ v\ measurable\ \Big|\ \int_\Omega |v(\mathbf{x})|^2 w(\mathbf{x})\,dx < \infty \right\}$$

is a Hilbert space with the inner product

$$(u,v)_w = \int_\Omega u(\mathbf{x})\,\overline{v(\mathbf{x})}\,w(\mathbf{x})\,dx.$$

This space is a weighted L^2 *space.*

Example 1.3.7 *Recall the Sobolev space* $W^{m,p}(a,b)$ *defined in Example 1.2.22. If we choose* $p = 2$, *then we obtain a Hilbert space. It is usually denoted by* $H^m(a,b) \equiv W^{m,2}(a,b)$. *The associated inner product is defined by*

$$(f,g)_{H^m} = \sum_{j=0}^m \left(f^{(j)}, g^{(j)} \right), \qquad f,g \in H^m(a,b)$$

using the standard inner product $(\cdot,\cdot)$ *of* $L^2(a,b)$. *Recall from Exercise 1.2.15 that* $H^1(a,b) \subseteq C[a,b]$.

1.3.2 Orthogonality

With the notion of an inner product at our disposal, we can define the angle between two vectors u and v as follows:

$$\theta = \arccos\left[\frac{(u,v)}{\|u\|\,\|v\|} \right].$$

This definition makes sense because, by the Schwarz inequality (Theorem 1.3.2), the argument of arccos is between -1 and 1. The case of a right angle is particularly important. We see that two vectors u and v form a right angle if and only if $(u,v) = 0$.

Definition 1.3.8 *Two vectors* u *and* v *are said to be* orthogonal *if* $(u,v) = 0$. *An element* $v \in V$ *is said to be orthogonal to a subset* $U \subseteq V$, *if* $(u,v) = 0$ *for any* $u \in U$.

Definition 1.3.9 *Let U be a subset of an inner product space V. We define its* orthogonal complement *to be the set*

$$U^\perp = \{v \in V \mid (v, u) = 0 \; \forall u \in U\}.$$

The orthogonal complement of any set is a closed subspace (cf. Exercise 1.3.7).

Definition 1.3.10 *Let V be an inner product space.*
(a) Suppose V is finite dimensional. A basis $\{v_1, \ldots, v_n\}$ of V is said to be an orthogonal basis *if*

$$(v_i, v_j) = 0, \qquad 1 \leq i \neq j \leq n.$$

If, additionally, $\|v_i\| = 1$, $1 \leq i \leq n$, then we say the basis is orthonormal, *and we combine these conditions as*

$$(v_i, v_j) = \delta_{ij} \equiv \begin{cases} 1, \, i = j, \\ 0, \, i \neq j. \end{cases}$$

(b) Suppose V is infinite dimensional normed space. We say V has a countably infinite basis if there is a sequence $\{v_i\}_{i \geq 1} \subseteq V$ for which the following is valid: For each $v \in V$, we can find scalars $\{\alpha_{n,i}\}_{i=1}^n$, $n = 1, 2, \ldots$, such that

$$\left\| v - \sum_{i=1}^n \alpha_{n,i} v_i \right\| \to 0 \qquad as \; n \to \infty.$$

The space V is also said to be separable. *The sequence $\{v_i\}_{i \geq 1}$ is called a basis if any finite subset of the sequence is linearly independent. If V is an inner product space, and if the sequence $\{v_i\}_{i \geq 1}$ also satisfies*

$$(v_i, v_j) = \delta_{ij}, \qquad i, j \geq 1, \tag{1.3.3}$$

then $\{v_i\}_{i \geq 1}$ is called an orthonormal basis *for V.*
(c) We say that an infinite dimensional normed space V has a Schauder basis *$\{v_n\}_{n \geq 1}$ if for each $v \in V$, it is possible to write*

$$v = \sum_{n=1}^\infty \alpha_n v_n \tag{1.3.4}$$

as a convergent series in V for a unique choice of scalars $\{\alpha_n\}_{n \geq 1}$.

For a discussion of the distinction between V having a Schauder basis and V being separable, see [103, p.68]. For V an inner product space, it is straightforward to show that an orthonormal basis $\{v_n\}_{n \geq 1}$ is also a Schauder basis, and therefore (1.3.4) is valid for an orthonormal basis. The advantage of using an orthogonal or an orthonormal basis is that it is easy to decompose a vector as a linear combination of the basis elements. Assuming $\{v_n\}_{n \geq 1}$ is an orthonormal basis of V, let us determine the coefficients

$\{\alpha_n\}_{n \geq 1}$ in the decomposition (1.3.4) for any $v \in V$. By the continuity of the inner product and the orthonormality condition (1.3.3), we have

$$(v, v_k) = \sum_{n=1}^{\infty} \alpha_n (v_n, v_k) = \alpha_k.$$

Thus

$$v = \sum_{n=1}^{\infty} (v, v_n) \, v_n. \tag{1.3.5}$$

In addition, by direct computation using (1.3.3),

$$\left\| \sum_{n=1}^{N} (v, v_n) \, v_n \right\|^2 = \sum_{n=1}^{N} |(v, v_n)|^2.$$

Using the convergence of (1.3.5) in V, we can let $N \to \infty$ to obtain

$$\|v\| = \sqrt{\sum_{n=1}^{\infty} |(v, v_n)|^2}. \tag{1.3.6}$$

A simple consequence of the identity (1.3.6) is the inequality

$$\sum_{n=1}^{N} |(v, v_n)|^2 \leq \|v\|^2, \qquad N \geq 1, \quad v \in V. \tag{1.3.7}$$

The decomposition (1.3.5) can be termed as the *generalized Fourier series*; then the identity (1.3.6) can be called the *generalized Parseval identity*, whereas (1.3.7) can be called the *generalized Bessel inequality*.

Example 1.3.11 *Let* $V = L^2(0, 2\pi)$ *with complex scalars. The complex exponentials*

$$v_n(x) = \frac{1}{\sqrt{2\pi}} \, e^{inx}, \qquad n = 0, \pm 1, \pm 2, \ldots \tag{1.3.8}$$

form an orthonormal basis. For any $v \in L^2(0, 2\pi)$, *we have the Fourier series expansion*

$$v(x) = \sum_{n=-\infty}^{\infty} \alpha_n v_n(x) \tag{1.3.9}$$

where

$$\alpha_n = (v, v_n) = \frac{1}{\sqrt{2\pi}} \int_0^{2\pi} v(x) \, e^{-inx} \, dx. \tag{1.3.10}$$

Also (1.3.6) and (1.3.7) reduce to the ordinary Parseval identity and Bessel inequality.

When a non-orthogonal basis for an inner product space is given, there is a standard procedure to construct an orthonormal basis.

Theorem 1.3.12 (GRAM-SCHMIDT METHOD) *Let $\{w_n\}_{n\geq 1}$ be a basis of the inner product space V. Then there is an orthonormal basis $\{v_n\}_{n\geq 1}$ with the property that*

$$\text{span}\{w_n\}_{n=1}^N = \text{span}\{v_n\}_{n=1}^N \qquad \forall N \geq 1.$$

Proof. The proof is done inductively. For $N = 1$, define

$$v_1 = \frac{w_1}{\|w_1\|},$$

which satisfies $\|v_1\| = 1$. For $N \geq 2$, assume $\{v_n\}_{n=1}^{N-1}$ have been constructed with $(v_n, v_m) = \delta_{nm}$, $1 \leq n, m \leq N - 1$, and

$$\text{span}\{w_n\}_{n=1}^{N-1} = \text{span}\{v_n\}_{n=1}^{N-1}.$$

Write

$$\tilde{v}_N = w_N + \sum_{n=1}^{N-1} \alpha_{N,n} v_n.$$

Now choose $\{\alpha_{N,n}\}_{n=1}^{N-1}$ by setting

$$(\tilde{v}_N, v_n) = 0, \qquad 1 \leq n \leq N - 1.$$

This implies

$$\alpha_{N,n} = -(w_N, v_n), \qquad 1 \leq n \leq N - 1.$$

This procedure "removes" from w_N the components in the directions of $v_1, \ldots, v_{N-1}$.

Finally, define

$$v_N = \frac{\tilde{v}_N}{\|\tilde{v}_N\|},$$

which is meaningful since $\tilde{v}_N \neq 0$. (Why?) Then the sequence $\{v_n\}_{n=1}^N$ satisfies

$$(v_n, v_m) = \delta_{nm}, \quad 1 \leq n, m \leq N$$

and

$$\text{span}\{w_n\}_{n=1}^N = \text{span}\{v_n\}_{n=1}^N.$$

∎

The Gram-Schmidt method can be used, e.g., to construct an orthonormal basis in $L^2(-1, 1)$ for a polynomial space of certain degrees. As a result we obtain the well-known Legendre polynomials (after a proper scaling), which play an important role in some numerical analysis problems.

Example 1.3.13 *Let us construct the first three orthonormal polynomials in $L^2(-1,1)$. For this purpose, we take*

$$w_1(x) = 1, \qquad w_2(x) = x, \qquad w_3(x) = x^2.$$

Then easily,

$$v_1(x) = \frac{w_1(x)}{\|w_1\|} = \frac{1}{\sqrt{2}}.$$

To find $v_2(x)$, we write

$$\tilde{v}_2(x) = w_2(x) + \alpha_{2,1}v_1(x) = x + \frac{1}{\sqrt{2}}\alpha_{2,1}$$

and choose

$$\alpha_{2,1} = -(x, \frac{1}{\sqrt{2}}) = -\int_{-1}^{1} \frac{1}{\sqrt{2}} x \, dx = 0.$$

So $\tilde{v}_2(x) = x$, and

$$v_2(x) = \frac{\tilde{v}_2(x)}{\|\tilde{v}_2\|} = \sqrt{\frac{3}{2}}\, x.$$

Finally, we write

$$\tilde{v}_3(x) = w_3(x) + \alpha_{3,1}v_1(x) + \alpha_{3,2}v_2(x) = x^2 + \frac{1}{\sqrt{2}}\alpha_{3,1} + \sqrt{\frac{3}{2}}\alpha_{3,2}x.$$

Then

$$\alpha_{3,1} = -(w_3, v_1) = -\int_{-1}^{1} x^2 \frac{1}{\sqrt{2}} \, dx = -\frac{\sqrt{2}}{3},$$

$$\alpha_{3,2} = -(w_3, v_2) = -\int_{-1}^{1} x^2 \sqrt{\frac{3}{2}}\, x \, dx = 0.$$

Hence

$$\tilde{v}_3(x) = x^2 - \frac{1}{3}.$$

Since $\|\tilde{v}_3\|^2 = \frac{8}{45}$, we have

$$v_3(x) = \frac{3}{2}\sqrt{\frac{5}{2}} \left(x^2 - \frac{1}{3}\right).$$

The fourth orthonormal polynomial is

$$v_4(x) = \sqrt{\frac{7}{8}} \left(5x^3 - 3x\right).$$

The graphs of these first four Legendre polynomials are given in Figure 1.2.

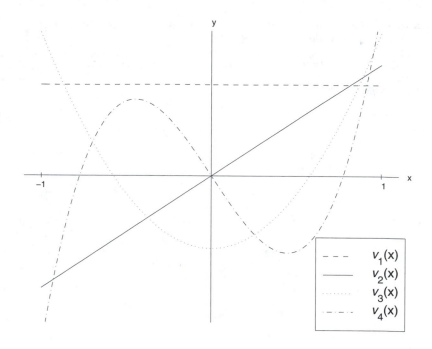

Figure 1.2. Graphs on $[-1, 1]$ of the orthonormal Legendre polynomials of degrees 0,1,2,3

As we see from Example 1.3.13, it is cumbersome to construct orthonormal (or orthogonal) polynomials directly. Fortunately, for many important cases of the weighted function $w(x)$ and integration interval (a, b), formulas of orthogonal polynomials in the weighted space $L^2_w(a, b)$ are known (see Section 3.4).

Exercise 1.3.1 *Given an inner product, show that the formula $\|u\| = \sqrt{(u, u)}$ defines a norm.*

Exercise 1.3.2 *Assume $f : \mathbb{R} \to \mathbb{R}$ is a continuous function, satisfying $f(\alpha) = f(\beta) + f(\alpha - \beta)$ for any $\alpha, \beta \in \mathbb{R}$, and $f(0) = 0$. Then $f(\alpha) = \alpha f(1)$.*
Solution: *From $f(\alpha) = f(\beta) + f(\alpha - \beta)$ and $f(0) = 0$, by an induction argument, we have $f(n\,\alpha) = n\,f(\alpha)$ for any integer n. Then from $f(\alpha) = 2\,f(\alpha/2)$, we have $f(1/2^n) = (1/2^n)\,f(1)$ for any integer $n \geq 0$. Finally, for any integer m, any non-negative integer n, $f(m\,2^{-n}) = m\,f(2^{-n}) = (m\,2^{-n})\,f(1)$. Now any rational can be represented as a finite sum $q = \sum_i m_i\,2^{-i}$. Hence, $f(q) = \sum_i f(m_i\,2^{-i}) = \sum_i m_i\,2^{-i}\,f(1) = q\,f(1)$. Since the set of the rational numbers is dense in $\mathbb{R}$ and f is a continuous function, we see that for any real ξ, $f(\xi) = \xi\,f(1)$.*

Exercise 1.3.3 *The norms* $\| \cdot \|_p$, $1 \leq p \leq \infty$, *over the space* $\mathbb{R}^d$ *are defined in Example 1.2.3. Find all the values of p for which the norm* $\| \cdot \|_p$ *is induced by an inner product.*
Hint: Apply Theorem 1.3.4.

Exercise 1.3.4 *Let* $w_1, \ldots, w_d$ *be positive constants. Show that the formula*

$$(\mathbf{x}, \mathbf{y}) = \sum_{i=1}^{d} w_i x_i y_i$$

defines an inner product on $\mathbb{R}^d$. *This is an example of a weighted inner product. What happens if we only assume* $w_i \geq 0$, $1 \leq i \leq d$?

Exercise 1.3.5 *Let* $A \in \mathbb{R}^{d \times d}$ *be a symmetric, positive definite matrix and let* $(\cdot, \cdot)$ *be the Euclidean inner product on* $\mathbb{R}^d$. *Show that the quantity* $(A\mathbf{x}, \mathbf{y})$ *defines an inner product on* $\mathbb{R}^d$.

Exercise 1.3.6 *Show that in an inner product space,* $\|u + v\| = \|u\| + \|v\|$ *for some* $u, v \in V$ *if and only if* u *and* v *are non-negatively linearly dependent (i.e., for some* $c_0 \geq 0$, *either* $u = c_0 v$ *or* $v = c_0 u$).

Exercise 1.3.7 *Prove that the orthogonal complement of a subset is a closed subspace.*

Exercise 1.3.8 *Let* V_0 *be a subset of a Hilbert space* V. *Show that the following statements are equivalent:*

(a) V_0 *is dense in* V; *i.e., for any* $v \in V$, *there exists* $\{v_n\}_{n \geq 1} \subseteq V_0$ *such that* $\|v - v_n\|_V \to 0$ *as* $n \to \infty$.

(b) $V_0^{\perp} = \{0\}$.

(c) *If* $u \in V$ *satisfies* $(u, v) = 0 \; \forall v \in V_0$, *then* $u = 0$.

(d) *For every* $0 \neq u \in V$, *there is a* $v \in V_0$ *such that* $(u, v) \neq 0$.

Exercise 1.3.9 *On* $C^1[a, b]$, *define*

$$(f, g)_* = f(a)g(a) + \int_0^1 f'(x)g'(x)\,dx, \qquad f, g \in C^1[a, b]$$

and $\|f\|_* = \sqrt{(f, f)_*}$. *Show that*

$$\|f\|_\infty \leq c \|f\|_* \qquad \forall f \in C^1[a, b]$$

for a suitable constant c.

Exercise 1.3.10 *Consider the Fourier series (1.3.9) for a function* $v \in C_p^m(2\pi)$ *with* $m \geq 2$. *Show that*

$$\left\| v - \sum_{n=-N}^{N} \alpha_n v_n \right\|_\infty \leq \frac{c_m(v)}{N^{m-1}}, \qquad N \geq 1.$$

Hint: *Use integration by parts in* (1.3.10).

1.4 Spaces of continuously differentiable functions

Spaces of continuous functions and continuously differentiable functions were introduced in Example 1.1.2. In this section, we provide a more detailed review of these spaces.

Let Ω be an open bounded subset of $\mathbb{R}^d$. A typical point in $\mathbb{R}^d$ is denoted by $\mathbf{x} = (x_1, \ldots, x_d)^T$. For multivariable functions, it is convenient to use the multi-index notation for partial derivatives. A multi-index is an ordered collection of d non-negative integers, $\alpha = (\alpha_1, \ldots, \alpha_d)$. The quantity $|\alpha| = \sum_{i=1}^{d} \alpha_i$ is said to be the *length* of α.

If v is an m-times differentiable function, then for any α with $|\alpha| \le m$,

$$D^\alpha v(\mathbf{x}) = \frac{\partial^{|\alpha|} v(\mathbf{x})}{\partial x_1^{\alpha_1} \cdots \partial x_d^{\alpha_d}}$$

is the αth order partial derivative. This is a handy notation for partial derivatives. Some examples are

$$\frac{\partial v}{\partial x_1} = D^\alpha v \quad \text{for } \alpha = (1, 0, \ldots, 0),$$

$$\frac{\partial^d v}{\partial x_1 \cdots \partial x_d} = D^\alpha v \quad \text{for } \alpha = (1, 1, \ldots, 1).$$

The set of all the derivatives of order m of a function v can be written as $\{D^\alpha v \mid |\alpha| = m\}$. For low-order partial derivatives, there are other commonly used notations; e.g., the partial derivative $\partial v / \partial x_i$ is also written as $\partial_{x_i} v$, or $\partial_i v$, or $v_{,x_i}$, or $v_{,i}$.

The space $C(\Omega)$ consists of all real-valued functions that are continuous on Ω. Since Ω is open, a function from the space $C(\Omega)$ is not necessarily bounded. For example, with $d = 1$ and $\Omega = (0, 1)$, the function $v(x) = 1/x$ is continuous but unbounded on $(0, 1)$. Indeed, a function from the space $C(\Omega)$ can behave "nastily" as the variable approaches the boundary of Ω. Usually, it is more convenient to deal with continuous functions that are continuous up to the boundary. Let $C(\overline{\Omega})$ be the space of functions that are *uniformly continuous* on Ω. Any function in $C(\overline{\Omega})$ is bounded. The notation $C(\overline{\Omega})$ is consistent with the fact that a uniformly continuous function on Ω has a unique continuous extension to $\overline{\Omega}$. The space $C(\overline{\Omega})$ is a Banach space with its canonical norm

$$\|v\|_{C(\overline{\Omega})} = \sup\{|v(\mathbf{x})| \mid \mathbf{x} \in \Omega\} \equiv \max\{|v(\mathbf{x})| \mid \mathbf{x} \in \overline{\Omega}\}.$$

We have $C(\overline{\Omega}) \subseteq C(\Omega)$, and the inclusion is proper; i.e., there are functions $v \in C(\Omega)$ that cannot be extended to a continuous function on $\overline{\Omega}$. A simple example is $v(x) = 1/x$ on $(0, 1)$.

Denote by $\mathbb{Z}_+$ the set of non-negative integers. For any $m \in \mathbb{Z}_+$, $C^m(\Omega)$ is the space of functions that, together with their derivatives of order less than or equal to m, are continuous on Ω; that is,

$$C^m(\Omega) = \{v \in C(\Omega) \mid D^\alpha v \in C(\Omega) \text{ for } |\alpha| \leq m\}.$$

This is a linear space. The notation $C^m(\overline{\Omega})$ denotes the space of functions which, together with their derivatives of order less than or equal to m, are continuous up to the boundary,

$$C^m(\overline{\Omega}) = \{v \in C(\overline{\Omega}) \mid D^\alpha v \in C(\overline{\Omega}) \text{ for } |\alpha| \leq m\}.$$

The space $C^m(\overline{\Omega})$ is a Banach space with the norm

$$\|v\|_{C^m(\overline{\Omega})} = \max_{|\alpha| \leq m} \|D^\alpha v\|_{C(\overline{\Omega})}.$$

Algebraically, $C^m(\overline{\Omega}) \subseteq C^m(\Omega)$. When $m = 0$, we usually write $C(\Omega)$ and $C(\overline{\Omega})$ instead of $C^0(\Omega)$ and $C^0(\overline{\Omega})$. We set

$$C^\infty(\Omega) = \bigcap_{m=0}^{\infty} C^m(\Omega) \equiv \{v \in C(\Omega) \mid v \in C^m(\Omega) \ \forall m \in \mathbb{Z}_+\},$$

$$C^\infty(\overline{\Omega}) = \bigcap_{m=0}^{\infty} C^m(\overline{\Omega}) \equiv \{v \in C(\overline{\Omega}) \mid v \in C^m(\overline{\Omega}) \ \forall m \in \mathbb{Z}_+\}.$$

These are spaces of infinitely differentiable functions.

Given a function v on Ω, its support is defined to be

$$\text{support } v = \overline{\{\mathbf{x} \in \Omega \mid v(\mathbf{x}) \neq 0\}}.$$

We say that v has a *compact support* if support v is a proper subset of Ω: support $v \subset \Omega$. Thus, if v has a compact support, then there is a neighboring open strip about the boundary $\partial\Omega$ such that v is zero on the part of the strip that lies inside Ω. Later on, we need the space

$$C_0^\infty(\Omega) = \{v \in C^\infty(\Omega) \mid \text{support } v \subset \Omega\}.$$

Obviously, $C_0^\infty(\Omega) \subseteq C^\infty(\overline{\Omega})$. In the case Ω is an interval such that $\Omega \supset (-1, 1)$, a standard example of a non-analytic $C_0^\infty(\Omega)$ function is

$$v(x) = \begin{cases} e^{1/(x^2-1)}, & |x| < 1, \\ 0, & \text{otherwise.} \end{cases}$$

1.4.1   Hölder spaces

A function v defined on Ω is said to be *Lipschitz continuous* if for some constant c, there holds the inequality

$$|v(\mathbf{x}) - v(\mathbf{y})| \leq c \|\mathbf{x} - \mathbf{y}\| \qquad \forall \mathbf{x}, \mathbf{y} \in \Omega.$$

In this formula, $\|\mathbf{x} - \mathbf{y}\|$ denotes the standard Euclidean distance between $\mathbf{x}$ and $\mathbf{y}$. The smallest possible constant in the above inequality is called the

Lipschitz constant of v, and is denoted by $\mathrm{Lip}(v)$. The Lipschitz constant is characterized by the relation

$$\mathrm{Lip}(v) = \sup \left\{ \frac{|v(\mathbf{x}) - v(\mathbf{y})|}{\|\mathbf{x} - \mathbf{y}\|} \,\middle|\, \mathbf{x}, \mathbf{y} \in \Omega,\ \mathbf{x} \neq \mathbf{y} \right\}.$$

More generally, a function v is said to be *Hölder continuous* with exponent $\beta \in (0, 1]$ if for some constant c,

$$|v(\mathbf{x}) - v(\mathbf{y})| \leq c \, \|\mathbf{x} - \mathbf{y}\|^{\beta} \qquad \text{for } \mathbf{x}, \mathbf{y} \in \Omega.$$

The Hölder space $C^{0,\beta}(\overline{\Omega})$ is defined to be the subspace of $C(\overline{\Omega})$ that consists of functions that are Hölder continuous with the exponent β. With the norm

$$\|v\|_{C^{0,\beta}(\overline{\Omega})} = \|v\|_{C(\overline{\Omega})} + \sup \left\{ \frac{|v(\mathbf{x}) - v(\mathbf{y})|}{\|\mathbf{x} - \mathbf{y}\|^{\beta}} \,\middle|\, \mathbf{x}, \mathbf{y} \in \Omega,\ \mathbf{x} \neq \mathbf{y} \right\}$$

the space $C^{0,\beta}(\overline{\Omega})$ becomes a Banach space. When $\beta = 1$, the Hölder space $C^{0,1}(\overline{\Omega})$ consists of all the Lipschitz continuous functions.

For $m \in \mathbb{Z}_+$ and $\beta \in (0, 1]$, we similarly define the Hölder space

$$C^{m,\beta}(\overline{\Omega}) = \left\{ v \in C^m(\overline{\Omega}) \mid D^\alpha v \in C^{0,\beta}(\overline{\Omega}) \text{ for all } \alpha \text{ with } |\alpha| = m \right\};$$

this is a Banach space with the norm

$$\|v\|_{C^{m,\beta}(\overline{\Omega})} = \|v\|_{C^m(\overline{\Omega})}$$
$$+ \sum_{|\alpha|=m} \sup \left\{ \frac{|D^\alpha v(\mathbf{x}) - D^\alpha v(\mathbf{y})|}{\|\mathbf{x} - \mathbf{y}\|^{\beta}} \,\middle|\, \mathbf{x}, \mathbf{y} \in \Omega,\ \mathbf{x} \neq \mathbf{y} \right\}.$$

Exercise 1.4.1 *Show that $C(\overline{\Omega})$ with the norm $\|v\|_{C(\overline{\Omega})}$ is a Banach space.*

Exercise 1.4.2 *Show that the space $C^1(\overline{\Omega})$ with the norm $\|v\|_{C(\overline{\Omega})}$ is not a Banach space.*

Exercise 1.4.3 *Let $v_n(x) = \frac{1}{n} \sin nx$. Show that $v_n \to 0$ in $C^{0,\beta}[0,1]$ for any $\beta \in (0, 1)$, but $v_n \not\to 0$ in $C^{0,1}[0,1]$.*

Exercise 1.4.4 *Discuss whether it is meaningful to use the Hölder space $C^{0,\beta}(\overline{\Omega})$ with $\beta > 1$.*

Exercise 1.4.5 *Consider $v(s) = s^\alpha$ for some $0 < \alpha < 1$. For which $\beta \in (0, 1]$ is it true that $v \in C^{0,\beta}[0,1]$?*

1.5 L^p spaces

In the study of $L^p(\Omega)$ spaces, we identify functions (i.e., such functions are considered identical) that are equal almost everywhere (a.e.) on Ω. For

$p \in [1, \infty)$, $L^p(\Omega)$ is the linear space of measurable functions $v : \Omega \to \mathbb{R}$ such that

$$\|v\|_{L^p(\Omega)} = \left\{ \int_\Omega |v(\mathbf{x})|^p dx \right\}^{1/p} < \infty. \tag{1.5.1}$$

The space $L^\infty(\Omega)$ consists of all essentially bounded measurable functions $v : \Omega \to \mathbb{R}$,

$$\|v\|_{L^\infty(\Omega)} = \inf_{\text{meas}\,(\Omega')=0} \sup_{\mathbf{x} \in \Omega \setminus \Omega'} |v(\mathbf{x})| < \infty. \tag{1.5.2}$$

Some basic properties of the L^p spaces are summarized in the following theorem.

Theorem 1.5.1 *Let Ω be an open bounded set in $\mathbb{R}^d$.*

(a) *For $p \in [1, \infty]$, $L^p(\Omega)$ is a Banach space.*

(b) *For $p \in [1, \infty]$, every Cauchy sequence in $L^p(\Omega)$ has a subsequence that converges pointwise a.e. on Ω.*

(c) *If $1 \le p \le q \le \infty$, then $L^q(\Omega) \subseteq L^p(\Omega)$,*

$$\|v\|_{L^p(\Omega)} \le \text{meas}\,(\Omega)^{\frac{1}{p}-\frac{1}{q}} \|v\|_{L^q(\Omega)} \qquad \forall\, v \in L^q(\Omega),$$

and

$$\|v\|_{L^\infty(\Omega)} = \lim_{p \to \infty} \|v\|_{L^p(\Omega)} \qquad \forall\, v \in L^\infty(\Omega).$$

(d) *If $1 \le p \le r \le q \le \infty$ and we choose $\theta \in [0, 1]$ such that*

$$\frac{1}{r} = \frac{\theta}{p} + \frac{(1-\theta)}{q},$$

then

$$\|v\|_{L^r(\Omega)} \le \|v\|_{L^p(\Omega)}^\theta \|v\|_{L^q(\Omega)}^{1-\theta} \qquad \forall\, v \in L^q(\Omega).$$

In (c), when $q = \infty$, $1/q$ is understood to be 0. The result (d) is called an *interpolation property* of the L^p spaces. To prove (c) and (d), we need to use the *Hölder inequality*. We first prove *Young's inequality*.

Lemma 1.5.2 (YOUNG'S INEQUALITY) *Let $a, b \ge 0$, $p, q > 1$, $1/p+1/q = 1$. Then*

$$ab \le \frac{a^p}{p} + \frac{b^q}{q}.$$

Proof. For any fixed $b \ge 0$, define a function

$$f(a) = \frac{a^p}{p} + \frac{b^q}{q} - ab$$

on $[0, \infty)$. From $f'(a) = 0$ we obtain $a = b^{1/(p-1)}$. We have $f(b^{1/(p-1)}) = 0$. Since $f(0) \ge 0$, $\lim_{a \to \infty} f(a) = \infty$ and f is continuous on $[0, \infty)$, we see that

$$\inf_{0 \leq a < \infty} f(a) = f(b^{1/(p-1)}) = 0.$$

Hence Young's inequality holds.   ∎

Lemma 1.5.3 (MODIFIED YOUNG'S INEQUALITY) *Let $a, b \geq 0$, $\varepsilon > 0$, $p, q > 1$, $1/p + 1/q = 1$. Then*

$$ab \leq \frac{\varepsilon \, a^p}{p} + \frac{\varepsilon^{1-q} b^q}{q}.$$

Lemma 1.5.4 (HÖLDER'S INEQUALITY) *Let $u \in L^p(\Omega)$, $v \in L^q(\Omega)$, $p, q \geq 1$, $1/p + 1/q = 1$. Then*

$$\int_\Omega |u(\mathbf{x}) \, v(\mathbf{x})| \, dx \leq \|u\|_{L^p(\Omega)} \|v\|_{L^q(\Omega)}.$$

Remark 1.5.5 *If $p = 1$, then $q = \infty$. Formally, we write $1/\infty = 0$.*

Proof. (HÖLDER'S INEQUALITY) The inequality is obviously true if $p = 1$ or ∞, or $\|u\|_{L^p(\Omega)} = 0$. For $p \in (1, \infty)$ and $\|u\|_{L^p(\Omega)} \neq 0$, we use the modified Young's inequality to obtain

$$\int_\Omega |u(\mathbf{x}) \, v(\mathbf{x})| \, dx \leq \frac{\varepsilon}{p} \|u\|_{L^p(\Omega)}^p + \frac{\varepsilon^{1-q}}{q} \|v\|_{L^q(\Omega)}^q \quad \forall \varepsilon > 0.$$

Then we set $\varepsilon = \|v\|_{L^q(\Omega)} / \|u\|_{L^p(\Omega)}^{p-1}$; this choice of ε minimizes the value of the right-hand side of the above inequality.   ∎

To show that (1.5.1) (and (1.5.2)) defines a norm, we need to verify the triangle inequality, which in this case is called the *Minkowski inequality*.

Lemma 1.5.6 (MINKOWSKI INEQUALITY)

$$\|u + v\|_{L^p(\Omega)} \leq \|u\|_{L^p(\Omega)} + \|v\|_{L^p(\Omega)} \quad \forall u, v \in L^p(\Omega), \ p \in [1, \infty].$$

Proof. The inequality is obviously true for $p = 1$ and ∞. Suppose $p \in (1, \infty)$. Applying the Hölder inequality, we have

$$\int_\Omega |u(\mathbf{x}) + v(\mathbf{x})|^p dx$$

$$\leq \int_\Omega |u(\mathbf{x}) + v(\mathbf{x})|^{p-1} |u(\mathbf{x})| \, dx + \int_\Omega |u(\mathbf{x}) + v(\mathbf{x})|^{p-1} |v(\mathbf{x})| \, dx$$

$$\leq \left(\int_\Omega |u(\mathbf{x}) + v(\mathbf{x})|^{(p-1)q} dx \right)^{1/q} \left(\|u\|_{L^p(\Omega)} + \|v\|_{L^p(\Omega)} \right)$$

$$= \left(\int_\Omega |u(\mathbf{x}) + v(\mathbf{x})|^p dx \right)^{1-\frac{1}{p}} \left(\|u\|_{L^p(\Omega)} + \|v\|_{L^p(\Omega)} \right).$$

Therefore, the Minkowski inequality holds.   ∎

Smooth functions are dense in $L^p(\Omega)$, $1 \leq p < \infty$.

Theorem 1.5.7 *Let $\Omega \subseteq \mathbb{R}^d$ be an open set, $1 \le p < \infty$. Then the space $C_0^\infty(\Omega)$ is dense in $L^p(\Omega)$; in other words, for any $v \in L^p(\Omega)$, there exists a sequence $\{v_n\} \subseteq C_0^\infty(\Omega)$ such that*

$$\|v_n - v\|_{L^p(\Omega)} \to 0 \quad \text{as } n \to \infty.$$

For any $m \in \mathbb{Z}_+$, by noting the inclusions $C_0^\infty(\Omega) \subseteq C^m(\overline{\Omega}) \subseteq L^p(\Omega)$, we see that the space $C^m(\overline{\Omega})$ is also dense in $L^p(\Omega)$.

Exercise 1.5.1 *Prove the modified Young's inequality by applying Young's inequality.*

Exercise 1.5.2 *Lemma 1.5.2 is a special case of the following general Young's inequality: Let $f : [0, \infty) \to [0, \infty)$ be a continuous, strictly increasing function such that $f(0)$, $\lim_{x \to \infty} f(x) = \infty$. Denote g as the inverse function of f. For $0 \le x < \infty$, define*

$$F(x) = \int_0^x f(t)\, dt, \qquad G(x) = \int_0^x g(t)\, dt.$$

Then

$$ab \le F(a) + F(b), \qquad \forall a, b \ge 0,$$

and the equality holds if and only if $b = f(a)$. Prove this result and deduce Lemma 1.5.2 from it.

Exercise 1.5.3 *Show the generalized Hölder inequality*

$$\left| \int_\Omega v_1 \cdots v_m\, dx \right| \le \|v_1\|_{L^{p_1}(\Omega)} \cdots \|v_m\|_{L^{p_m}(\Omega)} \quad \forall v_i \in L^{p_i}(\Omega),\ 1 \le i \le m,$$

where the exponents $p_i > 0$ satisfy the relation $\sum_{i=1}^m 1/p_i = 1$.

Exercise 1.5.4 *Consider the function*

$$f(x) = \begin{cases} e^{-1/x^2}, & x > 0 \\ 0, & x = 0. \end{cases}$$

Prove

$$\lim_{x \searrow 0} f^{(m)}(x) = 0$$

for all integers $m \ge 0$.

1.6 Compact sets

There are several definitions of the concept of compact set, most being equivalent in the setting of a normed linear space.

Definition 1.6.1 (a) *Let S be a subset of a normed linear space V. We say S has an* open covering *by a collection of open sets $\{U_\alpha \mid \alpha \in \Lambda\}$, Λ an index set, if*

$$S \subseteq \bigcup_{\alpha \in \Lambda} U_\alpha.$$

We say S is compact *if for every open covering $\{U_\alpha\}$ of S, there is a finite subcover $\{U_{\alpha_j} \mid j = 1, \dots, m\} \subseteq \{U_\alpha \mid \alpha \in \Lambda\}$ which also covers S.*

(b) *Equivalently, S is* compact *if every sequence $\{x_j\} \subseteq S$ contains a convergent subsequence $\{x_{j_k}\}$ that converges to a point $x \in S$.*

(c) *If S is a set for which $\overline{S}$ is compact, we say S is* precompact.

Part (a) of the definition is the general definition of compact set, valid in general topological spaces; and (b) is the usual form used in metric spaces (spaces with a distance function defining the topology of the space). In every normed linear space, a compact set is both closed and bounded.

For finite dimensional spaces, the compact sets are readily identified.

Theorem 1.6.2 (HEINE-BOREL THEOREM) *Let V be a finite-dimensional normed linear space, and let S be a subset of V. Then S is compact if and only if S is both closed and bounded.*

A proof can be found in most textbooks on advanced calculus; for example, see [4, Theorems 3–38, 3–40]. For infinite-dimensional normed linear spaces, it is more difficult to identify the compact sets. The results are dependent on the properties of the norm being used. We give an important result for the space of continuous functions $C(D)$ with the uniform norm $\|\cdot\|_\infty$, with some set $D \subseteq \mathbb{R}^d$. A proof is given in [88, p. 27].

Theorem 1.6.3 (ARZELA-ASCOLI THEOREM) *Let $S \subseteq C(D)$, with $D \subseteq \mathbb{R}^d$ closed and bounded. Suppose that the functions in S are uniformly bounded and equicontinuous over D, meaning that*

$$\sup_{f \in S} \|f\|_\infty < \infty$$

and

$$|f(\mathbf{x}) - f(\mathbf{y})| \le c_S(\varepsilon) \quad for \quad \|\mathbf{x} - \mathbf{y}\| \le \varepsilon \qquad \forall f \in S$$

with $c_S(\varepsilon) \to 0$ as $\varepsilon \to 0$. Then S is precompact in $C(D)$.

In Chapter 6, we review compact embedding results for Sobolev spaces, and these provide examples of compact sets in Sobolev spaces.

Suggestion for Further Readings

Detailed discussions of normed spaces, Banach spaces, Hilbert spaces, linear operators on normed spaces and their properties are found in most textbooks on Functional Analysis; see for example CONWAY [40], HUTSON

AND PYM [81], KANTOROVICH AND AKILOV [88], KESAVAN [94], KREYSZIG [103], ZEIDLER [174, 175].

In this work, we emphasize the ability to understand and correctly apply results from functional analysis in order to analyse various numerical methods and procedures. An important pioneering paper that advocated and developed this approach to numerical analysis is that of L.V. KANTOROVICH [87]. It was published in 1948; and much of the work of this paper appears in expanded form in KANTOROVICH AND AKILOV [88, chaps. 14–18]. Another associated and influential work is the text of KANTOROVICH AND KRYLOV [89], which appeared in several editions over a 30-year period. Other important general texts that set the study of numerical analysis within a framework of functional analysis include AUBIN [17], COLLATZ [38], CRYER [41], LEBEDEV [105], LINZ [108], and MOORE [121].

2
Linear Operators on Normed Spaces

Many of the basic problems of applied mathematics share the property of *linearity*, and linear spaces and linear operators provide a general and useful framework for the analysis of such problems. More complicated applications often involve nonlinear operators, and a study of linear operators also offers some useful tools for the analysis of nonlinear operators. In this chapter we review some basic results on linear operators, and we give some illustrative applications to obtain results in numerical analysis. Some of the results are quoted without proof; and usually the reader can find detailed proofs of the results in a standard textbook on functional analysis (e.g., see Conway [40], Kantorovich and Akilov [88], and Zeidler [174, 175]).

Linear operators are used in expressing mathematical problems, often leading to equations to be solved or to functions to be optimized. To examine the theoretical solvability of a mathematical problem and to develop numerical methods for its solution, we must know additional properties about the operators involved in our problem. The most important such properties in applied mathematics involve one of the following concepts or some mix of them.

- Closeness to a problem whose solvability theory is known. The *Geometric Series Theorem* given in Section 2.3 is the basis of most results for linear operator equations in this category.

- Closeness to a finite dimensional problem. One variant of this leads to the theory of *completely continuous* or *compact* linear operators, which is taken up in Section 2.8.

- Arguments based on finding the minimum of a function, with the point at which the minimum is attained being the solution to the problem being studied. The function being minimized is sometimes called an *energy* function. This is taken up in later chapters, but some of its framework is provided in the material of this chapter.

There are other important means of examining the solvability of mathematical problems in applied mathematics, based on Fourier analysis, complex analysis, positivity of an operator within the context of partially order linear spaces, and other techniques. However, we make only minimal use of such tools in this text.

2.1   Operators

Given two sets V and W, an *operator* T from V to W is a rule that assigns to each element in a subset of V a unique element in W. The *domain* $\mathcal{D}(T)$ of T is the subset of V where T is defined,

$$\mathcal{D}(T) = \{v \in V \mid T(v) \text{ is defined}\},$$

and the *range* $\mathcal{R}(T)$ of T is the set of the elements in W generated by T,

$$\mathcal{R}(T) = \{w \in W \mid w = T(v) \text{ for some } v \in \mathcal{D}(T)\}.$$

It is also useful to define the *null set*, the set of the zeros of the operator,

$$\mathcal{N}(T) = \{v \in V \mid T(v) = 0\}.$$

An operator is sometimes also called a mapping, a transformation, or a function. Usually the domain $\mathcal{D}(T)$ is understood to be the whole set V, unless it is stated explicitly to be otherwise.

Addition and scalar multiplication of operators are defined as they are for ordinary functions. Let S and T be operators mapping from V to W. Then $S + T$ is an operator from V to W with the domain $\mathcal{D}(S) \cap \mathcal{D}(T)$ and the rule

$$(S + T)(v) = S(v) + T(v) \qquad \forall\, v \in \mathcal{D}(S) \cap \mathcal{D}(T).$$

Let $\alpha \in \mathbb{K}$. Then αT is an operator from V to W with the domain $\mathcal{D}(T)$ and the rule

$$(\alpha T)(v) = \alpha T(v) \qquad \forall\, v \in \mathcal{D}(T).$$

Definition 2.1.1 *An operator* $T : V \to W$ *is said to be* one-to-one *or* injective *if*

$$v_1 \neq v_2 \quad \Longrightarrow \quad T(v_1) \neq T(v_2). \tag{2.1.1}$$

The operator is said to map V onto W or is called surjective *if $\mathcal{R}(T) = W$. If T is both injective and surjective, it is called a* bijection *from V to W.*

Evidently, when $T : V \to W$ is bijective, we can define its inverse $T^{-1} :$ $W \to V$ by the rule

$$v = T^{-1}(w) \quad \Longleftrightarrow \quad w = T(v).$$

More generally, if $T : V \to W$ is one-to-one, we can define its inverse from $\mathcal{R}(T) \subseteq W$ to V by using the above rule.

Example 2.1.2 *Let V be a linear space. The* identity operator $I : V \to V$ *is defined by*

$$I(v) = v \qquad \forall\, v \in V.$$

It is a bijection from V to V; and moreover, its inverse is also the identity operator.

Example 2.1.3 *Let $V = \mathbb{C}^n$, $W = \mathbb{C}^m$, and $L(\mathbf{v}) = A\mathbf{v}$, $\mathbf{v} \in \mathbb{C}^n$, where $A = (a_{ij}) \in \mathbb{C}^{m \times n}$ is a complex matrix and $A\mathbf{v}$ denotes matrix-vector multiplication. From results in finite-dimensional linear algebra, the operator L is injective if and only if $\mathrm{rank}(A) = n$; and L is surjective if and only if $\mathrm{rank}(A) = m$.*

Example 2.1.4 *We consider the differentiation operator d/dx from $V = C[0,1]$ to $W = C[0,1]$ defined by*

$$\frac{d}{dx} : v \mapsto v' \quad \text{for } v \in C^1[0,1].$$

We take the domain of the operator, $\mathcal{D}(d/dx)$, to be $C^1[0,1]$, which is a proper subspace of $C[0,1]$. It can be verified that the differentiation operator is a surjection, $\mathcal{R}(d/dx) = C[0,1]$. The differentiation operator is not injective, and its null set is the set of constant functions.

Example 2.1.5 *Although the differentiation operator d/dx is not injective from $C^1[0,1]$ to $C[0,1]$, the following operator*

$$D : v(x) \mapsto \begin{pmatrix} v'(x) \\ v(0) \end{pmatrix}$$

is a bijection between $V = C^1[0,1]$ and $W = C[0,1] \times \mathbb{R}$.

If both V and W are normed spaces, we can talk about the continuity and boundedness of the operators.

Definition 2.1.6 *Let V and W be two normed spaces. An operator $T : V \to W$ is* continuous *at $v \in \mathcal{D}(T)$ if*

$$\{v_n\} \subseteq \mathcal{D}(T) \quad \text{and} \quad v_n \to v \text{ in } V \quad \Longrightarrow \quad T(v_n) \to T(v) \text{ in } W.$$

T is said to be continuous *if it is continuous over its domain $\mathcal{D}(T)$. The operator is* bounded *if for any $r > 0$ there is an $R > 0$ such that*

$$v \in \mathcal{D}(T) \quad \text{and} \quad \|v\| \le r \quad \Longrightarrow \quad \|T(v)\| \le R.$$

We observe that an alternative definition of the boundedness is that for any set $B \subseteq \mathcal{D}(T)$,

$$\sup_{v \in B} \|v\|_V < \infty \quad \Longrightarrow \quad \sup_{v \in B} \|T(v)\|_W < \infty.$$

Example 2.1.7 *Let us consider the differentiation operator again. The spaces* $C[0,1]$ *and* $C^1[0,1]$ *are associated with their standard norms*

$$\|v\|_{C[0,1]} = \max_{0 \leq x \leq 1} |v(x)|$$

and

$$\|v\|_{C^1[0,1]} = \|v\|_{C[0,1]} + \|v'\|_{C[0,1]}. \tag{2.1.2}$$

Then the operator

$$T_1 = \frac{d}{dx} : C^1[0,1] \subseteq C[0,1] \to C[0,1]$$

is not continuous using the infinity norm of $C[0,1]$, *while the operator*

$$T_2 = \frac{d}{dx} : C^1[0,1] \to C[0,1]$$

is continuous using the norm of (2.1.2).

Exercise 2.1.1 *Consider Example 2.1.7. Show that* T_1 *is unbounded and* T_2 *is bounded, as asserted in the example.*

2.2 Continuous linear operators

This chapter is focused on the analysis of a particular type of operators called linear operators. From now on, when we write $T : V \to W$, we implicitly assume $\mathcal{D}(T) = V$, unless stated otherwise. As in Chapter 1, $\mathbb{K}$ denotes the set of scalars associated with the vector space under consideration.

Definition 2.2.1 *Let* V *and* W *be two linear spaces. An operator* $L : V \to W$ *is said to be* linear *if*

$$L(\alpha_1 v_1 + \alpha_2 v_2) = \alpha_1 L(v_1) + \alpha_2 L(v_2) \quad \forall\, v_1, v_2 \in V, \ \forall\, \alpha_1, \alpha_2 \in \mathbb{K},$$

or equivalently,

$$L(v_1 + v_2) = L(v_1) + L(v_2) \quad \forall\, v_1, v_2 \in V,$$
$$L(\alpha\, v) = \alpha\, L(v) \quad \forall\, v \in V, \ \forall\, \alpha \in \mathbb{K}.$$

For a linear operator L, *we usually write* $L(v)$ *as* Lv.

An important property of a linear operator is that the continuity and boundedness are equivalent. We state and prove this result in the form of a theorem after two preparatory propositions that are themselves important.

Proposition 2.2.2 *Let V and W be normed spaces, $L : V \to W$ a linear operator. Then continuity of L over the whole space is equivalent to its continuity at any one point, say, at $v = 0$.*

Proof. Assume

$$v_n \to 0 \implies Lv_n \to 0. \qquad (2.2.1)$$

Let $v \in V$ be arbitrarily given and $\{v_n\} \subseteq V$ be a sequence converging to v. Then $v_n - v \to 0$, and by (2.2.1), $L(v_n - v) = Lv_n - Lv \to 0$; i.e., $Lv_n \to Lv$. Hence L is continuous at v.   ∎

Proposition 2.2.3 *Let V and W be normed spaces, $L : V \to W$ a linear operator. Then L is bounded if and only if there exists a constant $\gamma \geq 0$ such that*

$$\|Lv\|_W \leq \gamma \|v\|_V \qquad \forall\, v \in V. \qquad (2.2.2)$$

Proof. Obviously (2.2.2) implies the boundedness. Conversely, suppose L is bounded, then

$$\gamma \equiv \sup_{v \in B_1} \|Lv\|_W < \infty,$$

where $B_1 = \{v \in V \mid \|v\|_V \leq 1\}$ is the unit ball centered at 0. Now for any $v \neq 0$, $v/\|v\|_V \in B_1$ and by the linearity of L,

$$\|Lv\|_W = \|v\|_V\, \|L(v/\|v\|_V)\|_W \leq \gamma \|v\|_V.$$

    ∎

Theorem 2.2.4 *Let V and W be normed spaces, $L : V \to W$ a linear operator. Then L is continuous on V if and only if it is bounded on V.*

Proof. Firstly we assume L is not bounded and prove that it is not continuous at 0. Since L is unbounded, we can find a bounded sequence $\{v_n\} \subseteq V$ such that $\|Lv_n\| \to \infty$. Without loss of generality, we may assume $Lv_n \neq 0$ for all n. Then we define a new sequence

$$\tilde{v}_n = \frac{v_n}{\|Lv_n\|_W}.$$

This sequence has the property that $\tilde{v}_n \to 0$ and $\|L\tilde{v}_n\|_W = 1$. Thus L is not continuous.

Secondly we assume L is bounded and show that it must be continuous. Indeed from (2.2.2) we have the Lipschitz inequality

$$\|Lv_1 - Lv_2\|_W \leq \gamma \|v_1 - v_2\|_V \qquad \forall\, v_1, v_2 \in V, \qquad (2.2.3)$$

which implies the continuity in an obvious fashion.   ∎

From (2.2.3), we see that for a linear operator, continuity and Lipschitz continuity are equivalent.

We use the notation $\mathcal{L}(V, W)$ for the set of all the continuous linear operators from a normed space V to another normed space W. In the special case $W = V$, we use $\mathcal{L}(V)$ to replace $\mathcal{L}(V, V)$. We see that for a linear operator, boundedness (2.2.2) is equivalent to continuity. Thus if $L \in \mathcal{L}(V, W)$, it is meaningful to define

$$\|L\|_{V,W} = \sup_{0 \neq v \in V} \frac{\|Lv\|_W}{\|v\|_V}. \tag{2.2.4}$$

Using the linearity of L, we have the following relations

$$\|L\|_{V,W} = \sup_{v \in B_1} \|Lv\|_W = \sup_{v : \|v\|_V = 1} \|Lv\|_W = \frac{1}{r} \sup_{v : \|v\|_V = r} \|Lv\|_W$$

for any $r > 0$. The norm $\|L\|_{V,W}$ is the maximum size in W of the image under L of the unit ball B_1 in V.

Theorem 2.2.5 *The set $\mathcal{L}(V, W)$ is a linear space, and (2.2.4) defines a norm over the space.*

We leave the proof of the theorem to the reader. The norm (2.2.4) is usually called the *operator norm* of L, which enjoys the following compatibility property

$$\|Lv\|_W \leq \|L\|_{V,W} \|v\|_V \qquad \forall\, v \in V. \tag{2.2.5}$$

If it is not stated explicitly, we always understand the norm of an operator as an operator norm defined by (2.2.4). Another useful inequality involving operator norms is given in the following result.

Theorem 2.2.6 *Let U, V and W be normed spaces, and let $S : U \to V$ and $T : V \to W$ be continuous linear operators. Then the composite operator $TS : U \to W$ defined by*

$$TS(v) = T(S(v)) \quad \forall\, v \in U$$

is a continuous linear mapping. Moreover,

$$\|TS\|_{U,W} \leq \|S\|_{U,V} \|T\|_{V,W}. \tag{2.2.6}$$

Proof. We only need to prove (2.2.6). By (2.2.5), for any $v \in U$,

$$\|TS(v)\|_W = \|T(S(v))\|_W \leq \|T\|_{V,W} \|Sv\|_V \leq \|T\|_{V,W} \|S\|_{U,V} \|v\|_U.$$

Hence, (2.2.6) is valid. ∎

As an important special case, if V is a normed space and if $L \in \mathcal{L}(V)$, then for any positive integer n,

$$\|L^n\| \leq \|L\|^n.$$

Both (2.2.5) and (2.2.6) are very useful relations for the error analysis of some numerical methods.

For a linear operator, the null set $\mathcal{N}(L)$ becomes a subspace of V, and we have the statement

$$L \text{ is one-to-one} \quad \Longleftrightarrow \quad \mathcal{N}(L) = \{0\}.$$

Example 2.2.7 *Let V be a linear space. Then the identity operator I : $V \rightarrow V$ belongs to $\mathcal{L}(V)$, and $\|I\| = 1$.*

Example 2.2.8 *Recall Example 2.1.3. Let $V = \mathbb{C}^n$, $W = \mathbb{C}^m$, and $L(\mathbf{v}) = A\mathbf{v}$, $\mathbf{v} \in \mathbb{C}^n$, where $A = (a_{ij}) \in \mathbb{C}^{m \times n}$ is a complex matrix. If the norms on V and W are $\| \cdot \|_\infty$, then the operator norm is the matrix ∞-norm,*

$$\|A\|_\infty = \max_{1 \le i \le m} \sum_{j=1}^{n} |a_{ij}|.$$

If the norms on V and W are $\| \cdot \|_1$, then the operator norm is the matrix 1-norm,

$$\|A\|_1 = \max_{1 \le j \le n} \sum_{i=1}^{m} |a_{ij}|.$$

If the norms on V and W are $\| \cdot \|_2$, then the operator norm is the spectral norm

$$\|A\|_2 = \sqrt{r_\sigma(A^*A)}.$$

For a square matrix B, $r_\sigma(B)$ denotes the spectral radius of the matrix B,

$$r_\sigma(B) = \max_{\lambda \in \sigma(B)} |\lambda|$$

and $\sigma(B)$ denotes the spectrum of B, the set of all the eigenvalues of B. Proofs of these results are given in [11, Section 7.3].

Example 2.2.9 *Let $V = W = C[a,b]$ with the norm $\| \cdot \|_\infty$. Let $k \in C([a,b]^2)$, and define $K : C[a,b] \rightarrow C[a,b]$ by*

$$(Kv)(x) = \int_a^b k(x,y)\, v(y)\, dy. \tag{2.2.7}$$

The mapping K in (2.2.7) is an example of a linear integral operator, *and the function $k(\cdot, \cdot)$ is called the* kernel function *of the integral operator. Under the continuity assumption on $k(\cdot, \cdot)$, the integral operator is continuous from $C[a,b]$ to $C[a,b]$. Furthermore,*

$$\|K\| = \max_{a \le x \le b} \int_a^b |k(x,y)|\, dy. \tag{2.2.8}$$

The linear integral operator (2.2.7) is later used extensively.

2.2.1 $\mathcal{L}(V, W)$ as a Banach space

In approximating integral and differential equations, the integral or differential operator is often approximated by a sequence of operators of a simpler form. In such cases, it is important to consider the limits of convergent sequences of bounded operators, and this makes it important to have $\mathcal{L}(V, W)$ be a complete space.

Theorem 2.2.10 *Let V be a normed space, and W be a Banach space. Then $\mathcal{L}(V, W)$ is a Banach space.*

Proof. Let $\{L_n\}$ be a Cauchy sequence in $\mathcal{L}(V, W)$. This means

$$\epsilon_n \equiv \sup_{p \geq 1} \|L_{n+p} - L_n\| \to 0 \quad \text{as } n \to \infty.$$

We must define a limit for $\{L_n\}$ and show that it belongs to $\mathcal{L}(V, W)$.
For each $v \in V$,

$$\|L_{n+p}v - L_n v\|_W \leq \epsilon_n \|v\|_V \to 0 \quad \text{as } n \to \infty. \tag{2.2.9}$$

Thus $\{L_n v\}$ is a Cauchy sequence in W. Since W is complete, the sequence has a limit, denoted by $L(v)$. This defines an operator $L : V \to W$. Let us prove that L is linear, bounded, and $\|L_n - L\|_{V,W} \to 0$ as $n \to \infty$.
For any $v_1, v_2 \in V$ and $\alpha_1, \alpha_2 \in \mathbb{K}$,

$$\begin{aligned}
L(\alpha_1 v_1 + \alpha_2 v_2) &= \lim_{n \to \infty} L_n(\alpha_1 v_1 + \alpha_2 v_2) \\
&= \lim_{n \to \infty} (\alpha_1 L_n v_1 + \alpha_2 L_n v_2) \\
&= \alpha_1 \lim_{n \to \infty} L_n v_1 + \alpha_2 \lim_{n \to \infty} L_n v_2 \\
&= \alpha_1 L(v_1) + \alpha_2 L(v_2).
\end{aligned}$$

Thus L is linear.
Now for any $v \in V$, we take the limit $p \to \infty$ in (2.2.9) to obtain

$$\|Lv - L_n v\|_W \leq \epsilon_n \|v\|_V.$$

Thus

$$\|L - L_n\|_{V,W} = \sup_{\|v\|_V \leq 1} \|Lv - L_n v\|_W \leq \epsilon_n \to 0 \quad \text{as } n \to \infty.$$

Hence $L \in \mathcal{L}(V, W)$ and $L_n \to L$ as $n \to \infty$. ∎

Exercise 2.2.1 *For a linear operator $L : V \to W$, show that $L(0) = 0$.*

Exercise 2.2.2 *Prove that $\mathcal{L}(V, W)$ is a linear space, and (2.2.4) defines a norm on the space.*

Exercise 2.2.3 *Assume $k \in C([a, b]^2)$. Show that the operator K defined by (2.2.7) is continuous, and its operator norm is given by the formula (2.2.8).*

Exercise 2.2.4 *A linear operator L is called nonsingular if $\mathcal{N}(L) = \{0\}$. Otherwise it is called singular. Show that if L is nonsingular, then a solution of the equation $Lu = f$ is unique.*

Exercise 2.2.5 *If a linear operator $L : V \to W$ is nonsingular and maps V onto W, then for any $f \in W$, the equation $Lu = f$ has a unique solution $u \in V$.*

2.3   The geometric series theorem and its variants

The following result is used commonly in numerical analysis and applied mathematics. It is also the means by which we can analyze the solvability of problems that are "close" to another problem known to be uniquely solvable.

Theorem 2.3.1 (GEOMETRIC SERIES THEOREM) *Let V be a Banach space, $L \in \mathcal{L}(V)$. Assume*

$$\|L\| < 1. \tag{2.3.1}$$

Then $I - L$ is a bijection on V, its inverse is a bounded linear operator, and

$$\|(I - L)^{-1}\| \le \frac{1}{1 - \|L\|}. \tag{2.3.2}$$

Proof. Define a sequence in $\mathcal{L}(V)$: $M_n = \sum_{i=0}^{n} L^i$, $n \ge 0$. For $p \ge 1$,

$$\|M_{n+p} - M_n\| = \left\|\sum_{i=n+1}^{n+p} L^i\right\| \le \sum_{i=n+1}^{n+p} \|L^i\| \le \sum_{i=n+1}^{n+p} \|L\|^i.$$

Using the assumption (2.3.1), we have

$$\|M_{n+p} - M_n\| \le \frac{\|L\|^{n+1}}{1 - \|L\|}. \tag{2.3.3}$$

Hence,

$$\sup_{p \ge 1} \|M_{n+p} - M_n\| \to 0 \quad \text{as } n \to \infty,$$

and $\{M_n\}$ is a Cauchy sequence in $\mathcal{L}(V)$. Since $\mathcal{L}(V)$ is complete, there is an $M \in \mathcal{L}(V)$ with

$$\|M_n - M\| \to 0 \quad \text{as } n \to \infty.$$

Using the definition of M_n and simple algebraic manipulation,

$$(I - L)\, M_n = M_n(I - L) = I - L^{n+1}.$$

Let $n \to \infty$ to get

$$(I - L)\, M = M\, (I - L) = I.$$

This relation implies $I - L$ is a bijection, and

$$M = (I - L)^{-1} = \lim_{n \to \infty} \sum_{i=0}^{n} L^i = \sum_{n=0}^{\infty} L^n.$$

To prove the bound (2.3.2), first note that

$$\|M_n\| \le \sum_{i=0}^{n} \|L\|^i \le \frac{1}{1 - \|L\|}.$$

Taking the limit $n \to \infty$, we obtain (2.3.2). ■

The theorem says that under the stated assumptions, for any $f \in V$, the equation $(I - L)\, u = f$ has a unique solution $u = (I - L)^{-1} f \in V$. Moreover, the solution depends continuously on the right-hand side f: Letting $(I - L)\, u_1 = f_1$ and $(I - L)\, u_2 = f_2$, it follows that

$$u_1 - u_2 = (I - L)^{-1}\, (f_1 - f_2),$$

and so

$$\|u_1 - u_2\| \le c\, \|f_1 - f_2\|$$

with $c = 1/(1 - \|L\|)$.

Example 2.3.2 *Consider the linear integral equation of the second kind*

$$\lambda\, u(x) - \int_a^b k(x, y)\, u(y)\, dy = f(x), \quad a \le x \le b \qquad (2.3.4)$$

with $\lambda \ne 0$, $k(x, y)$ continuous for $x, y \in [a, b]$, and $f \in C[a, b]$. Let $V = C[a, b]$ with the norm $\| \cdot \|_\infty$. Symbolically, we write

$$(\lambda I - K)\, u = f \qquad (2.3.5)$$

where K is the linear integral operator generated by the kernel function $k(\cdot, \cdot)$. We also will often write this as $(\lambda - K)\, u = f$, understanding it to mean the same as in (2.3.5).

This equation (2.3.5) can be converted into the form needed in the geometric series theorem:

$$(I - L)\, u = \frac{1}{\lambda}\, f, \qquad L = \frac{1}{\lambda}\, K.$$

Applying the geometric series theorem, we assert that if

$$\|L\| = \frac{1}{|\lambda|} \|K\| < 1,$$

then $(I - L)^{-1}$ exists and

$$\|(I - L)^{-1}\| \le \frac{1}{1 - \|L\|}.$$

Equivalently, if

$$\|K\| = \max_{a \le x \le b} \int_a^b |K(x,y)|\, dy < |\lambda|, \tag{2.3.6}$$

then $(\lambda I - K)^{-1}$ *exists and*

$$\|(\lambda I - K)^{-1}\| \le \frac{1}{|\lambda| - \|K\|}.$$

Hence under the assumption (2.3.6), for any $f \in C[a,b]$, *the integral equation (2.3.4) has a unique solution* $u \in C[a,b]$ *and*

$$\|u\|_\infty \le \|(\lambda I - K)^{-1}\|\, \|f\|_\infty \le \frac{\|f\|_\infty}{|\lambda| - \|K\|}.$$

We observe that the geometric series theorem is a straightforward generalization to linear continuous operators on a Banach space of the power series

$$(1-x)^{-1} = \sum_{n=0}^\infty x^n, \qquad |x| < 1$$

or its complex version

$$(1-z)^{-1} = \sum_{n=0}^\infty z^n, \qquad z \in \mathbb{C},\ |z| < 1.$$

From the proof of the theorem we see that for a linear operator $L \in \mathcal{L}(V)$ over a Banach space V, if $\|L\| < 1$, then the series $\sum_{n=0}^\infty L^n$ converges in $\mathcal{L}(V)$ and the value of the series is the operator $(I - L)^{-1}$. More generally, we can similarly define an operator-valued function $f(L)$ of an operator variable L from a real function $f(x)$ of a real variable x (or a complex-valued function $f(z)$ of a complex variable z), as long as $f(x)$ is analytic at $x = 0$; i.e.,

$$f(x) = \sum_{n=0}^\infty a_n x^n, \qquad |x| < \gamma$$

for some constant $\gamma > 0$, where $a_n = f^{(n)}(0)/n!$, $n \ge 0$. Now if V is a Banach space and $L \in \mathcal{L}(V)$ satisfies $\|L\| < \gamma$, then we define

$$f(L) = \sum_{n=0}^\infty a_n L^n.$$

The series on the right-hand side is a well-defined operator in $\mathcal{L}(V)$, thanks to the assumption $\|L\| < \gamma$. We now give some examples of operator-valued functions obtained by this approach, with $L \in \mathcal{L}(V)$ and V a Banach space:

$$e^L = \sum_{n=0}^{\infty} \frac{1}{n!} L^n,$$

$$\sin(L) = \sum_{n=0}^{\infty} \frac{(-1)^n}{(2n+1)!} L^{2n+1},$$

$$\arctan(L) = \sum_{n=0}^{\infty} \frac{(-1)^n}{2n+1} L^{2n+1}, \quad \|L\| < 1.$$

2.3.1 A generalization

To motivate a generalization of Theorem 2.3.1, consider the *Volterra integral equation of the second kind*

$$u(x) - \int_0^x \ell(x,y)u(y)\, dy = f(x), \qquad x \in [0, B]. \tag{2.3.7}$$

Here, $B > 0$ and we assume the kernel function $\ell(x, y)$ is continuous for $0 \leq y \leq x \leq B$, and $f \in C[0, B]$. We can use a variant of the geometric series theorem to show that this equation is uniquely solvable, irregardless of the size of the kernel function $\ell(x, y)$. Symbolically, we write this integral equation as $(I - L)\, u = f$.

Corollary 2.3.3 *Let V be a Banach space, $L \in \mathcal{L}(V)$. Assume for some $m \geq 2$ that*

$$\|L^m\| < 1. \tag{2.3.8}$$

Then $I - L$ is a bijection on V, its inverse is a bounded linear operator, and

$$\| (I - L)^{-1} \| \leq \frac{1}{1 - \|L^m\|} \sum_{i=0}^{m-1} \|L^i\| .$$

Proof. From Theorem 2.3.1, we know that $(I - L^m)^{-1}$ exists as a bounded bijective operator on V to V, with

$$\|(I - L^m)^{-1}\| \leq \frac{1}{1 - \|L^m\|}$$

and with the series

$$\sum_{j=0}^{\infty} L^{jm}$$

convergent in $\mathcal{L}(V)$. Then we can make use of the identity

$$\left(\sum_{i=0}^{m-1} L^i \right) \left(\sum_{j=0}^{\infty} L^{jm} \right) = \sum_{k=0}^{\infty} L^k$$

to prove that $(I - L)^{-1}$ exits and belongs to $\mathcal{L}(V)$. The details are similar to those in the proof of Theorem 2.3.1, and we omit them here.   ∎

Example 2.3.4 *Returning to (2.3.7), define*

$$Lv(x) = \int_0^x \ell(x, y)v(y)\, dy, \quad 0 \le x \le B, \quad v \in C[0, B].$$

Easily, L is a bounded linear operator on $C[0, B]$ to itself. The iterated operators L^k take the form

$$L^k v(x) = \int_0^x \ell_k(x, y)v(y)\, dy$$

for $k = 2, 3, \ldots$, and $\ell_1(x, y) \equiv \ell(x, y)$. It is straightforward to show

$$\ell_{k+1}(x, y) = \int_y^x \ell_k(x, z)\ell(z, y)\, dz, \qquad k = 1, 2, \ldots$$

Let

$$M(x) = \max_{0 \le y \le x} |\ell(x, y)|, \qquad 0 \le x \le B.$$

It is relatively straightforward, using induction, to show that

$$|\ell_k(x, y)| \le M(x)^k \frac{(x - y)^{k-1}}{(k - 1)!}, \qquad 0 \le y \le x \le B, \qquad k = 1, 2, 3, \ldots$$

Then

$$\|L^k\| \le \frac{M(B)^k B^k}{k!}, \qquad k = 1, 2, \ldots$$

It is clear that the right side converges to zero as $k \to \infty$, and thus (2.3.8) is satisfied for m large enough. We can also use this result to construct bounds for the solutions of (2.3.7), which we leave to Exercise 2.3.6.

2.3.2   A perturbation result

An important technique in applied mathematics is to study an equation by relating it to a "nearby" equation for which there is a known solvability result. One of the more popular tools is the following perturbation theorem.

Theorem 2.3.5 *Let V and W be normed spaces with at least one of them being complete. Assume $L \in \mathcal{L}(V, W)$ has a bounded inverse $L^{-1} : W \to V$. Assume $M \in \mathcal{L}(V, W)$ satisfies*

$$\|M - L\| \le \frac{1}{\|L^{-1}\|}. \tag{2.3.9}$$

Then $M : V \to W$ is a bijection, $M^{-1} \in \mathcal{L}(W, V)$ and

$$\|M^{-1}\| \le \frac{\|L^{-1}\|}{1 - \|L^{-1}\|\, \|L - M\|}. \tag{2.3.10}$$

Moreover,

$$\|L^{-1} - M^{-1}\| \leq \frac{\|L^{-1}\|^2 \|L - M\|}{1 - \|L^{-1}\| \|L - M\|}. \tag{2.3.11}$$

For solutions of the equations $Lv_1 = w$ and $Mv_2 = w$, we have the estimate

$$\|v_1 - v_2\| \leq \|M^{-1}\| \|(L - M) v_1\|. \tag{2.3.12}$$

Proof. We write M as a perturbation of L. If W is complete, we write

$$M = [I - (L - M) L^{-1}] L;$$

while if V is complete, we write

$$M = L [I - L^{-1}(L - M)].$$

Let us prove the result for the case W is complete.

The operator $(L - M) L^{-1} \in \mathcal{L}(W)$ satisfies

$$\|(L - M) L^{-1}\| \leq \|L - M\| \|L^{-1}\| < 1.$$

Thus applying the geometric series theorem, $[I - (L - M) L^{-1}]^{-1}$ exists and

$$\|[I - (L - M) L^{-1}]^{-1}\| \leq \frac{1}{1 - \|L^{-1}(L - M)\|} \leq \frac{1}{1 - \|L^{-1}\| \|L - M\|}.$$

So M^{-1} exists with

$$M^{-1} = L^{-1} [I - (L - M) L^{-1}]^{-1}$$

and

$$\|M^{-1}\| \leq \|L^{-1}\| \|[I - (L - M) L^{-1}]^{-1}\| \leq \frac{\|L^{-1}\|}{1 - \|L^{-1}\| \|L - M\|}.$$

To prove (2.3.11), we write

$$L^{-1} - M^{-1} = M^{-1}(M - L) L^{-1},$$

take norms and use (2.3.10). For the estimate (2.3.12), write

$$v_1 - v_2 = (L^{-1} - M^{-1}) w = M^{-1}(M - L) L^{-1} w = M^{-1}(M - L) v_1$$

and take norms and bounds. ∎

The above theorem can be paraphrased as follows: *An operator that is close to an operator with a bounded inverse will itself have a bounded inverse.* This is the framework for innumerable solvability results for linear differential and integral equations, and variations of it are also used with nonlinear operator equations.

The estimate (2.3.11) can be termed the local Lipschitz continuity of the operator inverse. The estimate (2.3.12) can be used both as an *a priori* and an *a posteriori* error estimate, depending on the way we use it. First, let us view the equation $Lv = w$ as the exact problem, and we take a sequence of approximation problems $L_n v_n = w$, $n = 1, 2, \ldots$. Assuming the sequence $\{L_n\}$ converges to L, we can apply the perturbation theorem to conclude that at least for sufficiently large n, the equation $L_n v_n = w$ has a unique solution v_n, and we have the error estimate

$$\|v - v_n\| \leq \|L_n^{-1}\| \, \|(L - L_n) \, v\|.$$

The *consistency* of the approximation is defined by the condition

$$\|(L - L_n) \, v\| \to 0,$$

while the *stability* is defined by the condition that $\{\|L_n^{-1}\|\}_{n \text{ large}}$ is uniformly bounded. We see that consistency plus stability implies *convergence*:

$$\|v - v_n\| \to 0.$$

The error estimate provides sufficient conditions for convergence (and order error estimate under regularity assumptions on the solution v) before we actually solve the approximation problem $L_n v_n = w$. Such an estimate is called an *a priori* error estimate. We notice that usually an *a priori* error estimate does not tell us quantitatively how small is the error.

Another way to use (2.3.12) is to view $Mv = w$ as the exact problem, and $L = M_n$ an approximation of M, $n = 1, 2, \ldots$. Denote v_n to be the solution of the approximation equation $M_n v_n = w$; the equation is uniquely solvable at least for sufficiently large n. Then we have the error estimate

$$\|v - v_n\| \leq \|M^{-1}\| \, \|(M - M_n) \, v_n\|.$$

Suppose we can estimate the term $\|M^{-1}\|$. Then after the approximate solution v_n is found, the above estimate offers a numerical upper bound for the error. Such an estimate is called an *a posteriori* error estimate.

Example 2.3.6 *We examine the solvability of the integral equation*

$$\lambda \, u(x) - \int_0^1 \sin(xy) \, u(y) \, dy = f(x), \qquad 0 \leq x \leq 1, \qquad (2.3.13)$$

with $\lambda \neq 0$. From the discussion of the Example 2.3.2, if

$$|\lambda| > \|K\| = \int_0^1 \sin(y) \, dy = 1 - \cos(1) \approx 0.4597 \qquad (2.3.14)$$

is satisfied, then for every $f \in C[0, 1]$, (2.3.13) admits a unique solution $u \in C[0, 1]$.

To extend the values of λ for which (2.3.13) has a unique solution, we apply the perturbation theorem. Since $\sin(xy) \approx xy$ for small values of $|xy|$, we compare (2.3.13) with

$$\lambda \, v(x) - \int_0^1 x \, y \, v(y) \, dy = f(x), \qquad 0 \leq x \leq 1. \qquad (2.3.15)$$

In the notation of the perturbation theorem, equation (2.3.13) is $Mu = f$ and (2.3.15) is $Lv = f$. The normed space is $V = C[0,1]$ with the norm $\|\cdot\|_\infty$, and $L, M \in \mathcal{L}(V)$.

The integral equation (2.3.15) can be solved explicitly. From (2.3.15), assuming $\lambda \neq 0$, we have that every solution v takes the form

$$v(x) = [f(x) + cx]$$

for some constant c. Substituting this back into (2.3.15) leads to a formula for c, and then

$$v(x) = \frac{1}{\lambda}\left[f(x) + \frac{1}{\lambda - \frac{1}{3}}\int_0^1 x\,y\,f(y)\,dy\right] \quad \text{if } \lambda \neq 0, \frac{1}{3}. \tag{2.3.16}$$

The relation (2.3.16) defines $L^{-1}f$ for all $f \in C[0,1]$.

To use the perturbation theorem, we need to measure several quantities. It can be computed that

$$\|L^{-1}\| \leq \frac{1}{|\lambda|}\left[1 + \frac{1}{2\,|\lambda - \frac{1}{3}|}\right]$$

and

$$\|L - M\| = \int_0^1 (y - \sin y)\,dy = \cos(1) - \frac{1}{2} \approx 0.0403.$$

The condition (2.3.9) is implied by

$$\frac{1}{|\lambda|}\left[1 + \frac{1}{2\,|\lambda - \frac{1}{3}|}\right] < \frac{1}{\cos(1) - \frac{1}{2}}. \tag{2.3.17}$$

A graph of the left side of this inequality is given in Figure 2.1. If λ is assumed to be real, then there are three cases to be considered: $\lambda > \frac{1}{3}$, $0 < \lambda < \frac{1}{3}$, and $\lambda < 0$. For the case $\lambda < 0$, (2.3.17) is true if and only if $\lambda < \lambda_0 \approx -0.0881$, the negative root of the equation

$$\lambda^2 - \left(\frac{5}{6} - \cos(1)\right)\lambda - \frac{5}{6}\left(\cos(1) - \frac{1}{2}\right) = 0.$$

As a consequence of the perturbation theorem, we have that if $\lambda < \lambda_0$, then (2.3.13) is uniquely solvable for all $f \in C[0,1]$. This is a significant improvement over the condition (2.3.14). Bounds can also be given on the solution u, but these are left to the reader, as are the remaining two cases for λ.

Exercise 2.3.1 *Consider the integral equation*

$$\lambda u(x) - \int_0^1 \frac{u(y)\,dy}{1 + x^2 y^2} = f(x), \qquad 0 \leq x \leq 1$$

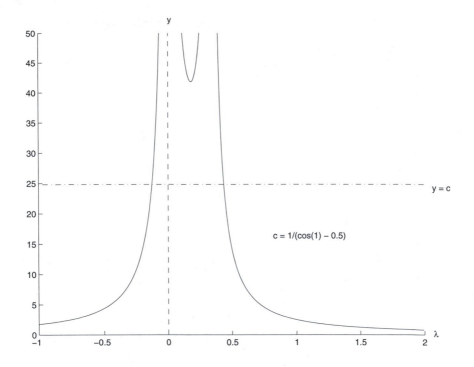

Figure 2.1. Graph of the left-hand side of inequality (2.3.17)

*for a given $f \in C[0, 1]$. Show this equation has a unique continuous solution
u if $|\lambda|$ is chosen sufficiently large. For such values of λ, bound the solution
u in terms of $\|f\|_\infty$.*

Exercise 2.3.2 *Complete the solvability analysis for Example 2.3.6.*

Exercise 2.3.3 *Repeat the solvability analysis of Example 2.3.6 for the
integral equation*

$$\lambda u(x) - \int_0^1 u(y) \tan^{-1}(xy)\, dy = f(x), \qquad 0 \le x \le 1.$$

Use the approximation based on the Taylor approximation

$$\tan^{-1} s \approx s$$

for small values of s.

Exercise 2.3.4 *Assume the conditions of the geometric series theorem are
satisfied. Then for any $f \in V$, the equation $(I - L)u = f$ has a unique
solution $u \in V$. Show that this solution can be approximated by a sequence*

$\{u_n\}$ defined by: $u_0 \in V$, $u_n = f + Lu_{n-1}$, $n = 1, 2, \ldots$. Derive an error estimate for $\|u - u_n\|$.

Exercise 2.3.5 Let $f \in C[0, 1]$. Show that the continuous solution of the boundary value problem

$$-u''(x) = f(x), \quad 0 < x < 1,$$
$$u(0) = u(1) = 0$$

is

$$u(x) = \int_0^1 k(x, y) \, f(y) \, dy,$$

where the kernel function $k(x, y) = \min(x, y) (1 - \max(x, y))$. Let $a \in C[0, 1]$. Apply the geometric series theorem to show that the boundary value problem

$$-u''(x) + a(x) \, u(x) = f(x), \quad 0 < x < 1,$$
$$u(0) = u(1) = 0$$

has a unique continuous solution u if $\max_{0 \le x \le 1} |a(x)| \le a_0$ is sufficiently small. Give an estimate of the value a_0.

Exercise 2.3.6 Recall the Volterra equation (2.3.7). Bound the solution u using Corollary 2.3.3. Separately, obtain a bound for u by examing directly the convergence of the series

$$u = \sum_{k=0}^{\infty} L^k f$$

and relating it to the Taylor series for $\exp(M(B)B)$.

2.4 Some more results on linear operators

In this section, we collect together several independent results that are important in working with linear operators.

2.4.1 An extension theorem

Bounded operators are often defined on a subspace of a larger space, and it is desirable to extend the domain of the original operator to the larger space, while retaining the boundedness of the operator.

Theorem 2.4.1 (EXTENSION THEOREM) Let V be a normed space, and let $\widehat{V}$ denote its completion. Let W be a Banach space. Assume $L \in \mathcal{L}(V, W)$. Then there is a unique operator $\widehat{L} \in \mathcal{L}(\widehat{V}, W)$ with

$$\widehat{L} v = L v \qquad \forall v \in V$$

and

$$\|\widehat{L}\|_{\widehat{V},W} = \|L\|_{V,W}.$$

The operator $\widehat{L}$ is called an extension *of L.*

Proof. Given $v \in \widehat{V}$, let $\{v_n\} \subseteq V$ with $v_n \to v$ in $\widehat{V}$. The sequence $\{Lv_n\}$ is a Cauchy sequence in W by the following inequality:

$$\|Lv_{n+p} - Lv_n\| \leq \|L\| \, \|v_{n+p} - v_n\|.$$

Since W is complete, there is a limit $\widehat{L}(v) \in W$. We must show that $\widehat{L}$ is well defined (i.e., $\widehat{L}(v)$ does not depend on the choice of the sequence $\{v_n\}$), linear, and bounded.

To show that $\widehat{L}$ is well defined, let $v_n \to v$ and $\tilde{v}_n \to v$ with $\{v_n\}, \{\tilde{v}_n\} \subseteq V$. Then as $n \to \infty$,

$$\|Lv_n - L\tilde{v}_n\| \leq \|L\| \, \|v_n - \tilde{v}_n\| \leq \|L\| \, (\|v_n - v\| + \|\tilde{v}_n - v\|) \to 0.$$

Thus $\{Lv_n\}$ and $\{L\tilde{v}_n\}$ must have the same limit.

To show the linearity, let $u_n \to u$ and $v_n \to v$, and let $\alpha, \beta \in \mathbb{K}$. Then

$$\widehat{L}(\alpha \, u + \beta \, v) = \lim_{n \to \infty} L(\alpha \, u_n + \beta \, v_n) = \lim_{n \to \infty} (\alpha \, Lu_n + \beta \, Lv_n) = \alpha \, \widehat{L}u + \beta \, \widehat{L}v.$$

To show the boundedness, let $v_n \to v$ and $\{v_n\} \subseteq V$. Then taking the limit $n \to \infty$ in

$$\|Lv_n\|_W \leq \|L\| \, \|v_n\|_V = \|L\| \, \|v_n\|_{\widehat{V}},$$

we obtain

$$\|\widehat{L}v\|_W \leq \|L\| \, \|v\|_{\widehat{V}}.$$

So $\widehat{L}$ is bounded and

$$\|\widehat{L}\| = \sup_{0 \neq v \in \widehat{V}} \frac{\|\widehat{L}v\|_W}{\|v\|_{\widehat{V}}} \leq \|L\|.$$

To see that $\|\widehat{L}\|_{\widehat{V},W} = \|L\|_{V,W}$, we note

$$\|L\|_{V,W} = \sup_{0 \neq v \in V} \frac{\|Lv\|_W}{\|v\|_V} = \sup_{0 \neq v \in V} \frac{\|\widehat{L}v\|_W}{\|v\|_{\widehat{V}}} \leq \sup_{0 \neq v \in \widehat{V}} \frac{\|\widehat{L}v\|_W}{\|v\|_{\widehat{V}}} = \|\widehat{L}\|_{\widehat{V},W}.$$

To show that $\widehat{L}$ is unique, let $\tilde{L}$ be another extension of L to $\widehat{V}$. Let $v \in \widehat{V}$ and let $v_n \to v$, $\{v_n\} \subseteq V$. Then

$$\|\tilde{L}v - Lv_n\|_W = \|\tilde{L}v - \tilde{L}v_n\|_W \leq \|\tilde{L}\| \, \|v - v_n\| \to 0 \quad \text{as } n \to \infty.$$

This shows $Lv_n \to \tilde{L}v$ as $n \to \infty$. On the other hand, $Lv_n \to \widehat{L}v$. So we must have $\tilde{L}v = \widehat{L}v$, for any $v \in \widehat{V}$. Therefore, $\tilde{L} = \widehat{L}$. ∎

There are a number of ways in which this theorem can be used. Often we wish to work with linear operators that are defined and bounded on some normed space, but the space is not complete with the given norm. Since most function space arguments require complete spaces, the above theorem allows us to proceed with our arguments on a larger complete space, with an operator that agrees with our original one on the original space.

Example 2.4.2 Let $V = C^1[0,1]$ with the inner product norm

$$\|v\|_{1,2} = (\|v\|_2^2 + \|v'\|_2^2)^{1/2}.$$

The completion of $C^1[0,1]$ with respect to $\|\cdot\|_{1,2}$ is the Sobolev space $H^1(0,1)$, which was introduced earlier in Example 1.3.7. (Details of Sobolev spaces are also given later in Chapter 6.) Let $W = L^2(0,1)$ with the standard norm $\|\cdot\|_2$.

Define the differentiation operator $D : C^1[0,1] \to L^2(0,1)$ by

$$(Dv)(x) = v'(x), \qquad 0 \le x \le 1, \ v \in C^1[0,1].$$

We have

$$\|Dv\|_2 = \|v'\|_2 \le \|v\|_{1,2},$$

and thus

$$\|D\|_{V,W} \le 1.$$

By the extension theorem, we can extend D to $\widehat{D} \in \mathcal{L}(H^1(0,1), L^2(0,1))$, the differentiation operator on $H^1(0,1)$. A more concrete realization of $\widehat{D}$ can be obtained using the theory of distributions. This is also discussed in Chapter 6.

2.4.2 Open mapping theorem

This theorem is widely used in obtaining boundedness of inverse operators. When considered in the context of solving an equation $Lv = w$, the theorem says that *existence* and *uniqueness* of solutions for all $w \in W$ implies the *stability* of the solution v; i.e., "small changes" in the given data w cause only "small changes" in the solution v. For a proof of this theorem, see [40, p. 91] or [175, p. 179].

Theorem 2.4.3 Let V and W be Banach spaces. If $L \in \mathcal{L}(V,W)$ is a bijection, then $L^{-1} \in \mathcal{L}(W,V)$.

To be more precise concerning the stability of the problem being solved, let $Lv = w$ and $L\hat{v} = \hat{w}$. We then have

$$v - \hat{v} = L^{-1}(w - \hat{w}),$$

and then

$$\|v - \hat{v}\| \le \|L^{-1}\| \, \|w - \hat{w}\|.$$

As $w - \hat{w}$ becomes small, so must $v - \hat{v}$. The term $\|L^{-1}\|$ gives a relationship between the size of the error in the data w and that of the error in the solution v. A more important way is to consider the relative changes in the two errors:

$$\frac{\|v - \hat{v}\|}{\|v\|} \leq \frac{\|L^{-1}\| \, \|w - \hat{w}\|}{\|v\|} = \|L^{-1}\| \|L\| \frac{\|w - \hat{w}\|}{\|L\| \, \|v\|}.$$

Applying $\|w\| \leq \|L\| \|v\|$, we obtain

$$\frac{\|v - \hat{v}\|}{\|v\|} \leq \|L^{-1}\| \|L\| \frac{\|w - \hat{w}\|}{\|w\|}. \qquad (2.4.1)$$

The quantity $\mathrm{cond}(L) \equiv \|L^{-1}\| \, \|L\|$ is called the *condition number* of the equation, and it relates the relative errors in the data w and the solution v. Note that we always have $\mathrm{cond}(L) \geq 1$ as

$$\|L^{-1}\| \, \|L\| \geq \|L^{-1}L\| = \|I\| = 1.$$

Problems with a small condition number are called *well-conditioned*, while those with a large condition number *ill-conditioned*.

In a related vein, consider a problem $Lv = w$, $L : V \to W$, in which L is bounded and injective, but not surjective. The inverse operator L^{-1} exists on the range $\mathcal{R}(L) \subseteq W$. If L^{-1} is unbounded on $\mathcal{R}(L)$ to V, we say the original problem $Lv = w$ is *ill-posed* or *unstable*. Such problems are not considered in this text, but there are a number of important applications (e.g., many indirect sensing devices) that fall into this category. Problems in which L^{-1} is bounded (along with L) are called *well-posed* or *stable*; they can still be ill-conditioned, as was discussed in the preceding paragraph.

2.4.3 Principle of uniform boundedness

Another widely used set of results refer to the collective boundedness of a set of linear operators.

Theorem 2.4.4 *Let $\{L_n\}$ be a sequence of bounded linear operators from a Banach space V to another Banach space W. Assume for every $v \in V$, the sequence $\{L_n v\}$ is bounded. Then*

$$\sup_n \|L_n\| < \infty.$$

This theorem is often called the *principle of uniform boundedness*; see [40, p. 95] or [175, p. 172] for a proof and a more extended development. We also have the following useful variant of this principle.

Theorem 2.4.5 (BANACH-STEINHAUS THEOREM) *Let V and W be Banach spaces, and let $L, L_n \in \mathcal{L}(V, W)$. Let V_0 be a dense subspace of V. Then in order for $L_n v \to Lv \; \forall v \in V$, it is necessary and sufficient that*
(a) $L_n v \to Lv$, $\forall v \in V_0$; and
(b) $\sup_n \|L_n\| < \infty$.

Proof. ($\Rightarrow$) Assume $L_n v \to L v$ for all $v \in V$. Then (a) follows trivially; and (b) follows from the principle of uniform boundedness.

($\Leftarrow$) Assume (a) and (b). Denote $B = \sup_n \|L_n\|$. Let $v \in V$ and $\epsilon > 0$. By the denseness of V_0 in V, there is an element $v_\epsilon \in V_0$ such that

$$\|v - v_\epsilon\| \le \frac{\epsilon}{3 \, \max\{\|L\|, B\}}.$$

Then

$$\|L v - L_n v\| \le \|L v - L v_\epsilon\| + \|L v_\epsilon - L_n v_\epsilon\| + \|L_n v_\epsilon - L_n v\|$$
$$\le \|L\| \, \|v - v_\epsilon\| + \|L v_\epsilon - L_n v_\epsilon\| + \|L_n\| \, \|v_\epsilon - v\|$$
$$\le \frac{2\,\epsilon}{3} + \|L v_\epsilon - L_n v_\epsilon\|.$$

Using (a), we can find a natural number n_ϵ such that

$$\|L v_\epsilon - L_n v_\epsilon\| \le \frac{\epsilon}{3}, \quad n \ge n_\epsilon.$$

Combining these results,

$$\|L v - L_n v\| \le \epsilon, \quad n \ge n_\epsilon.$$

Therefore, $L_n v \to L v$ as $n \to \infty$. ∎

Next, we apply Banach-Steinhaus theorem to discuss the convergence of numerical quadratures (i.e., numerical integration formulas).

2.4.4 Convergence of numerical quadratures

As an example, let us consider the convergence of numerical quadratures for the computation of the integral

$$L v = \int_0^1 w(x) \, v(x) \, dx,$$

where w is a weighted function, $w(x) \ge 0$, $w \in L^1(0,1)$. There are several approaches to constructing numerical quadratures. One popular approach is to replace the function v by some interpolant of it, denoted here by Πv, and then define the corresponding numerical quadrature by the formula

$$\int_0^1 w(x) \, \Pi v(x) \, dx.$$

The topic of function interpolation is discussed briefly in Section 3.1. If Πv is taken to be the Lagrange polynomial interpolant of v on a uniform partition of the integration interval, the resulting quadratures are called *Newton-Cotes integration formulas*. It is well known that high-degree polynomial interpolation on a uniform partition leads to strong oscillations near the boundary of the interval and hence divergence of the interpolation in many cases. Correspondingly, one cannot expect the convergence of the Newton-Cotes integration formulas. To guarantee the convergence, one may use the Lagrange polynomial interpolant Πv of v on a properly chosen partition with more nodes placed near the boundary, or one may use a piecewise polynomial interpolant on a uniform partition or a partition suitably refined in areas where the integrand $w\,v$ changes rapidly. With the use of piecewise polynomial interpolants of v, we get the celebrated *trapezoidal rule* (using piecewise linear interpolation) and *Simpson's rule* (using piecewise quadratic interpolation).

A second popular approach to constructing numerical quadratures is by the *method of undetermined parameters*. We approximate the integral by a sequence of finite sums, each of them being a linear combination of some function values (and more generally, derivative values can used in the sums as well). In other words, we let

$$Lv \approx L_n v = \sum_{i=0}^{n} w_i^{(n)} v(x_i^{(n)})$$

and choose the weights $\{w_i^{(n)}\}_{i=0}^{n}$ and the nodes $\{x_i^{(n)}\}_{i=0}^{n} \subseteq [0,1]$ by some specific requirements. Some of the weights and nodes may be prescribed *a priori* according to the context of the applications, and the remaining ones are usually determined by requiring the quadrature be exact for polynomials of degree as high as possible. If none of the weights and nodes is prescribed, then we may choose these $2n+2$ quantities so that the quadrature is exact for any polynomial of degree less than or equal to $2n+1$. The resulting numerical quadratures are called *Gaussian quadratures*.

Detailed discussions of numerical quadratures (for the case when the weight function $w \equiv 1$) can be found in [11, Section 5.3]. Here we study the convergence of numerical quadratures in an abstract framework.

Let there be given a sequence of quadratures

$$L_n v = \sum_{i=0}^{n} w_i^{(n)} v(x_i^{(n)}), \tag{2.4.2}$$

where $0 \le x_0^{(n)} < x_1^{(n)} < \cdots < x_n^{(n)} \le 1$ is a partition of $[0,1]$. We regard L_n as a linear functional defined on $C[0,1]$ with the standard uniform norm. It is straightforward to show that

$$\|L_n\| = \sum_{i=0}^{n} |w_i^{(n)}|, \tag{2.4.3}$$

and this is left as an exercise for the reader.

As an important special case, assume the quadrature scheme L_n is exact for polynomials of degree less than or equal to $d(n)$; i.e.,

$$L_n v = L v \qquad \forall\, v \in \mathcal{P}_{d(n)}.$$

Here $\mathcal{P}_{d(n)}$ is the space of all the polynomials of degree less than or equal to $d(n)$, and we assume $d(n) \to \infty$ as $n \to \infty$. Then an application of the Banach-Steinhaus theorem shows that $L_n v \to Lv$ for any $v \in C[0,1]$ if and only if

$$\sup_n \sum_{i=0}^{n} |w_i^{(n)}| < \infty.$$

Continuing the discussion on the convergence of numerical quadratures, we assume all the conditions stated in the previous paragraph are valid. Additionally, we assume the weights $w_i^{(n)} \geq 0$. Then it follows that $L_n v \to Lv$ for any $v \in C[0,1]$ (cf. Exercise 2.4.2).

From the point of view of numerical computations, it is considered somewhat important to have non-negative quadrature weights to avoid round-off accumulations. It can be shown that for the Gaussian quadratures, all the quadrature weights are non-negative; and if the weight function w is positive on $(0,1)$, then the quadrature weights are positive. See [11, Section 5.3] for an extended discussion of Gaussian quadrature.

Exercise 2.4.1 *Prove* (2.4.3).

Exercise 2.4.2 *Consider the quadrature formula* (2.4.2). *Assume all the weights $w_i^{(n)}$ are non-negative and the quadrature formula is exact for polynomials of degree less than or equal to $d(n)$ with $d(n) \to \infty$ as $n \to \infty$. Prove the convergence of the quadratures: $L_n v \to Lv$, for all $v \in C[0,1]$.*

Exercise 2.4.3 *A popular family of numerical quadratures is constructed by approximating the integrand by its piecewise polynomial interpolants. We take the composite trapezoidal rule as an example. The integral to be computed is*

$$Lv = \int_0^1 v(x)\, dx.$$

We divide the interval $[0,1]$ into n equal parts, and denote $x_i = i/n$, $0 \leq i \leq n$, as the nodes. Then we approximate v by its piecewise linear interpolant $\Pi_n v$ defined by

$$\Pi_n v(x) = n\,(x_i - x)\,v(x_{i-1}) + n\,(x - x_{i-1})\,v(x_i)$$

for $x_{i-1} \leq x \leq x_i$, $1 \leq i \leq n$.   Then the composite trapezoidal rule is

$$L_n v = \int_0^1 \Pi_n v(x)\, dx = \frac{1}{n}\left[\frac{1}{2}v(x_0) + \sum_{i=1}^{n-1} v(x_i) + \frac{1}{2}v(x_n)\right].$$

Show that $L_n v \to Lv$ for any $v \in C[0,1]$.

Using piecewise polynomials of higher degrees based on non-uniform partitions of the integration interval, we can develop other useful numerical quadratures.

Exercise 2.4.4 *In the formula (2.4.1), show that the inequality can be made as close as desired to equality for suitable choices of v and $\widetilde{v}$.*

2.5 Linear functionals

An important special case of linear operators is when they take on scalar values. Let V be a normed space, and $W = \mathbb{K}$, the set of scalars associated with V. The elements in $\mathcal{L}(V, \mathbb{K})$ are called *linear functionals*. Since $\mathbb{K}$ is complete, $\mathcal{L}(V, \mathbb{K})$ is a Banach space. This space is usually denoted as V' and it is called the *dual space* of V. Usually we use lowercase letters, such as ℓ, to denote a linear functional.

In some references, the term *linear functional* is used for the linear operators from a normed space to $\mathbb{K}$, without the functionals' being necessarily bounded. In this work, since we use exclusively linear functionals that are bounded, we use the term "linear functionals" to refer to only bounded linear functionals.

Example 2.5.1 *Let $\Omega \subseteq \mathbb{R}^d$ be a bounded open set. It is a well-known result that for $1 \leq p < \infty$, the dual space of $L^p(\Omega)$ can be identified with $L^{p'}(\Omega)$. Here p' is the conjugate exponent of p, defined by the relation*

$$\frac{1}{p} + \frac{1}{p'} = 1.$$

By convention, $p' = \infty$ when $p = 1$. In other words, given an $\ell \in (L^p(\Omega))'$, there is a function $u \in L^{p'}(\Omega)$, uniquely determined a.e., such that

$$\ell(v) = \int_\Omega u(\mathbf{x}) v(\mathbf{x}) \, dx \qquad \forall \, v \in L^p(\Omega). \tag{2.5.1}$$

Conversely, for any $u \in L^{p'}(\Omega)$, the rule

$$v \longmapsto \int_\Omega u(\mathbf{x}) v(\mathbf{x}) \, dx, \qquad v \in L^p(\Omega)$$

defines a bounded linear functional on $L^p(\Omega)$. It is convenient to identify $\ell \in (L^p(\Omega))'$ and $u \in L^{p'}(\Omega)$, related as in (2.5.1). Then we write

$$(L^p(\Omega))' = L^{p'}(\Omega), \qquad 1 \leq p < \infty.$$

The dual space of $L^\infty(\Omega)$, however, is larger than the space $L^1(0, 1)$.

All the results discussed in the previous sections for general linear operators apply immediately to linear functionals. In addition, there are useful results particular to linear functionals only.

2.5.1   An extension theorem for linear functionals

We have seen that a bounded linear operator can be extended to the closure of its domain. It is also possible to extend linear functionals defined on an arbitrary subspace to the whole space.

Theorem 2.5.2 (HAHN-BANACH THEOREM) *Let V_0 be a subspace of a normed space V, and $\ell : V_0 \to \mathbb{K}$ be linear and bounded. Then there exists an extension $\hat{\ell} \in V'$ of ℓ with $\hat{\ell}(v) = \ell(v) \ \forall v \in V_0$, and $\|\hat{\ell}\| = \|\ell\|$.*

A proof can be found in [40, p. 79] or [175, p. 4]. Note that if V_0 is not dense in V, then the extension need not be unique.

Example 2.5.3 *This example is important in the analysis of some numerical methods for solving integral equations. Let $V = L^\infty(0,1)$. This is the space of all cosets (or equivalence classes)*

$$\mathbf{v} = [v] = \{u \text{ Lebesgue measurable on } [0,1] \mid u = v \text{ a.e. in } [0,1]\}$$

for which

$$\|\mathbf{v}\|_\infty \equiv \|v\|_\infty = \operatorname*{ess\,sup}_{0 \le x \le 1} |v(x)| < \infty. \qquad (2.5.2)$$

With this norm, $L^\infty(0,1)$ is a Banach space.

Let V_0 be the set of all cosets $\mathbf{v} = [v]$, where $v \in C[0,1]$. It is a proper subspace of $L^\infty(0,1)$. When restricted to V_0, the norm (2.5.2) is equivalent to the usual norm $\|\cdot\|_\infty$ on $C[0,1]$. It is common to write $V_0 = C[0,1]$; but this is an abuse of notation, and it is important to keep in mind the distinction between V_0 and $C[0,1]$.

Let $c \in [0,1]$, and define

$$\ell_c([v]) = v(c) \qquad \forall v \in C[0,1]. \qquad (2.5.3)$$

The linear functional $\ell_c([v])$ is well defined on V_0. From

$$|\ell_c([v])| = |v(c)| \le \|v\|_\infty = \|\mathbf{v}\|_\infty, \qquad \mathbf{v} = [v],$$

we see that $\|\ell_c\| \le 1$. By choosing $v \in C[0,1]$ with $v(c) = \|v\|_\infty$, we then obtain

$$\|\ell_c\| = 1.$$

Using the Hahn-Banach theorem, we can extend ℓ_c to $\hat{\ell}_c : L^\infty(0,1) \to \mathbb{K}$ with

$$\|\hat{\ell}_c\| = \|\ell_c\| = 1.$$

The functional $\hat{\ell}_c$ extends to $L^\infty(0,1)$ the concept of point evaluation of a function, to functions that are only Lebesgue measurable and that are not precisely defined because of being members of a coset.

Somewhat surprisingly, many desirable properties of ℓ_c are carried over to $\hat{\ell}_c$. These include the following:

- *Assume* $[v] \in L^{\infty}(0,1)$ *satisfies* $m \le v(x) \le M$ *for almost all* x *in some open interval about* c. *Then* $m \le \hat{\ell}_c([v]) \le M$.

- *Assume* c *is a point of continuity of* v. *Then*

$$\hat{\ell}_c([v]) = v(c),$$

$$\lim_{a \to c} \hat{\ell}_a([v]) = v(c).$$

These ideas and properties carry over to $L^{\infty}(\Omega)$, *with* Ω *a closed, bounded set in* $\mathbb{R}^d$, $d \ge 1$, *and* $\mathbf{c} \in \Omega$. *For additional detail and the application of this extension to numerical integral equations, see [14].*

In some applications, a stronger form of the Hahn-Banach theorem is needed. We begin by introducing another useful concept for functionals.

Definition 2.5.4 *A functional* p *on a real vector space* V *is said to be* sublinear *if*

$$p(u + v) \le p(u) + p(v) \qquad \forall\, u, v \in V,$$
$$p(\alpha v) = \alpha\, p(v) \qquad \forall\, \alpha \ge 0.$$

We note that a semi-norm is a sublinear functional. A proof of the following result can be found in [40, p. 78] or [175, p. 2].

Theorem 2.5.5 (GENERALIZED HAHN-BANACH THEOREM) *Let* V *be a linear space,* $V_0 \subseteq V$ *a subspace. Suppose* $p : V \to \mathbb{R}$ *is a sublinear functional and* $\ell : V_0 \to \mathbb{R}$ *a linear functional such that* $\ell(v) \le p(v)$ *for all* $v \in V_0$. *Then* ℓ *can be extended to* V *such that* $\ell(v) \le p(v)$ *for all* $v \in V$.

Note that $p(v) = c\,\|v\|_V$, where c is a positive constant, is a sublinear functional on V. With this choice of p, we obtain the original Hahn-Banach theorem. A useful new consequence of the generalized Hahn-Banach theorem is the following.

Corollary 2.5.6 *Let* V *be a normed space. For any* $v_0 \in V$, *there exists* $\ell \in V'$ *such that* $\|\ell\| = 1$ *and* $\ell(v_0) = \|v_0\|$.

2.5.2   The Riesz representation theorem

On Hilbert spaces, linear functionals are limited in the forms they can take. The following theorem makes this more precise; and the result is one used in developing the solvability theory for some important partial differential equations and boundary integral equations. The theorem also provides a tool for introducing the concept of the adjoint of a linear operator in the next section.

Theorem 2.5.7 (RIESZ REPRESENTATION THEOREM) *Let* V *be a Hilbert space,* $\ell \in V'$. *Then there is a unique* $u \in V$ *for which*

$$\ell(v) = (v, u) \qquad \forall\, v \in V. \tag{2.5.4}$$

In addition,

$$\|\ell\| = \|u\|. \tag{2.5.5}$$

Proof. Assuming the existence of u, we first prove its uniqueness. Suppose $\tilde{u} \in V$ satisfies

$$\ell(v) = (v, u) = (v, \tilde{u}) \quad \forall\, v \in V.$$

Then

$$(v, u - \tilde{u}) = 0 \quad \forall\, v \in V.$$

Take $v = u - \tilde{u}$. Then $\|u - \tilde{u}\| = 0$, which implies $u = \tilde{u}$.

We give two derivations of the existence of u, both for the case of a real Hilbert space.

STANDARD PROOF OF EXISTENCE. Denote

$$N = \mathcal{N}(\ell) = \{v \in V \mid \ell(v) = 0\},$$

which is a subspace of V. If $N = V$, then $\|\ell\| = 0$, and we may take $u = 0$.

Now suppose $N \neq V$. Then there exists at least one $v_0 \in V$ such that $\ell(v_0) \neq 0$. It is possible to decompose V as the direct sum of N and $N^\perp$ (cf. Section 3.5). From this, we have the decomposition $v_0 = v_1 + v_2$ with $v_1 \in N$ and $v_2 \in N^\perp$. Then $\ell(v_2) = \ell(v_0) \neq 0$.

For any $v \in V$, we have the property

$$\ell\left(v - \frac{\ell(v)}{\ell(v_2)} v_2\right) = 0.$$

Thus

$$v - \frac{\ell(v)}{\ell(v_2)} v_2 \in N,$$

and in particular, it is orthogonal to v_2:

$$\left(v - \frac{\ell(v)}{\ell(v_2)} v_2, v_2\right) = 0;$$

i.e.,

$$\ell(v) = \left(v, \frac{\ell(v_2)}{\|v_2\|^2} v_2\right).$$

In other words, we may take u to be $(\ell(v_2)/\|v_2\|^2) v_2$.

PROOF USING A MINIMIZATION PRINCIPLE. From Theorem 3.2.12 in Chapter 3, we know the problem

$$\inf_{v \in V}\left[\frac{1}{2}\|v\|^2 - \ell(v)\right]$$

has a unique solution $u \in V$. The solution u is characterized by the relation (2.5.4).

We complete the proof of the theorem by showing (2.5.5). From (2.5.4) and the Cauchy-Schwarz inequality,

$$|\ell(v)| \leq \|u\| \, \|v\| \quad \forall \, v \in V.$$

Hence

$$\|\ell\| \leq \|u\|.$$

Let $v = u$ in (2.5.4). Then

$$\ell(u) = \|u\|^2$$

and

$$\|\ell\| = \sup_{v \neq 0} \frac{|\ell(v)|}{\|v\|} \geq \frac{|\ell(u)|}{\|u\|} \geq \|u\|.$$

Therefore, (2.5.5) holds. ∎

This theorem seems very straightforward, and its proof seems fairly simple. Nonetheless, this is a fundamental tool in the solvability theory for elliptic partial differential equations, as we see later in Chapter 7.

Example 2.5.8 *Let $\Omega \subseteq \mathbb{R}^d$ be open bounded. $V = L^2(\Omega)$ is a Hilbert space. By the Riesz representation theorem, there is a one-to-one correspondence between V' and V by the relation (2.5.4). We can identify $\ell \in V'$ with $u \in V$ related by (2.5.4). In this sense, $(L^2(\Omega))' = L^2(\Omega)$.*

For the space $L^2(\Omega)$, the element u in (2.5.4) is almost always immediately apparent. But for spaces such as the Sobolev space $H^1(a, b)$ introduced in Example 1.3.7 of Chapter 1, the determination of u of (2.5.4) is often not as obvious. As an example, define $\ell \in (H^1(a, b))'$ by

$$\ell(v) = v(c), \qquad v \in H^1(a, b) \tag{2.5.6}$$

for some $c \in [a, b]$. This linear functional can be shown to be well defined (cf. Exercise 2.5.2). From the Riesz representation theorem, there is a unique $u \in H^1(a, b)$ such that

$$\int_a^b [u'(x) \, v'(x) + u(x) \, v(x)] \, dx = v(c) \qquad \forall \, v \in H^1(a, b).$$

The element u is the generalized solution of the boundary value problem

$$- u'' + u = \delta(x - c) \text{ in } (a, b),$$
$$u'(a) = u'(b) = 0,$$

where $\delta(x - c)$ is the Dirac δ-function at c.

Exercise 2.5.1 *Prove Corollary 2.5.6.*

Exercise 2.5.2 *Show that the functional defined in (2.5.6) is linear and bounded on $H^1(a, b)$.*

Hint: Use the following results. For any $f \in H^1(a, b)$, f is continuous on $[a, b]$, and therefore,

$$\int_a^b f(x)\, dx = f(\zeta)$$

for some $\zeta \in (a, b)$. In addition,

$$f(c) = f(\zeta) + \int_\zeta^c f'(x)\, dx.$$

2.6 Adjoint operators

The notion of an adjoint operator is a generalization of the matrix transpose to infinite-dimensional spaces. First let us derive a defining property for the matrix transpose. Let $A \in \mathbb{R}^{m \times n}$, which is viewed as a linear continuous operator from $\mathbb{R}^n$ to $\mathbb{R}^m$. We use the regular Euclidean inner products for the spaces $\mathbb{R}^n$ and $\mathbb{R}^m$. Then

$$\mathbf{y}^T A \mathbf{x} = (A\mathbf{x}, \mathbf{y})_{\mathbb{R}^m}, \quad \mathbf{x}^T A^T \mathbf{y} = (\mathbf{x}, A^T \mathbf{y})_{\mathbb{R}^n} \quad \forall \mathbf{x} \in \mathbb{R}^n,\ \mathbf{y} \in \mathbb{R}^m.$$

Since $\mathbf{y}^T A \mathbf{x}$ is a real number, $\mathbf{y}^T A \mathbf{x} = (\mathbf{y}^T A \mathbf{x})^T = \mathbf{x}^T A^T \mathbf{y}$. We observe that the transpose (or adjoint) A^T is uniquely defined by the property

$$(A\mathbf{x}, \mathbf{y})_{\mathbb{R}^m} = (\mathbf{x}, A^T \mathbf{y})_{\mathbb{R}^n} \quad \forall \mathbf{x} \in \mathbb{R}^n,\ \mathbf{y} \in \mathbb{R}^m.$$

Turn to the general situation. Assume V and W are Hilbert spaces, $L \in \mathcal{L}(V, W)$. Let us use the Riesz representation theorem to define a new operator $L^* : W \to V$, called the *adjoint* of L. For simplicity, we assume in this section that $\mathbb{K} = \mathbb{R}$ for the set of scalars associated with W and V. Given $w \in W$, define a linear functional $\ell_w \in V'$ by

$$\ell_w(v) = (Lv, w)_W \quad \forall v \in V.$$

This linear functional is bounded because

$$|\ell_w(v)| \leq \|Lv\|\, \|w\| \leq \|L\|\, \|v\|\, \|w\|$$

and so

$$\|\ell_w\| \leq \|L\|\, \|w\|.$$

By the Riesz representation theorem, there is a uniquely determined element, denoted by $L^*(w) \in V$ such that

$$\ell_w(v) = (v, L^*(w)) \quad \forall v \in V.$$

We write

$$(Lv, w)_W = (v, L^*(w))_V \quad \forall v \in V,\, w \in W.$$

We first show that L^* is linear. Let $w_1, w_2 \in W$, and consider the linear functionals

$$\ell_1(v) = (Lv, w_1)_W = (v, L^*(w_1))_V,$$
$$\ell_2(v) = (Lv, w_2)_W = (v, L^*(w_2))_V$$

for any $v \in V$. Add these relations,

$$(Lv, w_1 + w_2)_W = (v, L^*(w_1) + L^*(w_2))_V \qquad \forall\, v \in V.$$

By definition,

$$(Lv, w_1 + w_2)_W = (v, L^*(w_1 + w_2))_V;$$

so

$$(v, L^*(w_1 + w_2))_V = (v, L^*(w_1) + L^*(w_2))_V \qquad \forall\, v \in V.$$

This implies

$$L^*(w_1 + w_2) = L^*(w_1) + L^*(w_2).$$

By a similar argument, for any $\alpha \in \mathbb{K}$, any $w \in W$,

$$L^*(\alpha\, w) = \alpha\, L^*(w).$$

Hence L^* is linear and we write $L^*(w) = L^*w$, and the defining relation is

$$(Lv, w)_W = (v, L^*w)_V \qquad \forall\, v \in V,\ w \in W. \tag{2.6.1}$$

Then we show the boundedness of L^*. We have

$$\|L^*w\| = \|\ell_w\| \le \|L\| \,\|w\| \quad \forall\, w \in W.$$

Thus

$$\|L^*\| \le \|L\| \tag{2.6.2}$$

and L^* is bounded. Let us show that actually the inequality in (2.6.2) can be replaced by an equality. For this, we consider the adjoint of L^*, defined by the relation

$$(L^*w, v)_V = (w, (L^*)^*v)_W \qquad \forall\, v \in V,\ w \in W.$$

Thus

$$(w, (L^*)^*v)_W = (w, Lv)_W \qquad \forall\, v \in V,\ w \in W.$$

By writing this as $(w, (L^*)^*v - Lv)_W = 0$ and letting $w = (L^*)^*v - Lv$, we obtain

$$(L^*)^*v = Lv \qquad \forall\, v \in V.$$

Hence

$$(L^*)^* = L. \tag{2.6.3}$$

We then apply (2.6.1) to L^* to obtain

$$\|L\| = \|(L^*)^*\| \leq \|L^*\|.$$

Combining this with (2.6.2), we have

$$\|L^*\| = \|L\|. \tag{2.6.4}$$

From the above derivation, we see that for a continuous linear operator between Hilbert spaces, the adjoint of its adjoint is the operator itself.

In the special situation $V = W$ and $L = L^*$, we say L is a *self-adjoint* operator. When L is a self-adjoint operator from $\mathbb{R}^n$ to $\mathbb{R}^n$, it is represented by a symmetric matrix in $\mathbb{R}^{n \times n}$. Equations of the form $Lv = w$ with L self-adjoint occur in many important physical settings, and the study of them forms a large and important area within functional analysis.

Example 2.6.1 *Let $V = W = L^2(a, b)$ with scalars $\mathbb{K}$ the real numbers and the standard norm $\| \cdot \|_2$. Consider the linear integral operator*

$$Kv(x) = \int_a^b k(x, y)\, v(y)\, dy, \qquad a \leq x \leq b,$$

where the kernel function satisfies the condition

$$B = \left[\int_a^b \int_a^b |k(x, y)|^2 dx\, dy \right]^{1/2} < \infty.$$

For any $v \in L^2(a, b)$,

$$\|Kv\|_2^2 = \int_a^b \left| \int_a^b k(x, y)\, v(y)\, dy \right|^2 dx$$

$$\leq \int_a^b \left[\int_a^b |k(x, y)|^2 dy \right] \left[\int_a^b |v(y)|^2 dy \right] dx$$

$$= B^2 \|v\|_2^2.$$

Thus,

$$\|Kv\|_2 \leq B\, \|v\| \quad \forall\, v \in L^2(a, b),$$

and then

$$\|K\| \leq B.$$

Hence we see that K is a continuous linear operator on $L^2(a, b)$.

Now let us find the adjoint of K. By the defining relation (2.6.1),

$$(K^*w, v) = (w, Kv)$$

$$= \int_a^b w(x) \left[\int_a^b k(x, y)\, v(y)\, dy \right] dx$$

$$= \int_a^b \left[\int_a^b k(x, y)\, w(x)\, dx \right] v(y)\, dy$$

for any $v, w \in L^2(a, b)$. This implies

$$K^* v(y) = \int_a^b k(x, y)\, v(x)\, dx \qquad \forall\, v \in L^2(a, b).$$

The integral operator K is self-adjoint if and only if $k(x, y) = k(y, x)$.

Given a Hilbert space V, the set of self-adjoint operators on V form a subspace of $\mathcal{L}(V)$. Indeed the following result is easy to verify.

Proposition 2.6.2 *If $L_1, L_2 \in \mathcal{L}(V)$ are self-adjoint, then for any real scalars α_1 and α_2, the operator $\alpha_1 L_1 + \alpha_2 L_2$ is self-adjoint.*

Proof. From Exercise 2.6.1, we have

$$(\alpha_1 L_1 + \alpha_2 L_2)^* = \alpha_1 L_1^* + \alpha_2 L_2^*.$$

Since L_1 and L_2 are self-adjoint,

$$(\alpha_1 L_1 + \alpha_2 L_2)^* = \alpha_1 L_1 + \alpha_2 L_2.$$

Hence $\alpha_1 L_1 + \alpha_2 L_2$ is self-adjoint. ∎

Proposition 2.6.3 *Assume $L_1, L_2 \in \mathcal{L}(V)$ are self-adjoint. Then $L_1 L_2$ is self-adjoint if and only if $L_1 L_2 = L_2 L_1$.*

Proof. Since L_1 and L_2 are self-adjoint, we have

$$(L_1 L_2 u, v) = (L_2 u, L_1 v) = (u, L_2 L_1 v) \qquad \forall\, u, v \in V.$$

Thus

$$(L_1 L_2)^* = L_2 L_1.$$

It follows that $L_1 L_2$ is self-adjoint if and only if $L_1 L_2 = L_2 L_1$ is valid. ∎

Corollary 2.6.4 *Suppose $L \in \mathcal{L}(V)$ is self-adjoint. Then for any non-negative integer n, L^n is self-adjoint (here by convention, $L^0 = I$, the identity operator). Consequently, for any polynomial $p(x)$ with real coefficients, the operator $p(L)$ is self-adjoint.*

We have a useful characterization of the norm of a self-adjoint operator.

Theorem 2.6.5 *Let $L \in \mathcal{L}(V)$ be self-adjoint. Then*

$$\|L\| = \sup_{\|v\|=1} |(Lv, v)|. \qquad (2.6.5)$$

Proof. Denote $M = \sup_{\|v\|=1} |(Lv, v)|$. First, for any $v \in V$, $\|v\| = 1$, we have

$$|(Lv, v)| \leq \|Lv\| \, \|v\| \leq \|L\|.$$

So

$$M \leq \|L\|. \qquad (2.6.6)$$

Now for any $u, v \in V$, we have the identity

$$(Lu, v) = \frac{1}{4} \left[(L(u+v), u+v) - (L(u-v), u-v) \right].$$

Thus

$$|(Lu, v)| \leq \frac{M}{4} \left(\|u+v\|^2 + \|u-v\|^2 \right) = \frac{M}{2} \left(\|u\|^2 + \|v\|^2 \right).$$

For $u \in V$ with $Lu \neq 0$, we take $v = (\|u\|/\|Lu\|) \, Lu$ in the above inequality to obtain

$$\|u\| \, \|Lu\| \leq M \, \|u\|^2;$$

i.e.,

$$\|Lu\| \leq M \, \|u\|.$$

Obviously, this inequality also holds if $Lu = 0$. Hence,

$$\|Lu\| \leq M \, \|u\| \quad \forall u \in V,$$

and we see that $\|L\| \leq M$. This inequality and (2.6.6) imply (2.6.5). ∎

Exercise 2.6.1 *Prove the following properties for adjoint operators.*

$$(\alpha_1 L_1 + \alpha_2 L_2)^* = \alpha_1 L_1^* + \alpha_2 L_2^*, \quad \alpha_1, \alpha_2 \text{ real,}$$
$$(L_1 L_2)^* = L_2^* L_1^*,$$
$$(L^*)^* = L.$$

Exercise 2.6.2 *Regard $C[0, 1]$ as an inner product space with the standard inner product*

$$(u, v) = \int_0^1 u(x)v(x) \, dx.$$

Define $K : C[0, 1] \to C[0, 1]$ by

$$Kf(x) = \int_0^x k(x, y) f(y) \, dy, \quad 0 \leq y \leq 1, \quad f \in C[0, 1],$$

with $k(x, y)$ continuous for $0 \leq y \leq x \leq 1$. Show K is a bounded operator. What is K^*? To what extent can the assumption of continuity of k be made less restrictive?

2.7 Types of convergence

We begin by introducing the concepts of strong convergence and weak convergence in a normed space.

Definition 2.7.1 *Let V be a normed space, V' its dual space. A sequence $\{u_n\} \subseteq V$ converges strongly to $u \in V$ if*

$$\lim_{n\to\infty} \|u - u_n\| = 0,$$

and we write $u_n \to u$ as $n \to \infty$. The sequence $\{u_n\}$ converges weakly to $u \in V$ if

$$\ell(u_n) \to \ell(u) \quad as \ n \to \infty, \ \forall \ell \in V'.$$

In this case, we write $u_n \rightharpoonup u$ as $n \to \infty$.

Example 2.7.2 *Let $f \in L^2(0, 2\pi)$. Then we know that the Fourier series of f converges in $L^2(0, 2\pi)$. Therefore the Fourier coefficients converge to zero, and in particular,*

$$\int_0^{2\pi} f(x) \sin(nx) \, dx \to 0 \qquad \forall f \in L^2(0, 2\pi).$$

This result is known as the Riemann-Lebesgue lemma. *Thus the sequence $\{\sin(nx) \mid n \geq 1\}$ converges weakly to 0 in $L^2(0, 2\pi)$. But certainly the sequence $\{\sin(nx) \mid n \geq 1\}$ does not converge strongly to 0 in $L^2(0, 2\pi)$.*

Strong convergence implies weak convergence, but not vice versa as Example 2.7.2 shows, unless the space V is finite dimensional. In a finite-dimensional space, it is well known that a bounded sequence has a convergent subsequence (cf. Theorem 1.6.2). In an infinite-dimensional space, we have only a weaker property; but it is still useful in proving many existence results.

Definition 2.7.3 *A normed space V is said to be* reflexive *if $(V')' = V$.*

An immediate consequence of this definition is that a reflexive normed space must be complete (i.e., a Banach space). By the Riesz representation theorem, it is relatively straightforward to show that any Hilbert space is reflexive.

The most important property of a reflexive Banach space is given in the next result. It is fundamental in the development of an existence theory for abstract optimization problems (cf. Section 3.2). A proof is given in [40, p. 132] and [175, p. 64].

Theorem 2.7.4 *Suppose V is a reflexive Banach space. Then any bounded sequence has a weakly convergent subsequence.*

Let $\Omega \subseteq \mathbb{R}^d$ be open and bounded. Recall from Example 2.5.1 that for $p \in (1, \infty)$, the dual space of $L^p(\Omega)$ is $L^{p'}(\Omega)$, where p' is the conjugate of

p defined by the relation $1/p' + 1/p = 1$. Therefore, $(L^p(\Omega)')' = L^{p'}(\Omega)' = L^p(\Omega)$; i.e., if $p \in (1, \infty)$, then the space $L^p(\Omega)$ is reflexive. Consequently, the above theorem implies the following: If $\{u_n\}$ is a bounded sequence in $L^p(\Omega)$, $\sup_n \|u_n\|_{L^p(\Omega)} < \infty$, then we can find a subsequence $\{u_{n'}\} \subseteq \{u_n\}$ and a function $u \in L^p(\Omega)$ such that

$$\lim_{n' \to \infty} \int_\Omega u_{n'}(\mathbf{x})\, v(\mathbf{x})\, dx = \int_\Omega u(\mathbf{x})\, v(\mathbf{x})\, dx \qquad \forall\, v \in L^{p'}(\Omega).$$

Finally, we introduce the concepts of strong convergence and *weak-** convergence of a sequence of linear operators.

Definition 2.7.5 *Let V and W be normed spaces. A sequence of linear operators $\{L_n\}$ from V to W is said to converge strongly to a linear operator $L : V \to W$ if*

$$\lim_{n \to \infty} \|L - L_n\| = 0.$$

In this case, we write $L_n \to L$ as $n \to \infty$. We say $\{L_n\}$ converges weak-* to L and write $L_n \rightharpoonup^* L$ if

$$\lim_{n \to \infty} L_n v = L v \qquad \forall\, v \in V.$$

Exercise 2.7.1 *Consider the linear operators from $C[a,b]$ to $\mathbb{R}$ defined by*

$$Lv = \int_a^b v(x)\, dx$$

and

$$L_n v = \frac{b-a}{n} \sum_{i=1}^{n} v\left(a + \frac{b-a}{n} i\right), \qquad n = 1, 2, \ldots.$$

We recognize that $\{L_n v\}$ is a sequence of Riemann sums for the integral Lv. Show that $L_n \rightharpoonup^* L$ but $L_n \not\to L$.

Exercise 2.7.2 *Show that in an inner product space,*

$$u_n \to u \quad \Longleftrightarrow \quad u_n \rightharpoonup u \text{ and } \|u_n\| \to \|u\|.$$

2.8 Compact linear operators

When V is a finite-dimensional linear space and $A : V \to V$ is linear, the equation $Au = w$ has a well-developed solvability theory. To extend these results to infinite-dimensional spaces, we introduce the concept of a *compact operator* K and then we give a theory for operator equations $Au = w$ in which $A = I - K$. Equations of the form

$$u - Ku = f \tag{2.8.1}$$

are called "equations of the second kind," and generally K is assumed to have special properties. The main idea is that compact operators are in some sense closely related to finite-dimensional operators, i.e., operators with a finite dimensional range. If K is truly finite dimensional, in a sense we define below, then (2.8.1) can be reformulated as a finite system of linear equations and solved exactly. If K is compact, then it is close to being finite dimensional; and the solvability theory of (2.8.1) is similar to that for the finite-dimensional case.

In the following, recall the discussion in Section 1.6.

Definition 2.8.1 *Let V and W be normed vector spaces, and let $K : V \to W$ be linear. Then K is* compact *if the set*

$$\{Kv \mid \|v\|_V \leq 1\}$$

has compact closure in W. This is equivalent to saying that for every bounded sequence $\{v_n\} \subseteq V$, the sequence $\{Kv_n\}$ has a subsequence that is convergent to some point in W. Compact operators are also called completely continuous operators.

There are other definitions for a compact operator, but the above is the one used most commonly. In the definition, the spaces V and W need not be complete; but in virtually all applications, they are complete. With completeness, some of the proofs of the properties of compact operators become simpler, and we will always assume V and W are complete (i.e., Banach spaces) when dealing with compact operators.

2.8.1   Compact integral operators on $C(D)$

Let D be a closed bounded set in $\mathbb{R}^d$. We define

$$Kv(\mathbf{x}) = \int_D k(\mathbf{x}, \mathbf{y})v(\mathbf{y})\, dy, \quad \mathbf{x} \in D, \quad v \in C(D). \tag{2.8.2}$$

The space $C(D)$ is to have the norm $\|\cdot\|_\infty$. We want to formulate conditions under which $K : C(D) \to C(D)$ is both bounded and compact. We assume $k(\mathbf{x}, \mathbf{y})$ is integrable as a function of $\mathbf{y}$, for all $\mathbf{x} \in D$, and further we assume the following.
A1. $\lim_{h \to 0} \omega(h) = 0$, with

$$\omega(h) \equiv \sup_{\substack{\mathbf{x}, \mathbf{z} \in D \\ \|\mathbf{x} - \mathbf{z}\| \leq h}} \int_D |k(\mathbf{x}, \mathbf{y}) - k(\mathbf{z}, \mathbf{y})|\, dy. \tag{2.8.3}$$

In this, $\|\mathbf{x} - \mathbf{z}\|$ denotes the Euclidean length of $\mathbf{x} - \mathbf{z}$.
A2.

$$\sup_{\mathbf{x} \in D} \int_D |k(\mathbf{x}, \mathbf{y})|\, dy < \infty. \tag{2.8.4}$$

Using A1, if $v(\mathbf{y})$ is bounded and integrable, then $Kv(\mathbf{x})$ is continuous, with

$$|Kv(\mathbf{x}) - Kv(\mathbf{y})| \le \omega(\|\mathbf{x} - \mathbf{y}\|)\|v\|_\infty. \tag{2.8.5}$$

Using A2, we have boundedness of K, with

$$\|K\| = \max_{\mathbf{x} \in D} \int_D |k(\mathbf{x}, \mathbf{y})| \, dy. \tag{2.8.6}$$

To discuss compactness of K, we first need to identify the compact sets in $C(D)$. To do this, we use Arzela-Ascoli theorem (Theorem 1.6.3). Now consider the set $S = \{Kv \mid v \in C(D) \text{ and } \|v\|_\infty \le 1\}$. This is uniformly bounded, since $\|Kv\|_\infty \le \|K\|\|v\|_\infty \le \|K\|$. In addition, S is equicontinuous from (2.8.5). Thus S has compact closure in $C(D)$, and K is a compact operator on $C(D)$ to $C(D)$.

What are the kernel functions k that satisfy A1–A2? Easily, these assumptions are satisfied if $k(\mathbf{x}, \mathbf{y})$ is a continuous function of $(\mathbf{x}, \mathbf{y}) \in D$. In addition, let $D = [a, b]$ and consider

$$Kv(x) = \int_a^b \log|x - y| \, v(y) \, dy \tag{2.8.7}$$

and

$$Kv(x) = \int_a^b \frac{1}{|x - y|^\beta} v(y) \, dy \tag{2.8.8}$$

with $\beta < 1$. These operators K can be shown to satisfy A1–A2, although we omit the proof. Later we show by other means that these are compact operators. An important and related example is

$$Kv(\mathbf{x}) = \int_D \frac{1}{|\mathbf{x} - \mathbf{y}|^\beta} v(\mathbf{y}) \, dy, \qquad \mathbf{x} \in D, \ v \in C(D).$$

The set $D \subseteq \mathbb{R}^d$ is assumed to be closed, bounded, and have a non-empty interior. This operator satisfies A1–A2 provided $\beta < d$, and therefore K is a compact operator from $C(D) \rightarrow C(D)$.

Still for the case $D = [a, b]$, another way to show that $k(x, y)$ satisfies A1 and A2 is to rewrite k in the form

$$k(x, y) = \sum_{i=0}^p h_i(x, y) l_i(x, y) \tag{2.8.9}$$

for some $p > 0$, with each $l_i(x, y)$ continuous for $a \le x, y \le b$ and each $h_i(x, y)$ satisfying A1–A2. It is left to the reader to show that in this case, k also satisfies A1–A2. The utility of this approach is that it is sometimes difficult to show directly that k satisfies A1–A2, whereas showing (2.8.9) with h_i, l_i satisfying the specified conditions may be easier.

Example 2.8.2 *Let $[a, b] = [0, \pi]$ and $k(x, y) = \log |\cos x - \cos y|$. Rewrite the kernel function as*

$$k(x, y) = \underbrace{|x - y|^{-\frac{1}{2}}}_{h(x,y)} \underbrace{|x - y|^{\frac{1}{2}} \log |\cos x - \cos y|}_{l(x,y)}. \qquad (2.8.10)$$

Easily, l is continuous; and from the discussion following (2.8.8), h satisfies A1–A2. Thus k is the kernel of a compact integral operator on $C[0, \pi]$ to $C[0, \pi]$.

2.8.2   Properties of compact operators

Another way of obtaining compact operators is to look at limits of simpler "finite-dimensional operators" in $\mathcal{L}(V, W)$, the Banach space of bounded linear operators from V to W. This gives another perspective on compact operators, one that leads to improved intuition by emphasizing their close relationship to operators on finite-dimensional spaces.

Definition 2.8.3 *Let V and W be linear spaces. The linear operator $K : V \to W$ is a* finite-rank *operator if $\mathcal{R}(K)$, the range of K, is finite dimensional.*

Proposition 2.8.4 *Let V and W be normed linear spaces, and let $K : V \to W$ be a bounded finite-rank operator. Then K is a compact operator.*

Proof. $\mathcal{R}(K)$ is a normed finite-dimensional space, and therefore it is complete. Consider the set

$$S = \{Kv \mid \|v\|_V \le 1\}.$$

The set S is bounded by $\|K\|$. Also $S \subseteq \mathcal{R}(K)$. Then S has compact closure, since all bounded closed sets in a finite-dimensional space are compact. This shows K is compact. ∎

Example 2.8.5 *Let $V = W = C[a, b]$ with $\| \cdot \|_\infty$. Consider the kernel function*

$$k(x, y) = \sum_{i=1}^{n} \beta_i(x) \, \gamma_i(y) \qquad (2.8.11)$$

with each β_i continuous on $[a, b]$ and each $\gamma_i(y)$ absolutely integrable on $[a, b]$. Then the associated integral operator K is a bounded, finite-rank operator on $C[a, b]$ to $C[a, b]$:

$$Kv(x) = \sum_{i=1}^{n} \beta_i(x) \int_a^b \gamma_i(y) \, v(y) \, dy, \qquad v \in C[a, b]. \qquad (2.8.12)$$

We have

$$\|K\| \le \sum_{i=1}^{n} \|\beta_i\|_\infty \int_a^b |\gamma_i(y)| \, dy.$$

From (2.8.12), $Kv \in C[a,b]$ and $\mathcal{R}(K) \subseteq \text{span}\{\beta_1, \ldots, \beta_n\}$, a finite-dimensional space.

Kernel functions of the form (2.8.11) are called *degenerate*. Below we see that the associated integral equation $(\lambda I - K)v = f$, $\lambda \neq 0$, is essentially a finite-dimensional equation.

Proposition 2.8.6 *Let $K \in \mathcal{L}(U,V)$ and $L \in \mathcal{L}(V,W)$; and let either K or L be compact. Then LK is a compact operator from U to W.*

The proof is left as Exercise 2.8.1 for the reader.

The following result gives the framework for using finite-rank operators to obtain similar, but more general compact operators.

Proposition 2.8.7 *Let V and W be normed spaces, with W complete. Let $K \in \mathcal{L}(V,W)$, let $\{K_n\}$ be a sequence of compact operators in $\mathcal{L}(V,W)$, and assume $K_n \to K$ in $\mathcal{L}(V,W)$. Then K is compact.*

This is a standard result in most books on functional analysis (e.g., see [40, p. 174] or [44, p. 486]).

For almost all function spaces V that occur in applied mathematics, the compact operators can be characterized as being the limit of a sequence of bounded finite-rank operators. This gives a further justification for the presentation of Proposition 2.8.7.

Example 2.8.8 *Let D be a closed and bounded set in $\mathbb{R}^d$. For example, D could be a region with a non-empty interior, a piecewise smooth surface, or a piecewise smooth curve. Let $k(\mathbf{x}, \mathbf{y})$ be a continuous function of $\mathbf{x}, \mathbf{y} \in D$. Suppose we can define a sequence of continuous degenerate kernel functions $k_n(\mathbf{x}, \mathbf{y})$ for which*

$$\max_{\mathbf{x} \in D} \int_D |k(\mathbf{x}, \mathbf{y}) - k_n(\mathbf{x}, \mathbf{y})| \, dy \to 0 \quad as \quad n \to \infty. \qquad (2.8.13)$$

Then for the associated integral operators, it easily follows that $K_n \to K$; and by Proposition 2.8.7, K is compact. The result (2.8.13) is true for general continuous functions $k(x, y)$, and we leave to the exercises the proof for various choices of D. Of course, we already knew that K was compact in this case, from the discussion following (2.8.8). But the present approach shows the close relationship of compact operators and finite-dimensional operators.

Example 2.8.9 *Let $V = W = C[a,b]$ with norm $\|\cdot\|_\infty$. Consider the kernel function*

$$k(x, y) = \frac{1}{|x - y|^\gamma} \qquad (2.8.14)$$

for some $0 < \gamma < 1$. Define a sequence of continuous kernel functions to approximate it:

$$k_n(x, y) = \begin{cases} \dfrac{1}{|x - y|^\gamma}, & |x - y| \geq \dfrac{1}{n}, \\ n^\gamma, & |x - y| \leq \dfrac{1}{n}. \end{cases} \qquad (2.8.15)$$

This merely limits the height to that of $k(x, y)$ when $|x - y| = \frac{1}{n}$. Easily, $k_n(x, y)$ is a continuous function for $a \leq x, y \leq b$, and thus the associated integral operator K_n is compact on $C[a, b]$. For the associated integral operators,

$$\|K - K_n\| = \frac{2\gamma}{1 - \gamma} \cdot \frac{1}{n^{1-\gamma}},$$

which converges to zero as $n \to \infty$. By Proposition 2.8.7, K is a compact operator on $C[a, b]$.

2.8.3 Integral operators on $L^2(a, b)$

Let $V = W = L^2(a, b)$, and let K be the integral operator associated with a kernel function $k(x, y)$. We first show that under suitable assumptions on k, the operator K maps $L^2(a, b)$ to $L^2(a, b)$ and is bounded. Let

$$M = \left[\int_a^b \int_a^b |k(x, y)|^2 dy\, dx \right]^{1/2} \qquad (2.8.16)$$

and assume $M < \infty$. For $v \in L^2(a, b)$, use the Cauchy-Schwarz inequality to obtain

$$\|Kv\|_2^2 = \int_a^b \left| \int_a^b k(x, y)v(y)dy \right|^2 dx$$

$$\leq \int_a^b \left[\int_a^b |K(x, y)|^2\, dy \right] \left[\int_a^b |v(y)|^2 dy \right] dx$$

$$= M^2 \|v\|_2^2.$$

This proves that $Kv \in L^2(a, b)$ and

$$\|K\| \leq M. \qquad (2.8.17)$$

This bound is comparable to the use of the Frobenius matrix norm to bound the operator norm of a matrix $A: \mathbb{R}^n \to \mathbb{R}^m$, when the vector norm $\|\cdot\|_2$ is being used. Recall that the Frobenius norm of a matrix A is given by

$$\|A\|_F = \sqrt{\sum_{i,j} |A_{i,j}|^2}.$$

Kernel functions K for which $M < \infty$ are called *Hilbert-Schmidt kernel functions*, and the quantity M in (2.8.16) is called the *Hilbert-Schmidt norm* of K.

For integral operators K with a degenerate kernel function as in (2.8.11), the operator K is bounded if all β_i, $\gamma_i \in L^2(a, b)$. This is a straightforward result that we leave as a problem for the reader. From Proposition 2.8.4, the integral operator is then compact.

To examine the compactness of K for more general kernel functions, we assume there is a sequence of kernel functions $k_n(x, y)$ for which (i) $K_n : L^2(a, b) \to L^2(a, b)$ is compact, and (ii)

$$M_n \equiv \left[\int_a^b \int_a^b |k(x, y) - k_n(x, y)|^2 dy\, dx \right]^{\frac{1}{2}} \to 0 \quad \text{as } n \to \infty. \quad (2.8.18)$$

For example, if K is continuous, then this follows from (2.8.13). The operator $K - K_n$ is an integral operator, and we apply (2.8.16)–(2.8.17) to it to obtain

$$\|K - K_n\| \le M_n \to 0 \quad \text{as } n \to \infty.$$

From Proposition 2.8.7, this shows K is compact. For any Hilbert-Schmidt kernel function, (2.8.18) can be shown to hold for a suitable choice of degenerate kernel functions k_n.

We leave it to the problems to show that $\log|x - y|$ and $|x - y|^{-\gamma}$, $\gamma < \frac{1}{2}$, are Hilbert-Schmidt kernel functions. For $\frac{1}{2} \le \gamma < 1$, the kernel function $|x - y|^{-\gamma}$ still defines a compact integral operator K on $L^2(a, b)$, but the above theory for Hilbert-Schmidt kernel functions does not apply. For a proof of the compactness of K in this case, see Mikhlin [119, p. 160].

2.8.4   The Fredholm alternative theorem

Integral equations were studied in the 19th century as one means of investigating boundary value problems for Laplace's equation, for example,

$$\begin{aligned} \Delta u(\mathbf{x}) &= 0, & \mathbf{x} \in \Omega, \\ u(\mathbf{x}) &= f(\mathbf{x}), & \mathbf{x} \in \partial\Omega, \end{aligned} \quad (2.8.19)$$

and other elliptic partial differential equations. In the early 1900s, Ivar Fredholm gave necessary and sufficient conditions for the solvability of a large class of Fredholm integral equations of the second kind; and with these results, he then was able to give much more general existence theorems for the solution of boundary value problems such as (2.8.19). In this subsection, we state and prove the most important results of Fredholm; and in the following subsection, we give additional results without proof.

The theory of integral equations has been developed by many people, with David Hilbert being among the most important popularizers of the area. The subject of integral equations continues as an important area of study in applied mathematics; and for an introduction that includes a review of much recent literature, see Kress [100]. For an interesting historical account of the development of functional analysis as it was affected by the development of the theory of integral equations, see Bernkopf [24]. From here on, to simplify notation, for a scalar λ and an operator $K : V \to V$, we use $\lambda - K$ for the operator $\lambda I - K$, where $I : V \to V$ is the identity operator.

Theorem 2.8.10 (FREDHOLM ALTERNATIVE) *Let V be a Banach space, and let $K : V \to V$ be compact. Then the equation $(\lambda - K)u = f$, $\lambda \neq 0$, has a unique solution $u \in V$ if and only if the homogeneous equation $(\lambda - K)v = 0$ has only the trivial solution $v = 0$. In such a case, the operator $\lambda - K : V \overset{1-1}{\underset{\text{onto}}{\to}} V$ has a bounded inverse $(\lambda - K)^{-1}$.*

Proof. The theorem is true for any compact operator K, but our proof is only for those compact operators that are the limit of a sequence of bounded finite-rank operators. For a more general proof, see Kress [100, Chap. 3] or Conway [40, p. 217]. We remark that the theorem is a generalization of the following standard result for finite-dimensional vector spaces V. For A a matrix of order n, with $V = \mathbb{R}^n$ or $\mathbb{C}^n$ (with A having real entries for the former case), the linear system $Au = w$ has a unique solution $u \in V$ for all $w \in V$ if and only if the homogeneous linear system $Az = 0$ has only the zero solution $z = 0$.

(**a**) We begin with the case where K is finite-rank and bounded. Let $\{\varphi_1, \dots, \varphi_n\}$ be a basis for $\mathcal{R}(K)$, the range of K. Rewrite the equation $(\lambda - K)u = f$ as

$$u = \frac{1}{\lambda}\left(f + Ku\right). \tag{2.8.20}$$

If this equation has a unique solution $u \in V$, then

$$u = \frac{1}{\lambda}\left(f + c_1\varphi_1 + \cdots + c_n\varphi_n\right) \tag{2.8.21}$$

for some uniquely determined set of constants $c_1, \dots, c_n$.

By substituting (2.8.21) into the equation, we have

$$\lambda\left\{\frac{1}{\lambda}f + \frac{1}{\lambda}\sum_{i=1}^{n}c_i\varphi_i\right\} - \frac{1}{\lambda}Kf - \frac{1}{\lambda}\sum_{j=1}^{n}c_jK\varphi_j = y.$$

Multiply by λ, and then simplify to obtain

$$\lambda\sum_{i=1}^{n}c_i\varphi_i - \sum_{j=1}^{n}c_jK\varphi_j = Kf. \tag{2.8.22}$$

Using the basis $\{\varphi_i\}$ for $\mathcal{R}(K)$, write

$$Kf = \sum_{i=1}^{n}\gamma_i\varphi_i$$

and

$$K\varphi_j = \sum_{i=1}^{n} a_{ij}\varphi_i, \quad 1 \le j \le n.$$

The coefficients $\{\gamma_i\}$ and $\{a_{ij}\}$ are uniquely determined. Substituting into (2.8.22) and rearranging,

$$\sum_{i=1}^{n} \left\{ \lambda c_i - \sum_{j=1}^{n} a_{ij} c_j \right\} \varphi_i = \sum_{i=1}^{n} \gamma_i \varphi_i.$$

By the independence of the basis elements φ_i, we obtain the linear system

$$\lambda c_i - \sum_{j=1}^{n} a_{ij} c_j = \gamma_i , \quad 1 \le i \le n. \tag{2.8.23}$$

Claim: This linear system and the equation $(\lambda - K)u = f$ are completely equivalent in their solvability, with (2.8.21) furnishing a one-to-one correspondence between the solutions of the two of them.

We have shown above that if u is a solution of $(\lambda - K)u = f$, then $(c_1, \dots, c_n)^{\mathrm{T}}$ is a solution of (2.8.23). In addition, suppose u_1 and u_2 are distinct solutions of $(\lambda - K)u = f$. Then

$$Ku_1 = \lambda u_1 - f \quad \text{and} \quad Ku_2 = \lambda u_2 - f, \quad \lambda \ne 0,$$

are also distinct vectors in $\mathcal{R}(K)$, and thus the associated vectors of coordinates $(c_1^{(1)}, \dots, c_n^{(1)})^T$ and $(c_1^{(2)}, \dots, c_n^{(2)})^T$,

$$K\varphi_i = \sum_{k=1}^{n} c_k^{(i)} \varphi_k , \quad i = 1, 2$$

must also be distinct.

For the converse statement, suppose $(c_1, \dots, c_n)^{\mathrm{T}}$ is a solution of (2.8.23). Define a vector $u \in V$ by using (2.8.21), and then check whether this u

satisfies the integral equation (2.8.20):

$$(\lambda - K)u = \lambda \left\{ \frac{1}{\lambda}f + \frac{1}{\lambda}\sum_{i=1}^{n} c_i \varphi_i \right\} - \frac{1}{\lambda}Kf - \frac{1}{\lambda}\sum_{j=1}^{n} c_j K\varphi_j$$

$$= f + \frac{1}{\lambda}\left\{ \lambda\sum_{i=1}^{n} c_i \varphi_i - Kf - \sum_{j=1}^{n} c_j K\varphi_j \right\}$$

$$= f + \frac{1}{\lambda}\left\{ \sum_{i=1}^{n} \lambda c_i \varphi_i - \sum_{i=1}^{n} \gamma_i \varphi_i - \sum_{j=1}^{n} c_j \sum_{i=1}^{n} a_{ij} \varphi_i \right\}$$

$$= f + \frac{1}{\lambda}\sum_{i=1}^{n} \underbrace{\left\{ \lambda c_i - \gamma_i - \sum_{j=1}^{n} a_{ij} c_j \right\}}_{=0,\ i=1,\dots,n} \varphi_i$$

$$= f.$$

Also, distinct coordinate vectors $(c_1, \dots, c_n)$ lead to distinct solutions u in (2.8.21), because of the linear independence of the basis vectors $\{\varphi_1, \dots, \varphi_n\}$. This completes the proof of the claim given above.

Now consider the Fredholm alternative theorem for $(\lambda - K)u = f$ with this finite-rank operator K. Suppose

$$\lambda - K : V \overset{1-1}{\underset{onto}{\to}} V.$$

Then trivially, the null space $\mathcal{N}(\lambda - K) = \{0\}$. For the converse, assume $(\lambda - K)v = 0$ has only the solution $v = 0$; and note that we want to show that $(\lambda - K)u = f$ has a unique solution for every $f \in V$.

Consider the associated linear system (2.8.23). It can be shown to have a unique solution for all right-hand sides $(\gamma_1, \dots, \gamma_n)$ by showing that the homogeneous linear system has only the zero solution. The latter is done by means of the equivalence of the homogeneous linear system to the homogeneous equation $(\lambda - K)v = 0$, which implies $v = 0$. But since (2.8.23) has a unique solution, so must $(\lambda - K)u = f$, and it is given by (2.8.21).

We must also show that $(\lambda - K)^{-1}$ is bounded. This can be done directly by a further examination of the consequences of K's being a bounded and finite-rank operator; but it is simpler to just cite the open mapping theorem (cf. Theorem 2.4.3).

(b) Assume now that $\|K - K_n\| \to 0$, with K_n finite rank and bounded. Rewrite $(\lambda - K)u = f$ as

$$[\lambda - (K - K_n)]u = f + K_n u, \quad n \geq 1. \tag{2.8.24}$$

Pick an index $m > 0$ for which

$$\|K - K_m\| < |\lambda| \tag{2.8.25}$$

and fix it. By the geometric series theorem (cf. Theorem 2.3.1),

$$Q_m \equiv [\lambda - (K - K_m)]^{-1}$$

exists and is bounded, with

$$\|Q_m\| \leq \frac{1}{|\lambda| - \|K - K_m\|}.$$

The equation (2.8.24) can now be written in the equivalent form

$$u - Q_m K_m u = Q_m f. \tag{2.8.26}$$

The operator $Q_m K_m$ is bounded and finite rank. The boundedness follows from that of Q_m and K_m. To show it is finite rank, let $\mathcal{R}(K_m) = \text{span}\{\varphi_1, \ldots, u_m\}$. Then

$$\mathcal{R}(Q_m K_m) = \text{span}\{Q_m \varphi_1, \ldots, Q_m u_m\}$$

is a finite-dimensional space.

The equation (2.8.26) is one to which we can apply part (a) of this proof. Assume $(\lambda - K)v = 0$ implies $v = 0$. By the above equivalence, this yields

$$(I - Q_m K_m)v = 0 \quad \implies \quad v = 0.$$

But from part (a), this says $(I - Q_m K_m)u = w$ has a unique solution u for every $w \in V$, and in particular, for $w = Q_m f$ as in (2.8.26). By the equivalence of (2.8.26) and $(\lambda - K)u = f$, we have that the latter is uniquely solvable for every $f \in V$. The boundedness of $(\lambda - K)^{-1}$ follows from part (a) and the boundedness of Q_m; or the open mapping theorem can be cited, as earlier in part (a). ∎

For many practical problems in which K is not compact, it is important to note what makes this proof work. It is *not* necessary to have a sequence of bounded and finite-rank operators $\{K_n\}$ for which $\|K - K_n\| \to 0$. Rather, it is necessary to satisfy the inequality (2.8.25) for one finite-rank operator K_m; and in applying the proof to other operators K, it is necessary only that K_m be compact. In such a case, the proof following (2.8.25) remains valid, and the Fredholm alternative still applies to such an equation:

$$(\lambda - K)u = f.$$

2.8.5 Additional results on Fredholm integral equations

In this subsection, we give additional results on the solvability of compact equations of the second kind, $(\lambda - K)u = f$, with $\lambda \neq 0$. No proofs are given, and the reader is referred to a standard text on integral equations (e.g., see Kress [100] or Mikhlin [118]).

Definition 2.8.11 *Let $K : V \to V$. If there is a scalar λ and an associated vector $u \neq 0$ for which $Ku = \lambda u$, then λ is called an* eigenvalue *and u an associated* eigenvector *of the operator K. (When dealing with compact* *operators K, we generally are interested in only the non-zero eigenvalues of K.)*

In the following, recall that $\mathcal{N}(A)$ denotes the null space of A.

Theorem 2.8.12 *Let $K : V \to V$ be compact, and let V be a Banach space. Then—*

1. *The eigenvalues of K form a discrete set in the complex plane $\mathbb{C}$, with 0 as the only possible limit point.*

2. *For each non-zero eigenvalue λ of K, there are only a finite number of linearly independent eigenvectors.*

3. *Each non-zero eigenvalue λ of K has finite index $\nu(\lambda) \geq 1$. This means*

$$\mathcal{N}(\lambda - K) \underset{\neq}{\subseteq} \mathcal{N}((\lambda - K)^2) \underset{\neq}{\subseteq} \cdots$$
$$\underset{\neq}{\subseteq} \mathcal{N}((\lambda - K)^{\nu(\lambda)}) = \mathcal{N}((\lambda - K)^{\nu(\lambda)+1}). \qquad (2.8.27)$$

In addition, $\mathcal{N}((\lambda - K)^{\nu(\lambda)})$ is finite dimensional. The elements of the subspace $\mathcal{N}((\lambda - K)^{\nu(\lambda)})$ are called generalized eigenvectors of K.

4. *For all $\lambda \neq 0$, $\mathcal{R}(\lambda - K)$ is closed in V.*

5. *For each non-zero eigenvalue λ of K,*

$$V = \mathcal{N}((\lambda - K)^{\nu(\lambda)}) \oplus \mathcal{R}((\lambda - K)^{\nu(\lambda)}) \qquad (2.8.28)$$

is a decomposition of V into invariant subspaces. This implies that every $u \in V$ can be written as $u = u_1 + u_2$ with unique choices

$$u_1 \in \mathcal{N}((\lambda - K)^{\nu(\lambda)}) \quad and \quad u_2 \in \mathcal{R}((\lambda - K)^{\nu(\lambda)}).$$

Being invariant means that

$$K : \mathcal{N}((\lambda - K)^{\nu(\lambda)}) \to \mathcal{N}((\lambda - K)^{\nu(\lambda)}),$$
$$K : \mathcal{R}((\lambda - K)^{\nu(\lambda)}) \to \mathcal{R}((\lambda - K)^{\nu(\lambda)}).$$

6. *The Fredholm alternative theorem and the above results (1)–(5) remain true if K^m is compact for some $m > 1$.*

For results on the speed with which the eigenvalues $\{\lambda_n\}$ of compact integral operators K converge to zero, see Hille and Tamarkin [77] and Fenyö and Stolle [50, Section 8.9]. Generally, as the differentiability of the kernel function $k(x, y)$ increases, the speed of convergence to zero of the eigenvalues also increases.

For the following results, recall from Section 2.6 the concept of an adjoint operator.

Lemma 2.8.13 *Let V be a Hilbert space with scalars the complex numbers $\mathbb{C}$, let $K : V \to V$ be a compact operator. Then $K^* : V \to V$ is also a compact operator.*

This implies that the operator K^* also shares the properties stated above for the compact operator K. There is, however, a closer relationship between the operators K and K^*, which is given in the following theorem.

Theorem 2.8.14 *Let V be a Hilbert space with scalars the complex numbers $\mathbb{C}$, let $K : V \to V$ be a compact operator, and let λ be a non-zero eigenvalue of K. Then—*

1. *$\bar{\lambda}$ is an eigenvalue of the adjoint operator K^*. In addition, $\mathcal{N}(\lambda - K)$ and $\mathcal{N}(\bar{\lambda} - K^*)$ have the same dimension.*

2. *The equation $(\lambda - K)u = f$ is solvable if and only if*

$$(f, v) = 0 \quad \forall v \in \mathcal{N}(\bar{\lambda} - K^*). \tag{2.8.29}$$

An equivalent way of writing this is

$$\mathcal{R}(\lambda - K) = \mathcal{N}(\bar{\lambda} - K^*)^\perp,$$

the subspace orthogonal to $\mathcal{N}(\bar{\lambda} - K^*)$. With this, we can write the decomposition

$$V = \mathcal{N}(\bar{\lambda} - K^*) \oplus \mathcal{R}(\lambda - K). \tag{2.8.30}$$

Theorem 2.8.15 *Let V be a Hilbert space with scalars the complex numbers $\mathbb{C}$, and let $K : V \to V$ be a self-adjoint compact operator. Then all eigenvalues of K are real and of index $\nu(\lambda) = 1$. In addition, the corresponding eigenvectors can be chosen as an orthonormal set. Order the nonzero eigenvalues as follows:*

$$|\lambda_1| \geq |\lambda_2| \geq \cdots \geq |\lambda_n| \geq \cdots > 0 \tag{2.8.31}$$

with each eigenvalue repeated according to its multiplicity (i.e., the dimension of $\mathcal{N}(\lambda - K)$). Then we write

$$Ku_i = \lambda_i u_i , \quad i \geq 1 \tag{2.8.32}$$

with

$$(u_i, u_j) = \delta_{ij}.$$

Also, the eigenvectors $\{u_i\}$ form an orthonormal basis for $\overline{\mathcal{R}(\lambda - K)}$.

Much of the theory of self-adjoint boundary value problems for ordinary and partial differential equations is based on theorems 2.8.14 and 2.8.15. Moreover, the completeness in $L^2(D)$ of many families of functions is proven by showing they are the eigenfunctions to a self-adjoint differential equation or integral equation problem.

Example 2.8.16 *Let* $D = \{\mathbf{x} \in \mathbb{R}^3 \mid \|\mathbf{x}\| = 1\}$, *the unit sphere in* $\mathbb{R}^3$, *and let* $V = L^2(D)$. *In this,* $\|\mathbf{x}\|$ *denotes the Euclidean length of* $\mathbf{x}$. *Define*

$$Kv(\mathbf{x}) = \int_D \frac{v(\mathbf{y})}{\|\mathbf{x} - \mathbf{y}\|} dS_{\mathbf{y}}, \qquad \mathbf{x} \in D. \qquad (2.8.33)$$

This is a compact operator, a proof of which is given in Mikhlin [119, p. 160]. The eigenfunctions of K *are called* spherical harmonics, *a much-studied set of functions (e.g., see [55], [111]). For each integer* $k \geq 0$, *there are* $2k + 1$ *independent spherical harmonics of degree* k; *and for each such spherical harmonic* φ_k, *we have*

$$K\varphi_k = \frac{4\pi}{2k+1} \varphi_k \qquad k = 0, 1, \ldots \qquad (2.8.34)$$

Letting $\mu_k = 4\pi/(2k + 1)$, *we have* $\mathcal{N}(\mu_k - K)$ *has dimension* $2k + 1$, $k \geq 0$. *It is well known that the set of all spherical harmonics form a basis for* $L^2(D)$, *in agreement with Theorem 2.8.15.*

Exercise 2.8.1 *Prove Proposition 2.8.6.*

Exercise 2.8.2 *Suppose* k *is a degenerate kernel function given by (2.8.11) with all* $\beta_i, \gamma_i \in L^2(a, b)$. *Show that the integral operator* K, *defined by*

$$Kv(x) = \int_a^b k(x, y)v(y) \, dy$$

is bounded from $L^2(a, b)$ *to* $L^2(a, b)$.

Exercise 2.8.3 *Consider the integral operator (2.8.2). Assume the kernel function* k *has the form (2.8.9) with each* $l_i(x, y)$ *continuous for* $a \leq x, y \leq b$ *and each* $h_i(x, y)$ *satisfying A1–A2. Prove that* k *also satisfies A1–A2.*

Exercise 2.8.4 *Show that* $\log |x - y|$ *and* $|x - y|^{-\gamma}$, $\gamma < \frac{1}{2}$, *are Hilbert-Schmidt kernel functions.*

Exercise 2.8.5 *Consider the integral equation*

$$\lambda f(x) - \int_0^1 e^{x-y} f(y) \, dy = g(x), \qquad 0 \leq x \leq 1,$$

with $g \in C[0, 1]$. *Denote the integral operator in the equation by* K, *and consider* K *as a mapping on* $C[0, 1]$ *into itself, and use the uniform norm* $\|\cdot\|_\infty$. *Find a bound for the condition number*

$$\text{cond}(\lambda - K) \equiv \|\lambda - K\| \, \|(\lambda - K)^{-1}\|$$

within the framework of the space $C[0, 1]$. *Do this for all values of* λ *for which* $(\lambda - K)^{-1}$ *exists as a bounded operator on* $C[0, 1]$. *Comment on how the condition number varies with* λ.

Exercise 2.8.6 *Recall Example 2.3.6 of Section 2.3. Use the approximation*

$$e^{xy} \approx 1 + xy$$

to examine the solvability of the integral equation

$$\lambda u(x) - \int_0^1 e^{xy} u(y)\, dy = f(x), \qquad 0 \le x \le 1.$$

To solve the integral equation associated with the kernel $1 + xy$, use the method developed in the proof of Theorem 2.8.10.

Exercise 2.8.7 *For any $f \in C[0,1]$, define*

$$\mathcal{A}f(x) = \begin{cases} \displaystyle\int_0^x \frac{f(y)}{\sqrt{x^2 - y^2}}\, dy, & 0 < x \le 1, \\[2ex] \dfrac{\pi}{2} f(0), & x = 0. \end{cases}$$

This is called an Abel integral operator. Show that $f(x) = x^\alpha$ is an eigenfunction of $\mathcal{A}$ for every $\alpha \ge 0$. What is the corresponding eigenvalue? Can $\mathcal{A}$ be a compact operator?

2.9 The resolvent operator

Let V be a complex Banach space; e.g., let $V = C(D)$ be the set of continuous complex-valued functions on a closed set D with the uniform norm $\|\cdot\|_\infty$; and let $L : V \to V$ be a bounded linear operator. From the geometric series theorem (Theorem 2.3.1), we know that if $|\lambda| > \|L\|$, then $(\lambda - L)^{-1}$ exists as a bounded linear operator from V to V. It is useful to consider the set of all complex numbers λ for which such a $(\lambda - L)^{-1}$ exists on V to V.

Definition 2.9.1 (a) *Let V be a complex Banach space, and let $L : V \to V$ be a bounded linear operator. We say $\lambda \in \mathbb{C}$ belongs to the* resolvent set *of L if $(\lambda - L)^{-1}$ exists as a bounded linear operator from V to V. The resolvent set of L is denoted by $\rho(L)$. The operator $(\lambda - L)^{-1}$ is called the* resolvent operator.
(b) *The set $\sigma(L) = \mathbb{C} - \rho(L)$ is called the* spectrum *of L.*

From the remarks preceding the definition,

$$\{\lambda \in \mathbb{C} : |\lambda| > \|L\|\} \subseteq \rho(L).$$

In addition, we have the following.

Lemma 2.9.2 *$\rho(L)$ is an open set in $\mathbb{C}$; and consequently, $\sigma(L)$ is a closed set.*

Proof. Let $\lambda_0 \in \rho(L)$. We use the perturbation result in Theorem 2.3.5 to show that all points λ in a sufficiently small neighborhood of λ_0 also are in $\rho(L)$; this is sufficient for showing $\rho(L)$ is open. Since $(\lambda_0 - L)^{-1}$ is a bounded linear operator on V to V, consider all $\lambda \in \mathbb{C}$ for which

$$|\lambda - \lambda_0| < \frac{1}{\left\| (\lambda_0 - L)^{-1} \right\|}. \tag{2.9.1}$$

Using Theorem 2.3.5, we have that $(\lambda - L)^{-1}$ also exists as a bounded operator from V to V, and moreover,

$$\left\| (\lambda - L)^{-1} - (\lambda_0 - L)^{-1} \right\| \leq \frac{|\lambda - \lambda_0| \left\| (\lambda_0 - L)^{-1} \right\|^2}{1 - |\lambda - \lambda_0| \left\| (\lambda_0 - L)^{-1} \right\|}. \tag{2.9.2}$$

This shows

$$\{ \lambda \in \mathbb{C} : |\lambda - \lambda_0| < \varepsilon \} \subseteq \rho(L),$$

provided ε is chosen sufficiently small, which shows $\rho(L)$ is an open set.

The inequality (2.9.2) shows that $R(\lambda) \equiv (\lambda - L)^{-1}$ is a continuous function of λ from $\mathbb{C}$ to $\mathcal{L}(V)$. ∎

A complex number λ can belong to $\sigma(L)$ for several different reasons. Following is a standard classification scheme.

1. *Point spectrum.* $\lambda \in \sigma_P(L)$ means that λ is an eigenvalue of L. Thus there is a non-zero eigenvector $u \in V$ for which $Lu = \lambda u$. Such cases were explored in Section 2.8 with L a compact operator. In this latter case, the non-zero portion of $\sigma(L)$ consists entirely of eigenvalues, and moreover, 0 is the only possible point in $\mathbb{C}$ to which sequences of eigenvalues can converge.

2. *Continuous spectrum.* $\lambda \in \sigma_C(L)$ means that $\lambda - L$ is one-to-one, $\mathcal{R}(\lambda - L) \neq V$, and $\overline{\mathcal{R}(\lambda - L)} = V$. Note that if $\lambda \neq 0$, then L cannot be compact. (Why?) This type of situation, $\lambda \in \sigma_C(L)$, occurs in solving equations $(\lambda - L)u = f$ that are ill-posed. In the case $\lambda = 0$, such equations can often be written as an integral equation of the first kind

$$\int_a^b \ell(x, y) u(y) \, dy = f(x), \qquad a \leq x \leq b$$

with $\ell(x, y)$ continuous and smooth.

3. *Residual spectrum.* $\lambda \in \sigma_R(L)$ means $\lambda \in \sigma(L)$ and that it is in neither the point spectrum nor continuous spectrum. This case can be further subdivided, into cases with $\mathcal{R}(\lambda - L)$ closed and not closed. The latter case consists of ill-posed problems, much as with the case of continuous spectrum. For the former case, the equation $(\lambda - L)u = f$

is usually a well-posed problem; but some change in it is often needed when developing practical methods of solution.

If L is a compact operator on V to V, and if V is infinite dimensional, then it can be shown that $0 \in \sigma(L)$. In addition in this case, if 0 is not an eigenvalue of L, then L^{-1} can be shown to be unbounded on $\mathcal{R}(L)$. Equations $Lu = f$ with L compact make up a significant proportion of ill-posed problems.

2.9.1 $R(\lambda)$ as a holomorphic function

Let $\lambda_0 \in \rho(L)$. Returning to the proof of Lemma 2.9.2, we can write $R(\lambda) \equiv (\lambda - L)^{-1}$ as

$$R(\lambda) = \sum_{k=0}^{\infty} (-1)^k (\lambda - \lambda_0)^k R(\lambda_0)^{k+1} \tag{2.9.3}$$

for all λ satisfying (2.9.1). Thus we have a power series expansion of $R(\lambda)$ about the point λ_0. This can be used to introduce the idea that R is an analytic (or holomorphic) function from $\rho(L) \subseteq \mathbb{C}$ to the vector space $\mathcal{L}(V)$. Many of the definitions, ideas, and results of *complex analysis* can be extended to analytic vector-valued functions. See [44, p. 566] for an introduction to these ideas.

In particular, we can introduce line integrals. We are especially interested in line integrals of the form

$$g_\Gamma(L) = \frac{1}{2\pi i} \int_\Gamma (\mu - L)^{-1} g(\mu) \, d\mu, \tag{2.9.4}$$

Note that whereas $g : \rho(L) \to \mathbb{C}$, the quantity $g_\Gamma(L) \in \mathcal{L}(V)$. In this integral, Γ is a piecewise smooth curve of finite length in $\rho(L)$; and Γ can consist of several finite disjoint curves. In complex analysis, such integrals occur in connection with studying *Cauchy's theorem*.

Let $\mathcal{F}(L)$ denote the set of all functions g that are analytic on some open set U containing $\sigma(L)$, with the set U dependent on the function g (U need not be connected). For functions in $\mathcal{F}(L)$, a number of important results can be shown for the operators $g(L)$ of (2.9.4) with $g \in \mathcal{F}(L)$. For a proof of the following, see [44, p. 568]

Theorem 2.9.3 *Let $f, g \in \mathcal{F}(L)$, and let $f_\Gamma(L)$, $g_\Gamma(L)$ be defined using (2.9.4), assuming Γ is located within the domain of analyticity of both f and g. Then*
(a) *$f \cdot g \in \mathcal{F}(L)$, and $f_\Gamma(L) \cdot g_\Gamma(L) = (f \cdot g)_\Gamma(L)$;*
(b) *if f has a power series expansion*

$$f(\lambda) = \sum_{n=0}^{\infty} a_n \lambda^n$$

that is valid in some open disk about $\sigma(L)$, then

$$f_\Gamma(L) = \sum_{n=0}^{\infty} a_n L^n.$$

In numerical analysis, such integrals (2.9.4) become a means for studying the convergence of algorithms for approximating the eigenvalues of L.

Theorem 2.9.4 *Let L be a compact operator from V to V, and let λ_0 be a nonzero eigenvalue of L. Introduce*

$$E(\lambda_0, L) = \frac{1}{2\pi i} \int_{|\lambda - \lambda_0| = \varepsilon} (\lambda - L)^{-1} \, d\lambda \qquad (2.9.5)$$

with ε less than the distance from λ_0 to the remaining portion of $\sigma(L)$. Then—
(a) $E(\lambda_0, L)$ is a projection operator on V to V.
(b) $E(\lambda_0, L)V$ is the set of all ordinary and generalized eigenvectors associated with λ_0; i.e.,

$$E(\lambda_0, L)V = \mathcal{N}((\lambda - K)^{\nu(\lambda_0)}),$$

with the latter taken from (2.8.27) and $\nu(\lambda_0)$ the index of λ_0.

For a proof of these results, see Dunford and Schwartz [44, pp. 566–580].
When L is approximated by a sequence of operators $\{L_n\}$, we can examine the convergence of the eigenspaces of L_n to those of L by means of tools fashioned from (2.9.5). Examples of such analyses can be found in [7], [9], and Chatelin [33].

Exercise 2.9.1 *Let $\lambda \in \rho(L)$. Define $d(\lambda)$ to be the distance from λ to $\sigma(L)$,*

$$d(\lambda) = \min_{\kappa \in \sigma(L)} |\lambda - \kappa|.$$

Show that

$$\left\| (\lambda - L)^{-1} \right\| \geq \frac{1}{d(\lambda)}.$$

This shows $\left\| (\lambda - L)^{-1} \right\| \to \infty$ as $\lambda \to \sigma(L)$.

Exercise 2.9.2 *Let $V = C[0,1]$, and let L be the Volterra integral operator*

$$Lu(x) = \int_0^x k(x,y)u(y) \, dy, \qquad 0 \leq x \leq 1, \qquad u \in C[0,1].$$

with $k(x,y)$ continuous for $0 \leq y \leq x \leq 1$. What is $\sigma(L)$?

Exercise 2.9.3 *Derive (2.9.3).*

Exercise 2.9.4 *Let $F \subseteq \rho(L)$ be closed and bounded. Show $(\lambda - L)^{-1}$ is a continuous function of $\lambda \in F$, with*

$$\max_{\lambda \in F} \left\| (\lambda - L)^{-1} \right\| < \infty.$$

Exercise 2.9.5 *Let L be a bounded linear operator on a Banach space V to V; and let $\lambda_0 \in \sigma(L)$ be an isolated nonzero eigenvalue of L. Let $\{L_n\}$ be a sequence of bounded linear operators on V to V with $\|L - L_n\| \to 0$ as $n \to \infty$. Let $F \subseteq \rho(L)$ be closed and bounded. Show that there exists N such that*

$$n \geq N \quad \Rightarrow \quad F \subseteq \rho(L_n).$$

This shows that approximating sequences $\{L_n\}$ cannot produce extraneous convergent sequences of approximating eigenvalues.
Hint: *use the preceding Exercise 2.9.4 as a lemma.*

Exercise 2.9.6 *Assume L is a compact operator on V to V, a complex Banach space, and let $\{L_n\}$ be a sequence of approximating bounded linear compact operators with $\|L - L_n\| \to 0$ as $n \to \infty$. Referring to the curve $\Gamma = \{\lambda : |\lambda - \lambda_0| = \varepsilon\}$ of (2.9.5), we have from Exercise 2.9.5 that we can define*

$$E(\sigma_n, L_n) = \frac{1}{2\pi i} \int_{|\lambda - \lambda_0| = \varepsilon} (\lambda - L_n)^{-1} \, d\lambda, \qquad n \geq N,$$

with σ_n denoting the portion of $\sigma(L_n)$ located within Γ. Prove

$$\|E(\sigma_n, L_n) - E(\lambda_0, L)\| \to 0 \quad as \quad n \to \infty.$$

It can be shown that $\mathcal{R}(E(\sigma_n, L_n))$ consists of combinations of the simple and generalized eigenvectors of L_n corresponding to the eigenvalues of L_n within σ_n. In addition, prove that for every $u \in \mathcal{N}((\lambda - K)^{\nu(\lambda_0)})$,

$$E(\sigma_n, L_n)u \to u \quad as \quad n \to \infty.$$

This shows convergence of approximating simple and generalized eigenfunctions of L_n to those of L.

Suggestion for Further Readings

See "Suggestion for Further Readings" in Chapter 1.

3

Approximation Theory

In this chapter, we deal with the problem of approximation of functions. A prototype problem can be described as follows: For some function f, known exactly or approximately, find an approximation that has a more simply computable form, with the error of the approximation within a given error tolerance. Often the function f is not known exactly. For example, if the function comes from a physical experiment, we usually have a table of function values only. Even when a closed-form expression is available, it may happen that the expression is not easily computable, for example,

$$f(x) = \int_0^x e^{-t^2} dt.$$

The approximating functions need to be of simple form so that it is easy to make calculations with them. The most commonly used classes of approximating functions are the polynomials, piecewise polynomial functions, and trigonometric polynomials.

We give several approaches to the construction of approximating functions. In Section 3.1, we define and analyze the use of interpolation functions. In Section 3.2 we define the concept of best uniform approximation, and in Section 3.3 we look at best approximation in the sense of mean-square or L^2 error. Section 3.4 discusses the important special case of approximations using orthogonal polynomials, and Section 3.5 introduces a more abstract framework for approximations, one using finite-rank projection operators. The chapter concludes with a discussion in Section 3.6 of the uniform error in polynomial and trigonometric approximations of smooth functions.

3.1 Interpolation theory

We begin by discussing the interpolation problem in an abstract setting. Let V be a normed vector space over a field $\mathbb{K}$ of numbers ($\mathbb{R}$ or $\mathbb{C}$). Recall that the space of all the linear continuous functionals on V is called the *dual space* of V and is denoted by V' (cf. Section 2.5).

An abstract interpolation problem can be stated in the following form. Suppose V_n is an n-dimensional subspace of V, with a basis $\{v_1, \ldots, v_n\}$. Let $L_i \in V'$, $1 \le i \le n$, be n linear continuous functionals. Given n numbers $b_i \in \mathbb{K}$, $1 \le i \le n$, find $u_n \in V_n$ such that the interpolation conditions

$$L_i u_n = b_i, \qquad 1 \le i \le n$$

are satisfied.

Some questions arise naturally: Does the interpolation problem have a solution? If so, is it unique? If the interpolation function is used to approximate a given function $f(x)$, what can be said about error in the approximation?

Definition 3.1.1 *We say that the functionals L_i, $1 \le i \le n$, are linearly independent over V_n if*

$$\sum_{i=1}^{n} a_i L_i(v) = 0, \quad \forall v \in V_n \implies a_i = 0, \quad 1 \le i \le n.$$

Lemma 3.1.2 *The linear functionals $L_1, \ldots, L_n$ are linearly independent over V_n if and only if*

$$\det(L_i v_j) = \det \begin{bmatrix} L_1 v_1 & \cdots & L_1 v_n \\ \vdots & \ddots & \vdots \\ L_n v_1 & \cdots & L_n v_n \end{bmatrix} \neq 0.$$

Proof. By definition,

$L_1, \ldots, L_n$ are linearly independent over V_n

$$\iff \quad \sum_{i=1}^{n} a_i L_i(v_j) = 0, \ 1 \le j \le n \implies a_i = 0, \ 1 \le i \le n$$

$$\iff \quad \det(L_i v_j) \neq 0.$$

∎

Theorem 3.1.3 *The following statements are equivalent:*

1. *The interpolation problem has a unique solution.*

2. *The functionals $L_1, \ldots, L_n$ are linearly independent over V_n.*

3. The only element $u_n \in V_n$ satisfying

$$L_i u_n = 0, \qquad 1 \leq i \leq n,$$

is $u_n = 0$.

4. For any data $\{b_i\}_{i=1}^n$, there exists one $u_n \in V_n$ such that

$$L_i u_n = b_i, \qquad 1 \leq i \leq n.$$

Proof. From linear algebra, for a square matrix $A \in \mathbb{K}^{n \times n}$, the following statements are equivalent:

1. The system $A\mathbf{x} = \mathbf{b}$ has a unique solution $\mathbf{x} \in \mathbb{K}^n$ for any $\mathbf{b} \in \mathbb{K}^n$.

2. $\det(A) \neq 0$.

3. If $A\mathbf{x} = \mathbf{0}$, then $\mathbf{x} = \mathbf{0}$.

4. For any $\mathbf{b} \in \mathbb{K}^n$, the system $A\mathbf{x} = \mathbf{b}$ has a solution $\mathbf{x} \in \mathbb{K}^n$.

The results of the theorem now follow from these statements and the previous lemma. ∎

Now given $u \in V$, its interpolant $u_n = \sum_{i=1}^n a_i v_i$ in V_n is defined by the interpolation conditions

$$L_i u_n = L_i u, \qquad 1 \leq i \leq n.$$

The coefficients $\{a_i\}_{i=1}^n$ can be found from the linear system

$$\begin{pmatrix} L_1 v_1 & \cdots & L_1 v_n \\ \vdots & \ddots & \vdots \\ L_n v_1 & \cdots & L_n v_n \end{pmatrix} \begin{pmatrix} a_1 \\ \vdots \\ a_n \end{pmatrix} = \begin{pmatrix} L_1 u \\ \vdots \\ L_n u \end{pmatrix}$$

which has a unique solution if the functionals $L_1, \ldots, L_n$ are linearly independent over V_n.

The question of an error analysis in the abstract framework is difficult. For a general discussion of such error analysis, see Davis [42, Chap. 3]. Here we only give error analysis results for certain concrete situations.

3.1.1 Lagrange polynomial interpolation

Let f be a continuous function defined on a finite closed interval $[a, b]$. Let

$$\Delta : a \leq x_0 < x_1 < \cdots < x_n \leq b$$

be a partition of the interval $[a, b]$. Choose $V = C[a, b]$, the space of continuous functions $f : [a, b] \to \mathbb{K}$; and choose V_{n+1} to be $\mathcal{P}_n$, the space of the polynomials of degree less than or equal to n. Then the Lagrange interpolant of degree n of f is defined by the conditions

$$p_n(x_i) = f(x_i), \qquad 0 \leq i \leq n, \quad p_n \in \mathcal{P}_n. \tag{3.1.1}$$

Here the interpolation linear functionals are

$$L_i f = f(x_i), \qquad 0 \le i \le n.$$

If we choose the regular basis $v_j(x) = x^j$ $(0 \le j \le n)$ for $\mathcal{P}_n$, then it can be shown that

$$\det(L_i v_j)_{(n+1) \times (n+1)} = \prod_{j>i}(x_j - x_i) \ne 0. \qquad (3.1.2)$$

Thus there exists a unique Lagrange interpolation polynomial.

Furthermore, we have the representation formula

$$p_n(x) = \sum_{i=0}^{n} f(x_i)\, \phi_i(x), \qquad \phi_i(x) \equiv \prod_{j \ne i} \frac{x - x_j}{x_i - x_j}, \qquad (3.1.3)$$

called *Lagrange's formula* for the interpolation polynomial. The functions ϕ_i satisfy the special interpolation conditions

$$\phi_i(x_j) = \delta_{ij} = \begin{cases} 0, \, i \ne j, \\ 1, \, i = j. \end{cases}$$

The functions $\{\phi_i\}_{i=0}^{n}$ form a basis for $\mathcal{P}_n$, and they are often called *Lagrange basis functions*. See Figure 3.1 for graphs of $\{\phi_i(x)\}$ for $n = 3$, the case of cubic interpolation, with even spacing.

Outside of the framework of Theorem 3.1.3, the formula (3.1.3) shows directly the existence of a solution to the Lagrange interpolation problem (3.1.1). The uniqueness result can also be proven by showing that the interpolant corresponding to the homogeneous data is zero. Let us show this. Let $p_n \in \mathcal{P}_n$ with $p_n(x_i) = 0$, $0 \le i \le n$. Then the polynomial p_n must contain the factors $(x-x_i)$, $1 \le i \le n$. Since $\deg(p_n) \le n$ and $\deg \Pi_{i=1}^{n}(x-x_i) = n$, we have

$$p_n(x) = c \prod_{i=1}^{n}(x - x_i)$$

for some constant c. Using the condition $p_n(x_0) = 0$, we see that $c = 0$ and therefore, $p_n \equiv 0$. We note that by Theorem 3.1.3, this result on the uniqueness of the solvability of the homogeneous problem also implies the existence of a solution.

In the above, we have indicated three methods for showing the existence and uniqueness of a solution to the interpolation problem (3.1.1). The method based on showing the determinant of the coefficient is nonzero, as in (3.1.2), can be done easily only in simple situations such as Lagrange polynomial interpolation. Usually it is simpler to show that the interpolant corresponding to the homogeneous data is zero, even for complicated interpolation conditions. For practical calculations, it is also useful to have a representation formula that is the analogue of (3.1.3), but such a formula is sometimes difficult to find.

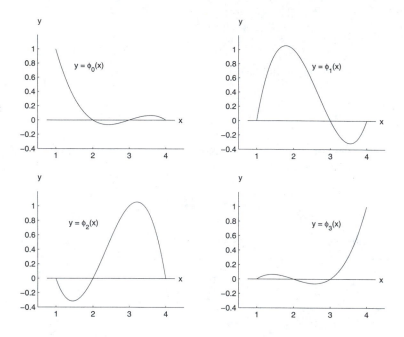

Figure 3.1. The Lagrange basis functions for $n = 3$, with nodes $\{1, 2, 3, 4\}$

These results on the existence and uniqueness of polynomial interpolation extend to the case that $\{x_0, \ldots, x_n\}$ are any $n + 1$ distinct points in the complex plane $\mathbb{C}$. The proofs remain the same.

For the interpolation error in Lagrange polynomial interpolation, we have the following.

Proposition 3.1.4 *Assume $f \in C^{n+1}[a, b]$. Then there exists a ξ_x between $\min_i\{x_i, x\}$ and $\max_i\{x_i, x\}$ such that*

$$f(x) - p_n(x) = \frac{\omega_n(x)}{(n+1)!} f^{(n+1)}(\xi_x), \qquad \omega_n(x) = \prod_{i=0}^{n}(x - x_i). \qquad (3.1.4)$$

Proof. The result is obvious if $x = x_i$, $0 \le i \le n$. Suppose $x \ne x_i$, $0 \le i \le n$, and denote

$$E(x) = f(x) - p_n(x).$$

Consider the function

$$g(t) = E(t) - \frac{\omega_n(t)}{\omega_n(x)} E(x).$$

We see that $g(t)$ has $(n + 2)$ distinct roots, namely, $t = x$ and $t = x_i$, $0 \le i \le n$. By the Mean Value Theorem, $g'(t)$ has $n + 1$ distinct roots. Applying

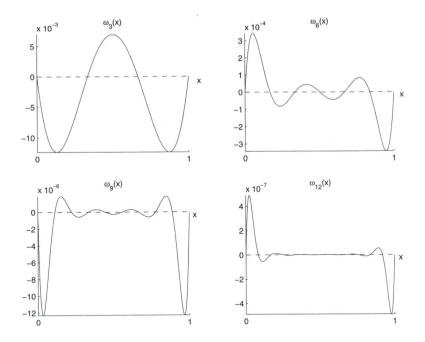

Figure 3.2. Examples of the polynomials $\omega_n(x)$ occurring in the interpolation error formulas (3.1.4) and (3.1.5)

repeatedly the Mean Value Theorem to derivatives of g, we conclude that $g^{(n+1)}(t)$ has a root $\xi_x \in (\min_i\{x_i, x\}, \max_i\{x_i, x\})$. Then

$$0 = g^{(n+1)}(\xi_x) = f^{(n+1)}(\xi_x) - \frac{(n+1)!}{\omega_n(x)} E(x),$$

and the result is proved. ∎

There are other ways of looking at polynomial interpolation error. Using Newton divided differences, we can show

$$f(x) - p_n(x) = \omega_n(x) \, f[x_0, x_1, \ldots, x_n, x] \tag{3.1.5}$$

with $f[x_0, x_1, \ldots, x_n, x]$ a divided difference of f of order $n + 1$. See [11, Section 3.2] for a development of this approach, together with a general discussion of divided differences and their use in interpolation.

We should note that high-degree polynomial interpolation with a uniform mesh is likely to lead to problems. Figure 3.2 contains graphs of $\omega_n(x)$ for various degrees n. From these graphs, it is clear that the error behavior is worse near the endpoint nodes than near the center node points. This leads to $p_n(x)$ failing to converge for such simple functions as $f(x) = \left(1 + x^2\right)^{-1}$ on $[-5, 5]$, a famous example due to Carl Runge. A further discussion of this can be found in [11, Section 3.5]. In contrast, interpolation using the zeros of Chebyshev polynomials leads to excellent results. This is discussed further in Section 3.6 of this chapter; and a further discussion is given in [11, p. 228].

3.1.2 Hermite polynomial interpolation

The main idea is to use values of both $f(x)$ and $f'(x)$ as interpolation conditions. Assume f is a continuously differentiable function on a finite interval $[a, b]$. Let

$$\Delta : a \le x_1 < \cdots < x_n \le b$$

be a partition of the interval $[a, b]$. Then the Hermite interpolant $p_{2n-1} \in \mathcal{P}_{2n-1}$ of degree less than or equal to $2n - 1$ of f is chosen to satisfy

$$p_{2n-1}(x_i) = f(x_i), \qquad p'_{2n+1}(x_i) = f'(x_i), \qquad 1 \le i \le n. \qquad (3.1.6)$$

We have results on Hermite interpolation similar to those for Lagrange interpolation, as given in Exercise 3.1.6.

More generally, for a given set of non-negative integers $\{m_i\}_{i=0}^n$, one can define a general Hermite interpolation problem as follows. Find $p_N \in \mathcal{P}_N(a, b)$, $N = \sum_{i=1}^n (m_i + 1) - 1$, to satisfy the interpolation conditions

$$p_N^{(j)}(x_i) = f^{(j)}(x_i), \qquad 0 \le j \le m_i, \quad 1 \le i \le n.$$

Again it can be shown that the interpolant with the homogeneous data is zero so that the interpolation problem has a unique solution. Also if $f \in C^{N+1}[a, b]$, then the error satisfies

$$f(x) - p_N(x) = \frac{1}{(N+1)!} \prod_{i=0}^n (x - x_i)^{m_i+1} f^{(N+1)}(\xi_x)$$

for some $\xi_x \in [a, b]$. For an illustration of an alternative error formula for the Hermite interpolation problem (3.1.6) that involves only the Newton divided difference of f, see [11, p. 161].

3.1.3 Piecewise polynomial interpolation

For simplicity, we focus our discussion on piecewise linear interpolation. Let $f \in C[a, b]$, and let

$$\Delta : a = x_0 < x_1 < \cdots < x_n = b$$

be a partition of the interval $[a, b]$. Denote $h_i = x_i - x_{i-1}$, $1 \le i \le n$, and $h = \max_{1 \le i \le n} h_i$. The piecewise linear interpolant $\Pi_\Delta f$ of f is defined through the following two requirements:

- For each $i = 1, \ldots, n$, $\Pi_\Delta f|_{[x_{i-1}, x_i]}$ is linear.

- For $i = 0, 1, \ldots, n$, $\Pi_\Delta f(x_i) = f(x_i)$.

It is easy to see that $\Pi_\Delta f$ exists and is unique, and

$$\Pi_\Delta f(x) = \frac{x_i - x}{h_i} f(x_{i-1}) + \frac{x - x_{i-1}}{h_i} f(x_i), \qquad x \in [x_{i-1}, x_i], \quad (3.1.7)$$

for $1 \le i \le n$.

For general $f \in C[a, b]$, it is relatively straightforward to show

$$\max_{x \in [a,b]} |f(x) - \Pi_\Delta f(x)| \le \omega(f, h) \qquad (3.1.8)$$

with $\omega(f, h)$ the *modulus of continuity* of f on $[a, b]$:

$$\omega(f, h) = \max_{\substack{|x-y| \le h \\ a \le x, y \le b}} |f(x) - f(y)| .$$

Now suppose $f \in C^2[a, b]$. By using (3.1.4) and (3.1.7), it is straightforward to show that

$$\max_{x \in [a,b]} |f(x) - \Pi_\Delta f(x)| \le \frac{h^2}{8} \max_{x \in [a,b]} |f''(x)|. \qquad (3.1.9)$$

Now instead of $f \in C^2[a, b]$, assume $f \in H^2(a, b)$ so that

$$\|f\|_{H^2(a,b)}^2 = \int_a^b \left[|f(x)|^2 + |f'(x)|^2 + |f''(x)|^2 \right] dx < \infty.$$

Here $H^2(a, b)$ is an example of Sobolev spaces. An introductory discussion was given in Examples 1.2.22 and 1.3.7. The space $H^2(a, b)$ consists of continuously differentiable functions f whose second derivative exists a.e. and belongs to $L^2(a, b)$. A detailed discussion of Sobolev spaces is given in Chapter 6. We are interested in estimating the error in the piecewise linear interpolant $\Pi_\Delta f$ and its derivative $(\Pi_\Delta f)'$ under the assumption $f \in H^2(a, b)$.

We consider the error in the L^2 sense,

$$\|f - \Pi_\Delta f\|_{L^2(a,b)}^2 = \int_a^b |f(x) - \Pi_\Delta f(x)|^2 dx$$

$$= \sum_{i=1}^n \int_{x_{i-1}}^{x_i} |f(x) - \Pi_\Delta f(x)|^2 dx.$$

For a function $\widehat{f} \in H^2(0, 1)$, let $\widehat{\Pi}\widehat{f}$ be its linear interpolant:

$$\widehat{\Pi}\widehat{f}(\xi) = \widehat{f}(0)(1 - \xi) + \widehat{f}(1)\xi, \quad 0 \le \xi \le 1.$$

By Taylor's theorem,

$$\widehat{f}(0) = \widehat{f}(\xi) - \xi \, \widehat{f}'(\xi) - \int_{\xi}^{0} t \, \widehat{f}''(t) \, dt,$$

$$\widehat{f}(1) = \widehat{f}(\xi) + (1 - \xi) \, \widehat{f}(\xi) + \int_{\xi}^{1} (1 - t) \, \widehat{f}''(t) \, dt.$$

Thus

$$\widehat{f}(\xi) - \widehat{\Pi}\widehat{f}(\xi) = -\xi \int_{\xi}^{1} (1 - t) \, \widehat{f}''(t) \, dt - (1 - \xi) \int_{0}^{\xi} t \, \widehat{f}''(t) \, dt,$$

and therefore

$$\int_{0}^{1} |\widehat{f}(\xi) - \widehat{\Pi}\widehat{f}(\xi)|^2 d\xi \leq c \int_{0}^{1} |\widehat{f}''(\xi)|^2 d\xi \tag{3.1.10}$$

for some constant c independent of $\widehat{f}$. Using (3.1.10),

$$\int_{x_{i-1}}^{x_i} |f(x) - \Pi_{\Delta}f(x)|^2 dx$$

$$= h_i \int_{0}^{1} |f(x_{i-1} + h_i\xi) - \widehat{\Pi}f(x_{i-1} + h_i\xi)|^2 d\xi$$

$$\leq c\,h_i \int_{0}^{1} \left| \frac{d^2 f(x_{i-1} + h_i\xi)}{d\xi^2} \right|^2 d\xi$$

$$= c\,h_i^5 \int_{0}^{1} |f''(x_{i-1} + h_i\xi)|^2 d\xi$$

$$= c\,h_i^4 \int_{x_{i-1}}^{x_i} |f''(x)|^2 dx.$$

Therefore,

$$\|f - \Pi_{\Delta}f\|_{L^2(a,b)}^2 = \sum_{i=1}^{n} \int_{x_{i-1}}^{x_i} |f(x) - \Pi_{\Delta}f(x)|^2 dx \leq c\,h^4\|f''\|_{L^2(a,b)}^2;$$

i.e.,

$$\|f - \Pi_{\Delta}f\|_{L^2(a,b)} \leq c\,h^2\|f''\|_{L^2(a,b)}. \tag{3.1.11}$$

A similar argument shows

$$\|f' - (\Pi_{\Delta}f)'\|_{L^2(a,b)} \leq \widetilde{c}\,h\,\|f''\|_{L^2(a,b)}. \tag{3.1.12}$$

for another constant $\widetilde{c} > 0$.

In the theory of finite element interpolation, the above argument is generalized to error analysis of piecewise polynomial interpolation of any degree.

3.1.4 Trigonometric interpolation

Another important and widely used class of approximating functions are
the trigonometric polynomials

$$p_n(x) = a_0 + \sum_{j=1}^{n} [a_j \cos(jx) + b_j \sin(jx)]. \tag{3.1.13}$$

If $|a_n| + |b_n| \neq 0$, we say $p_n(x)$ is a trigonometric polynomial of degree n.
The function $p_n(x)$ is often considered as a function on the unit circle, in
which case $p_n(\theta)$ would be a more sensible notation, with θ the central angle
for a point on the unit circle. The set of the trigonometric polynomials of
degree less than or equal to n is denoted by $\mathcal{T}_n$.

An equivalent way of writing such polynomials is as

$$p_n(x) = \sum_{j=-n}^{n} c_j e^{ijx}. \tag{3.1.14}$$

The equivalence is given by

$$a_0 = c_0, \qquad a_j = c_j + c_{-j}, \qquad b_j = i(c_j - c_{-j}).$$

Many computations with trigonometric polynomials are easier with (3.1.14)
than with (3.1.13). With (3.1.14), we also can write

$$p_n(x) = \sum_{j=-n}^{n} c_j z^j = z^{-n} \sum_{k=0}^{2n} c_{k-n} z^k, \qquad z = e^{ix}, \tag{3.1.15}$$

which brings us back to something involving polynomials.

The trigonometric polynomials of (3.1.13) are periodic with period 2π,
and thus we choose our interpolation nodes from the interval $[0, 2\pi)$ or any
interval of length of 2π:

$$0 \leq x_0 < x_1 < \cdots < x_{2n} < 2\pi.$$

Often we use an even spacing, with

$$x_j = jh, \qquad j = 0, 1, \ldots, 2n, \qquad h = \frac{2\pi}{2n+1}. \tag{3.1.16}$$

The interpolation problem is to find a trigonometric polynomial $p_n(x)$ of
degree less than or equal to n for which

$$p_n(x_j) = b_j, \qquad j = 0, 1, \ldots, 2n \tag{3.1.17}$$

for given data values $\{b_j \mid 0 \leq j \leq 2n\}$. The existence and uniqueness of a
solution of this problem can be reduced to that for Lagrange polynomial
interpolation by means of the final formula in (3.1.15). Using it, we introduce the distinct complex nodes $z_j = e^{ix_j}$, $j = 0, 1, \ldots, 2n$. Then (3.1.17)

n	$\|f - p_n\|_\infty$	n	$\|f - p_n\|_\infty$
1	$1.16E + 00$	8	$2.01E - 07$
2	$2.99E - 01$	9	$1.10E - 08$
3	$4.62E - 02$	10	$5.53E - 10$
4	$5.67E - 03$	11	$2.50E - 11$
5	$5.57E - 04$	12	$1.04E - 12$
6	$4.57E - 05$	13	$4.01E - 14$
7	$3.24E - 06$	14	$2.22E - 15$

Table 3.1. Trigonometric interpolation errors for (3.1.18)

can be rewritten as the polynomial interpolation problem

$$\sum_{k=0}^{2n} c_{k-n} z_j^k = z_j^n b_j, \qquad j = 0, 1, \ldots, 2n.$$

All results from the polynomial interpolation problem with complex nodes can be applied to the trigonometric interpolation problem. For additional detail, see [11, Section 3.8]. Error results are given in Section 3.6 for the interpolation of a periodic function using trigonometric polynomials.

Example 3.1.5 *Consider the periodic function*

$$f(x) = e^{\sin x} \sin x. \qquad (3.1.18)$$

Table 3.1 contains the maximum errors in the trigonometric interpolation polynomial $p_n(x)$ for varying values of n.

Exercise 3.1.1 *Show that there is a unique quadratic function p_2 satisfying the conditions*

$$p_2(0) = a_0, \quad p_2(1) = a_1, \quad \int_0^1 p_2(x)\,dx = \bar{a}$$

with given a_0, a_1, and $\bar{a}$.

Exercise 3.1.2 *Given a function f on $C[a, b]$, the moment problem is to find $p_n \in \mathcal{P}_n(a, b)$ such that*

$$\int_a^b x^i p_n(x)\,dx = \int_a^b x^i f(x)\,dx, \quad 0 \le i \le n.$$

Show that the problem has a unique solution.

Exercise 3.1.3 *Let $x_0 < x_1 < x_2$ be three real numbers. Consider finding a polynomial $p(x)$ of degree ≤ 3 for which*

$$p(x_0) = y_0, \qquad p(x_2) = y_2,$$
$$p'(x_1) = y_1', \qquad p''(x_1) = y_1''$$

with given data $\{y_0, y_2, y_1', y_1''\}$. Show there exists a unique such polynomial.

Exercise 3.1.4 *Derive the formula (3.1.2) for the Vandermonde determinant of order $n + 1$.*
Hint: *Introduce*

$$V_n(x) = \det \begin{bmatrix} 1 & x_0 & x_0^2 & \cdots & x_0^n \\ 1 & x_1 & x_1^2 & \cdots & x_1^n \\ \vdots & \vdots & \vdots & \ddots & \vdots \\ 1 & x_{n-1} & x_{n-1}^2 & \cdots & x_{n-1}^n \\ 1 & x & x^2 & \cdots & x^n \end{bmatrix}$$

Show

$$V_n(x) = V_{n-1}(x_{n-1})(x - x_0) \cdots (x - x_{n-1})$$

and use this to prove (3.1.2).

Exercise 3.1.5 *Show that the Lagrange formula (3.1.3) can be rewritten in the form*

$$p_n(x) = \frac{\displaystyle\sum_{j=0}^{n} \frac{w_j f(x_j)}{x - x_j}}{\displaystyle\sum_{j=0}^{n} \frac{w_j}{x - x_j}}$$

for x not a node point, for suitable values of $\{w_j\}$ that are dependent on only the nodes $\{x_j\}$. This formula is called the barycentric representation *of $p_n(x)$.*

Exercise 3.1.6 *Show that the Hermite interpolation problem (3.1.6) admits a unique solution. Find a representation formula for the interpolant. Derive the error relation*

$$f(x) - p_{2n-1}(x) = \frac{f^{(2n)}(\xi)}{(2n)!} \prod_{i=1}^{n} (x - x_i)^2$$

for some $\xi \in (\min_i\{x_i, x\}, \max_i\{x_i, x\})$, if $f \in C^{2n}[a, b]$.

Exercise 3.1.7 *Let us derive an error estimate for the composite trapezoidal rule, the convergence of which was discussed in Exercise 2.4.3. A standard error estimate is*

$$|L_n v - L v| \leq c\, h^2 \max_{0 \leq x \leq 1} |v''(x)|$$

with $h = 1/n$. Assume $v'' \in L^1(0, 1)$. Show that

$$|L_n v - L v| \leq c\, h^2 \|v''\|_{L^1(0,1)};$$

i.e., the smoothness requirement on the integrand can be weakened while the same order error estimate is kept. Improved estimates of this kind are valid for errors of more general numerical quadratures.

Exercise 3.1.8 *An elementary argument for the improved error estimate of the preceding exercise is possible, under additional smoothness assumption on the integrand. Suppose $v \in C^2[a, b]$. For the composite trapezoidal rule, show that*

$$L_n v - L v = \int_a^b K_T(x)\, v''(x)\, dx,$$

where the "Peano kernel function" K_T is defined by

$$K_T(x) = \frac{1}{2}\, (x - x_{k-1})\, (x_k - x), \qquad x_{k-1} \le x \le x_k$$

for $k = 1, 2, \ldots, n$. Use this relation to prove the quadrature error bound

$$|L_n v - L v| \le c\, h^2 \|v''\|_{L^1(0,1)}.$$

Exercise 3.1.9 *As another example of similar nature, show that for the composite Simpson's rule, the following error representation is valid:*

$$L_n v - L v = \int_a^b K_S(x)\, v^{(4)}(x)\, dx,$$

where the Peano kernel function K_S is defined by

$$K_T(x) = \begin{cases} \frac{h}{18}\, (x - x_{k-2})^3 - \frac{1}{24}\, (x - x_{k-2})^4, & x_{k-2} \le x \le x_{k-1}, \\ \frac{h}{18}\, (x_k - x)^3 - \frac{1}{24}\, (x_k - x)^4, & x_{k-1} \le x \le x_k \end{cases}$$

for $k = 2, 4, \ldots, n$. Use this relation to prove the quadrature error bound

$$|L_n v - L v| \le c\, h^4 \|v^{(4)}\|_{L^1(0,1)}.$$

Exercise 3.1.10 (a) *For the nodes $\{x_j\}$ of (3.1.16), show the identity*

$$\sum_{j=0}^{2n} e^{ikx_j} = \begin{cases} 2n + 1, & e^{ix_k} = 1 \\ 0, & e^{ix_k} \ne 1 \end{cases}$$

for $k \in \mathbb{Z}$ and $x_k = 2\pi k / (2n + 1)$.
(b) *Find the trigonometric interpolation polynomial that solves the problem (3.1.17) with the evenly spaced nodes of (3.1.16). Consider finding the interpolation polynomial in the form of (3.1.14). Show that the coefficients $\{c_j\}$ are given by*

$$c_\ell = \frac{1}{2n+1} \sum_{j=0}^{2n} b_j e^{-i\ell x_j}, \qquad \ell = -n, \ldots, n.$$

Hint: Use the identity in part (a) to solve the linear system

$$\sum_{k=-n}^{n} c_k e^{ikt_j} = b_j, \qquad j = 0, 1, \ldots, 2n.$$

Begin by multiplying equation j by $e^{-i\ell x_j}$, and then sum the equations over j.

Exercise 3.1.11 *Prove the error bound* (3.1.8).

3.2   Best approximation

We approximate a function $f(x)$ by choosing some member of a restricted class of functions. For example, a polynomial was selected from $\mathcal{P}_n$ by using interpolation to $f(x)$. It is useful to consider the best that can be done with such a class of approximating functions: How small an error is possible when selecting an approximation from the given class of approximating functions? This is known as the *best approximation problem*. The solution depends on the function f, on the class of approximating functions, and on the norm by which the error is being measured. The best known cases use the uniform norm $\| \cdot \|_\infty$, the L^1-norm, and the L^2-norm (and other Hilbert space norms). We examine the best approximation problem in this section and some of the following sections.

Throughout this section, V is allowed to be either a real or complex linear space.

3.2.1   Convexity, lower semicontinuity

A best approximation problem can be described by the minimization of a certain functional, and some rather general results can be given within such a framework. We begin by introducing some useful concepts.

Definition 3.2.1 *Let V be a real or complex linear space, $K \subseteq V$. The set K is said to be* convex *if*

$$u, v \in K \quad \Longrightarrow \quad \lambda u + (1 - \lambda) v \in K \quad \forall \lambda \in (0, 1).$$

Informally, the line segment joining any two elements of K is also contained in K.

If K is convex, by induction we can show

$$u_i \in K, \quad 1 \leq i \leq n \quad \Longrightarrow \quad \sum_{i=1}^{n} \lambda_i u_i \in K \quad \forall \lambda_i \geq 0 \text{ with } \sum_{i=1}^{n} \lambda_i = 1.$$

$$(3.2.1)$$

Such an expression $\sum_{i=1}^{n} \lambda_i u_i$ is called a *convex combination* of $\{u_i\}_{i=1}^{n}$.

Definition 3.2.2 *Let K be a convex set in a linear space V. A function $f : K \to \mathbb{R}$ is said to be* convex *if*

$$f(\lambda u + (1 - \lambda) v) \leq \lambda f(u) + (1 - \lambda) f(v) \quad \forall u, v \in K, \quad \forall \lambda \in [0, 1].$$

The function f is strictly convex *if the above inequality is strict for $u \neq v$ and $\lambda \in (0, 1)$.*

To obtain a more intuitive sense of what it means for a function f to be convex, interpret it geometrically for the graph of a real-valued convex function f over $\mathbb{R}^2$. If any two points u and v in $\mathbb{R}^2$ are connected by a straight line segment L, then any point on the line segment joining $(u, f(u))$ and $(v, f(v))$ is located above the function value for the corresponding point on L. The reader should note that the term "strictly convex" has another meaning in the literature on approximation theory, related somewhat to our definition but still distinct from it.

Definition 3.2.3 *Let V be a normed space. A set $K \subseteq V$ is* closed *if $\{v_n\} \subseteq K$ and $v_n \to v$ imply $v \in K$. The set K is* weakly closed *if $\{v_n\} \subseteq K$ and $v_n \rightharpoonup v$ imply $v \in K$.*

Definition 3.2.4 *Let V be a normed space, $K \subseteq V$. A function $f : K \to \mathbb{R}$ is* (sequentially) lower semicontinuous (l.s.c.) *if $\{v_n\} \subseteq K$ and $v_n \to v \in K$ imply*

$$f(v) \le \liminf_{n \to \infty} f(v_n).$$

The function f is weakly sequentially lower semicontinuous *or* weakly lower semicontinuous (w.l.s.c.) *if the above inequality is valid for any sequence $\{v_n\} \subseteq K$ with $v_n \rightharpoonup v \in K$.*

Obviously continuity implies lower semicontinuity. The converse statement is not true, as lower semicontinuity allows discontinuity in a function. It is easily seen that if f is w.l.s.c., then it is l.s.c. The notion of weak lower semicontinuity is very useful in a number of topics with applied and computational mathematics, including the study of boundary value problems for elliptic partial differential equations.

Example 3.2.5 *We examine an example of a w.l.s.c. function. Let V be a normed space and let us show that the norm function is w.l.s.c. For this, let $\{v_n\} \subseteq V$ be a weakly convergent sequence, $v_n \rightharpoonup v \in K$. By Corollary 2.5.6, there is an $\ell \in V'$ such that $\ell(v) = \|v\|$ and $\|\ell\| = 1$. We notice that*

$$\ell(v_n) \le \|\ell\| \, \|v_n\| = \|v_n\|.$$

Therefore,

$$\|v\| = \ell(v) = \lim_{n \to \infty} \ell(v_n) \le \liminf_{n \to \infty} \|v_n\|.$$

So $\| \cdot \|$ is w.l.s.c.

In an inner product space, a simpler proof is possible to show the norm function is w.l.s.c. Indeed, assume V is an inner product space, and let $\{v_n\} \subseteq V$ be a weakly convergent sequence, $v_n \rightharpoonup v$. Then

$$\|v\|^2 = (v, v) = \lim_{n \to \infty} (v, v_n) \le \liminf_{n \to \infty} \|v\| \, \|v_n\|,$$

and we easily obtain

$$\|v\| \le \liminf_{n \to \infty} \|v_n\|.$$

We now present a useful result on geometric functional analysis derived from the generalized Hahn-Banach theorem, concerning separation of convex sets.

Definition 3.2.6 *Let V be a real normed space, and A and B non-empty sets in V. The sets A and B are said to be* separated *if there is a non-zero linear continuous functional ℓ on V and a number $\alpha \in \mathbb{R}$ such that*

$$\ell(u) \leq \alpha \quad \forall u \in A, \quad \ell(v) \geq \alpha \quad \forall v \in B.$$

If the inequalities are strict, then we say the sets A and B are strictly *separated.*

The next result follows from Theorem 2.5.5; a proof of the result can be found in, e.g., [46].

Theorem 3.2.7 *Let V be a real normed space, and let A and B be two non-empty disjoint convex subsets of V such that one of them is compact, and the other is closed. Then the sets A and B can be strictly separated.*

This result is used later in Section 10.4.

3.2.2 Some abstract existence results

Given a real space V, a subset $K \subseteq V$, and a functional $f : K \to \mathbb{R}$, we consider the problem of finding a minimizer $v = u$ for the expression

$$\inf_{v \in K} f(v). \tag{3.2.2}$$

A general reference for the results of this subsection is [173], including proofs of most of the results given here.

Before we present a general result on the existence of a solution to the problem (3.2.2), let us recall the classical result of Weierstraß: A real-valued continuous function f on a bounded closed interval $[a, b]$ $(-\infty < a < b < \infty)$ has a maximum and a minimum. We review main steps in the proof of the result for the part regarding a minimum. Denote

$$\alpha = \inf_{x \in [a,b]} f(x).$$

Then by the definition of infimum, there is a sequence $\{x_n\} \subseteq [a, b]$ such that $f(x_n) \to \alpha$ as $n \to \infty$. Since the bounded closed interval $[a, b]$ is compact, we have a subsequence $\{x_{n'}\} \subseteq \{x_n\}$ and some $x_0 \in [a, b]$ such that

$$x_{n'} \to x_0 \quad \text{as } n' \to \infty.$$

Now the function f is assumed to be continuous, so

$$f(x_0) = \lim_{n' \to \infty} f(x_{n'}) = \alpha,$$

i.e., x_0 is a minimizer of f on $[a, b]$. When we try to extend the result and the proof to the problem (3.2.2) on a general setting, we notice the following points.

- The continuity of f is too restrictive. From the above argument, we observe that it is enough to assume the lower semicontinuity

$$f(x_0) \leq \liminf_{n' \to \infty} f(x_{n'}). \tag{3.2.3}$$

  This condition allows f to be discontinuous.

- In an infinite-dimensional Banach space V, a bounded sequence does not necessarily contain a convergent subsequence (cf. Example 2.7.2). Nevertheless, if V is a reflexive Banach space, then Theorem 2.7.4 states that a bounded sequence in V contains a weakly convergent subsequence. Therefore, for the problem (3.2.2), we assume V is reflexive, K is bounded and weakly closed. This last condition ensures that the weak limit of a weakly convergent subsequence in K lies in K. Relatedly, the condition (3.2.3) is assumed for any subsequence $\{x_{n'}\}$ that converges weakly to x_0.

With the above consideration, we see that the conditions of the next result are quite natural.

Theorem 3.2.8 *Assume V is a reflexive Banach space, and assume $K \subseteq V$ is bounded and weakly closed. If $f : K \to \mathbb{R}$ is weakly sequentially l.s.c., then the problem (3.2.2) has a solution.*

Proof. Denote

$$\alpha = \inf_{v \in K} f(v).$$

By the definition of infimum, there exists a sequence $\{u_n\} \subseteq K$ with

$$f(u_n) \to \alpha \quad \text{as } n \to \infty.$$

Since K is bounded, $\{u_n\}$ is a bounded sequence in the space V. Since V is reflexive, Theorem 2.7.4 implies that there exists a subsequence $\{u_{n'}\} \subseteq \{u_n\}$ that converges weakly to $u \in V$. Since K is weakly closed, we have $u \in K$; and since f is weakly sequentially l.s.c., we have

$$f(u) \leq \liminf_{n' \to \infty} f(u_{n'}).$$

Therefore, $f(u) = \alpha$, and u is a solution of the minimization problem (3.2.2). Note that this proof also shows α is finite, $\alpha > -\infty$. ∎

In the above theorem, K is assumed to be bounded. Often we have the situation where K is unbounded (a subspace, for example). We can drop the boundedness assumption on K, and as a compensation we assume f to be coercive over K.

Definition 3.2.9 *Let V be a normed space, $K \subseteq V$. A real-valued functional f on V is said to be* coercive *over K if*

$$f(v) \to \infty \quad \text{as } \|v\| \to \infty, \ v \in K.$$

Theorem 3.2.10 *Assume V is a reflexive Banach space, $K \subseteq V$ is weakly closed. If $f : K \to \mathbb{R}$ is weakly sequentially l.s.c. and coercive on K, then the problem (3.2.2) has a solution.*

Proof. Pick any $v_0 \in K$ and define

$$K_0 = \{ v \in K \mid f(v) \le f(v_0) \}.$$

Since f is coercive, K_0 is bounded. Since K is weakly closed and f is weakly sequentially l.s.c., we see that K_0 is weakly closed. The problem (3.2.2) is equivalent to

$$\inf_{v \in K_0} f(v),$$

which has at least one solution from the Theorem 3.2.8. ∎

These results are rather general in nature. In applications, it is usually not convenient to verify the conditions associated with weakly convergent sequences. We replace these conditions by ones easier to verify. First we record a result of fundamental importance in convex analysis. A proof is given in [47, p. 6].

Theorem 3.2.11 (MAZUR LEMMA) *Assume V is a normed space, and assume $\{v_n\}_{n \geq 1}$ is a sequence converging weakly to u. Then there is a sequence $\{u_n\}_{n \geq 1}$ of convex combinations of $\{v_n\}_{n \geq 1}$,*

$$u_n = \sum_{i=n}^{N(n)} \lambda_i^{(n)} v_i, \quad \sum_{i=n}^{N(n)} \lambda_i^{(n)} = 1, \ \lambda_i^{(n)} \geq 0, \ n \leq i \leq N(n),$$

which converges strongly to u.

It is left as Exercise 3.2.2 to prove the following corollaries of the Mazur lemma.

- If K is convex and closed, then it is weakly closed.

- If f is convex and l.s.c. (or continuous), then it is weakly sequentially l.s.c.

Now we have the following variants of the existence results, and they are sufficient for our applications.

Theorem 3.2.12 *Assume V is a reflexive Banach space, $K \subseteq V$ is convex and closed, and $f : K \to \mathbb{R}$ is convex and l.s.c. If either*
(a) K is bounded
or
(b) f is coercive on K,

then the minimization problem (3.2.2) has a solution. Furthermore, if f is strictly convex, then a solution to the problem (3.2.2) is unique.

Proof. It remains to show that if f is strictly convex, then a minimizer of f over K is unique. Let us argue by contradiction. Assume there were two minimizers $u_1 \neq u_2$, with $f(u_1) = f(u_2)$ the minimal value of f on K. Since K is convex, $(u_1 + u_2)/2 \in K$. By the strict convexity of f, we would have

$$f\left(\frac{u_1 + u_2}{2}\right) < \frac{1}{2}\left(f(u_1) + f(u_2)\right) = f(u_1).$$

This relation contradicts the assumption that u_1 is a minimizer. ∎

In certain applications, the space V is not reflexive (e.g., $V = C[a,b]$). In such a case the above theorems are not applicable. Nevertheless, we notice that the reflexivity of V is used only to extract a weakly convergent subsequence from a bounded sequence in K. Also notice that we only need the completeness of the subset K, not that of the space V. Hence, we may modify the above theorem as follows.

Theorem 3.2.13 *Assume V is a normed space, $K \subseteq V$ is a convex and closed finite-dimensional subset, and $f : K \to \mathbb{R}$ is convex and l.s.c. If either*

(a) *K is bounded*

or

(b) *f is coercive on K,*

then the minimization problem (3.2.2) has a solution. Furthermore, if f is strictly convex, then a solution to the problem (3.2.2) is unique.

3.2.3   Existence of best approximation

Let us apply the above results to a best approximation problem. Let $u \in V$—we are interested in finding elements from $K \subseteq V$ that are closest to u among the elements in K. More precisely, we are interested in the minimization problem

$$\inf_{v \in K} \|u - v\|. \tag{3.2.4}$$

Obviously (3.2.4) is a problem of the form (3.2.2) with

$$f(v) = \|u - v\|.$$

Certainly $f(v)$ is convex and continuous (and hence l.s.c.). Furthermore, $f(v)$ is coercive if K is unbounded. We thus have the following existence theorems on best approximations.

Theorem 3.2.14 *Assume $K \subseteq V$ is a closed, convex subset of a reflexive Banach space V. Then there is an element $\widehat{u} \in K$ such that*

$$\|u - \widehat{u}\| = \inf_{v \in K} \|u - v\|.$$

Theorem 3.2.15 *Assume $K \subseteq V$ is a convex and closed finite-dimensional subset of a normed space V. Then there is an element $\widehat{u} \in K$ such that*

$$\|u - \widehat{u}\| = \inf_{v \in K} \|u - v\|.$$

In particular, a finite-dimensional subspace is both convex and closed.

Theorem 3.2.16 *Assume K is a finite-dimensional subspace of the normed space V. Then there is an element $\widehat{u} \in K$ such that*

$$\|u - \widehat{u}\| = \inf_{v \in K} \|u - v\|.$$

Example 3.2.17 *Let $V = C[a, b]$ (or $L^p(a, b)$) and $K = \mathcal{P}_n$, the space of all the polynomials of degree less than or equal to n. Associated with the space V, we may use $L^p(a, b)$ norms, $1 \le p \le \infty$. The previous results ensure that for any $f \in C[a, b]$ (or $L^p(a, b)$), there exists a polynomial $f_n \in \mathcal{P}_n$ such that*

$$\|f - f_n\|_{L^p(a,b)} = \inf_{q_n \in \mathcal{P}_n} \|f - q_n\|_{L^p(a,b)}.$$

Certainly, for a different value of p, we have a different best approximation f_n. When $p = \infty$, f_n is called a "best uniform approximation of f."

The existence of a best approximation from a finite-dimensional subspace can also be proven directly. To do so, reformulate the minimization problem as a problem of minimizing a non-negative continuous real-valued function over a closed bounded subset of $\mathbb{R}^n$ or $\mathbb{C}^n$, and then appeal to the Heine-Borel theorem from elementary analysis (cf. Theorem 1.6.2). This is left as Exercise 3.2.3.

3.2.4   Uniqueness of best approximation

Showing uniqueness requires greater attention to the properties of the norm or to the characteristics of the approximating subset K.

Arguing as in the proof for the uniqueness part in Theorem 3.2.12, we can easily show the next result.

Theorem 3.2.18 *Assume V is a normed space, and further assume that the function $f(v) \equiv \|v\|^p$ is strictly convex for some $p \ge 1$. Let K be a convex subset of V. Then for any $u \in V$, a best approximation $\widehat{u}$ from K is unique.*

If V is an inner product space, then $f(v) \equiv \|v\|^2$ is a strictly convex function on V (cf. Exercise 3.3.3), and therefore a solution to the best approximation problem in an inner product space is unique (provided it exists). Other approaches to proving the uniqueness of a best approximation can be found in Davis [42], giving uniqueness results for approximation in $L^p(a, b)$ for $1 < p < \infty$.

Notice that the strict convexity of the norm is a sufficient condition for the uniqueness of a best approximation, but the condition is not necessary. For example, the norm $\| \cdot \|_{L^\infty (a,b)}$ is not strictly convex, yet there are classical results stating that a best uniform approximation is unique for important classes of approximating functions. The following is the best known of such results.

Theorem 3.2.19 (CHEBYSHEV EQUI-OSCILLATION THEOREM) *Let $f \in C[a, b]$ for a finite interval $[a, b]$, and let $n \geq 0$ be an integer. Then there is a unique solution $\widehat{p}_n \in \mathcal{P}_n$ to the minimization*

$$\rho_n (f) \equiv \min_{p \in \mathcal{P}_n} \|f - p\|_\infty .$$

It is characterized uniquely as follows. There is a set of $n + 2$ numbers

$$a \leq x_0 < x_1 < \cdots < x_{n+1} \leq b,$$

not necessarily unique, for which

$$f(x_j) - \widehat{p}_n(x_j) = \sigma (-1)^j \rho_n (f), \qquad j = 0, 1, \ldots, n + 1,$$

with $\sigma = +1$ or -1.

Theorem 3.2.20 *Let g be a continuous 2π-periodic function on $\mathbb{R}$, and let $n \geq 0$ be an integer. Then there is a unique trigonometric polynomial $\widehat{q}_n \in \mathcal{T}_n$ of degree $\leq n$ (cf. (3.1.13) in Section 3.1.4) satisfying the minimization*

$$\rho_n (g) \equiv \min_{q \in \mathcal{T}_n} \|g - q\|_\infty .$$

Proofs of these two theorems are given in Meinardus [117, Section 3] and Davis [42, Chap. 7]. We return to these best uniform approximations in Section 3.6, where we look at the size of $\rho_n (f)$ as a function of the smoothness of f.

Exercise 3.2.1 *Let $g \in C[0, 1]$ and let $n \geq 0$ be an integer. Define*

$$E(g) \equiv \inf_{\deg(p) \leq n} \left[\max_{0 \leq x \leq 1} \left(1 + x^2\right) |g(x) - p(x)| \right]$$

with $p(x)$ denoting a polynomial. Consider the minimization problem of finding at least one polynomial $\widehat{p}(x)$ of degree at most n for which

$$E(g) \equiv \max_{0 \leq x \leq 1} \left(1 + x^2\right) |g(x) - \widehat{p}(x)|.$$

What can you say about the solvability of this problem?

Exercise 3.2.2 *Apply the Mazur lemma to show that in a normed space a convex closed set is weakly closed, and a convex l.s.c. function is w.l.s.c.*

Exercise 3.2.3 *Give a direct proof of Theorem 3.2.16, as discussed following Example 3.2.17.*

Exercise 3.2.4 *Prove (3.2.1).*

3.3   Best approximations in inner product spaces

In an inner product space V, the norm $\|\cdot\|$ is induced by an associated inner product. The square of such a norm is strictly convex (Exercise 3.3.3), so the best approximation is unique from Theorem 3.2.18. Alternatively, the uniqueness of the best approximation can be verified using the following characterization of a best approximation when the norm is induced by an inner product.

Throughout this section, we assume V is a real inner product space. Many of the results generalize, and in some cases, they are stated in a more general form. A general reference for the results of this section is [173], including proofs of most of the results given here.

Lemma 3.3.1 *Let K be a convex subset of a real inner product space V. For any $u \in V$, $\widehat{u} \in K$ is its best approximation in K if and only if it satisfies*

$$(u - \widehat{u}, v - \widehat{u}) \leq 0 \qquad \forall v \in K. \tag{3.3.1}$$

Proof. Suppose $\widehat{u} \in K$ is a best approximation of u. Let $v \in K$ be arbitrary. Then, since K is convex, $\widehat{u} + \lambda (v - \widehat{u}) \in K$, $\lambda \in [0, 1]$. Hence the function

$$\varphi(\lambda) = \|u - [\widehat{u} + \lambda (v - \widehat{u})]\|^2, \qquad \lambda \in [0, 1],$$

has its minimum at $\lambda = 0$. We then have

$$0 \leq \varphi'(0) = -2 (u - \widehat{u}, v - \widehat{u});$$

i.e., (3.3.1) holds.

Conversely, assume (3.3.1) is valid. Then for any $v \in K$,

$$\begin{aligned}
\|u - v\|^2 &= \|(u - \widehat{u}) + (\widehat{u} - v)\|^2 \\
&= \|u - \widehat{u}\|^2 + 2 (u - \widehat{u}, \widehat{u} - v) + \|\widehat{u} - v\|^2 \\
&\geq \|u - \widehat{u}\|^2;
\end{aligned}$$

i.e., $\widehat{u}$ is a best approximation of u in K.   ∎

The geometric meaning of this lemma is that the angle between the two vectors $u - \widehat{u}$ and $v - \widehat{u}$ is in the range $[\pi/2, \pi]$.

Corollary 3.3.2 *Let K be a convex set of an inner product space V. Then for any $u \in V$, its best approximation is unique.*

Proof. Assume both $\widehat{u}_1, \widehat{u}_2 \in K$ are best approximations. Then from the lemma,

$$(u - \widehat{u}_1, v - \widehat{u}_1) \leq 0 \quad \forall v \in K.$$

In particular, we choose $v = \widehat{u}_2$ to obtain

$$(u - \widehat{u}_1, \widehat{u}_2 - \widehat{u}_1) \leq 0.$$

Similarly,

$$(u - \widehat{u}_2, \widehat{u}_1 - \widehat{u}_2) \leq 0.$$

Adding the last two inequalities, we get

$$-\|\widehat{u}_1 - \widehat{u}_2\|^2 \leq 0.$$

Therefore, $\widehat{u}_1 = \widehat{u}_2$.   ∎

Now combining the above uniqueness result and the existence results from the last subsection, we can state the following theorems.

Theorem 3.3.3 *Assume $K \subseteq V$ is a closed, convex subset of a Hilbert space V. Then there is a unique element $\widehat{u} \in K$ such that*

$$\|u - \widehat{u}\| = \inf_{v \in K} \|u - v\|.$$

The element $\widehat{u}$ is also characterized by the inequality (3.3.1).

We call $\widehat{u}$ the projection of u onto K, and write $\widehat{u} = P_K(u)$. In general, P_K is a nonlinear operator, called the *projection operator*. It is not difficult to prove the following properties of the projection operator by using the characterization (3.3.1).

Proposition 3.3.4 *Assume $K \subseteq V$ is a closed, convex subset of a Hilbert space V. Then the projection operator is monotone,*

$$(P_K(u) - P_K(v), u - v) \geq 0 \qquad \forall\, u, v \in V,$$

and it is non-expansive,

$$\|P_K(u) - P_K(v)\| \leq \|u - v\| \qquad \forall\, u, v \in V.$$

Theorem 3.3.5 *Assume $K \subseteq V$ is a convex and closed finite-dimensional subset of an inner product space V. Then there is a unique element $\widehat{u} \in K$ such that*

$$\|u - \widehat{u}\| = \inf_{v \in K} \|u - v\|.$$

Theorem 3.3.6 *Assume K is a complete subspace of a real or complex inner product space V. Then there is a unique element $\widehat{u} \in K$ such that*

$$\|u - \widehat{u}\| = \inf_{v \in K} \|u - v\|.$$

In the situation described in Theorem 3.3.6, since K is a subspace, the best approximation is characterized by the property (cf. Lemma 3.3.1)

$$(u - \widehat{u}, v) = 0 \quad \forall v \in K.$$

In other words, the "error" $u - \widehat{u}$ is orthogonal to the subspace K. The projection mapping P_K is then called an *orthogonal projection operator*. Its main properties are summarized in the next theorem. For a detailed discussion, see [88, pp. 147, 172–174].

Theorem 3.3.7 *Assume K is a complete subspace of a real or complex inner product space V. Then the orthogonal projection operator $P_K : V \to V$ is linear, self-adjoint; i.e.,*

$$(P_K u, v) = (u, P_K v) \qquad \forall\, u, v \in K. \tag{3.3.2}$$

In addition,

$$\|v\|^2 = \|P_K v\|^2 + \|v - P_K v\|^2 \qquad \forall\, v \in V; \tag{3.3.3}$$

and as a consequence,

$$\|P_K\| = 1. \tag{3.3.4}$$

An important special situation arises when we know an orthonormal basis $\{\phi_n\}_{n\geq 1}$ of the space V, and $K = V_n = \mathrm{span}\,\{\phi_1, \ldots, \phi_n\}$. The element $P_n u \in V_n$ is the minimizer of

$$\min_{v \in V_n} \|u - v\|.$$

We find this minimizer by considering the minimization of the non-negative function

$$f(b_1, \ldots, b_n) = \left\| u - \sum_{i=1}^{n} b_i \phi_i \right\|^2,$$

which is equivalent to minimizing over V_n. It is straightforward to obtain the identity

$$f(b_1, \ldots, b_n) = \|u\|^2 - \sum_{i=1}^{n} |(u, \phi_i)|^2 + \sum_{i=1}^{n} |b_i - (u, \phi_i)|^2$$

the verification of which is left to the reader. Clearly, the minimum of f is attained by letting $b_i = (u_i, \phi_i)$, $i = 1, \ldots, n$. Thus the orthogonal projection of u into V_n is given by

$$P_n u = \sum_{i=1}^{n} (u, \phi_i)\, \phi_i. \tag{3.3.5}$$

Since

$$\|u - P_n u\| = \inf_{v \in V_n} \|u - v\| \to 0 \quad \text{as} \quad n \to \infty,$$

we have the expansion

$$u = \lim_{n\to\infty} \sum_{i=1}^{n} (u, \phi_i)\, \phi_i = \sum_{i=1}^{\infty} (u, \phi_i)\, \phi_i,$$

where the limit is understood in the sense of the norm $\|\cdot\|$.

Example 3.3.8 *A very important application is the least squares approximation of continuous functions by polynomials. Let $V = L^2(-1, 1)$, and*

$V_n = \mathcal{P}_n(-1, 1)$, *the space of polynomials of degree less than or equal to* n. *Note that the dimension of* V_n *is* $n + 1$ *instead of* n; *this fact does not have any essential influence in the above discussions. An orthonormal polynomial basis for* V *is known,* $\{\phi_n \equiv L_n\}_{n\geq 1}$ *consists of the Legendre polynomials,*

$$L_0(x) = \frac{1}{\sqrt{2}}, \quad L_n(x) = \sqrt{\frac{2n+1}{2}} \frac{1}{2^n n!} \frac{d^n}{dx^n} \left[(x^2 - 1)^n \right], \quad n \geq 1.$$

$$(3.3.6)$$

For any $u \in V$, *its least squares best approximation from* $\mathcal{P}_n(-1, 1)$ *is given by the formula*

$$P_n u(x) = \sum_{i=0}^{n} (u, L_i)_{L^2(-1,1)} L_i(x).$$

We have the convergence

$$\lim_{n\to\infty} \|u - P_n u\|_{L^2(-1,1)} = 0.$$

Therefore,

$$\|u\|_{L^2(-1,1)}^2 = \lim_{n\to\infty} \|P_n u\|_{L^2(-1,1)}^2$$

$$= \lim_{n\to\infty} \sum_{i=0}^{n} |(u, L_i)_{L^2(-1,1)}|^2$$

$$= \sum_{i=0}^{\infty} |(u, L_i)_{L^2(-1,1)}|^2$$

known as Parseval's equality. We also have

$$u = \lim_{n\to\infty} \sum_{i=0}^{n} (u, L_i)_{L^2(-1,1)} L_i = \sum_{i=0}^{\infty} (u, L_i)_{L^2(-1,1)} L_i$$

in the sense of $L^2(-1, 1)$ *norm.*

Example 3.3.9 *An equally important example is the least squares approximation of a function* $f \in L^2(0, 2\pi)$ *by trigonometric polynomials (cf. (3.1.13)). Let* $V_n = \mathcal{T}_n$, *the set of all trigonometric polynomials of degree* $\leq n$. *Then the least squares approximation is given by*

$$p_n(x) = \frac{1}{2} a_0 + \sum_{j=1}^{n} [a_j \cos(jx) + b_j \sin(jx)], \qquad (3.3.7)$$

where

$$a_j = \frac{1}{\pi} \int_0^{2\pi} f(x) \cos(jx) \, dx, \qquad j \geq 0,$$

$$b_j = \frac{1}{\pi} \int_0^{2\pi} f(x) \sin(jx) \, dx, \qquad j \geq 1.$$

$$(3.3.8)$$

As with Example 3.3.8, we can look at the convergence of (3.3.7). This leads to the well-known Fourier series expansion

$$f(x) = a_0 + \sum_{j=1}^{\infty} [a_j \cos(jx) + b_j \sin(jx)].$$

Further development of this example is left as Exercise 3.3.4.

Exercise 3.3.1 *Let $V = L^2(\Omega)$, and*

$$K = \{v \in V \mid \|v\|_{L^2(\Omega)} \leq 1\}.$$

For any $u \in V$, find its projection on K.

Exercise 3.3.2 *Prove Theorem 3.3.7.*

Exercise 3.3.3 *Show that in an inner product space V, the function $f(v) \equiv \|v\|^2$ is strictly convex.*

Exercise 3.3.4 *Given a function $f \in L^2(0, 2\pi)$, show that its best approximation in the space $\mathcal{T}_n$ with respect to the norm of $L^2(0, 2\pi)$ is given by the partial sum*

$$\frac{a_0}{2} + \sum_{j=1}^{n} (a_j \cos jx + b_j \sin jx)$$

of the Fourier series of f with the Fourier coefficients

$$a_j = \frac{1}{\pi} \int_0^{2\pi} f(x) \cos jx \, dx, \quad b_j = \frac{1}{\pi} \int_0^{2\pi} f(x) \sin jx \, dx.$$

Derive Parseval's equality for this case.

Exercise 3.3.5 *Repeat Exercise 3.3.4, but use the basis*

$$\{e^{ijx} \mid -n \leq j \leq n\}$$

for $\mathcal{T}_n$. Find a formula for the least squares approximation of $f(x)$ in $L^2(0, 2\pi)$. Give Parseval's equality and give a formula for $\|u - P_n u\|$ in terms of the Fourier coefficients of f when using this basis.

3.4 Orthogonal polynomials

The discussion of Example 3.3.8 at the end of the previous section can be extended in a more general framework of weighted L^2-spaces. As in Example 3.3.8, we use the interval $[-1, 1]$. Let $w(x)$ be a weight function on $[-1, 1]$; i.e., it is positive almost everywhere and it is integrable on $[-1, 1]$. Then we can introduce a weighted function space

$$L_w^2(-1, 1) = \left\{ v \text{ is measurable on } [-1, 1] \; \middle| \; \int_{-1}^{1} |v(x)|^2 w(x) \, dx < \infty \right\}.$$

This is a Hilbert space with the inner product

$$(u, v)_{0,w} = \int_{-1}^{1} u(x)\, v(x)\, w(x)\, dx$$

and the corresponding norm

$$\|v\|_{0,w} = \sqrt{(v, v)_{0,w}}\,.$$

Two functions $u, v \in L_w^2(-1, 1)$ are said to be orthogonal if $(u, v)_{0,w} = 0$.

Starting with the monomials $\{1, x, x^2, \dots\}$, we can apply the Gram-Schmidt procedure described in Section 1.3 to construct a system of orthogonal polynomials $\{p_n(x)\}_{n=0}^{\infty}$ such that the degree of p_n is n. For any $u \in L_w^2(-1, 1)$, the best approximating polynomial of degree less than or equal to N is

$$P_N u(x) = \sum_{n=0}^{N} \xi_n p_n(x), \qquad \xi_n = \frac{(u, p_n)_{0,w}}{\|p_n\|_{0,w}^2}, \quad 0 \le n \le N.$$

This can be verified directly. The best approximation $P_N u$ is characterized by the property that it is the orthogonal projection of u onto the polynomial space $P_N(-1, 1)$ with respect to the inner product $(\cdot, \cdot)_{0,w}$.

A family of well-known orthogonal polynomials, called the *Jacobi polynomials*, are related to the weight function

$$w^{(\alpha,\beta)}(x) = (1 - x)^{\alpha}(1 + x)^{\beta}, \quad -1 < \alpha, \beta < 1. \tag{3.4.1}$$

A detailed discussion of these polynomials can be found in the reference [156]. Here we mention some results for two of the most important special cases.

When $\alpha = \beta = 0$, the Jacobi polynomials become Legendre polynomials, which were discussed in Example 3.3.8. Conventionally, the Legendre polynomials are defined to be

$$L_0(x) = 1, \quad L_n(x) = \frac{1}{2^n n!} \frac{d^n}{dx^n}\left[(x^2 - 1)^n\right], \ n \ge 1. \tag{3.4.2}$$

These polynomials are orthogonal, and

$$(L_m, L_n)_0 = \frac{2}{2n + 1}\, \delta_{mn}.$$

The Legendre polynomials satisfy the differential equation

$$[(1 - x^2)\, L_n'(x)]' + n\,(n + 1)\, L_n(x) = 0, \quad n = 0, 1, \dots,$$

and the recursion formula

$$L_{n+1}(x) = \frac{2n + 1}{n + 1}\, x\, L_n(x) - \frac{n}{n + 1}\, L_{n-1}(x), \quad n = 1, 2, \dots$$

with $L_0(x) = 1$ and $L_1(x) = x$. Graphs of orthonormalized Legendre polynomials of degrees $n = 0, 1, 2, 3$ were given earlier in Figure 1.2 of Subsection 1.3.2 in Chapter 1.

To present some error estimates related to orthogonal projection polynomials, we need to use the notion of Sobolev spaces as reviewed in Chapter 6. A reader without prior knowledge on Sobolev spaces may skip the following error estimates in a first-time reading.

For any $u \in L^2(-1,1)$, its N-th degree $L^2(-1,1)$-projection polynomial $P_N u$ is

$$P_N u(x) = \sum_{n=0}^{N} \xi_n L_n(x), \qquad \xi_n = \frac{2n+1}{2}(u, L_n)_0, \quad 0 \le n \le N.$$

It is shown in [31] that if $u \in H^s(-1,1)$ with $s > 0$, then the following error estimates hold,

$$\|u - P_N u\|_0 \le c\, N^{-s}\|u\|_s,$$
$$\|u - P_N u\|_1 \le c\, N^{3/2-s}\|u\|_s.$$

Here $\|\cdot\|_s$ denotes the $H^s(-1,1)$-norm, and below we use $(\cdot, \cdot)_1$ for the inner product in $H^1(-1,1)$.

Notice that the error estimate in the $L^2(-1,1)$-norm is of optimal order as expected, yet the error estimate in the $H^1(-1,1)$-norm is not of optimal order. In order to improve the approximation order also in the $H^1(-1,1)$-norm, another orthogonal projection operator $P_{1,N} : H^1(-1,1) \to \mathcal{P}_N$ can be introduced: For $u \in H^1(-1,1)$, its projection $P_{1,N} u \in \mathcal{P}_N$ is defined by

$$(P_{1,N} u, v)_1 = (u, v)_1 \quad \forall v \in \mathcal{P}_N.$$

It is shown in [112] that

$$\|u - P_{1,N} u\|_k \le c\, N^{k-s}\|u\|_s, \quad k = 0, 1, \ s \ge 1. \tag{3.4.3}$$

Notice that the error is of optimal order in both the $L^2(-1,1)$-norm and the $H^1(-1,1)$-norm.

Another important special case of (3.4.1) is when $\alpha = \beta = -1/2$. The weight function here is

$$w(x) = \frac{1}{\sqrt{1-x^2}}$$

and the weighted inner product is

$$(u, v)_{0,w} = \int_{-1}^{1} \frac{u(x)\, v(x)}{\sqrt{1-x^2}}\, dx.$$

The corresponding orthogonal polynomials are called *Chebyshev polynomials of the first kind*,

$$T_n(x) = \cos(n \arccos x), \quad n = 0, 1, \ldots. \tag{3.4.4}$$

These functions are orthogonal,

$$(T_m, T_n)_{0,w} = \frac{\pi}{2} c_n \delta_{mn}, \qquad n, m \ge 0$$

with $c_0 = 2$ and $c_n = 1$ for $n \geq 1$. The Chebyshev polynomials satisfy the differential equation

$$-[\sqrt{1-x^2}\,T_n'(x)]' = n^2 \frac{T_n(x)}{\sqrt{1-x^2}}, \quad n = 0, 1, \ldots$$

and the recursion formula

$$T_{n+1}(x) = 2\,x\,T_n(x) - T_{n-1}(x), \quad n \geq 1, \tag{3.4.5}$$

with $T_0(x) = 1$ and $T_1(x) = x$.

Above, we considered orthogonal polynomials defined on the interval $[-1, 1]$. On a general finite interval $[a, b]$ ($b > a$), we can use a simple linear transformation of the independent variables, and reduce the study of orthogonal polynomials on the interval $[a, b]$ to that on the interval $[-1, 1]$. It is also possible to study orthogonal polynomials defined on unbounded intervals.

Orthogonal polynomials are important in the derivation and analysis of Gaussian numerical integration (cf., e.g., [11, Section 5.3]), and in the study of a family of powerful numerical methods, called spectral methods, for solving differential equations (cf., e.g., [23, 30, 62]). For a more extended introduction to orthogonal polynomials, see [42, Chap. 10].

Exercise 3.4.1 *Use (3.4.2) and integration by parts to show $(L_n, L_m) = 0$ for $m \neq n$, $m, n \geq 0$.*

Exercise 3.4.2 *Derive formulas for the Legendre polynomials of (3.4.2) and the Chebyshev polynomials of (3.4.4) over a general interval $[a, b]$. For the Chebyshev polynomials, what is the appropriate weight function over $[a, b]$?*

Exercise 3.4.3 *Derive (3.4.5) from (3.4.4).*

Exercise 3.4.4 *Find the zeros of $T_n(x)$ for $n \geq 1$. Find the points at which*

$$\max_{-1 \leq x \leq 1} |T_n(x)|$$

is attained.

Exercise 3.4.5 *Using the Gram-Schmidt process, compute orthogonal polynomials of degrees 0,1,2 for the weight function $w(x) = -\log x$ on $[0, 1]$.*

Exercise 3.4.6 *For $n \geq 0$, define*

$$S_n(x) = \frac{1}{n+1} T_{n+1}'(x)$$

using the Chebyshev polynomials of (3.4.4). These new polynomials $\{S_n(x)\}$ are called Chebyshev polynomials of the second kind.

(a) Show that $\{S_n(x)\}$ is an orthogonal family on $[-1, 1]$ with respect to the

weight function $w(x) = \sqrt{1 - x^2}$.
(b) *Show that* $\{S_n(x)\}$ *also satisfies the triple recursion relation* (3.4.5).

3.5 Projection operators

Projection operators are useful in discussing many approximation methods. Intuitively we are approximating elements of a vector space V using elements of a subspace W. Originally, this generalized the construction of an orthogonal projection from Euclidean geometry, finding the orthogonal projection of an element $v \in V$ in the subspace W. This has since been extended to general linear spaces that do not possess an inner product; and hence our discussion approaches the definition of projection operators from another perspective.

Definition 3.5.1 *Let V be a linear space, V_1 and V_2 subspaces of V. We say V is the* direct sum *of V_1 and V_2 and write $V = V_1 \oplus V_2$, if any element $v \in V$ can be uniquely decomposed as*

$$v = v_1 + v_2, \quad v_1 \in V_1, \ v_2 \in V_2. \tag{3.5.1}$$

Furthermore, if V is an inner product space, and $(v_1, v_2) = 0$ for any $v_1 \in V_1$ and any $v_2 \in V_2$, then V is called the orthogonal direct sum *of V_1 and V_2.*

There exists a one-to-one correspondence between direct sums and linear operators P satisfying $P^2 = P$.

Proposition 3.5.2 *Let V be a linear space. Then $V = V_1 \oplus V_2$ if and only if there is a linear operator $P : V \to V$ with $P^2 = P$ such that in the decomposition (3.5.1), $v_1 = Pv$ and $v_2 = (I - P)v$, and also $V_1 = P(V)$ and $V_2 = (I - P)(V)$.*

Proof. Let $V = V_1 \oplus V_2$. Then $Pv = v_1$ defines an operator from V to V. It is easy to verify that P is linear and maps V onto V_1 ($Pv_1 = v_1 \, \forall \, v_1 \in V_1$), and so $V_1 = P(V)$. Obviously $v_2 = (I - P)v$ and $(I - P)v_2 = v_2 \, \forall \, v_2 \in V_2$.

Conversely, with the operator P, for any $v \in V$ we have the decomposition $v = Pv + (I - P)v$. We must show this decomposition is unique. Suppose $v = v_1 + v_2$, $v_1 \in V_1$, $v_2 \in V_2$. Then $v_1 = Pw$ for some $w \in V$. This implies $Pv_1 = P^2w = Pw = v_1$. Similarly, $Pv_2 = 0$. Hence, $Pv = v_1$, and then $v_2 = v - v_1 = (I - P)v$. ∎

Definition 3.5.3 *Let V be a Banach space. An operator $P \in \mathcal{L}(V)$ with the property $P^2 = P$ is called a* projection operator. *The subspace $P(V)$ is called the corresponding* projection space. *The direct sum*

$$V = P(V) \oplus (I - P)(V)$$

is called a topological direct sum.

If V is a Hilbert space, P is a projection operator, and $V = P(V) \oplus (I - P)(V)$ is an orthogonal direct sum, then we call P an orthogonal projection operator.

It is easy to see that a projection operator P is orthogonal if and only if

$$(Pv, (I - P)w) = 0 \qquad \forall\, v, w \in V. \tag{3.5.2}$$

Example 3.5.4 *Figure 3.3 illustrates the orthogonal direct decomposition of an arbitrary vector in $\mathbb{R}^2$ which defines an orthogonal projection operator P from $\mathbb{R}^2$ to V_1. In particular, when V_1 is the x_1-axis, we have*

$$Pv = \begin{pmatrix} v_1 \\ 0 \end{pmatrix} \quad for \quad v = \begin{pmatrix} v_1 \\ v_2 \end{pmatrix}.$$

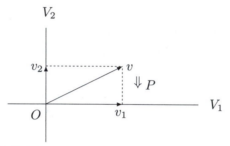

FIGURE 3.3. Orthogonal projection in $\mathbb{R}^2$

Example 3.5.5 (LAGRANGE INTERPOLATION) *Let $V = C[a, b]$, $V_1 = \mathcal{P}_n$ the space of the polynomials of degree less than or equal to n, and let $\Delta : a = x_0 < x_1 < \cdots < x_n = b$ be a partition of the interval $[a, b]$. For $v \in C[a, b]$, we define $Pv \in \mathcal{P}_n$ to be the Lagrange interpolant of v corresponding to the partition Δ; i.e., Pv satisfies the interpolation conditions: $Pv(x_i) = v(x_i)$, $0 \le i \le n$. From the discussion of Section 3.1, the interpolant Pv is uniquely determined. The uniqueness of the interpolant implies that P is a projection operator. Explicitly,*

$$Pv(x) = \sum_{i=0}^{n} \left(\prod_{j \neq i} \frac{x - x_j}{x_i - x_j} \right) v(x_i),$$

using the Lagrange formula for the interpolant.

Example 3.5.6 (PIECEWISE LINEAR INTERPOLATION) *Again we let $V = C[a, b]$ and $\Delta : a = x_0 < x_1 < \cdots < x_n = b$ a partition of the interval $[a, b]$. This time, we take V_1 to be the space of continuous piecewise linear functions:*

$$V_1 = \{ v \in C[a, b] \mid v|_{[x_{i-1}, x_i]} \text{ is linear, } 1 \le i \le n \}.$$

Then for any $v \in C[a, b]$, Pv is the piecewise linear function uniquely determined by the interpolation conditions $Pv(x_i) = v(x_i)$, $0 \le i \le n$. This is an example of a finite element space.

Example 3.5.7 *Generally, let V_n be an n-dimensional subspace of a Hilbert space V. Suppose $\{u_1, \ldots, u_n\}$ is an orthonormal basis of V_n. For any $v \in V$, the formula*

$$Pv = \sum_{i=1}^{n} (u_i, v)\, u_i$$

defines an orthogonal projection from V onto V_n.

Example 3.5.8 *Recall the least squares approximation using trigonometric polynomials, in (3.3.7)–(3.3.8). This defines an orthogonal projection from $L^2(0, 2\pi)$ to $\mathcal{T}_n$. Denote it by $\mathcal{F}_n f$. In the following section, we discuss $\mathcal{F}_n$ as a projection from $C_p(2\pi)$ to $\mathcal{T}_n$. Recall from examples 1.1.2(g), 1.2.4(a) that the space $C_p(2\pi)$ consists of all continuous functions g on $\mathbb{R}$ for which*

$$g(x + 2\pi) \equiv g(x)$$

and the norm is $\|\cdot\|_\infty$. Proposition 3.5.9, given below, also applies to $\mathcal{F}_n$ and the linear space $L^2(0, 2\pi)$.

If V is an inner product space, then we define the orthogonal complement of a subspace V_1 as

$$V_1^{\perp} = \{v \in V \mid (v, v_1) = 0 \quad \forall\, v_1 \in V_1\}.$$

The proof of the following is left as Exercise 3.5.3.

Proposition 3.5.9 (ORTHOGONAL PROJECTION) *Let V_1 be a closed linear subspace of the Hilbert space V, with its orthogonal complement $V_1^{\perp}$. Let $P : V \to V_1$. Then*
(a) The operator P is an orthogonal projection if and only if it is a self-adjoint projection.
(b) $V = V_1 \oplus V_1^{\perp}$.
(c) There exists exactly one orthogonal projection operator P from V onto V_1. We have

$$\|v - Pv\| = \inf_{w \in V_1} \|v - w\| \quad \forall\, v \in V.$$

The operator $I - P$ is the orthogonal projection onto $V_1^{\perp}$.
(d) If $P : V \to V$ is an orthogonal projection operator, then $P(V)$ is a closed subspace of V, and we have the orthogonal direct sum

$$V = P(V) \oplus (I - P)(V).$$

Exercise 3.5.1 *Show that if P is a projection operator (or an orthogonal projection operator), then so is $I - P$.*

Exercise 3.5.2 *Let V be a Hilbert space, let V_1 be a finite-dimensional subspace with basis $\{\varphi_1, \ldots, \varphi_n\}$, and let P be an orthogonal projection of V onto V_1. Show that $Px = 0$ if and only if $P\varphi_j = 0$ for $j = 1, \ldots, n$.*

Exercise 3.5.3 *Prove Proposition 3.5.9.*

Exercise 3.5.4 *Let P be a bounded projection on the Banach space V. Show that $\|P\| \geq 1$. If V is a Hilbert space, and if P is an orthogonal projection, show that $\|P\| = 1$.*

Exercise 3.5.5 (a) *Find a formula for $\|P\|$ in Example 3.5.5.*
(b) *Find a formula for $\|P\|$ in Example 3.5.6.*

Exercise 3.5.6 *Extend Example 3.5.6 to piecewise quadratic interpolation. Use evenly spaced node points. What is $\|P\|$ in this case?*

3.6   Uniform error bounds

Approximation in the uniform norm is quite important in numerical analysis and applied mathematics, and polynomials are the most important type of approximants. However, most convergence analyses on the uniform approximation of continuous functions by polynomials must be shown directly by special arguments, rather than by appealing to some more general theory. The first important such result is due to Weierstrass, and it is much stronger than first might be expected.

Theorem 3.6.1 (WEIERSTRASS) *Let $f \in C[a, b]$, let $\varepsilon > 0$. Then there exists a polynomial $p(x)$ for which*

$$\|f - p\|_\infty \leq \varepsilon.$$

This result says that any continuous function f can be approximated uniformly by polynomials, no matter how badly behaved f may be on $[a, b]$. Several proofs of this seminal result are given in [42, Chap. 6], including an interesting constructive result using Bernstein polynomials.

For many uses of approximation theory in numerical analysis, we need error bounds for the best uniform approximation of a function $f(x)$ on an interval $[a, b]$. We are interested in two such problems: the uniform approximation of a smooth function by polynomials and the uniform approximation of a smooth 2π-periodic function by trigonometric polynomials. These problems were discussed previously in Section 3.2.4, with theorems 3.2.19 and 3.2.20 giving the uniqueness of the best approximants for these two forms of approximation. Initially, we study the polynomial approximation problem on the special interval $[-1, 1]$, and then the results obtained extend easily to an arbitrary interval $[a, b]$ by a simple linear change of variables. We consider the approximation of a 2π-periodic function by trigonometric polynomials as taking place on the interval $[-\pi, \pi]$

in most cases, as it aids in dealing with the special cases of even and odd 2π-periodic functions.

An important first step is to note that these two problems are closely connected. Given a function $f \in C^m[-1, 1]$, introduce the function

$$g(\theta) = f(\cos \theta). \tag{3.6.1}$$

The function g is an even 2π-periodic function and it is also m-times continuously differentiable. If we examine the best approximating trigonometric polynomial $q_n(\theta)$ for any even 2π-periodic function $g(\theta)$, the uniqueness result (Theorem 3.2.20) can be used to show that $q_n(\theta)$ has the form

$$q_n(\theta) = a_0 + \sum_{j=1}^{n} a_j \cos(j\theta). \tag{3.6.2}$$

The proof is based on using the property that g is even and that the best approximation is unique; cf. Exercise 3.6.2.

For the best uniform trigonometric approximation of (3.6.1), given in (3.6.2), use the substitution $x = \cos \theta$ and trigonometric identities to show the existence of $p_n \in \mathcal{P}_n$ with

$$q_n(\theta) = p_n(\cos \theta)$$

and with $p_n(x)$ having the same degree as $q_n(\theta)$. Conversely, if $p \in \mathcal{P}_n$ is given, then $q_n(\theta) \equiv p_n(\cos \theta)$ can be shown to have the form (3.6.2).

Using these results,

$$\max_{0 \le \theta \le \pi} |g(\theta) - q_n(\theta)| = \max_{-1 \le x \le 1} |f(x) - p_n(x)|.$$

In addition, it is straightforward to show that $p_n(x)$ must be the best uniform approximation to $f(x)$ on $[-1, 1]$. If it were not, we could produce a better uniform approximation, call it $r_n(x)$; and then $r_n(\cos \theta)$ would be a better uniform approximation to $g(\theta)$ on $[0, \pi]$, a contradiction. This equivalence allows us to concentrate on only one of our two approximating problems, that of approximating a 2π-periodic function $g(\theta)$ by a trigonometric polynomial $q_n(\theta)$. The results then transfer immediately to the ordinary polynomial approximation problem for a function $f \in C^m[-1, 1]$.

As a separate result, it can also be shown that when given any 2π-periodic function $g(\theta)$, there is a corresponding function $f(x)$ of equal smoothness for which there is an equivalence between their best uniform approximations in the respective approximating spaces $\mathcal{T}_n$ and $\mathcal{P}_n$. For this construction, see [117, page 46].

We state without proof the main results. For proofs, see Meinardus [117, Section 5.5]. Recall that the notation $C_p(2\pi)$ denotes the Banach space of 2π-periodic functions, with the uniform norm as the norm.

Theorem 3.6.2 (JACKSON'S THEOREM) *Suppose the 2π-periodic function $g(\theta)$ possesses continuous derivatives up to order k. Further assume that* *the kth derivative satisfies a Hölder condition:*

$$\left| g^{(k)}(\theta_1) - g^{(k)}(\theta_2) \right| \le M_k \left| \theta_1 - \theta_2 \right|^\alpha, \qquad -\infty < \theta_1, \theta_2 < \infty,$$

for some $M_k > 0$ and some $\alpha \in (0, 1]$. (We say that $g \in C_p^{k,\alpha}(2\pi)$.) Then the error in the best approximation $q_n(\theta)$ to $g(\theta)$ satisfies

$$\max_{-\infty < \theta < \infty} |g(\theta) - q_n(\theta)| \le c^{k+1} \frac{M_k}{n^{k+\alpha}} \tag{3.6.3}$$

with $c = 1 + \frac{1}{2}\pi^2$.

Theorem 3.6.3 (JACKSON'S THEOREM) *Suppose $f \in C^k[-1, 1]$ and that the kth derivative satisfies a Hölder condition:*

$$\left| f^{(k)}(x_1) - f^{(k)}(x_2) \right| \le M_k \left| x_1 - x_2 \right|^\alpha, \qquad -1 \le x_1, x_2 \le 1,$$

for some $M_k > 0$ and some $\alpha \in (0, 1]$. Then the error in the best approximation $p_n(x)$ to $f(x)$ satisfies

$$\max_{-1 \le x \le 1} |f(x) - p_n(x)| \le d_k \, c^{k+1} \frac{M_k}{n^{k+\alpha}} \tag{3.6.4}$$

with $c = 1 + \frac{1}{2}\pi^2$ and d_k any number satisfying

$$d_k \ge \frac{n^{k+\alpha}}{n(n-1)\cdots(n-k+1)(n-k)^\alpha}, \qquad n \ge 1.$$

Note that the right-hand fraction tends to 1 as $n \to \infty$, and therefore a finite bound d_k does exist for each $k \ge 0$.

3.6.1 Uniform error bounds for L^2-approximations

The Fourier series of a function $f \in L^2(-\pi, \pi)$ is a widely used tool in applied and computational mathematics, and as such, error bounds are needed for the convergence of the series. We return to this topic in later chapters, deriving additional error bounds for the error in the context of Sobolev spaces (cf. Section 6.5). But here we look at bounds based on the above Theorem 3.6.2.

The Fourier series for a function $f \in L^2(-\pi, \pi)$ was given in Example 3.3.9, with the formulas (3.3.7)–(3.3.8). As introduced in the preceding section, we also use the notation $\mathcal{F}_n f$ to denote the partial Fourier series of terms of degree $\le n$. It is straightforward to obtain bounds for $f - \mathcal{F}_n f$ in $L^2(-\pi, \pi)$, once uniform error bounds are known. Simply use

$$\|g\|_2 \le \sqrt{2\pi}\, \|g\|_\infty, \qquad g \in C[-\pi, \pi],$$

and therefore

$$\|f - \mathcal{F}_n f\|_2 \le \sqrt{2\pi}\, \|f - \mathcal{F}_n f\|_\infty, \qquad f \in C[-\pi, \pi]. \tag{3.6.5}$$

The error bounds follow immediately. An alternative set of L^2-bounds are introduced in Section 6.5.

Obtaining results in the uniform norm is more difficult. Begin by using standard trigonometric identities to rewrite the formulas (3.3.7)–(3.3.8) for $\mathcal{F}_n f$ as

$$\mathcal{F}_n f(x) = \frac{1}{\pi} \int_{-\pi}^{\pi} D_n(x - y)\, f(y)\, dy \qquad (3.6.6)$$

where

$$D_n(\theta) = \frac{1}{2} + \sum_{j=1}^{n} \cos(j\theta)\,; \qquad (3.6.7)$$

and if $x \notin \{2j\pi \mid j = 0, \pm 1, \pm 2, \dots\}$,

$$D_n(\theta) = \frac{\sin\left(n + \frac{1}{2}\right) x}{2 \sin \frac{1}{2} x}.$$

The function D_n is called the *Dirichlet kernel function*. Many results on the behavior of the partial Fourier sums $\mathcal{F}_n f(x)$ are obtained by an examination of the formula (3.6.6).

For $f \in C_p(2\pi)$, use this formula to obtain

$$\max_{x} |\mathcal{F}_n f(x)| \leq \frac{1}{\pi} \max_{x} \int_{-\pi}^{\pi} |D_n(x - y)|\, dy\, \|f\|_{\infty}$$

$$= \frac{2}{\pi} \int_{0}^{\pi} |D_n(y)|\, dy\, \|f\|_{\infty}\,.$$

The last step uses the facts that $D_n(\theta)$ is even and 2π-periodic. From this, we see

$$\mathcal{F}_n : C_p(2\pi) \to \mathcal{T}_n \subseteq C_p(2\pi)$$

is a bounded projection operator with

$$\|\mathcal{F}_n\| \leq L_n \equiv \frac{2}{\pi} \int_{0}^{\pi} |D_n(y)|\, dy. \qquad (3.6.8)$$

By regarding (3.6.6) as defining an integral operator from $C_p(2\pi)$ to itself, it can be seen that $\|\mathcal{F}_n\| = L_n$ (cf. (2.2.8)).

The numbers $\{L_n\}$ are called *Lebesgue constants*, and a great deal is known about them. In particular, it is shown in Zygmund [176, Chap. 2, p. 67] that

$$\|\mathcal{F}_n\| = L_n = \frac{4}{\pi^2} \log n + O(1), \qquad n \geq 1. \qquad (3.6.9)$$

Thus $\{\|\mathcal{F}_n\|\}$ is an unbounded sequence. This implies the existence of a function $f \in C_p(2\pi)$ for which $\mathcal{F}_n f$ does not converge uniformly to f.

To prove the last statement, begin by noting that $\mathcal{F}_n f = f$ for any $f \in \mathcal{T}_n$; and moreover, note that the trigonometric polynomials are dense in $C_p(2\pi)$. It then follows from the Banach-Steinhaus theorem (Theorem 2.4.5) that there exist functions $f \in C_p(2\pi)$ for which $\mathcal{F}_n f$ does not converge uniformly to f. Note, however, that since such an f is in $L^2(-\pi, \pi)$, $\mathcal{F}_n f$ does converge to f in the L^2-norm.

In contrast to the above results in $C_p(2\pi)$, recall that $\mathcal{F}_n$ is an orthogonal projection operator with respect to $L^2(-\pi, \pi)$; and therefore $\|\mathcal{F}_n\| = 1$.

Uniform error bounds for the Fourier series

For a given function f and a given integer $n \geq 0$, let q_n denote the best approximation of f from the approximating subspace $\mathcal{T}_n$. Note that $\mathcal{F}_n(q_n) = q_n$. Then using the linearity of $\mathcal{F}_n$,

$$f - \mathcal{F}_n(f) = (f - q_n) - \mathcal{F}_n(f - q_n).$$

Taking norms of both sides,

$$\|f - \mathcal{F}_n(f)\|_\infty \leq (1 + \|\mathcal{F}_n\|) \|f - q_n\|_\infty.$$

Assuming $f \in C_p^{k,\alpha}(2\pi)$ for some $k \geq 0$ and some $\alpha \in (0, 1]$, we have

$$\|f - \mathcal{F}_n(f)\|_\infty \leq (1 + \|\mathcal{F}_n\|) c^{k+1} \frac{M_k}{n^{k+\alpha}} \qquad (3.6.10)$$

and

$$\|f - \mathcal{F}_n(f)\|_\infty \leq c_k \frac{\log n}{n^{k+\alpha}} \quad \text{for } n \geq 2. \qquad (3.6.11)$$

Here $c = 1 + \pi^2/2$, M_k is the Hölder constant for $f^{(k)}$ and c_k is a constant linearly dependent on M_k and otherwise independent of f. Combining this with (3.6.9), we see that if $f \in C_p^{0,\alpha}(2\pi)$ for some α, then $\mathcal{F}_n(f)$ converges uniformly to f. For $\mathcal{F}_n(f)$ to fail to converge uniformly to f, the function f must be fairly badly behaved.

3.6.2 Interpolatory projections and their convergence

Recall the trigonometric interpolation discussion of Section 3.1.4. Let $f \in C_p(2\pi)$, let $n \geq 0$ be a given integer, and let the interpolation nodes be the evenly spaced points in (3.1.16). Denote the resulting interpolation formula by $\mathcal{I}_n(f)$. It is straightforward to show that this is a linear operator; and by the uniqueness of such trigonometric polynomial interpolation, it also follows that $\mathcal{I}_n$ is a projection operator on $C_p(2\pi)$ to $\mathcal{T}_n$. To discuss the convergence of $\mathcal{I}_n(f)$ to f, we can proceed in the same manner as when examining the convergence of $\mathcal{F}_n(f)$.

Begin by obtaining the Lagrange interpolation formula

$$\mathcal{I}_n f(x) = \frac{2}{2n+1} \sum_{j=0}^{2n} D_n(x - x_j) f(x_j). \qquad (3.6.12)$$

Its proof is left as Exercise 3.6.4. Using this formula,

$$\|\mathcal{I}_n\| = \frac{2}{2n+1} \max_x \sum_{j=0}^{2n} |D_n(x - x_j)|.$$

In Rivlin [138, p. 13], it is shown that

$$\|\mathcal{I}_n\| \le 1 + \frac{2}{\pi} \log n, \qquad n \ge 1 \qquad (3.6.13)$$

and it is also shown that $\|\mathcal{I}_n\|$ is exactly of order $\log n$ as $n \to \infty$. This result can be combined with an argument such as the one leading to (3.6.11) to obtain analogous results for the convergence of $\mathcal{I}_n f$. In fact, assuming $f \in C_p^{k,\alpha}(2\pi)$ for some $k \ge 0$ and some $\alpha \in (0,1]$, we have

$$\|f - \mathcal{I}_n f\|_\infty \le (1 + \|\mathcal{I}_n\|) c^{k+1} \frac{M_k}{n^{k+\alpha}} \qquad (3.6.14)$$

for any $n \ge 1$, and

$$\|f - \mathcal{I}_n f\|_\infty \le c_k \frac{\log n}{n^{k+\alpha}} \quad \text{for } n \ge 2 \qquad (3.6.15)$$

for c_k a constant linearly dependent on M_k and otherwise independent of f.

Exercise 3.6.1 *Show that $\cos(j\theta)$ can be written as $p_j(\cos\theta)$, with $p_j(x)$ a polynomial of degree j.*

Exercise 3.6.2 *Show that if $g(\theta)$ is an even 2π-periodic function, then its best approximation of degree n must take the form (3.6.2).*

Exercise 3.6.3 *Derive (3.6.6).*

Exercise 3.6.4 *Show that the functions*

$$\phi_j(x) \equiv \frac{2}{2n+1} D_n(x - x_j), \qquad 0 \le j \le 2n,$$

belong to $\mathcal{T}_n$ and that they satisfy

$$\phi_j(x_k) = \delta_{jk}$$

Thus show that (3.6.12) can be considered a "Lagrange interpolation formula."

Exercise 3.6.5 *Show that $\mathcal{I}_n(f)$ can be obtained from $\mathcal{F}_n(f)$ by a suitably chosen numerical integration.*

Suggestion for Further Readings

Interpolation is a standard topic found in every textbook on numerical analysis. The reader is referred to ATKINSON [11, Chap. 3], KRESS [101], and other numerical analysis textbooks for a more detailed discussion of polynomial interpolation, as well as other interpolation topics not touched upon in this work, such as interpolation with spline functions. The classic introduction to the theory of spline functions is DE BOOR [26]. For an introduction to the use of *wavelets*, see CHUI [34] or KAISER [85]. MEINARDUS [117] is an excellent reference for approximations by polynomials and trigonometric polynomials. DAVIS [42] contains an extensive discussion of interpolation and approximation problems in a very general framework.

A best approximation problem is a minimization problem, with or without constraints, and some of them are best studied within the framework of *optimization theory*. Some abstract minimization problems are best studied in the framework of *convex analysis*, and some excellent references on convex analysis include EKELAND AND TEMAM [47], ROCKAFELLAR [140] and ZEIDLER [173].

4
Nonlinear Equations and Their Solution by Iteration

Nonlinear functional analysis is the study of operators lacking the property of linearity. In this chapter, we consider nonlinear operator equations and their numerical solution. We begin the consideration of operator equations that take the form

$$u = T(u), \qquad u \in K. \qquad (4.0.1)$$

Here, V is a Banach space, K is a subset of V, and $T : K \to V$. The solutions of this equation are called *fixed points* of the operator T, as they are left unchanged by T. The most important method for analyzing the solvability theory for such equations is the *Banach fixed-point theorem*. We present the Banach fixed-point theorem and then discuss its application to the study of various iterative methods in numerical analysis.

We then consider an extension of the well-known Newton method to the more general setting of Banach spaces. For this purpose, we introduce the differential calculus for nonlinear operators on normed spaces.

We conclude the chapter with a brief introduction to another means of studying (4.0.1), using the concept of the *rotation of a completely continuous vector field*. There are many generalizations of the ideas of this chapter, and we intend this material as only a brief introduction.

4.1 The Banach fixed-point theorem

Let V be a Banach space with the norm $\| \cdot \|_V$, and let K be a subset of V. Let $T : K \to V$ be an operator defined on K. We are interested in the existence of a solution of the operator equation (4.0.1) and the possibility of approximating the solution u by the following iterative method. Pick an initial guess $u_0 \in K$, and define a sequence $\{u_n\}$ by the iteration formula

$$u_{n+1} = T(u_n), \qquad n = 0, 1, \ldots \tag{4.1.1}$$

To have this make sense, we see another requirement that must be imposed upon T:

$$T(v) \in K \qquad \text{for all } v \in K \tag{4.1.2}$$

The problem of solving an equation

$$f(u) = 0 \tag{4.1.3}$$

for some operator $f : K \subseteq V \to V$ can be reduced to an equivalent fixed-point problem of the form (4.0.1) by setting $T(v) = v - f(v)$, or more generally, $T(v) = v - c_0 f(v)$ for some constant scalar $c_0 \neq 0$. Thus any result on the fixed-point problem (4.0.1) can be translated into a result for an equation (4.1.3). In addition, the iterative method (4.1.1) then provides a possible approximation procedure for solving the equation (4.1.3). In the following Section 4.2, we look at such applications for solving equations in a variety of settings.

For the iterative method to work, we must assume something more than (4.1.2). To build some insight as to what further assumptions are needed on the operator T, consider the following simple example.

Example 4.1.1 *Take V to be the real line $\mathbb{R}$, and T an affine operator,*

$$Tx = a\,x + b, \qquad x \in \mathbb{R},$$

for some constants a and b. Now define the iterative method induced by the operator T. Let $x_0 \in \mathbb{R}$, and for $n = 0, 1, \ldots$, define

$$x_{n+1} = a\,x_n + b.$$

It is easy to see that

$$x_n = x_0 + n\,b \qquad \text{if } a = 1,$$

and

$$x_n = a^n x_0 + \frac{1 - a^n}{1 - a}\,b \qquad \text{if } a \neq 1.$$

Thus in the non-trivial case $a \neq 1$, the iterative method is convergent if and only if $|a| < 1$. Notice that the number $|a|$ occurs in the property

$$|Tx - Ty| \leq |a|\,|x - y| \quad \forall\, x, y \in \mathbb{R}.$$

Definition 4.1.2 *For an operator $T : K \subseteq V \to V$, we say it is contractive with contractivity constant $\alpha \in [0, 1)$ if*

$$\|T(u) - T(v)\|_V \leq \alpha\,\|u - v\|_V \qquad \forall\, u, v \in K.$$

The operator T is called non-expansive *if*

$$\|T(u) - T(v)\|_V \le \|u - v\|_V \qquad \forall\, u, v \in K,$$

and Lipschitz continuous *if there exists a constant $L \ge 0$ such that*

$$\|T(u) - T(v)\|_V \le L\,\|u - v\|_V \qquad \forall\, u, v \in K.$$

We see the following implications:

$$\text{contractivity} \implies \text{non-expansiveness}$$
$$\implies \text{Lipschitz continuity}$$
$$\implies \text{continuity.}$$

Theorem 4.1.3 (BANACH FIXED-POINT THEOREM) *Assume that K is a non-empty closed set in a Banach space V, and further, that $T : K \to K$ is a contractive mapping with contractivity constant α, $0 \le \alpha < 1$. Then the following results hold.*

1. *Existence and uniqueness: There exists a unique $u \in K$ such that*

$$u = T(u).$$

2. *Convergence and error estimates of the iteration: For any $u_0 \in K$, the sequence $\{u_n\} \subseteq K$ defined by $u_{n+1} = T(u_n)$, $n = 0, 1, \ldots$, converges to u:*

$$\|u_n - u\|_V \to 0 \quad \text{as } n \to \infty.$$

For the error, the following bounds are valid:

$$\|u_n - u\|_V \le \frac{\alpha^n}{1 - \alpha}\,\|u_0 - u_1\|_V, \tag{4.1.4}$$

$$\|u_n - u\|_V \le \frac{\alpha}{1 - \alpha}\,\|u_{n-1} - u_n\|_V, \tag{4.1.5}$$

$$\|u_n - u\|_V \le \alpha\,\|u_{n-1} - u\|_V. \tag{4.1.6}$$

Proof. Since $T : K \to K$, the sequence $\{u_n\}$ is well-defined. Let us first prove that $\{u_n\}$ is a Cauchy sequence. Using the contractivity of the mapping T, we have

$$\|u_{n+1} - u_n\|_V \le \alpha\,\|u_n - u_{n-1}\|_V \le \cdots \le \alpha^n\|u_1 - u_0\|_V.$$

Then for any $m \ge n \ge 1$,

$$\|u_m - u_n\|_V \le \sum_{j=0}^{m-n-1} \|u_{n+j+1} - u_{n+j}\|_V$$
$$\le \sum_{j=0}^{m-n-1} \alpha^{n+j}\|u_1 - u_0\|_V$$
$$\le \frac{\alpha^n}{1 - \alpha}\,\|u_1 - u_0\|_V. \tag{4.1.7}$$

Since $\alpha \in [0,1)$, $\|u_m - u_n\|_V \to 0$ as $m, n \to \infty$. Thus $\{u_n\}$ is a Cauchy sequence; and since K is a closed set in the Banach space V, $\{u_n\}$ has a limit $u \in K$. We take the limit $n \to \infty$ in $u_{n+1} = T(u_n)$ to see that $u = T(u)$ by the continuity of T; i.e., u is a fixed point of T.

Suppose $u_1, u_2 \in K$ are fixed points of T. Then from $u_1 = T(u_1)$ and $u_2 = T(u_2)$, we obtain

$$u_1 - u_2 = T(u_1) - T(u_2).$$

Hence

$$\|u_1 - u_2\|_V = \|T(u_1) - T(u_2)\|_V \leq \alpha \|u_1 - u_2\|_V,$$

which implies $\|u_1 - u_2\|_V = 0$ since $\alpha \in [0,1)$. So a fixed point of a contractive mapping is unique.

Now we prove the error estimates. Letting $m \to \infty$ in (4.1.7), we get the estimate (4.1.4). From

$$\|u_n - u\|_V = \|T(u_{n-1}) - T(u)\|_V \leq \alpha \|u_{n-1} - u\|_V$$

we obtain the estimate (4.1.6). This estimate together with

$$\|u_{n-1} - u\|_V \leq \|u_{n-1} - u_n\|_V + \|u_n - u\|_V$$

implies the estimate (4.1.5). ∎

This theorem is called by a variety of names in the literature, with the *contractive mapping theorem* another popular choice. It is also called *Picard iteration* in some settings.

Exercise 4.1.1 *In the Banach fixed-point theorem, we assume (1) V is a complete space, (2) K is a non-empty closed set in V, (3) $T : K \to K$, and (4) T is contractive. Find examples to show that each of these assumptions is necessary for the result of the theorem; in particular, the result fails to hold if all the other assumptions are kept except that T is only assumed to satisfy the inequality*

$$\|T(u) - T(v)\|_V < \|u - v\|_V \quad \forall u, v \in V, \ u \neq v.$$

Exercise 4.1.2 *Assume K is a non-empty closed set in a Banach space V, and that $T : K \to K$. Suppose T^m is a contraction for some positive integer m. Prove that T has a unique fixed point in K. Moreover, prove the iteration method*

$$u_{n+1} = T(u_n), \qquad n = 0, 1, 2, \ldots$$

coverges.

Exercise 4.1.3 *Let T be a contractive mapping on V to V. Using Theorem 4.1.3, the equation $v = T(v) + y$ has a unique solution, call it $u(y)$, for every $y \in V$. Show that $u(y)$ is a continuous function of y. Weaken the assumption of contractiveness as much as possible, still obtaining the same conclusion.*

Exercise 4.1.4 *Let V be a Banach space, and let T be a contractive mapping on $K \subseteq V$ to K, with $K = \{v \in V \mid \|v\| \le r\}$ for some $r > 0$. Assume $T(0) = 0$. Show that $v = T(v) + y$ has a unique solution in K for all sufficiently small choices of $y \in V$.*

4.2 Applications to iterative methods

The Banach fixed-point theorem presented in the preceding section contains the most desirable properties of a numerical method. Under the stated conditions, the approximation sequence is well-defined, and it is convergent to the unique solution of the problem. Furthermore, we know the convergence rate is linear (cf. (4.1.6)), we have an *a priori* error estimate (4.1.4) that can be used to determine the number of iterations needed to achieve a prescribed solution accuracy before actual computations take place, and we also have an *a posteriori* error estimate (4.1.5) that gives a computable error bound once some numerical solutions are calculated.

In this section, we apply the Banach fixed-point theorem to the analysis of numerical approximations of several problems. Later in Chapter 10, we use the Banach fixed-point theorem to show the unique solvability of some variational problems.

4.2.1 Nonlinear equations

Given a real-valued function of a real variable, $f : \mathbb{R} \to \mathbb{R}$, we are interested in computing its real roots; i.e., we are interested in solving the equation

$$f(x) = 0, \qquad x \in \mathbb{R}. \tag{4.2.1}$$

There are a variety of ways to reformulate this equation as an equivalent fixed-point problem of the form

$$x = T(x), \qquad x \in \mathbb{R}. \tag{4.2.2}$$

Some examples are $T(x) \equiv x - f(x)$ or more generally $T(x) \equiv x - c_0 f(x)$ for some constant $c_0 \ne 0$. A more sophisticated example is $T(x) = x - f(x)/f'(x)$, in which case the iterative method becomes the celebrated Newton's method. For this last example, we generally use Newton's method only for finding simple roots of $f(x)$, which means we need to assume $f'(x) \ne 0$ when $f(x) = 0$. We return to a study of the Newton's method later in Section 4.4. Specializing the Banach fixed-point theorem to the problem (4.2.2), we have the following well-known result.

Theorem 4.2.1 *Let $-\infty < a < b < \infty$ and $T : [a, b] \to [a, b]$ be a contractive function with contractivity constant $\alpha \in [0, 1)$. Then the following results hold.*

1. *Existence and uniqueness: There exists a unique solution $x \in [a, b]$ to the equation $x = T(x)$.*

2. *Convergence and error estimates of the iteration: For any $x_0 \in [a, b]$, the sequence $\{x_n\} \subseteq [a, b]$ defined by $x_{n+1} = T(x_n)$, $n = 0, 1, \ldots$, converges to x:*

$$x_n \to x \quad as \ n \to \infty.$$

   For the error, there hold the bounds

$$|x_n - x| \leq \frac{\alpha^n}{1 - \alpha} |x_0 - x_1|,$$

$$|x_n - x| \leq \frac{\alpha}{1 - \alpha} |x_{n-1} - x_n|,$$

$$|x_n - x| \leq \alpha |x_{n-1} - x|.$$

The contractiveness of the function T is guaranteed from the assumption that

$$\sup_{a \leq x \leq b} |T'(x)| < 1.$$

Indeed, using the Mean Value Theorem, we then see that T is contractive with the contractivity constant $\alpha = \sup_{a \leq x \leq b} |T'(x)|$.

4.2.2 Linear systems

Let $A \in \mathbb{R}^{m \times m}$ be an m by m matrix, and let us consider the linear system

$$A\mathbf{x} = \mathbf{b}, \qquad \mathbf{x} \in \mathbb{R}^m \tag{4.2.3}$$

where $\mathbf{b} \in \mathbb{R}^m$ is given. It is well known that (4.2.3) has a unique solution $\mathbf{x}$ for any $\mathbf{b}$ if and only if A is nonsingular, $\det(A) \neq 0$. We reformulate (4.2.3) as a fixed-point problem $\mathbf{x} = T(\mathbf{x})$.

A common practice for devising iterative methods of solving (4.2.3) is by using a matrix splitting

$$A = N - M$$

with N chosen in such a way that the system $N\mathbf{x} = \mathbf{k}$ is easily and uniquely solvable for any right side $\mathbf{k}$. Then the linear system (4.2.3) is rewritten as

$$N\mathbf{x} = M\mathbf{x} + \mathbf{b}.$$

Using this leads naturally to an iterative method for solving (4.2.3):

$$N\mathbf{x}_n = M\mathbf{x}_{n-1} + \mathbf{b}, \qquad n = 1, 2, \ldots \tag{4.2.4}$$

with $\mathbf{x}_0$ a given initial guess of the solution $\mathbf{x}$.

To more easily analyze the iteration, we rewrite these last two equations as

$$\mathbf{x} = N^{-1}M\mathbf{x} + N^{-1}\mathbf{b},$$
$$\mathbf{x}_n = N^{-1}M\mathbf{x}_{n-1} + N^{-1}\mathbf{b}.$$

The matrix $N^{-1}M$ is called the iteration matrix. Subtracting the two equations, we obtain the error equation

$$\mathbf{x} - \mathbf{x}_n = N^{-1}M\left(\mathbf{x} - \mathbf{x}_{n-1}\right).$$

Inductively,

$$\mathbf{x} - \mathbf{x}_n = \left(N^{-1}M\right)^n\left(\mathbf{x} - \mathbf{x}_0\right), \qquad n = 0, 1, 2, \ldots$$

We see that the iterative method converges if $\|N^{-1}M\| < 1$, where $\|\cdot\|$ is some matrix operator norm; i.e., it is a norm induced by some vector norm $\|\cdot\|$:

$$\|A\| = \sup_{\mathbf{x}\neq 0} \frac{\|A\mathbf{x}\|}{\|\mathbf{x}\|}. \tag{4.2.5}$$

Note that a necessary and sufficient condition for convergence of the iterative method (4.2.4) is $r_\sigma(N^{-1}M) < 1$, where $r_\sigma(\cdot)$ denotes the *spectral radius*:

$$r_\sigma(A) = \max_i |\lambda_i(A)|,$$

with $\{\lambda_i(A)\}$ the set of all eigenvalues of A.

The spectral radius of a matrix is an intrinsic property of the matrix, while a matrix norm is not. It is thus not surprising that a necessary and sufficient condition for convergence of the iterative method is described in terms of the spectral radius of the iteration matrix. We would also expect something of this kind since in finite-dimensional spaces, convergence of $\{\mathbf{x}_n\}$ in one norm is equivalent to convergence in every other norm (cf. Theorem 1.2.12 from Chapter 1).

We have the following relations between the spectral radius and norms of a matrix $A \in \mathbb{R}^{m\times m}$.

1. $r_\sigma(A) \leq \|A\|$ for any operator matrix norm $\|\cdot\|$.
   This result follows immediately from the definition of $r_\sigma(A)$, the defining relation of an eigenvalue, and the fact that the matrix norm $\|\cdot\|$ is generated by a vector norm.

2. For any $\varepsilon > 0$, there exists a matrix operator norm $\|\cdot\|_{A,\varepsilon}$ such that

   $$r_\sigma(A) \leq \|A\|_{A,\varepsilon} \leq r_\sigma(A) + \varepsilon.$$

   For a proof, see [82, p. 12].

3. $r_\sigma(A) = \lim_{n\to\infty} \|A^n\|^{1/n}$ for any matrix norm $\|\cdot\|$.

Notice that here the norm can be any matrix norm, not necessarily the ones generated by vector norms as in (4.2.5). This can be proven by using the Jordan canonical form; see [11, p. 490].

For applications to the solution of discretizations of Laplace's equation and some other elliptic partial differential equations, it is useful to write

$$A = D + L + U,$$

where D is the diagonal part of A, L and U are the strict lower and upper triangular parts. If we take $N = D$, then (4.2.4) reduces to

$$D\mathbf{x}_n = \mathbf{b} - (L + U)\mathbf{x}_{n-1},$$

which is the vector representation of the Jacobi method; the corresponding componentwise representation is

$$x_{n,i} = \frac{1}{a_{i,i}}\left(b_i - \sum_{j \neq i} a_{ij} x_{n-1,j}\right), \qquad 1 \leq i \leq m.$$

If we take $N = D + L$, then we obtain the Gauss-Seidel method,

$$(D + L)\mathbf{x}_n = \mathbf{b} - U\mathbf{x}_{n-1},$$

or equivalently,

$$x_{n,i} = \frac{1}{a_{i,i}}\left(b_i - \sum_{j=1}^{i-1} a_{ij} x_{n,j} - \sum_{j=i+1}^{m} a_{ij} x_{n-1,j}\right), \qquad 1 \leq i \leq m.$$

A more sophisticated splitting is obtained by setting

$$N = \frac{1}{\omega}D + L, \qquad M = \frac{1-\omega}{\omega}D - U,$$

where ω is an acceleration parameter. The corresponding iterative method with the (approximate) optimal choice of ω is called the SOR (successive overrelaxation) method. The componentwise representation of the SOR method is

$$x_{n,i} = x_{n-1,i} + \omega\left(b_i - \sum_{j=1}^{i-1} a_{ij} x_{n,j} - \sum_{j=i+1}^{m} a_{ij} x_{n-1,j}\right), \qquad 1 \leq i \leq m.$$

In vector form, we write it in the somewhat more intuitive form

$$\mathbf{z}_n = D^{-1}\left[b - L\mathbf{x}_n - U\mathbf{x}_{n-1}\right],$$
$$\mathbf{x}_n = \omega\mathbf{z}_n + (1 - \omega)\mathbf{x}_{n-1}.$$

With the equations to which this is usually applied, there is a well-understood theory for the choice of an optimal value of ω; and with that optimal value, the iteration converges much more rapidly than does the original Gauss-Seidel method on which it is based. Additional discussion of the framework (4.2.4) for iteration methods is given in [11, Section 8.6], [82].

4.2.3 Linear and nonlinear integral equations

Recall Example 2.3.2 from Chapter 2, in which we discussed solvability of the integral equation

$$\lambda u(x) - \int_a^b k(x, y)\, u(y)\, dy = f(x), \qquad a \le x \le b \qquad (4.2.6)$$

by means of the geometric series theorem. For simplicity, we assume $k \in C([a, b] \times [a, b])$ and let $f \in C[a, b]$, although these assumptions can be weakened considerably. In Example 2.3.2, we established that within the framework of the function space $C[a, b]$ with the uniform norm, the equation (4.2.6) was uniquely solvable if

$$\max_{a \le t \le b} \int_a^b |k(x, y)|\, dy < |\lambda|. \qquad (4.2.7)$$

If we rewrite the equation (4.2.6) as

$$u(x) = \frac{1}{\lambda} \int_a^b k(x, y)\, u(y)\, dy + \frac{1}{\lambda} f(x), \qquad a \le x \le b,$$

which has the form $u = T(u)$, then we can apply the Banach fixed-point theorem. Doing so, it is straightforward to derive a formula for the contractivity constant:

$$\alpha = \frac{1}{|\lambda|} \max_{a \le x \le b} \int_a^b |k(x, y)|\, dy.$$

The requirement that $\alpha < 1$ is exactly the assumption (4.2.7). Moreover, the fixed-point iteration

$$u_n(x) = \frac{1}{\lambda} \int_a^b k(x, y)\, u_{n-1}(y)\, dy + \frac{1}{\lambda} f(x), \qquad a \le x \le b, \qquad (4.2.8)$$

for $n = 1, 2, \dots$ can be shown to be equivalent to a truncation of the geometric series for solving (4.2.6). This is left as Exercise 4.2.4.

Nonlinear integral equations of the second kind

Nonlinear integral equations lack the property of linearity. Consequently, we must assume other properties in order to be able to develop a solvability theory for them. We discuss the major form of such equations. The integral equation

$$u(x) = \mu \int_a^b k(x, y, u(y))\, dy + f(x), \qquad a \le x \le b, \qquad (4.2.9)$$

is said to be a *Urysohn integral equation*. Here we assume that

$$f \in C[a,b] \quad \text{and} \quad k \in C([a,b] \times [a,b] \times \mathbb{R}). \qquad (4.2.10)$$

Moreover, we assume k satisfies a *uniform Lipschitz condition* with respect to its third argument:

$$|k(x,y,u_1) - k(x,y,u_2)| \leq M |u_1 - u_2|, \quad a \leq x, y \leq b, \quad u_1, u_2 \in \mathbb{R}. \qquad (4.2.11)$$

Since (4.2.9) is of the form $v = T(v)$, we can introduce the fixed-point iteration

$$u_n(x) = \mu \int_a^b k(x,y,u_{n-1}(y)) \, dy + f(x), \quad a \leq x \leq b, \quad n \geq 1. \qquad (4.2.12)$$

Theorem 4.2.2 *Assume f and k satisfy (4.2.10), (4.2.11). Moreover, assume*

$$|\mu| \, M \, (b - a) < 1.$$

Then the integral equation (4.2.9) has a unique solution $u \in C[a,b]$, and it can be approximated by the iteration method of (4.2.12).

Another well-studied nonlinear integral equation is

$$u(x) = \mu \int_a^b k(x,y) \, h(y,u(y)) \, dy + f(x), \qquad a \leq x \leq b,$$

with $k(x,y), h(y,u)$, and $f(x)$ given. This is called a *Hammerstein integral equation*. These equations are often derived as reformulations of boundary value problems for nonlinear ordinary differential equations. Multivariate generalizations of this equation are obtained as reformulations of boundary value problems for nonlinear elliptic partial differential equations. An interesting nonlinear integral equation that does not fall into the above categories is *Nekrasov's equation*:

$$\theta(x) = \lambda \int_0^\pi L(x,t) \frac{\sin \theta(t)}{1 + 3\lambda \int_0^t \sin \theta(s) \, ds} \, dt, \qquad 0 \leq x \leq \pi, \qquad (4.2.13)$$

$$L(x,t) = \frac{1}{\pi} \log \frac{\sin \frac{1}{2}(x+t)}{\sin \frac{1}{2}(x-t)}.$$

One solution is $\theta(x) \equiv 0$, and it is the non-zero solutions that are of interest. This arises in the study of the profile of water waves on liquids of infinite depth; and the equation involves interesting questions of solutions that bifurcate (see [120, p. 415]).

Nonlinear Volterra integral equations of the second kind

An equation of the form

$$u(t) = \int_a^t k(t, s, u(s))\, ds + f(t), \qquad t \in [a, b] \tag{4.2.14}$$

is called a nonlinear Volterra integral equation of the second kind. When $k(t, s, u)$ depends linearly on u, we get a linear Volterra integral equation, and such equations were investigated earlier in Example 2.3.4 of Section 2.3. The form of the equation (4.2.14) leads naturally to the iterative method

$$u_n(t) = \int_a^t k(t, s, u_{n-1}(s))\, ds + f(t), \qquad t \in [a, b], \quad n \geq 1. \tag{4.2.15}$$

Theorem 4.2.3 *Assume $k(t, s, u)$ is continuous for $a \leq s \leq t \leq b$ and $u \in \mathbb{R}$; and let $f \in C[a, b]$. Furthermore, assume*

$$|k(t, s, u_1) - k(t, s, u_2)| \leq M\, |u_1 - u_2|, \qquad a \leq s \leq t \leq b, \quad u_1, u_2 \in \mathbb{R}$$

for some constant M. Then the integral equation (4.2.14) has a unique solution $x \in C[a, b]$. Moreover, the iterative method (4.2.15) converges for any initial function $u_0 \in C[a, b]$.

Proof. There are at least two approaches to applying the Banach fixed-point theorem to prove the existence of a unique solution of (4.2.14). We give a sketch of the two approaches below, assuming the conditions stated in Theorem 4.2.3. We define the nonlinear integral operator

$$T : C[a, b] \to C[a, b], \qquad Tu(t) \equiv \int_a^t k(t, s, u(s))\, ds + f(t).$$

APPROACH 1. Let us show that for m sufficiently large, the operator T^m is a contraction on $C[a, b]$. For $x, y \in C[a, b]$,

$$Tu(t) - Tv(t) = \int_a^t [k(t, s, u(s)) - k(t, s, v(s))]\, ds.$$

Then

$$|Tu(t) - Tv(t)| \leq M \int_a^t |u(s) - v(s)|\, ds \tag{4.2.16}$$

and

$$|Tu(t) - Tv(t)| \leq M \|u - v\|_\infty (t - a).$$

Since

$$T^2 u(t) - T^2 v(t) = \int_a^t [k(t, s, Tu(s)) - k(t, s, Tv(s))]\, ds,$$

we get

$$|T^2 u(t) - T^2 v(t)| \leq M \int_a^t |Tu(s) - Tv(s)| \, ds$$

$$\leq \frac{[M(t-a)]^2}{2!} \|u - v\|_\infty.$$

By a mathematical induction, we obtain

$$|T^m u(t) - T^m v(t)| \leq \frac{[M(t-a)]^m}{m!} \|u - v\|_\infty.$$

Thus

$$\|T^m u - T^m v\|_\infty \leq \frac{M(b-a)]^m}{m!} \|u - v\|_\infty.$$

Since

$$\frac{[M(b-a)]^m}{m!} \to 0 \quad \text{as} \quad m \to \infty,$$

the operator T^m is a contraction on $C[a, b]$ when m is chosen sufficiently large. By the result in Exercise 4.1.2, the operator T has a unique fixed point in $C[a, b]$ and the iteration sequence converges to the solution. Derivation of error bounds is left as an exercise.

APPROACH 2. Over the space $C[a, b]$, let us introduce the norm

$$|||u||| = \max_{a \leq t \leq b} e^{-\beta t} |u(t)|,$$

which is equivalent to the standard norm $\|u\|_\infty$ on $C[a, b]$. The parameter β is to be chosen such that $\beta > M$. We modify the relation (4.2.16) as follows:

$$e^{-\beta t} |Tu(t) - Tv(t)| \leq M e^{-\beta t} \int_a^t e^{\beta s} e^{-\beta s} |u(s) - v(s)| \, ds.$$

Hence,

$$e^{-\beta t} |Tu(t) - Tv(t)| \leq M e^{-\beta t} |||u - v||| \int_a^t e^{\beta s} \, ds$$

$$= \frac{M}{\beta} e^{-\beta t} (e^{\beta t} - e^{\beta a}) |||u - v|||.$$

Therefore,

$$|||Tu - Tv||| \leq \frac{M}{\beta} |||u - v|||.$$

Since $\beta > M$, the operator T is a contraction on the Banach space $(V, |||\cdot|||)$. Then T has a unique fixed point that is the unique solution of the linear integral equation (4.2.14) and the iteration sequence converges. ∎

We observe that if the stated assumptions are valid over the interval $[a, \infty)$, then the conclusions of Theorem 4.2.3 remain true on $[a, \infty)$. This implies the equation

$$u(t) = \int_a^t k(t, s, u(s)) \, ds + f(t), \qquad t \geq a,$$

has a unique solution $u \in C[a, \infty)$; and for any $b > a$, we have the convergence $\|u - u_n\|_{C[a,b]} \to 0$ as $n \to \infty$ with $\{u_n\} \subseteq C[a, \infty)$ being defined by

$$u_n(t) = \int_a^t k(t, s, u_{n-1}(s)) \, ds + f(t), \qquad t \geq a.$$

Note that although the value of M may increase as $b - a$ increases, the result will remain valid.

4.2.4 Ordinary differential equations in Banach spaces

Let V be a Banach space and consider the initial value problem

$$\begin{cases} u'(t) = f(t, u(t)), & |t - t_0| < c, \\ u(t_0) = z. \end{cases} \qquad (4.2.17)$$

Here $z \in V$ and $f : [t_0 - c, t_0 + c] \times V \to V$ is continuous. For example, f could be an integral operator; and then (4.2.17) would be an "integro-differential equation." The differential equation problem (4.2.17) is equivalent to the integral equation

$$u(t) = z + \int_{t_0}^t f(s, u(s)) \, ds, \qquad |t - t_0| < c, \qquad (4.2.18)$$

which is of the form $u = T(u)$. This leads naturally to the fixed-point iteration method

$$u_n(t) = z + \int_{t_0}^t f(s, u_{n-1}(s)) \, ds, \qquad |t - t_0| < c, \quad n \geq 1. \qquad (4.2.19)$$

Denote, for $b > 0$,

$$Q_b \equiv \{(t, u) \in \mathbb{R} \times V \mid |t - t_0| \leq c, \ \|u - z\| \leq b\}.$$

We have the following existence and solvability theory for (4.2.17). The proof is a straightforward application of Theorem 4.1.3 and the ideas incorporated in the proof of Theorem 4.2.3.

Theorem 4.2.4 (GENERALIZED PICARD-LINDELÖF THEOREM) *Assume $f : Q_b \to V$ is continuous and is uniformly Lipschitz continuous with respect to its second argument:*

$$\|f(t, u) - f(t, v)\| \leq L \|u - v\|, \qquad \forall \, (t, u), (t, v) \in Q_b,$$

where L is a constant independent of t. Let

$$M = \max_{(t,u)\in Q_b} \|f(t,u)\|$$

and

$$c_0 = \min\left\{c, \frac{b}{M}\right\}.$$

Then the initial value problem (4.2.17) has a unique continuously differentiable solution $x(\cdot)$ on $[t_0 - c_0, t_0 + c_0]$; and the iterative method (4.2.19) converges for any initial value u_0 for which $\|z - u_0\| < b$,

$$\max_{|t-t_0|\leq c_0} \|u_n(t) - u(t)\| \to 0 \quad \text{as } n \to \infty.$$

Moreover, with $\alpha = 1 - e^{-L c_0}$, the error

$$\max_{|t-t_0|\leq c_0} \|u_n(t) - u(t)\|e^{-L|t-t_0|}$$

is bounded by each of the following:

$$\frac{\alpha^n}{1-\alpha} \max_{|t-t_0|\leq c_0} \|u_1(t) - u_0(t)\|e^{-L|t-t_0|},$$

$$\frac{\alpha}{1-\alpha} \max_{|t-t_0|\leq c_0} \|u_{n-1}(t) - u_n(t)\|e^{-L|t-t_0|},$$

$$\alpha \max_{|t-t_0|\leq c_0} \|u_{n-1}(t) - u(t)\|e^{-L|t-t_0|}.$$

Exercise 4.2.1 *This exercise illustrates the effect of the reformulation of the equation on the convergence of the iterative method. As an example, we compute the positive square root of 2, which is a root of the equation $x^2 - 2 = 0$. First, reformulating the equation as $x = 2/x$, we obtain an iterative method $x_n = 2/x_{n-1}$. Show that unless $x_0 = \sqrt{2}$, the method is not convergent.*

(Hint: *Compare x_{n+1} with x_{n-1}.*)

Then let us consider another reformulation. Notice that $\sqrt{2} \in [1,2]$ and is a fixed point of the equation

$$x = T(x) \equiv \frac{1}{4}(2 - x^2) + x.$$

Verify that $T : [1,2] \to [1,2]$ and $\max_{1\leq x\leq 2} |T'(x)| = 1/2$. Thus with any $x_0 \in [1,2]$, the iterative method

$$x_n = \frac{1}{4}(2 - x_{n-1}^2) + x_{n-1}, \quad n \geq 1$$

is convergent.

Exercise 4.2.2 *A matrix $A = (a_{ij})$ is called* **diagonally dominant** *if*

$$\sum_{j\neq i} |a_{ij}| < |a_{ii}| \quad \forall i.$$

Apply the Banach fixed-point theorem to show that if A is diagonally dominant, then both the Jacobi method and the Gauss-Seidel method converge.

Exercise 4.2.3 *Prove Theorem 4.2.2. In addition, state error bounds for the iteration (4.2.12) based on (4.1.4)–(4.1.6).*

Exercise 4.2.4 *Show that the iteration (4.2.8) is equivalent to some truncation of the geometric series for (4.2.6). Apply the fixed-point theorem to derive error bounds for the iteration based on (4.1.4)–(4.1.6).*

Exercise 4.2.5 *Derive error bounds for the iteration of Theorem 4.2.3.*

Exercise 4.2.6 *Generalize Theorem 4.2.3 to a system of d Volterra integral equations. Specifically, consider the equation*

$$\mathbf{u}(t) = \int_a^t k(t, s, \mathbf{u}(s)) \, ds + \mathbf{f}(t), \qquad t \in [a, b].$$

In this equation, $\mathbf{u}(t) \in \mathbb{R}^d$ and $k : \mathbb{R} \times \mathbb{R} \times \mathbb{R}^d \to \mathbb{R}^d$. Include error bounds for the corresponding iterative method.

Exercise 4.2.7 *Prove the generalized Picard-Lindelöf theorem.*

Exercise 4.2.8 *Gronwall's inequality provides an upper bound for a continuous function f on $[a, b]$ that satisfies the relation*

$$f(t) \le g(t) + \int_a^t h(s) \, f(s) \, ds, \quad t \in [a, b],$$

where g is continuous and $h \in L^1(a, b)$. Show that

$$f(t) \le g(t) + \int_a^t g(s) \, h(s) \exp\left(\int_s^t h(\tau) \, d\tau\right) ds \quad \forall \, t \in [a, b].$$

Moreover, if g is non-decreasing, then

$$f(t) \le g(t) \exp\left(\int_a^t h(s) \, ds\right) \quad \forall \, t \in [a, b].$$

In the special case when $h(s) = c > 0$, these inequalities reduce to

$$f(t) \le g(t) + c \int_a^t g(s) \, e^{c \, (t-a)} \, ds \quad \forall \, t \in [a, b]$$

and

$$f(t) \le g(t) \, e^{c \, (t-a)} \quad \forall \, t \in [a, b],$$

respectively.

Exercise 4.2.9 *Gronwall's inequality is useful in stability analysis. Let $f : [t_0 - a, t_0 + a] \times V \to V$ be continuous and Lipschitz continuous with respect to u,*

$$\|f(t, u) - f(t, v)\| \le L \, \|u - v\| \quad \forall \, t \in [t_0 - a, t_0 + a], \; u, v \in V.$$

Let r_1 and r_2 be continuous mappings from $[t_0 - a, t_0 + a]$ to V. Let u_1 and u_2 satisfy

$$u_1(t) = f(t, u_1(t)) + r_1(t),$$
$$u_2(t) = f(t, u_2(t)) + r_2(t).$$

Show that

$$\|u_1(t) - u_2(t)\| \leq e^{L\,|t-t_0|} \Big\{ \|u_1(t_0) - u_2(t_0)\|$$

$$+ \max_{|s-t_0| \leq |t-t_0|} \|r_1(s) - r_2(s)\| \Big\}.$$

Thus, the solution of the differential equation depends continuously on the source term r and the initial value.

Gronwall's inequality and its discrete analog are useful also in error estimates of some numerical methods.

4.3   Differential calculus for nonlinear operators

In this section, we generalize the notion of derivatives of real functions to that of operators. General references for this material are [21, Section 2.1], [88, Chap. 17].

4.3.1   Fréchet and Gâteaux derivatives

We first recall the definition of the derivative of a real function. Let I be an interval on $\mathbb{R}$, and x_0 an interior point of I. A function $f : I \to \mathbb{R}$ is differentiable at x_0 if and only if

$$f'(x_0) \equiv \lim_{h \to 0} \frac{f(x_0 + h) - f(x_0)}{h} \text{ exists;} \qquad (4.3.1)$$

or equivalently, for some number a,

$$f(x_0 + h) = f(x_0) + a\,h + o(|h|) \text{ as } h \to 0. \qquad (4.3.2)$$

where we let $f'(x_0) = a$ denote the derivative.

From the eyes of a first-year calculus student, for a real-valued real-variable function, the definition (4.3.1) looks simpler than (4.3.2), though the two definitions are equivalent. Nevertheless, the definition (4.3.2) clearly indicates that the nature of differentiation is (local) linearization. Moreover, the form (4.3.2) can be directly extended to define the derivative of a general operator, while the form (4.3.1) is useful for defining directional or partial derivatives of the operator. We illustrate this by looking at a vector-valued function of several real variables.

Let K be a subset of the space $\mathbb{R}^d$, with $\mathbf{x}_0$ as an interior point. Let $\mathbf{f} : K \to \mathbb{R}^m$. Following (4.3.2), we say $\mathbf{f}$ is differentiable at $\mathbf{x}_0$ if there exists a matrix (linear operator) $A \in \mathbb{R}^{m \times d}$ such that

$$\mathbf{f}(\mathbf{x_0} + \mathbf{h}) = \mathbf{f}(\mathbf{x_0}) + A\mathbf{h} + o(|\mathbf{h}|) \text{ as } \mathbf{h} \to \mathbf{0}, \qquad \mathbf{h} \in \mathbb{R}^d. \qquad (4.3.3)$$

We can show that $A = \nabla \mathbf{f}(\mathbf{x_0})$, the gradient or Jacobian of $\mathbf{f}$ at $\mathbf{x_0}$:

$$A_{i,j} = \frac{\partial f_i}{\partial x_j}.$$

There is a difficulty in extending (4.3.1) for the differentiability: how to extend the meaning of the divided difference $[f(x_0 + h) - f(x_0)]/h$ when h is a vector? On the other hand, (4.3.1) can be extended directly to provide the notion of a directional derivative: We do not linearize the function in all the possible directions of the variable $\mathbf{x}$ approaching $\mathbf{x_0}$; rather, we linearize the function along a fixed direction towards $\mathbf{x_0}$. In this way, we will only need to deal with a vector-valued function of one real variable, and then the divided difference in (4.3.1) makes sense. More precisely, let $\mathbf{h}$ be a fixed vector in $\mathbb{R}^d$, and we consider the function $\mathbf{f}(\mathbf{x_0} + t\,\mathbf{h})$, for $t \in \mathbb{R}$ in a neighborhood of 0. We then say $\mathbf{f}$ is differentiable at $\mathbf{x_0}$ with respect to $\mathbf{h}$, if there is a matrix A such that

$$\lim_{t \to 0} \frac{\mathbf{f}(\mathbf{x_0} + t\,\mathbf{h}) - \mathbf{f}(\mathbf{x_0})}{t} = A\mathbf{h}. \qquad (4.3.4)$$

In case $\|\mathbf{h}\| = 1$, we call the quantity $A\mathbf{h}$ the directional derivative of $\mathbf{f}$ at $\mathbf{x_0}$ along the direction $\mathbf{h}$. We notice that if $\mathbf{f}$ is differentiable at $\mathbf{x_0}$ following the definition (4.3.3), then (4.3.4) is also valid. But the converse is not true: The relation (4.3.4) for any $\mathbf{h} \in \mathbb{R}^d$ does not imply the relation (4.3.3) (see Exercise 4.3.1).

We now turn to the case of an operator $f : K \subseteq V \to W$ between two Banach spaces V and W. Let us adopt the convention that whenever we discuss the differentiability at a point u_0, implicitly we assume u_0 is an interior point of K; by this, we mean there is an $r > 0$ such that

$$B(u_0, r) \equiv \{u \in V \mid \|u - u_0\| \leq r\} \subseteq K.$$

Definition 4.3.1 *The operator f is* Fréchet differentiable *at u_0 if and only if there exists $A \in \mathcal{L}(V, W)$ such that*

$$f(u_0 + h) = f(u_0) + Ah + o(\|h\|), \quad h \to 0. \qquad (4.3.5)$$

The map A is called the Fréchet derivative of f at u_0, and we write $A = f'(u_0)$. The quantity $df(u_0; h) = f'(u_0)h$ is called the Fréchet differential of f at u_0. If f is Fréchet differentiable at all points in $K_0 \subseteq K$, we call $f' : K_0 \subseteq V \to \mathcal{L}(V, W)$ the Fréchet derivative of f on K_0.

Definition 4.3.2 *The operator f is* Gâteaux differentiable *at u_0 if and only if there exists $A \in \mathcal{L}(V, W)$ such that*

$$\lim_{t \to 0} \frac{f(u_0 + t\,h) - f(u_0)}{t} = Ah \quad \forall\, h \in V, \ \|h\| = 1. \qquad (4.3.6)$$

The map A is called the Gâteaux derivative of f at u_0, and we write $A = f'(u_0)$. The quantity $df(u_0; h) = f'(u_0)h$ is called the Gâteaux differential of f at u_0. If f is Gâteaux differentiable at all points in $K_0 \subseteq K$, we call $f' : K_0 \subseteq V \to \mathcal{L}(V, W)$ the Gâteaux derivative of f on K_0.

From the defining relation (4.3.5), we immediately obtain the next result.

Proposition 4.3.3 *If $f'(u_0)$ exists as a Fréchet derivative, then f is continuous at u_0.*

Evidently, the relation (4.3.6) is equivalent to

$$f(u_0 + t h) = f(u_0) + t Ah + o(|t|) \quad \forall h \in V, \ \|h\| = 1.$$

Thus a Fréchet derivative is also the Gâteaux derivative. The converse of this statement is not true, as is shown in Exercise 4.3.1. However, we have the following result.

Proposition 4.3.4 *A Fréchet derivative is also a Gâteaux derivative. Conversely, if the limit in (4.3.6) is uniform with respect to h with $\|h\| = 1$ or if the Gâteaux derivative is continuous at u_0, then the Gâteaux derivative at u_0 is also the Fréchet derivative at u_0.*

Now we present some differentiation rules. If we do not specify the type of derivative, then the result is valid for both the Fréchet derivative and the Gâteaux derivative.

Proposition 4.3.5 (SUM RULE) *If $f, g : K \subseteq V \to W$ are differentiable at u_0, then for any scalars α and β, $\alpha f + \beta g$ is differentiable at u_0 and*

$$(\alpha f + \beta g)'(u_0) = \alpha f'(u_0) + \beta g'(u_0).$$

Proposition 4.3.6 (PRODUCT RULE) *If $f_1 : K \subseteq V \to V_1$ and $f_2 : K \subseteq V \to V_2$ are differentiable at u_0, and $b : V_1 \times V_2 \to W$ is a bounded bilinear form, then the operator $B(u) = b(f_1(u), f_2(u))$ is differentiable at u_0, and*

$$B'(u_0)h = b(f_1'(u)h, f_2(u)) + b(f_1(u), f_2'(u)h) \quad h \in V.$$

Proposition 4.3.7 (CHAIN RULE) *Let $f : K \subseteq U \to V$, $g : L \subseteq V \to W$ be given with $f(K) \subseteq L$. Assume u_0 is an interior point of K, $f(u_0)$ is an interior point of L If $f'(u_0)$ and $g'(f(u_0))$ exist as Fréchet derivatives, then $g \circ f$ is Fréchet differentiable at u_0 and*

$$(g \circ f)'(u_0) = g'(f(u_0))f'(u_0).$$

If $f'(u_0)$ exists as a Gâteaux derivative and $g'(f(u_0))$ exists as a Fréchet derivative, then $g \circ f$ is Gâteaux differentiable at u_0 and the above formula holds.

Let us look at some examples.

Example 4.3.8 *Let $f : V \to W$ be a continuous affine operator,*

$$f(v) = Lv + b,$$

where $L \in \mathcal{L}(V,W)$, $b \in W$, and $v \in V$. Then f is Fréchet differentiable, and $f'(v) = L$ is constant.

Example 4.3.9 *For functions $T : K \subseteq \mathbb{R}^m \to \mathbb{R}^n$, the Fréchet derivative is the $n \times m$ Jacobian matrix evaluated at $v_0 = (x_1, \ldots, x_m)^T$:*

$$T'(v_0) = \left(\frac{\partial T_i(v_0)}{\partial x_j} \right)_{\substack{i=1:n \\ j=1:m}}.$$

Example 4.3.10 *Let $V = W = C[a,b]$ with the maximum norm. Assume $g \in C[a,b]$, $k \in C([a,b] \times [a,b] \times \mathbb{R})$. Then we can define the operator $T : V \to W$ by the formula*

$$T(u)(t) = g(t) + \int_a^b k(t, s, u(s)) \, ds.$$

The integral operator in this is called a Urysohn *integral operator. Let $u_0 \in C[a,b]$ be such that*

$$\frac{\partial k}{\partial u}(t, s, u_0(s)) \in C([a,b]^2).$$

Then T is Fréchet differentiable at u_0, and

$$(T'(u_0)h)(t) = \int_a^b \frac{\partial k}{\partial u}(t, s, u_0(s)) \, h(s) \, ds, \qquad h \in V.$$

The restriction that $k \in C([a,b] \times [a,b] \times \mathbb{R})$ can be relaxed in a number of ways, with the definition of $T'(u_0)$ still valid.

It is possible to introduce Fréchet and Gâteaux derivatives of higher order. For example, the second Fréchet derivative is the derivative of the Fréchet derivative. For $f : K \subseteq V \to W$ differentiable on $K_0 \subseteq K$, the Fréchet derivative is a mapping $f' : K_0 \subseteq V \to W$. If f' is Fréchet differentiable on K_0, then the second Fréchet derivative

$$f'' = (f')' : K_0 \subseteq V \to \mathcal{L}(V, \mathcal{L}(V, W)).$$

At each point $v \in K_0$, the second derivative $f''(v)$ can also be viewed as a bilinear mapping from $V \times V$ to W, and

$$f'' : K_0 \subseteq V \to \mathcal{L}(V \times V, W),$$

and this is generally the way f'' is regarded. Detailed discussions on Fréchet and Gâteaux derivatives, including higher-order derivatives, are given in [88, Section 17.2] and [170, Section 4.5].

4.3.2 Mean value theorems

We need to generalize the Mean Value Theorem for differentiable functions of a real variable. This then allows us to consider the effect on a nonlinear function of perturbations in its argument.

Proposition 4.3.11 *Assume U and V are real linear spaces. Let $F : K \subseteq U \to V$ with K an open set. Assume F is differentiable on K and that $F'(u)$ is a continuous function of u on K to $\mathcal{L}(U, V)$. Let $u, w \in K$ and assume the line segment joining them is also contained in K. Then*

$$\|F(u) - F(w)\|_V \leq \sup_{0 \leq \theta \leq 1} \|F'((1 - \theta)\, u + \theta w)\| \, \|u - w\|_U. \qquad (4.3.7)$$

Proof. Denote $y = F(u) - F(w)$. Using the Hahn-Banach theorem in the form of Corollary 2.5.6, justify the existence of a linear functional $T : V \to \mathbb{R}$ with $\|T\| = 1$ and $T(y) = \|y\|_V$. Introduce the real-valued function

$$g(t) = T(F(tu + (1 - t)\, w)), \qquad 0 \leq t \leq 1.$$

Note that $T(y) = g(1) - g(0)$.

We show g is continuously differentiable on $[0, 1]$ using the chain rule of Proposition 4.3.7. Introduce

$$g_1(t) = tu + (1 - t)\, w, \qquad g_1 : [0, 1] \to V,$$
$$g_2(v) = T(F(v)), \qquad g_2 : K \subseteq V \to \mathbb{R}.$$

For the interval $0 \leq t \leq 1$,

$$g(t) = g_2(g_1(t)),$$
$$g'(t) = g_2'(g_1(t))g_1'(t)$$
$$= [T \circ F'(tu + (1 - t)\, w)]\, (u - w)$$
$$= T\, [F'(tu + (1 - t)\, w)\, (u - w)].$$

According to the ordinary Mean Value Theorem, there is $\theta \in [0, 1]$ for which

$$\|F(u) - F(w)\|_V = g(1) - g(0) = g'(\theta)$$
$$= T\, [F'(\theta u + (1 - \theta)\, w)\, (u - w)]$$
$$\leq \|T\|\, \|F'(\theta u + (1 - \theta)\, w)\, (u - w)\|_W$$
$$\leq \|F'(\theta u + (1 - \theta)\, w)\|\, \|u - w\|_U.$$

The formula (4.3.7) follows immediately. ∎

The following result provides an error bound for the linear Taylor approximation for a nonlinear function. A proof similar to the above can be given for this lemma.

Proposition 4.3.12 *Assume U and V are real linear spaces. Let $F : K \subseteq U \to V$ with K an open set. Assume F is twice continuously differentiable on K, with $F'' : K \to \mathcal{L}(U \times U, V)$. Let $u_0, u_0 + h \in K$ along with the line segment joining them. Then*

$$\|F(u_0 + h) - [F(u_0) + F'(u_0)h]\|_V \leq \frac{1}{2} \sup_{0 \leq \theta \leq 1} \|F''(u_0 + \theta h)\|\, \|h\|_U^2.$$

4.3.3 Partial derivatives

The following definition is for either type of derivatives (Fréchet or Gâteaux).

Definition 4.3.13 *Let U, V, and W be Banach spaces, $f : \mathcal{D}(f) \subseteq U \times V \to W$. For fixed $v_0 \in V$, $f(u, v_0)$ is a function of u whose derivative at u_0, if it exists, is called the partial derivative of f with respect to u, and is denoted by $f_u(u_0, v_0)$. The partial derivative $f_v(u_0, v_0)$ is defined similarly.*

We explore the relation between the Fréchet derivative and partial Fréchet derivatives.

Proposition 4.3.14 *If f is Fréchet differentiable at (u_0, v_0), then the partial Fréchet derivatives $f_u(u_0, v_0)$ and $f_v(u_0, v_0)$ exist, and*

$$f'(u_0, v_0)(h, k) = f_u(u_0, v_0)\, h + f_v(u_0, v_0)\, k, \quad h \in U, \ k \in V. \qquad (4.3.8)$$

Conversely, if $f_u(u, v)$ and $f_v(u, v)$ exist in a neighborhood of (u_0, v_0) and are continuous at (u_0, v_0), then f is Fréchet differentiable at (u_0, v_0), and (4.3.8) holds.

Proof. Assume f is Fréchet differentiable at (u_0, v_0); then

$$f(u_0 + h, v_0 + k) = f(u_0, v_0) + f'(u_0, v_0)(h, k) + o(\|(h, k)\|).$$

Setting $k = 0$, we obtain

$$f(u_0 + h, v_0) = f(u_0, v_0) + f'(u_0, v_0)(h, 0) + o(\|h\|).$$

Therefore, $f_u(u_0, v_0)$ exists and

$$f_u(u_0, v_0)\, h = f'(u_0, v_0)(h, 0).$$

Similarly, $f_v(u_0, v_0)$ exists and

$$f_v(u_0, v_0)\, k = f'(u_0, v_0)(0, k).$$

Adding the two relations, we get (4.3.8).

Now assume $f_u(u, v)$ and $f_v(u, v)$ exist in a neighborhood of (u_0, v_0) and are continuous at (u_0, v_0). We have

$$\|f(u_0 + h, v_0 + k) - [f(u_0, v_0) + f_u(u_0, v_0)\, h + f_v(u_0, v_0)\, k]\|$$
$$\leq \|f(u_0 + h, v_0 + k) - [f(u_0, v_0 + k) + f_u(u_0, v_0 + k)\, h]\|$$
$$+ \|f_u(u_0, v_0 + k)\, h - f_u(u_0, v_0)\, h\|$$
$$+ \|f(u_0, v_0 + k) - [f(u_0, v_0) + f_v(u_0, v_0)\, k]\|$$
$$= o(\|(h, k)\|).$$

Hence, f is Fréchet differentiable at (u_0, v_0). ∎

Corollary 4.3.15 *A mapping $f(u, v)$ is continuously Fréchet differentiable in a neighborhood of (u_0, v_0) if and only if $f_u(u, v)$ and $f_v(u, v)$ are continuous in a neighborhood of (u_0, v_0).*

The above discussion can be extended straightforward to maps of several variables.

4.3.4 The Gâteaux derivative and convex minimization

Let us first use the notion of Gâteaux derivative to characterize the convexity of Gâteaux differentiable functionals.

Theorem 4.3.16 *Let V be a normed space and $K \subseteq V$ be a non-empty convex subset. Assume $f : K \to \mathbb{R}$ is Gâteaux differentiable. Then the following three statements are equivalent.*
(a) *f is convex.*
(b) *$f(v) \geq f(u) + \langle f'(u), v - u \rangle \ \forall \, u, v \in K$.*
(c) *$\langle f'(v) - f'(u), v - u \rangle \geq 0 \ \forall \, u, v \in K$.*

Proof. (a) $\Longrightarrow$ (b). For any $t \in [0, 1]$, by the convexity of f,

$$f(u + t\,(v - u)) \leq t\,f(v) + (1 - t)\,f(u).$$

Then

$$\frac{f(u + t\,(v - u)) - f(u)}{t} \leq f(v) - f(u), \quad t \in (0, 1].$$

Taking the limit $t \to 0+$, we obtain

$$\langle f'(u), v - u \rangle \leq f(v) - f(u).$$

(b) $\Longrightarrow$ (a). For any $u, v \in K$, any $\lambda \in [0, 1]$, we have

$$f(v) \geq f(u + \lambda\,(v - u)) + (1 - \lambda)\,\langle f'(u + \lambda\,(v - u)), v - u \rangle,$$
$$f(u) \geq f(u + \lambda\,(v - u)) + \lambda\,\langle f'(u + \lambda\,(v - u)), u - v \rangle.$$

Multiplying the first inequality by λ, the second inequality by $1 - \lambda$, and adding the two relations, we obtain

$$\lambda\,f(v) + (1 - \lambda)\,f(u) \geq f(u + \lambda\,(v - u)).$$

So f is a convex function.
(b) $\Longrightarrow$ (c). For any $u, v \in K$, we have

$$f(v) \geq f(u) + \langle f'(u), v - u \rangle,$$
$$f(u) \geq f(v) + \langle f'(v), u - v \rangle.$$

Add the two inequalities to obtain

$$0 \geq -\langle f'(v) - f'(u), v - u \rangle.$$

(c) $\Longrightarrow$ (b). Define a real function

$$\phi(t) = f(u + t\,(v - u)), \quad t \in [0, 1].$$

Using Taylor's theorem, we have

$$\phi(1) = \phi(0) + \phi'(\theta) \quad \text{for some } \theta \in (0, 1).$$

Notice that

$$\phi(1) = f(v), \quad \phi(0) = f(u).$$

Also

$$\phi'(\theta) = \langle f'(u + \theta(v - u)), v - u \rangle$$
$$= \frac{1}{\theta} \langle f'(u + \theta(v - u)) - f'(u), v - u \rangle + \langle f'(u), v - u \rangle$$
$$\geq \langle f'(u), v - u \rangle,$$

where in the last step we used the condition (c). Thus (b) holds. ∎

Let us then characterize minimizers of Gâteaux differentiable convex functionals.

Theorem 4.3.17 *Let V be a normed space and $K \subseteq V$ be a non-empty convex subset. Assume $f : K \to \mathbb{R}$ is Gâteaux differentiable. Then*

$$u \in K : \quad f(u) = \inf_{v \in K} f(v) \tag{4.3.9}$$

if and only if

$$u \in K : \quad \langle f'(u), v - u \rangle \geq 0 \quad \forall v \in K. \tag{4.3.10}$$

When K is a subspace, the inequality (4.3.10) reduces to an equality:

$$u \in K : \quad \langle f'(u), v \rangle = 0 \quad \forall v \in K. \tag{4.3.11}$$

Proof. Assume u satisfies (4.3.9). Then for any $t \in (0, 1)$, since $u + t(v - u) \in K$, we have

$$f(u) \leq f(u + t(v - u)).$$

Then we have (4.3.10) by an argument similar to the one used in the proof of Theorem 4.3.16 for the part "(a) $\Longrightarrow$ (b)."

Now assume u satisfies (4.3.10). Then since f is convex,

$$f(v) \geq f(u) + \langle f'(u), v - u \rangle \geq f(u).$$

When K is a subspace, we can take v in (4.3.10) to be $v + u$ for any $v \in K$ to obtain

$$\langle f'(u), v \rangle \geq 0 \quad \forall v \in K.$$

Since K is a subspace, $-v \in K$ and

$$\langle f'(u), -v \rangle \geq 0 \quad \forall v \in K.$$

Therefore, we have the equality (4.3.11). ∎

Exercise 4.3.1 *Let $f : \mathbb{R}^2 \to \mathbb{R}$ be defined by*

$$f(x_1, x_2) = \begin{cases} \dfrac{x_1 x_2^2}{x_1^2 + x_2^4}, & \text{if } (x_1, x_2) \neq (0, 0), \\ 0, & \text{if } (x_1, x_2) = (0, 0). \end{cases}$$

Show that the function is Gâteaux differentiable at $(0,0)$, is discontinuous at $(0,0)$ and hence is not Fréchet differentiable at $(0,0)$ by Proposition 4.3.3.

Exercise 4.3.2 *Prove Proposition 4.3.5.*

Exercise 4.3.3 *Prove Proposition 4.3.6.*

Exercise 4.3.4 *Prove Proposition 4.3.7.*

Exercise 4.3.5 *Prove Proposition 4.3.12*

Exercise 4.3.6 *Let V be a normed space, and $K \subseteq V$ be a convex subset. Assume $f : K \to \mathbb{R}$ is Gâteaux differentiable. Show that if*

$$\langle f'(v) - f'(u), v - u \rangle > 0, \quad u, v \in K, \ v \neq u,$$

then f is strictly convex on K.

Exercise 4.3.7 *Show that for $p \geq 2$, the real-valued function*

$$f(\xi) = \frac{1}{p}(1 + |\xi|^2)^p, \quad \xi \in \mathbb{R}^d$$

is strictly convex (cf. Definition 3.2.2 in Section 3.2).

Exercise 4.3.8 *Let $f : C^1[0,1] \to C[0,1]$ be defined by*

$$f(u) = \left(\frac{du}{dx}\right)^2, \quad u \in C^1[0,1].$$

Calculate $f'(u)$.

Exercise 4.3.9 *Let $A : V \to V$, with V a real Hilbert space. Define*

$$f(u) = \frac{1}{2}(Av, v).$$

Then $f : V \to \mathbb{R}$. Show the existence of the Frechet derivative $f'(u)$ and calculate it.

Exercise 4.3.10 (a) *Find the derivative of the nonlinear operator given in the right-hand side of (4.2.13);*
(b) *Let $u(t) = \sin \theta(t)$, and reformulate (4.2.13) as a new fixed-point problem $u = K(u)$. Find $K'(u)$ for $u = 0$.*

4.4   Newton's method

Let $f : \mathbb{R} \to \mathbb{R}$ be continuously differentiable and consider the equation

$$f(x) = 0.$$

Suppose we know an approximate solution x_n near a root of the equation x^*. Then by the Taylor's expansion,

$$\begin{aligned}
0 = f(x^*) \\
= f(x_n) + f'(x_n)(x^* - x_n) + o(|x^* - x_n|) \\
\approx f(x_n) + f'(x_n)(x^* - x_n).
\end{aligned}$$

Thus

$$x^* \approx x_n - [f'(x_n)]^{-1} f(x_n).$$

This leads to the well-known Newton method for solving the equation $f(x) = 0$:

$$x_{n+1} = x_n - [f'(x_n)]^{-1} f(x_n), \quad n = 0, 1, \dots.$$

In this section, we generalize the Newton method to solving operators equations.

4.4.1 Newton's method in a Banach space

Let U and V be two Banach spaces, $F : U \to V$ be Fréchet differentiable. We are interested in solving the equation

$$F(u) = 0. \tag{4.4.1}$$

The Newton method reads as follows: Choose an initial guess $u_0 \in U$; for $n = 0, 1, \dots$, compute

$$u_{n+1} = u_n - [F'(u_n)]^{-1} F(u_n). \tag{4.4.2}$$

Theorem 4.4.1 (LOCAL CONVERGENCE) *Assume u^* is a root of the equation (4.4.1) such that $[F'(u^*)]^{-1}$ exists and is a continuous linear map from V to U. Assume further that $F'(u)$ is locally Lipschitz continuous at x^*,*

$$\|F'(u) - F'(v)\| \leq L \|u - v\| \quad \forall u, v \in N(u^*),$$

where $N(u^*)$ is a neighborhood of u^*, and $L > 0$ is a constant. Then there exists a $\delta > 0$ such that if $\|u_0 - u^*\| \leq \delta$, the Newton's sequence $\{u_n\}$ is well-defined and converges to u^*. Furthermore, for some constant M we have the error bounds

$$\|u_{n+1} - u^*\| \leq M \|u_n - u^*\|^2$$

and

$$\|u_n - u^*\| \leq (M\delta)^{2^n} / M.$$

Proof. Upon redefining the neighborhood $N(u^*)$ if necessary, we may assume $[F'(u)]^{-1}$ exists on $N(u^*)$ and $c_0 = \sup_{u \in N(u^*)} \|[F'(u)]^{-1}\| < \infty$. Let us define

$$T(u) = u - [F'(u)]^{-1} F(u), \quad u \in N(u^*).$$

Notice that $T(u^*) = u^*$. For $u \in N(u^*)$, we have

$$T(u) - T(u^*) = u - u^* - [F'(u)]^{-1} F(u)$$
$$= [F'(u)]^{-1} \{F(u^*) - F(u) - F'(u)(u^* - u)\}$$
$$= [F'(u)]^{-1} \int_0^1 [F'(u + t(u^* - u)) - F'(u)] dt (u^* - u);$$

and by taking the norm,

$$\|T(u) - T(u^*)\| \leq \|[F'(u)]^{-1}\| \int_0^1 \|F'(u + t(u^* - u)) - F'(u)\| dt \|u^* - u\|$$

$$\leq \|[F'(u)]^{-1}\| \int_0^1 L t \|u^* - u\| dt \|u^* - u\|.$$

Hence,

$$\|T(u) - T(u^*)\| \leq \frac{c_0 L}{2} \|u - u^*\|^2. \tag{4.4.3}$$

Thus, if we choose $\delta < 2/(c_0 L)$ with the property $\overline{B}(u^*, \delta) \subseteq N(u^*)$, then $T : \overline{B}(u^*, \delta) \to \overline{B}(u^*, \delta)$ is an α-contraction with $\alpha = (c_0 L \delta)/2 < 1$. Therefore, by the Banach fixed-point theorem, T has a unique fixed point u^* in $\overline{B}(u^*, \delta)$ and the sequence $\{u_n\}$ converges to u^*. Denote $M = (c_0 L)/2$. From (4.4.3) we get the estimate

$$\|u_{n+1} - u^*\| \leq M \|u_n - u^*\|^2.$$

An inductive application of this inequality leads to

$$M \|u_n - u^*\| \leq (M \|u_0 - u^*\|)^{2^n}.$$

Hence both estimates of the theorem hold. ∎

The theorem clearly shows that the Newton method is locally convergent with quadratic convergence. The main drawback of the result is the dependence of the assumptions on a root of the equation, which is the quantity to be computed. The Kantorovich theory overcomes this difficulty. A proof of the following theorem can be found in [170, p. 210].

Theorem 4.4.2 (KANTOROVICH) *Suppose that*
(a) $F : D(F) \subseteq U \to V$ *is differentiable on an open convex set* $D(F)$, *and the derivative is Lipschitz continuous*

$$\|F'(u) - F'(v)\| \leq L \|u - v\| \quad \forall u, v \in D(F).$$

(b) *For some* $u_0 \in D(F)$, $[F'(u_0)]^{-1}$ *exists and is a continuous operator from* V *to* U, *and such that* $h = a b L \leq 1/2$ *for some* $a \geq \|[F'(u_0)]^{-1}\|$ *and* $b \geq \|[F'(u_0)]^{-1} F(u_0)\|$. *Denote*

$$t^* = \frac{1 - (1 - 2h)^{1/2}}{a L}, \quad t^{**} = \frac{1 + (1 - 2h)^{1/2}}{a L}.$$

(c) u_1 is chosen so that $\overline{B}(u_1, r) \subseteq D(F)$, where $r = t^* - b$.
Then the equation (4.4.1) has a solution $u^* \in \overline{B}(u_1, r)$ and the solution is
unique in $\overline{B}(u_0, t^{**}) \cap D(F)$; the sequence $\{u_n\}$ converges to u^*, and we
have the error estimate

$$\|u_n - u^*\| \leq \frac{(1 - (1 - 2h)^{1/2} 2^n}{2^n a L}, \quad n = 0, 1, \ldots.$$

The Kantorovich theorem provides sufficient conditions for the convergence of the Newton method. These conditions are usually difficult to verify.
Nevertheless, at least theoretically, the result is of great importance. For
other related discussions of Newton's method, see [21, pp. 116–118] and
[88, Chap. 18].

4.4.2 Applications

Nonlinear systems

Let $F : \mathbb{R}^d \to \mathbb{R}^d$ be a continuously differentiable function. A nonlinear
system is of the form

$$\mathbf{x} \in \mathbb{R}^d, \quad F(\mathbf{x}) = \mathbf{0}.$$

Then the Newton method is

$$\mathbf{x}_{n+1} = \mathbf{x}_n - [F'(\mathbf{x}_n)]^{-1} F(\mathbf{x}_n), \tag{4.4.4}$$

which can also be written in the form

$$F'(\mathbf{x}_n)\, \boldsymbol{\delta}_n = -F(\mathbf{x}_n), \quad \mathbf{x}_{n+1} = \mathbf{x}_n + \boldsymbol{\delta}_n.$$

So at each step, we solve a linear system. The method breaks down when
$F'(\mathbf{x}_n)$ is singular or nearly singular.

Nonlinear integral equations

Consider the nonlinear integral equation

$$u(t) = \int_0^1 k(t, s, u(s))\, ds \tag{4.4.5}$$

over the space $U = C[0, 1]$. Assume $k \in C([0, 1] \times [0, 1] \times \mathbb{R})$ and is continuously differentiable with respect to its third argument. Introducing an
operator $F : U \to U$ through the formula

$$F(u)(t) = u(t) - \int_0^1 k(t, s, u(s))\, ds, \quad t \in [0, 1],$$

the integral equation can be written in the form

$$F(u) = 0.$$

Newton's method for the problem is

$$u_{n+1} = u_n - [F'(u_n)]^{-1} F(u_n),$$

or equivalently,

$$F'(u_n)(u_{n+1} - u_n) = -F(u_n). \tag{4.4.6}$$

Let us compute the derivative of F.

$$F'(u)(v)(t) = \lim_{h \to 0} \frac{1}{h} \left[F(u + hv)(t) - F(u)(t) \right]$$

$$= \lim_{h \to 0} \frac{1}{h} \left[h\,v(t) - \int_0^1 (k(t, s, u(s) + h\,v(s)) - k(t, s, u(s)))\, ds \right]$$

$$= v(t) + \int_0^1 \frac{\partial k(t, s, u(s))}{\partial u}\, v(s)\, ds.$$

Therefore, the Newton iteration formula is

$$\delta_{n+1}(t) - \int_0^1 \frac{\partial k(t, s, u_n(s))}{\partial u} \delta_{n+1}(s)\, ds$$

$$= -u_n(t) + \int_0^1 k(t, s, u_n(s))\, u_n(s)\, ds, \tag{4.4.7}$$

$$u_{n+1}(t) = u_n(t) + \delta_{n+1}(t).$$

At each step, we solve a linear integral equation.

It is often faster computationally to use a modification of (4.4.6), using an fixed value of the derivative:

$$F'(u_0)(u_{n+1} - u_n) = -F(u_n). \tag{4.4.8}$$

The iteration formula is now

$$\delta_{n+1}(t) - \int_0^1 \frac{\partial k(t, s, u_0(s))}{\partial u} \delta_{n+1}(s)\, ds$$

$$= -u_n(t) + \int_0^1 k(t, s, u_n(s))\, u_n(s)\, ds, \tag{4.4.9}$$

$$u_{n+1}(t) = u_n(t) + \delta_{n+1}(t).$$

This converges more slowly; but the lack of change in the integral equation (since only the right side is varying) often leads to less computation than with (4.4.7).

Nonlinear differential equations

As a sample problem, we consider

$$\begin{cases} u''(t) = f(t, u(t)), & t \in (0, 1), \\ u(0) = u(1) = 0. \end{cases}$$

Here $f : [0, 1] \times \mathbb{R}$ is assumed to be continuous and continuously differentiable with respect to u. We take

$$U = C_0^2[0, 1] = \{ v \in C^2[0, 1] \mid v(0) = v(1) = 0 \}$$

with the norm $\| \cdot \|_{C^2[0,1]}$. Define

$$F(u)(t) = u''(t) - f(t, u(t)), \qquad t \in [0, 1].$$

It can be shown that

$$F'(u)(y)(t) = y''(t) - \frac{\partial f(t, u(t))}{\partial u} \, y(t).$$

Thus at each step, we solve a linearized boundary value problem

$$\begin{cases} u''_{n+1}(t) - \dfrac{\partial f}{\partial u}(t, u_n(t)) \, u_{n+1}(t) = f(t, u_n(t)) - \dfrac{\partial f}{\partial u}(t, u_n(t)) \, u_n(t), \\ \qquad\qquad t \in (0, 1), \\ u_{n+1}(0) = u_{n+1}(1) = 0. \end{cases}$$

Exercise 4.4.1 *Explore sufficient conditions for the convergence of the Newton method* (4.4.4).

Exercise 4.4.2 *Explore sufficient conditions for the convergence of the Newton method* (4.4.7).

4.5   Completely continuous vector fields

There are other means of asserting the existence of a solution to an equation. For example, if $f \in C[a, b]$, and if $f(a) \, f(b) < 0$, then the intermediate value theorem asserts the existence of a solution in $[a, b]$ to the equation $f(x) = 0$. We convert this to an existence theorem for fixed points as follows. Let $T : [a, b] \to [a, b]$ be continuous. Then $x = T(x)$ has a solution in $[a, b]$. This can be proved by reducing it to the earlier case, letting $f(x) \equiv x - T(x)$.

It is natural to try to extend this to multivariate functions f or T.

Theorem 4.5.1 (BROUWER'S FIXED-POINT THEOREM) *Let $K \subseteq \mathbb{R}^d$ be bounded, closed, and convex. Let $T : K \to K$ be continuous. Then T has at least one fixed point in the set K.*

For a proof, see [98, p. 94]; [88, pp. 636–639]; or [53, p. 232].

We would like to generalize this further, to operators on infinite-dimensional Banach spaces. There are several ways of doing this, and we describe two approaches, both based on the assumption that T is a *continuous compact operator*. This is a concept we generalize to nonlinear operators T from Definition 2.8.1 for linear operators. We begin with an example to show that some additional hypotheses are needed, in addition to those assumed in Theorem 4.5.1.

Example 4.5.2 *Let V be a Hilbert space with an orthonormal basis $\{\varphi_j\}_{j\geq 1}$. Then for every $v \in V$, we can write*

$$v = \sum_{j=1}^{\infty} \alpha_j \varphi_j, \qquad \|v\|_V = \sqrt{\sum_{j=1}^{\infty} |\alpha_j|^2}.$$

Let K be the unit ball in V,

$$K = \{v \mid \|v\|_V \leq 1\}.$$

Introduce a parameter $k > 1$, and then choose a second parameter $t < 1$ satisfying

$$0 < t \leq \sqrt{k^2 - 1}.$$

Define $T : K \to K$ by

$$T(v) = t(1 - \|v\|_V)\varphi_1 + \sum_{j=1}^{\infty} \alpha_j \varphi_{j+1}, \qquad v \in K. \tag{4.5.1}$$

This can be shown to be Lipschitz continuous on K, with

$$\|T(v) - T(w)\|_V \leq k \|v - w\|_V, \quad v, w \in K. \tag{4.5.2}$$

Moreover, the domain K is convex, closed, and bounded. However, T does not have a fixed point. This example is a modification of one given in [86].

Definition 4.5.3 *Let $T : K \subseteq V \to W$, with V and W Banach spaces. We say T is compact if for every bounded set $B \subseteq K$, the set $T(B)$ has compact closure in W. If T is both compact and continuous, we call T a completely continuous operator.*

When T is a linear operator, T compact implies T is bounded and hence continuous. This is not true in general when T is nonlinear; continuity of T must be assumed separately. Some authors include a requirement of continuity in their definition of T being compact, e.g., [21, p. 89]. With the above definition, we can state one generalization of Theorem 4.5.1. For proofs, see [21, p. 90], [88, p. 482], or [98, p. 124].

Theorem 4.5.4 (SCHAUDER'S FIXED-POINT THEOREM) *Let V be a Banach space and let $K \subseteq V$ be bounded, closed, and convex. Let $T : K \to K$ be a completely continuous operator. Then T has at least one fixed point in the set K.*

When dealing with equations involving differentiable nonlinear functions, say,

$$v = T(v), \tag{4.5.3}$$

a common approach is to "linearize the problem." This generally means we replace the nonlinear function by a linear Taylor series approximation,

$$T(v) \approx T(v_0) + T'(v_0)(v - v_0) \qquad (4.5.4)$$

for some suitably chosen point v_0. Then the equation (4.5.3) can be rewritten as

$$(I - T'(v_0))(v - v_0) \approx T(v_0) - v_0. \qquad (4.5.5)$$

This linearization procedure is a commonly used approach for the convergence analysis of approximation methods for solving (4.5.3), and it is used to this end in Section 11.6.

This leads us to consider the properties of $T'(v_0)$ and motivates consideration of the following result. A proof is given in [99, p. 77]. As a consequence, we can apply the Fredholm alternative theorem to the operator $I - T'(v_0)$.

Proposition 4.5.5 *Let V be a Banach space and let $K \subseteq V$ be an open set. Let $T : K \to V$ be a completely continuous operator that is differentiable at $v_0 \in K$. Then $T'(v_0)$ is a compact operator from V to V.*

4.5.1   The rotation of a completely continuous vector field

The concept of the rotation of a nonlinear mapping is a fairly deep and sophisticated consequence of topology, and a complete development of it is given in [98, Chap. 2]. We describe here the main properties of this "rotation."

Let $T : K \subseteq V \to V$, with V a Banach space, and assume T is completely continuous on K. We call the function

$$\Phi(v) \equiv v - T(v), \qquad v \in K$$

the *completely continuous vector field generated by T* (or Φ). Let B be a bounded, open subset of K and let S denote its boundary, and assume $\overline{B} \equiv B \cup S \subseteq K$. *Assume T has no fixed points on the boundary S.* Under the above assumptions, it is possible to define *the rotation of T* (or Φ) *over S.* This is an integer, denoted here by $\mathrm{Rot}(\Phi)$ with the following properties.

P1 If $\mathrm{Rot}(\Phi) \neq 0$, then T has at least one fixed point within the set B. (See [98, p. 123].)

P2 Assume there is a function $X(v, t)$ defined for $v \in \overline{B}$ and $0 \leq t \leq 1$, and assume it has the following properties.

(a) $X(v, 0) \equiv \Phi(v)$, $v \in \overline{B}$.
(b) $X(\cdot, t)$ is completely continuous on $\overline{B}$ for each $t \in [0, 1]$.
(c) For every $v \in S$, $X(v, t)$ is uniformly continuous in t.
(d) $v - X(v, t) \neq 0$ for all $v \in S$ and for $0 \leq t \leq 1$.

Then $\text{Rot}(\Phi) = \text{Rot}(\Psi)$, where $\Psi(v) \equiv v - X(v, 1)$. The mapping X is called a *homotopy*, and this property says *the rotation of a completely continuous vector field is invariant under homotopy.* (See [98, p. 108].)

P3 Let v_0 be an isolated fixed point of T in K. Then for all sufficiently small neighborhoods of v_0, $\text{Rot}(\Phi)$ over that neighborhood is constant; it is called the *index of the fixed point v_0*. If all fixed points of T on B are isolated, then the number of such fixed points is finite; call them $v_1, \ldots, v_r$. Moreover, $\text{Rot}(\Phi)$ equals the sum of the indexes of the individual fixed points $v_1, \ldots, v_r$. (See [98, p. 109].)

P4 Let v_0 be a fixed point of T and suppose that T has a continuous Fréchet derivative $T'(v)$ for all v in some neighborhood of v_0. In addition, assume 1 is not an eigenvalue of $T'(v_0)$. Then the index of v_0 in non-zero. More precisely, it equals $(-1)^\beta$ with β equal to the number of positive real eigenvalues of $T'(v_0)$ that are greater than 1, counted according to their multiplicity. Also, the fixed point v_0 is isolated. (See [98, p. 136].)

P5 Let v_0 be an isolated fixed point of T in B. Then the index of v_0 is zero if and only if there exists some open neighborhood N of v_0 such that for every $\delta > 0$, there exists completely continuous T_δ defined on $\overline{N}$ to V with

$$\|T(v) - T_\delta(v)\|_V \leq \delta, \qquad v \in \overline{N},$$

and with T_δ having no fixed points in $\overline{N}$. This says that isolated fixed points have index zero if and only if they are unstable with respect to completely continuous perturbations.

These ideas give a framework for the error analysis of numerical methods for solving some nonlinear integral equations and other problems. Such methods are examined in Subsection 11.6.2 of Chapter 11.

Exercise 4.5.1 (a) *Show that $T : B \to B$ in Example 4.5.2.*
(b) *Show the inequality (4.5.2).*
(c) *Show T does not have a fixed point in B.*

4.6   Conjugate Gradient Iteration

The *conjugate gradient method* is an iteration method that was originally devised for solving finite linear systems that were symmetric and positive definite. The method has since been generalized in a number of directions, and in this section, we consider its generalization to operator equations

$$Au = f. \tag{4.6.1}$$

In this, A is a bounded, positive-definite, self-adjoint, and invertible linear operator on a Hilbert space V. With these assumptions, (4.6.1) has a unique solution $u^* = A^{-1}f$. For simplicity in this section, we assume V is a real Hilbert space; and we further assume that V is *separable*, implying that it has a countable orthogonal basis.

The conjugate gradient method for solving $Au = f$ in V is defined as follows. Let u_0 be an initial guess for the solution u^*. Define $r_0 = f - Au_0$ and $s_0 = r_0$. For $k \geq 0$, define

$$
\begin{aligned}
u_{k+1} &= u_k + \alpha_k s_k & \alpha_k &= \frac{\|r_k\|^2}{(As_k, s_k)}, \\
r_{k+1} &= f - Au_{k+1} & & \\
s_{k+1} &= r_{k+1} + \beta_k s_k & \beta_k &= \frac{\|r_{k+1}\|^2}{\|r_k\|^2}.
\end{aligned}
\tag{4.6.2}
$$

The norm and inner product are those of V. There are several other equivalent formulations of the method. An introduction to the conjugate gradient method for finite-dimensional systems, together with some other equivalent ways of writing it, is given in [11, Section 8.9] and [61, Section 10.2].

The following theorem is taken from [129, p. 159]; and we omit the proof, as it calls for a more detailed investigation of the method than we wish to consider here. The proof also follows quite closely the proof for the finite-dimensional case, which is well known in the literature (e.g., see [106, p. 250]). In stating the theorem, we also use the following alternative inner product and norm:

$$
(v, u)_A = (Av, u), \qquad \|v\|_A = \sqrt{(v, v)_A}.
$$

Theorem 4.6.1 *Let A be a bounded, self-adjoint, linear operator satisfying*

$$
\sqrt{m}\, \|v\| \leq \|v\|_A \leq \sqrt{M}\, \|v\|, \qquad v \in V,
\tag{4.6.3}
$$

with $m, M > 0$ (which means that $\|\cdot\|_A$ and $\|\cdot\|$ are equivalent norms). Then the sequence $\{u_k\}$ from (4.6.2) converges to u^*, and

$$
\|u^* - u_{k+1}\|_A \leq \frac{M - m}{M + m} \|u^* - u_k\|_A, \qquad k \geq 0.
\tag{4.6.4}
$$

This shows $u_k \to u^*$ linearly.

Patterson [129, p. 163] also derives the improved result

$$
\|u^* - u_k\|_A \leq 2 \left(\frac{\sqrt{M} - \sqrt{m}}{\sqrt{M} + \sqrt{m}} \right)^k \|u^* - u_0\|_A, \qquad k \geq 0.
\tag{4.6.5}
$$

It follows that this is a more rapid rate of convergence by showing

$$
\frac{\sqrt{M} - \sqrt{m}}{\sqrt{M} + \sqrt{m}} \leq \frac{M - m}{M + m},
\tag{4.6.6}
$$

which we leave to the reader.

In the case that $A = I - K$, with K a compact operator, more can be said about the rate of convergence of u_k to u^*. For the remainder of this section, we assume K is a compact, self-adjoint operator on V to V.

The discussion of the convergence requires results on the eigenvalues of K. From Theorem 2.8.15 in Section 2.8.5, the eigenvalues of the self-adjoint compact operator K are real and the associated eigenvectors can be so chosen as to form an orthonormal basis for V:

$$K\phi_j = \lambda_j \phi_j, \qquad j = 1, 2, \dots$$

with $(\phi_i, \phi_j) = \delta_{i,j}$. Without any loss of generality, let the eigenvalues be ordered as follows:

$$|\lambda_1| \geq |\lambda_2| \geq |\lambda_3| \geq \cdots \geq 0. \tag{4.6.7}$$

We permit the number of non-zero eigenvalues to be finite or infinite. From Theorem 2.8.12 of Section 2.8.5,

$$\lim_{j \to \infty} \lambda_j = 0.$$

The eigenvalues of A are $\{1 - \lambda_j\}$ with $\{\phi_j\}$ as the corresponding orthogonal eigenvectors. The self-adjoint operator $A = I - K$ is positive definite if and only if

$$\delta \equiv \inf_{j \geq 1} (1 - \lambda_j) = 1 - \sup_{j \geq 1} \lambda_j > 0, \tag{4.6.8}$$

or equivalently, $\lambda_j < 1$ for all $j \geq 1$. For later use, also introduce

$$\Delta \equiv \sup_{j \geq 1} (1 - \lambda_j).$$

We note that

$$\|A\| = \Delta, \qquad \|A^{-1}\| = \frac{1}{\delta}. \tag{4.6.9}$$

With respect to (4.6.3), $M = \Delta$ and $m = \delta$. The result (4.6.4) becomes

$$\|u^* - u_{k+1}\|_A \leq \left(\frac{\Delta - \delta}{\Delta + \delta}\right) \|u^* - u_k\|_A, \qquad k \geq 0. \tag{4.6.10}$$

It is possible to improve on this geometric rate of convergence when dealing with equations

$$Au \equiv (I - K)u = f$$

to show a "superlinear" rate of convergence. The following result is due to Winther [166].

Theorem 4.6.2 *Let K be a self-adjoint compact operator on the Hilbert space V. Assume $A = I - K$ is a self-adjoint positive definite operator (with the notation used above). Let $\{u_k\}$ be generated by the conjugate gradient* *iteration (4.6.2). Then* $u_k \to u^*$ *superlinearly:*

$$\|u^* - u_k\| \le (c_k)^k \|u^* - u_0\|, \qquad k \ge 0 \qquad (4.6.11)$$

with $\lim\limits_{k \to \infty} c_k = 0.$

Proof. It is a standard result (cf. [106, p. 246]) that (4.6.2) implies that

$$u_k = u_0 + \widetilde{P}_{k-1}(A)r_0 \qquad (4.6.12)$$

with $\widetilde{P}_{k-1}(\lambda)$ a polynomial of degree $\le k - 1$. Letting $A = I - K$, this can be rewritten in the equivalent form

$$u_k = u_0 + \widehat{P}_{k-1}(K)r_0$$

for some other polynomial $\widehat{P}(\lambda)$ of degree $\le k - 1$. The conjugate gradient iterates satisfy an optimality property: If $\{y_k\}$ is another sequence of iterates, generated by another sequence of the form

$$y_k = u_0 + P_{k-1}(K)r_0, \qquad k \ge 1, \qquad (4.6.13)$$

for some sequence of polynomials $\{P_{k-1} : \deg(P_{k-1}) \le k - 1, \; k \ge 1\}$, then

$$\|u^* - u_k\|_A \le \|u^* - y_k\|_A, \qquad k \ge 0. \qquad (4.6.14)$$

For a proof, see Luenberger [106, p. 247].

Introduce

$$Q_k(\lambda) = \prod_{j=1}^{k} \frac{\lambda - \lambda_j}{1 - \lambda_j}, \qquad (4.6.15)$$

and note that $Q_k(1) = 1$. Define P_{k-1} implicitly by

$$Q_k(\lambda) = 1 - (1 - \lambda)P_{k-1}(\lambda),$$

and note that $\text{degree}(P_{k-1}) = k - 1$. Let $\{y_k\}$ be defined using (4.6.13). Define $\widetilde{e}_k = u^* - y_k$ and

$$\widetilde{r}_k = b - Ay_k = A\widetilde{e}_k.$$

We first bound $\widetilde{r}_k$, and then

$$\|\widetilde{e}_k\| \le \|A^{-1}\| \, \|\widetilde{r}_k\| = \frac{1}{\delta} \|\widetilde{r}_k\|.$$

Moreover,

$$\begin{aligned}
\|u^* - u_k\|_A &\le \|u^* - y_k\|_A \\
&\le \sqrt{\Delta} \, \|u^* - y_k\| \\
&\le \frac{\sqrt{\Delta}}{\delta} \|\widetilde{r}_k\|,
\end{aligned}$$

$$\|u^* - u_k\| \le \frac{1}{\sqrt{\delta}} \|u^* - u_k\|_A \le \frac{1}{\delta}\sqrt{\frac{\Delta}{\delta}} \|\widetilde{r}_k\|. \qquad (4.6.16)$$

From (4.6.13),

$$\begin{aligned}
\widetilde{r}_k &= b - A\left[y_0 + P_{k-1}(K)r_0\right] \\
&= \left[I - AP_{k-1}(K)\right]r_0 \\
&= Q_k(A)r_0.
\end{aligned} \tag{4.6.17}$$

Expand r_0 using the eigenfunction basis $\{\phi_1, \phi_2, \dots\}$:

$$r_0 = \sum_{j=1}^{\infty}(r_0, \phi_j)\phi_j.$$

Note that

$$Q_k(A)\phi_j = Q_k(\lambda_j)\phi_j, \qquad j \geq 1,$$

and thus $Q_k(A)\phi_j = 0$ for $j = 1, \dots, k$. Then (4.6.17) implies

$$\widetilde{r}_k = \sum_{j=1}^{\infty}(r_0, \phi_j)Q_k(A)\phi_j = \sum_{j=k+1}^{\infty}(r_0, \phi_j)Q_k(\lambda_j)\phi_j$$

and

$$\|\widetilde{r}_k\| \leq \alpha_k \sqrt{\sum_{j=k+1}^{\infty}(r_0, \phi_j)^2} \leq \alpha_k \|r_0\| \tag{4.6.18}$$

with

$$\alpha_k = \sup_{j \geq k+1}|Q_k(\lambda_j)|.$$

Examining $Q_k(\lambda_j)$ and using (4.6.7), we have

$$\alpha_k \leq \prod_{j=1}^{k}\frac{|\lambda_{k+1}| + |\lambda_j|}{1 - \lambda_j} \leq \prod_{j=1}^{k}\frac{2|\lambda_j|}{1 - \lambda_j}. \tag{4.6.19}$$

Recall the well known inequality

$$\left[\prod_{j=1}^{k}b_j\right]^{\frac{1}{k}} \leq \frac{1}{k}\sum_{j=1}^{k}b_j,$$

which relates the arithmetic and geometric means of k positive numbers $b_1, \dots, b_k$. Applying this, we have

$$\alpha_k \leq \left[\frac{2}{k}\sum_{j=1}^{k}\frac{|\lambda_j|}{1 - \lambda_j}\right]^k.$$

Since $\lim\limits_{j \to \infty} \lambda_j = 0$, it is a straightforward argument to show that

$$\lim_{k \to \infty} \frac{2}{k} \sum_{j=1}^{k} \frac{|\lambda_j|}{1 - \lambda_j} = 0. \tag{4.6.20}$$

We leave the proof to the reader.

Returning to (4.6.18) and (4.6.16), we have

$$\|u^* - u_k\| \le \frac{1}{\delta} \sqrt{\frac{\Delta}{\delta}} \, \|\widetilde{r}_k\|$$

$$\le \frac{1}{\delta} \sqrt{\frac{\Delta}{\delta}} \alpha_k \, \|r_0\|$$

$$\le \left(\frac{\Delta}{\delta}\right)^{\frac{3}{2}} \alpha_k \, \|u^* - u_0\|.$$

To obtain (4.6.11), define

$$c_k = \left(\frac{\Delta}{\delta}\right)^{\frac{3}{2k}} \left[\frac{2}{k} \sum_{j=1}^{k} \frac{|\lambda_j|}{1 - \lambda_j}\right]. \tag{4.6.21}$$

From (4.6.20), $c_k \to 0$ as $k \to \infty$.   ∎

It is of interest to know how rapidly c_k converges to zero. For this we restrict ourselves to compact integral operators. For simplicity, we consider only the single-variable case:

$$Kv(x) = \int_a^b k(x, y) v(y) \, dy, \quad x \in [a, b], \quad v \in L^2(a, b), \tag{4.6.22}$$

and $V = L^2(a, b)$. From (4.6.21), the speed of convergence of $c_k \to 0$ is essentially the same as that of

$$\tau_k \equiv \frac{1}{k} \sum_{j=1}^{k} \frac{|\lambda_j|}{1 - \lambda_j}. \tag{4.6.23}$$

In turn, the convergence of τ_k depends on the rate at which the eigenvalues λ_j converge to zero. We give two results from Flores [52] in the following theorem. In all cases, we also assume the operator $A = I - K$ is positive definite, which is equivalent to the assumption (4.6.8).

Theorem 4.6.3 (a) *Assume the integral operator K of (4.6.22) is a self-adjoint Hilbert-Schmidt integral operator, i.e.,*

$$\|K\|_{HS}^2 \equiv \int_D \int_D |k(t, s)|^2 \, ds \, dt < \infty.$$

Then

$$\frac{1}{\ell} \cdot \frac{|\lambda_1|}{1 - \lambda_1} \leq \tau_\ell \leq \frac{1}{\sqrt{\ell}} \|K\|_{HS} \|(I - K)^{-1}\|. \tag{4.6.24}$$

(b) *Assume $k(t, s)$ is a symmetric kernel with continuous partial derivatives of order up to p for some $p \geq 1$. Then there is a constant $M \equiv M(p)$ with*

$$\tau_\ell \leq \frac{M}{\ell} \zeta(p + \tfrac{1}{2}) \|(I - K)^{-1}\|, \qquad \ell \geq 1 \tag{4.6.25}$$

with $\zeta(z)$ the Riemann zeta function.

Proof. (a) It can be proven from Theorem 2.8.15 of Section 2.8.5 that

$$\sum_{j=1}^{\infty} \lambda_j^2 = \|K\|_{HS}^2.$$

From this, the eigenvalues λ_j can be shown to converge to zero with a certain speed. Namely,

$$j\lambda_j^2 \leq \sum_{i=1}^{j} \lambda_i^2 \leq \|K\|_{HS}^2,$$

$$\lambda_j \leq \frac{1}{\sqrt{j}} \|K\|_{HS}, \qquad j \geq 1.$$

This leads to

$$\begin{aligned}
\tau_\ell &= \frac{1}{\ell} \sum_{j=1}^{\ell} \frac{|\lambda_j|}{1 - \lambda_j} \\
&\leq \frac{1}{\delta} \frac{\|K\|_{HS}}{\ell} \sum_{j=1}^{\ell} \frac{1}{\sqrt{j}} \\
&\leq \frac{1}{\sqrt{\ell}} \frac{\|K\|_{HS}}{\delta}.
\end{aligned}$$

Recalling that $\delta^{-1} = \|A^{-1}\|$ proves the upper bound in (4.6.24). The lower bound in (4.6.24) is immediate from the definition of τ_ℓ.
(b) From Fenyö and Stolle [50, Sec. 8.9], the eigenvalues $\{\lambda_j\}$ satisfy

$$\lim_{j \to \infty} j^{p+0.5} \lambda_j = 0.$$

Let

$$\beta \equiv \sup j^{p+0.5} |\lambda_j|,$$

so that

$$|\lambda_j| \leq \frac{\beta}{j^{p+0.5}}, \qquad j \geq 1. \tag{4.6.26}$$

With this bound on the eigenvalues,

$$\tau_\ell \le \frac{\beta}{\ell\delta} \sum_{j=1}^{\ell} \frac{1}{j^{p+0.5}} \le \frac{\beta}{\ell\delta}\zeta(p+0.5).$$

This completes the proof of (4.6.25).   ∎

We see that the speed of convergence of $\{\tau_\ell\}$ (or equivalently, $\{c_\ell\}$) is no better than $O(\ell^{-1})$, regardless of the differentiability of the kernel function k. Moreover, for most cases of practical interest, it is no worse than $O(\ell^{-0.5})$. The result (4.6.11) was only a bound for the speed of convergence of the conjugate gradient method, although we expect that the convergence speed is no better than this. For additional discussion of the convergence of $\{\tau_\ell\}$, see [52].

The result (4.6.12) says that the vectors $u_k - u_0$ belong to the *Krylov subspace*

$$\mathcal{K}(A) \equiv \left\{r_0, Ar_0, A^2 r_0, \ldots, A^{k-1} r_0\right\};$$

and in fact, u_k is an optimal choice in the following sense:

$$\|u^* - u_k\|_A = \min_{y \in u_0 + \mathcal{K}(A)} \|u^* - y\|_A.$$

Other iteration methods have been based on choosing particular elements from $\mathcal{K}(A)$ using a different sense of optimality. For a general discussion of such generalizations to nonsymmetric finite linear systems, see [56]. The conjugate gradient method has also be extended to the solution of nonlinear systems and nonlinear optimization problems.

Exercise 4.6.1 *Prove the inequality* (4.6.6).

Exercise 4.6.2 *Derive the relation* (4.6.12).

Exercise 4.6.3 *Prove the result in* (4.6.20).

Suggestion for Further Readings

Many books and articles cover the convergence issue of iterative methods for finite linear systems; see, e.g., AXELSSON [18], DEMMEL [43], FREUND, GOLUB, AND NACHTIGAL [56], GOLUB AND VAN LOAN [61], KELLEY [91, 92], STEWART [152], and TREFETHEN AND BAU [160]. For finite-dimensional nonlinear systems, see the comprehensive work of ORTEGA AND RHEINBOLDT [127]. For optimization problems, a comprehensive reference is LUENBERGER [106].

Portions of this chapter follow ZEIDLER [170]. A more theoretical look at iteration methods for solving linear equations is given in NEVANLINNA [124]. The iterative solution of linear integral equations of the second kind

is treated in several books, including ATKINSON [13, Chap. 6] and KRESS
[100]. There are many other tools for the analysis of nonlinear problems,
and we refer the reader to BERGER [21], FRANKLIN [53], KANTOROVICH
AND AKILOV [88, Chaps. 16–18], KRASNOSELSKII [98], KRASNOSELSKII
AND ZABREYKO [99], and ZEIDLER [170].

5
Finite Difference Method

The finite difference method is a universally applicable numerical method for the solution of differential equations. In this chapter, for a sample parabolic partial differential equation, we introduce some difference schemes and analyze their convergence. We present the well-known Lax equivalence theorem and related theoretical results, and we apply them to the convergence analysis of difference schemes.

The finite difference method can be difficult to analyze, in part because it is quite general in its applicability. Much of the existing stability and convergence analysis is restricted to special cases, particularly to linear differential equations with constant coefficients. These results are then used to predict the behavior of difference methods for more complicated equations.

5.1   Finite difference approximations

The basic idea of the finite difference method is to approximate differential quotients by appropriate difference quotients, thus reducing a differential equation to an algebraic system. There are a variety of ways to do the approximation.

Suppose f is a differentiable real-valued function on $\mathbb{R}$. Let $x \in \mathbb{R}$ and $h > 0$. Then we have the following three popular difference approximations:

$$f'(x) \approx \frac{f(x+h) - f(x)}{h} \tag{5.1.1}$$

$$\frac{f(x) - f(x-h)}{h} \tag{5.1.2}$$

$$\frac{f(x+h) - f(x-h)}{2h}. \tag{5.1.3}$$

These differences are called a *forward difference*, a *backward difference*, and a *centered difference*, respectively. Supposing f has a second derivative, it is easy to verify that the approximation errors for the forward and backward differences are both $O(h)$. If the third derivative of f exists, then the approximation error for the centered difference is $O(h^2)$. We see that if the function is smooth, the centered difference is a more accurate approximation to the derivative.

The second derivative of the function f is usually approximated by a second-order centered difference:

$$f''(x) \approx \frac{f(x+h) - 2\,f(x) + f(x-h)}{h^2}. \tag{5.1.4}$$

It can be verified that when f has a fourth derivative, the approximation error is $O(h^2)$.

Now let us use these difference formulas to formulate some difference schemes for a sample initial-boundary value problem for a heat equation.

Example 5.1.1 *Let us consider the problem*

$$u_t = \nu\,u_{xx} + f(x,t) \quad \text{in } [0, \pi] \times [0, T], \tag{5.1.5}$$
$$u(0, t) = u(\pi, t) = 0, \quad 0 \le t \le T, \tag{5.1.6}$$
$$u(x, 0) = u_0(x). \tag{5.1.7}$$

The differential equation (5.1.5) can be used to model a variety of physical processes such as heat conduction (see, e.g., [133]). Here $\nu > 0$ is a constant; f and u_0 are given functions. To develop a finite difference method, we need to introduce grid points. Let N_x and N_t be positive integers, $h_x = \pi/N_x$, $h_t = T/N_t$ and define the partition points

$$x_j = j\,h_x, \quad j = 0, 1, \ldots, N_x,$$
$$t_m = m\,h_t, \quad m = 0, 1, \ldots, N_t.$$

A point of the form (x_j, t_m) is called a grid point *and we are interested in computing approximate solution values at the grid points. We use the notation v_j^m for an approximation to $u_j^m \equiv u(x_j, t_m)$ computed from a finite difference scheme. Write $f_j^m = f(x_j, t_m)$ and*

$$r = \nu\,h_t/h_x^2.$$

Then we can bring in several schemes.

The first scheme is

$$\frac{v_j^{m+1} - v_j^m}{h_t} = \nu \, \frac{v_{j+1}^m - 2v_j^m + v_{j-1}^m}{h_x^2} + f_j^m,$$

$$1 \le j \le N_x - 1, \ 0 \le m \le N_t - 1, \qquad (5.1.8)$$

$$v_0^m = v_{N_x}^m = 0, \qquad 0 \le m \le N_t, \qquad (5.1.9)$$

$$v_j^0 = u_0(x_j), \qquad 0 \le j \le N_x. \qquad (5.1.10)$$

This scheme is obtained by replacing the time derivative with a forward difference and the second spatial derivative with a second-order centered difference. Hence it is called a forward-time centered-space scheme. The difference equation (5.1.8) can be written as

$$v_j^{m+1} = (1 - 2r) \, v_j^m + r \, (v_{j+1}^m + v_{j-1}^m) + h_t f_j^m,$$

$$1 \le j \le N_x - 1, \ 0 \le m \le N_t - 1. \qquad (5.1.11)$$

Thus once the solution at the time level $t = t_m$ is computed, the solution at the next time level $t = t_{m+1}$ can be found explicitly. The forward scheme (5.1.8)–(5.1.10) is an explicit *method.*

Alternatively, we may replace the time derivative with a backward difference and still use the second spatial derivative with a second-order centered difference. The resulting scheme is a backward-time centered-space scheme:

$$\frac{v_j^m - v_j^{m-1}}{h_t} = \nu \, \frac{v_{j+1}^m - 2v_j^m + v_{j-1}^m}{h_x^2} + f_j^m,$$

$$1 \le j \le N_x - 1, \ 1 \le m \le N_t, \qquad (5.1.12)$$

$$v_0^m = v_{N_x}^m = 0, \qquad 0 \le m \le N_t, \qquad (5.1.13)$$

$$v_j^0 = u_0(x_j), \qquad 0 \le j \le N_x. \qquad (5.1.14)$$

The difference equation (5.1.12) can be written as

$$(1 + 2r) \, v_j^m - r \, (v_{j+1}^m + v_{j-1}^m) = v_j^{m-1} + h_t f_j^m,$$

$$1 \le j \le N_x - 1, \ 1 \le m \le N_t, \qquad (5.1.15)$$

which is supplemented by the boundary condition from (5.1.13). Thus in order to find the solution at the time level $t = t_m$ from the solution at $t = t_{m-1}$, we need to solve a tridiagonal linear system of order $N_x - 1$. The backward scheme (5.1.12)–(5.1.14) is an implicit *method.*

In the above two methods, we approximate the differential equation at $x = x_j$ and $t = t_m$. We can also consider the differential equation at $x = x_j$ and $t = t_{m-1/2}$, approximating the time derivative by a centered difference:

$$u_t(x_j, t_{m-1/2}) \approx \frac{u(x_j, t_m) - u(x_j, t_{m-1})}{h_t}.$$

Further, approximate the second spatial derivative by the second-order centered difference:

$$u_{xx}(x_j, t_{m-1/2}) \approx \frac{u(x_{j+1}, t_{m-1/2}) - 2\,u(x_j, t_{m-1/2}) + u(x_{j-1}, t_{m-1/2})}{h_t^2},$$

and then approximate the half-time values by averages:

$$u(x_j, t_{m-1/2}) \approx (u(x_j, t_m) + u(x_j, t_{m-1}))/2,$$

etc. As a result we arrive at the Crank-Nicolson *scheme:*

$$\frac{v_j^m - v_j^{m-1}}{h_t} = \nu \frac{(v_{j+1}^m - 2v_j^m + v_{j-1}^m) + (v_{j+1}^{m-1} - 2v_j^{m-1} + v_{j-1}^{m-1})}{2\,h_x^2}$$

$$+ f_j^{m-1/2}, \qquad 1 \le j \le N_x - 1, \ 1 \le m \le N_t, \quad (5.1.16)$$

$$v_0^m = v_{N_x}^m = 0, \qquad 0 \le m \le N_t, \tag{5.1.17}$$

$$v_j^0 = u_0(x_j), \qquad 0 \le j \le N_x. \tag{5.1.18}$$

Here $f_j^{m-1/2} = f(x_j, t_{m-1/2})$, *which is replaced by* $(f_j^m + f_j^{m-1})/2$ *sometimes. The difference equation* (5.1.16) *can be rewritten as*

$$\left(1 + \frac{r}{2}\right) v_j^m - \frac{r}{2} (v_{j+1}^m + v_{j-1}^m)$$

$$= \left(1 - \frac{r}{2}\right) v_j^{m-1} + \frac{r}{2} (u_{j+1}^{m-1} + u_{j-1}^{m-1}) + h_t f_j^{m-1/2}. \quad (5.1.19)$$

So we see that the Crank-Nicolson scheme is also an implicit method and at each time step we need to solve a tridiagonal linear system of order $N_x - 1$.

The three schemes derived above all seem reasonable approximations for the initial-boundary value problem (5.1.5)–(5.1.7). Let us do some numerical experiments to see if these schemes indeed produce useful results. Let us use the forward scheme (5.1.8)–(5.1.10) and the backward scheme (5.1.12)–(5.1.14) to solve the problem (5.1.5)–(5.1.7) with $\nu = 1$, $f(x,t) = 0$ and $u_0(x) = \sin x$. Results from the Crank-Nicolson scheme are qualitatively similar to those from the backward scheme but magnitudes are smaller, and are thus omitted. It can be verified that the exact solution is $u(x,t) = e^{-t} \sin x$. We consider numerical solution errors at $t = 1$.

Figure 5.1 shows solution errors of the forward scheme corresponding to several combinations of the values N_x and N_t (or equivalently, h_x and h_t). Convergence is observed only when N_x is substantially smaller than N_t (i.e. when h_t is substantially smaller than h_x). In the next two sections, we explain this phenomenon theoretically.

Figure 5.2 demonstrates solution errors of the backward scheme corresponding to the same values of N_x and N_t. We observe a good convergence pattern. The maximum solution error decreases as N_x and N_t increase. In Section 5.3, we prove that the maximum error at $t = 1$ is bounded by a constant times $(h_x^2 + h_t)$. This result explains the phenomenon in Figure

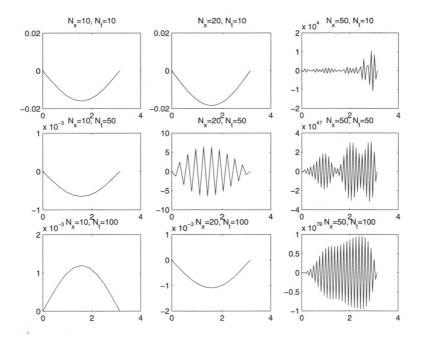

Figure 5.1. The forward scheme: errors at $t = 1$

5.2 that the error seems to decrease more rapidly with a decreasing h_t than h_x.

Naturally, a difference scheme is useful only if the scheme is convergent, i.e., if it can provide numerical solutions that approximate the exact solution. A necessary requirement for convergence is consistency of the scheme; that is, the difference scheme must be close to the differential equation in some sense. However, consistency alone does not guarantee the convergence, as we see from the numerical examples above. From the viewpoint of theoretical analysis, at each time level, some error is brought in, representing the discrepancy between the difference scheme and the differential equation. From the viewpoint of computer implementation, numerical values and numerical computations are subject to roundoff errors. Thus it is important to be able to control the propagation of errors. The ability to control the propagation of errors is termed *stability* of the scheme. We expect to have convergence for consistent, stable schemes. The well-known Lax theory for finite difference methods goes beyond this. The theory states that with properly defined notions of consistency, stability and convergence for a well-posed partial differential equation problem, a consistent scheme is convergent if and only if it is stable. In the next section, we present

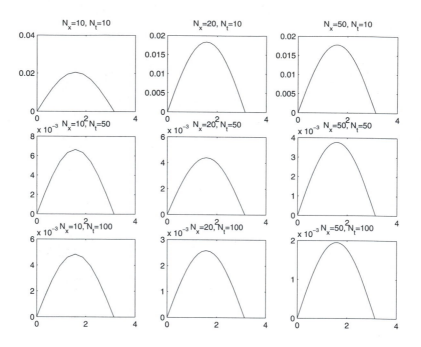

Figure 5.2. The backward scheme: errors at $t = 1$

one version of the Lax equivalence theory on the convergence of difference schemes. In the third section, we present and illustrate a variant of
the Lax equivalence theory that is usually more easily applicable to yield
convergence and convergence order results of difference schemes.

Exercise 5.1.1 *One approach for deriving difference formulas to approximate derivatives is the* method of undetermined coefficients. *Suppose f is
a smooth function on* $\mathbb{R}$. *Let* $h > 0$. *Determine coefficients a, b, and c so
that*

$$a\, f(x + h) + b\, f(x) + c\, f(x - h)$$

is an approximation of $f'(x)$ *with an order as high as possible; i.e., choose
a, b, and c such that*

$$|a\, f(x + h) + b\, f(x) + c\, f(x - h) - f'(x)| \leq O(h^p)$$

with a largest possible exponent p.

Exercise 5.1.2 *Do the same problem as Exercise 5.1.1 with* $f'(x)$ *replaced
by* $f''(x)$.

Exercise 5.1.3 *Is it possible to use*

$$a\,f(x+h) + b\,f(x) + c\,f(x-h)$$

with suitably chosen coefficients to approximate $f'''(x)$? How many function values are needed to approximate $f'''(x)$?

Exercise 5.1.4 *For the initial value problem of the one-way wave equation*

$$u_t + a\,u_x = f \quad \text{in } \mathbb{R} \times \mathbb{R}_+, \tag{5.1.20}$$
$$u(\cdot, 0) = u_0(\cdot) \quad \text{in } \mathbb{R}, \tag{5.1.21}$$

where $a \in \mathbb{R}$ is a constant, derive some difference schemes based on various combinations of difference approximations of the time derivative and spatial derivative.

Exercise 5.1.5 *The idea of the Lax-Wendroff scheme for solving the initial value problem of Exercise 5.1.4 is the following. Start with the Taylor expansion*

$$u(x_j, t_{m+1}) \approx u(x_j, t_m) + h_t u_t(x_j, t_m) + \frac{h_t^2}{2} u_{tt}(x_j, t_m). \tag{5.1.22}$$

From the differential equation, we have

$$u_t = -a\,u_x + f$$

and

$$u_{tt} = a^2 u_{xx} - a\,f_x + f_t.$$

Use these relations to replace the time derivatives in the right side of (5.1.22). Then replace the first and the second spatial derivatives by central differences. Finally replace f_x by a central difference and f_t by a forward difference.

Follow the above instructions to derive the Lax-Wendroff *scheme for solving (5.1.20)–(5.1.21).*

5.2   Lax equivalence theorem

In this section, we follow [67] to present one version of the Lax equivalence theorem for analyzing difference methods in solving initial value or initial-boundary value problems. The rigorous theory is developed in an abstract setting. To help understand the theory, we use the sample problem (5.1.5)–(5.1.7) with $f(x,t) = 0$ to illustrate the notation, assumptions, definitions and the equivalence result.

We first introduce an abstract framework. Let V be a Banach space, $V_0 \subseteq V$ a dense subspace of V. Let $L : V_0 \subseteq V \to V$ be a linear operator.

The operator L is usually unbounded and can be thought of as a differential operator. Consider the initial value problem

$$\begin{cases} \dfrac{du(t)}{dt} = Lu(t), & 0 \le t \le T, \\ u(0) = u_0. \end{cases} \tag{5.2.1}$$

This problem also represents an initial-boundary value problem with homogeneous boundary value conditions when they are included in definitions of the space V and the operator L. The next definition gives the meaning of a solution of the problem (5.2.1).

Definition 5.2.1 *A function* $u : [0,T] \to V$ *is a solution of the initial value problem* (5.2.1) *if for any* $t \in [0,T]$, $u(t) \in V_0$,

$$\lim_{\Delta t \to 0} \left\| \frac{1}{\Delta t} \left(u(t + \Delta t) - u(t) \right) - Lu(t) \right\| = 0, \tag{5.2.2}$$

and $u(0) = u_0$.

In the above definition, the limit in (5.2.2) is understood to be the right limit at $t = 0$ and the left limit at $t = T$.

Definition 5.2.2 *The initial value problem* (5.2.1) *is* well-posed *if for any* $u_0 \in V_0$, *there is a unique solution* $u = u(t)$ *and the solution depends continuously on the initial value: There exists a constant* $c_0 > 0$ *such that if* $u(t)$ *and* $\bar{u}(t)$ *are the solutions for the initial values* $u_0, \bar{u}_0 \in V_0$; *then*

$$\sup_{0 \le t \le T} \|u(t) - \bar{u}(t)\|_V \le c_0 \|u_0 - \bar{u}_0\|_V. \tag{5.2.3}$$

From now on, we assume the initial value problem (5.2.1) is well-posed. We denote the solution as

$$u(t) = S(t)\, u_0, \quad u_0 \in V_0.$$

Using the linearity of the operator L, it is easy to see that the solution operator $S(t)$ is linear. From the continuous dependence property (5.2.3), we have

$$\sup_{0 \le t \le T} \|S(t)\,(u_0 - \bar{u}_0)\|_V \le c_0 \|u_0 - \bar{u}_0\|_V,$$

$$\sup_{0 \le t \le T} \|S(t)u_0\|_V \le c_0 \|u_0\|_V \quad \forall\, u_0 \in V_0.$$

By Theorem 2.4.1, the operator $S(t) : V_0 \subseteq V \to V$ can be uniquely extended to a linear continuous operator $S(t) : V \to V$ with

$$\sup_{0 \le t \le T} \|S(t)\|_V \le c_0.$$

Definition 5.2.3 *For* $u_0 \in V \backslash V_0$, *we call* $u(t) = S(t)u_0$ *the generalized solution of the initial value problem* (5.2.1).

Example 5.2.4 *We use the following problem and its finite difference approximations to illustrate the use of the abstract framework of the section.*

$$\begin{cases} u_t = \nu\, u_{xx} & \text{in } [0, \pi] \times [0, T], \\ u(0, t) = u(\pi, t) = 0 & 0 \le t \le T, \\ u(x, 0) = u_0(x) & 0 \le x \le \pi. \end{cases} \tag{5.2.4}$$

We take $V = C_0[0, \pi] = \{v \in C[0, \pi] \mid v(0) = v(\pi) = 0\}$ with the norm $\|\cdot\|_{C[0,\pi]}$. We choose

$$V_0 = \left\{ v \;\middle|\; v(x) = \sum_{j=1}^{n} a_j \sin(jx),\; a_j \in \mathbb{R},\; n = 1, 2, \dots \right\}. \tag{5.2.5}$$

The verification that V_0 is dense in V is left as an exercise.

If $u_0 \in V_0$, then for some integer $n \ge 1$ and $b_1, \dots, b_n \in \mathbb{R}$,

$$u_0(x) = \sum_{j=1}^{n} b_j \sin(jx). \tag{5.2.6}$$

For this u_0, it can be verified directly that the solution is

$$u(x, t) = \sum_{j=1}^{n} b_j e^{-\nu j^2 t} \sin(jx). \tag{5.2.7}$$

By using the maximum principle for the heat equation (see, e.g., [48] or other textbooks on partial differential equations),

$$\min\{0, \min_{0 \le x \le \pi} u_0(x)\} \le u(x, t) \le \max\{0, \max_{0 \le x \le \pi} u_0(x)\},$$

we see that

$$\max_{0 \le x \le \pi} |u(x, t)| \le \max_{0 \le x \le \pi} |u_0(x)| \quad \forall t \in [0, T].$$

Thus the operator $S(t) : V_0 \subseteq V \to V$ is bounded.

Then for a general $u_0 \in V$, the problem (5.2.4) has a unique solution. If $u_0 \in V$ has a piecewise continuous derivative in $[0, \pi]$, then from the theory of Fourier series,

$$u_0(x) = \sum_{j=1}^{\infty} b_j \sin(jx)$$

and the solution $u(t)$ can be expressed as

$$u(x, t) = S(t)u_0(x) = \sum_{j=1}^{\infty} b_j e^{-\nu j^2 t} \sin(jx).$$

Return to the abstract problem (5.1.5)–(5.1.7). We present two results; the first one is on the time continuity of the generalized solution and the second one shows the solution operator $S(t)$ forms a semigroup.

Proposition 5.2.5 *For any $u_0 \in V_0$, the generalized solution of the initial value problem (5.1.5)–(5.1.7) is continuous in t.*

Proof. Choose a sequence $\{u_{0,n}\} \subseteq V_0$ that converges to u_0 in V:

$$\|u_{0,n} - u_0\|_V \to 0 \quad \text{as } n \to \infty.$$

Let $t_0 \in [0, T]$ be fixed, and $t \in [0, T]$. We write

$$u(t) - u(t_0) = S(t)u_0 - S(t_0)u_0$$
$$= S(t)(u_0 - u_{0,n}) + (S(t) - S(t_0))u_{0,n} - S(t_0)(u_0 - u_{0,n}).$$

Then

$$\|u(t) - u(t_0)\|_V \leq 2\,c_0\|u_{0,n} - u_0\|_V + \|(S(t) - S(t_0))u_{0,n}\|_V.$$

Given any $\varepsilon > 0$, we choose n sufficiently large such that

$$2\,c_0\|u_{0,n} - u_0\|_V < \frac{\varepsilon}{2}.$$

For this n, using (5.2.2) of the definition of the solution, we have a $\delta > 0$ such that

$$\|(S(t) - S(t_0))u_{0,n}\|_V < \frac{\varepsilon}{2} \quad \text{for } |t - t_0| < \delta.$$

Then for $t \in [0, T]$ with $|t - t_0| < \delta$, we have $\|u(t) - u(t_0)\|_V \leq \varepsilon$. ∎

Proposition 5.2.6 *Assume the problem (5.2.1) is well-posed. Then for all $t_1, t_0 \in [0, T]$ such that $t_1 + t_0 \leq T$, we have $S(t_1 + t_0) = S(t_1)\,S(t_0)$.*

Proof. The solution of the problem (5.2.1) is $u(t) = S(t)u_0$. We have $u(t_0) = S(t_0)u_0$ and $S(t)u(t_0)$ is the solution of the differential equation on $[t_0, T]$ with the initial condition $u(t_0)$ at t_0. By the uniqueness of the solution,

$$S(t)u(t_0) = u(t + t_0),$$

i.e.,

$$S(t_1)S(t_0)u_0 = S(t_1 + t_0)u_0.$$

Therefore, $S(t_1 + t_0) = S(t_1)\,S(t_0)$. ∎

Now we introduce a finite difference method defined by a one-parameter family of uniformly bounded linear operators

$$C(\Delta t) : V \to V, \quad 0 < \Delta t \leq \Delta t_0.$$

Here $\Delta t_0 > 0$ is a fixed number. The family $\{C(\Delta t)\}_{0 < \Delta t \leq \Delta t_0}$ is said to be uniformly bounded if there is a constant c such that

$$\|C(\Delta t)\| \leq c \quad \forall\, \Delta t \in (0, \Delta t_0].$$

The approximate solution is then defined by

$$u_{\Delta t}(m\,\Delta t) = C(\Delta t)^m u_0, \quad m = 1, 2, \ldots$$

Definition 5.2.7 (CONSISTENCY) *The difference method is consistent if there exists a dense subspace V_c of V such that for all $u_0 \in V_c$, for the corresponding solution u of the initial value problem (5.2.1), we have*

$$\lim_{\Delta t \to 0} \left\| \frac{1}{\Delta t} \left(C(\Delta t)u(t) - u(t + \Delta t) \right) \right\| = 0 \quad \text{uniformly in } [0, T].$$

Assume $V_c \cap V_0 \neq \emptyset$. For $u_0 \in V_c \cap V_0$, we write

$$\frac{1}{\Delta t} \left(C(\Delta t)u(t) - u(t + \Delta t) \right)$$

$$= \left(\frac{C(\Delta t) - I}{\Delta t} - L \right) u(t) - \left(\frac{u(t + \Delta t) - u(t)}{\Delta t} - Lu(t) \right).$$

Since

$$\frac{u(t + \Delta t) - u(t)}{\Delta t} - Lu(t) \to 0 \quad \text{as } \Delta t \to 0$$

by the definition of the solution, we see that

$$\left(\frac{C(\Delta t) - I}{\Delta t} - L \right) u(t) \to 0 \quad \text{as } \Delta t \to 0;$$

so $(C(\Delta t) - I)/\Delta t$ is a convergent approximation of the operator L.

Example 5.2.8 (continuation of Example 5.2.4) *Let us now consider the forward method and the backward method from Example 5.1.1 for the sample problem (5.2.4). For the forward method, we define the operator $C(\Delta t)$ by the formula*

$$C(\Delta t)v(x) = (1 - 2r)\, v(x) + r\, (v(x + \Delta x) + v(x - \Delta x)),$$

where $\Delta x = \sqrt{\nu \Delta t / r}$ and if $x \pm \Delta x \notin [0, \pi]$; then the function v is extended by oddness with period 2π. We will identify Δt with h_t and Δx with h_x. Then $C(\Delta t) : V \to V$ is a linear operator and it can be shown that

$$\|C(\Delta t)v\|_V \leq (|1 - 2r| + 2r)\, \|v\|_V \quad \forall\, v \in V.$$

So

$$\|C(\Delta t)\| \leq |1 - 2r| + 2r, \tag{5.2.8}$$

and the family $\{C(\Delta t)\}$ is uniformly bounded. The difference method is

$$u_{\Delta t}(t_m) = C(\Delta t)u_{\Delta t}(t_{m-1}) = C(\Delta t)^m u_0$$

or

$$u_{\Delta t}(\cdot, t_m) = C(\Delta t)^m u_0(\cdot).$$

Notice that in this form, the difference method generates an approximate solution $u_{\Delta t}(x, t)$ that is defined for $x \in [0, \pi]$ and $t = t_m$, $m = 0, 1, \ldots, N_t$.

Since

$$u_{\Delta t}(x_j, t_{m+1}) = (1 - 2r)\, u_{\Delta t}(x_j, t_m)$$
$$+ r\left(u_{\Delta t}(x_{j-1}, t_m) + u_{\Delta t}(x_{j+1}, t_m)\right),$$
$$1 \le j \le N_x - 1,\ 0 \le m \le N_t - 1,$$
$$u_{\Delta t}(0, t_m) = u_{\Delta t}(N_x, t_m) = 0,\quad 0 \le m \le N_t,$$
$$u_{\Delta t}(x_j, 0) = u_0(x_j),\quad 0 \le j \le N_x,$$

we see that the relation between the approximate solution $u_{\Delta t}$ and the so-lution v defined by the ordinary difference scheme (5.1.8)–(5.1.10) (with $f_j^m = 0$) is

$$u_{\Delta t}(x_j, t_m) = v_j^m. \tag{5.2.9}$$

As for the consistency, we take $V_c = V_0$. For the initial value function (5.2.6), we have the formula (5.2.7) for the solution which is obviously infinitely smooth. Now using Taylor expansions at (x, t), we have

$$C(\Delta t) u(x, t) - u(x, t + \Delta t)$$
$$= (1 - 2r)u(x, t) + r\left(u(x + \Delta x, t) + u(x - \Delta x, t)\right) - u(x, t + \Delta t)$$
$$= (1 - 2r)u(x, t) + r\left(2u(x, t) + u_{xx}(x, t)\right)$$
$$+ \frac{r}{4!}\left(u_{xxxx}(x + \theta_1 \Delta x, t) + u_{xxxx}(x - \theta_2 \Delta x, t)\right)(\Delta x)^4$$
$$- u(x, t) - u_t(x, t)\,\Delta t - \frac{1}{2}u_{tt}(x, t + \theta_3 \Delta t)(\Delta t)^2$$
$$= -\frac{1}{2}u_{tt}(x, t + \theta_3 \Delta t)(\Delta t)^2$$
$$- \frac{\nu^2}{24\, r}\left(u_{xxxx}(x + \theta_1 \Delta x, t) + u_{xxxx}(x - \theta_2 \Delta x, t)\right)(\Delta t)^2,$$

where, $\theta_1, \theta_2, \theta_3 \in (0, 1)$. Thus,

$$\left\| \frac{1}{\Delta t}\left(C(\Delta t)u(t) - u(t + \Delta t)\right) \right\| \le c\,\Delta t,$$

and we have the consistency of the scheme.
   For the backward method, $u_{\Delta t}(\cdot, t + \Delta t) = C(\Delta t)u_{\Delta t}(\cdot, t)$ is defined by

$$(1 + 2r)\,u_{\Delta t}(x, t + \Delta t) - r\left(u_{\Delta t}(x - \Delta x, t + \Delta t)\right.$$
$$\left. + u_{\Delta t}(x + \Delta x, t + \Delta t)\right) = u_{\Delta t}(x, t)$$

with $\Delta x = \sqrt{\nu \Delta t / r}$. Again, for $x \pm \Delta x \notin [0, \pi]$, the function u is extended by oddness with period 2π. Rewrite the relation in the form

$$u_{\Delta t}(x, t + \Delta t)$$
$$= \frac{r}{1 + 2r}\left(u_{\Delta t}(x - \Delta x, t + \Delta t) + u_{\Delta t}(x + \Delta x, t + \Delta t)\right) + \frac{u_{\Delta t}(x, t)}{1 + 2r}.$$

Let $\|u_{\Delta t}(\cdot, t + \Delta t)\|_V = |u_{\Delta t}(x_0, t + \Delta t)|$ for some $x_0 \in [0, \pi]$. Then

$$\|u_{\Delta t}(\cdot, t + \Delta t)\|_V \leq \frac{r}{1 + 2r}(|u_{\Delta t}(x - \Delta x, t + \Delta t)| + |u_{\Delta t}(x + \Delta x, t + \Delta t)|)$$
$$+ \frac{|u_{\Delta t}(x, t)|}{1 + 2r};$$

i.e.,

$$\|u_{\Delta t}(\cdot, t + \Delta t)\|_V \leq \frac{2r}{1 + 2r}\|u_{\Delta t}(\cdot, t + \Delta t)\|_V + \frac{\|u_{\Delta t}(\cdot, t)\|_V}{1 + 2r}.$$

So

$$\|u_{\Delta t}(\cdot, t + \Delta t)\|_V \leq \|u_{\Delta t}(\cdot, t)\|_V$$

and the family $\{C(\Delta t)\}_{0 < \Delta t \leq \Delta t_0}$ is uniformly bounded.

Showing consistency of the backward scheme is more involved, and the argument is similar to that in Example 5.3.4 where the definition of the consistency is slightly different but is essentially the same.

Let us return to the general case.

Definition 5.2.9 (CONVERGENCE) *The difference method is* convergent *if for any fixed $t \in [0, T]$, any $u_0 \in V$, we have*

$$\lim_{\Delta t_i \to 0} \|(C(\Delta t_i)^{m_i} - S(t))u_0\| = 0$$

where $\{m_i\}$ is a sequence of integers and $\{\Delta t_i\}$ a sequence of step sizes such that $\lim_{i \to \infty} m_i \Delta t_i = t$.

Definition 5.2.10 (STABILITY) *The difference method is* stable *if the operators*

$$\{C(\Delta t)^m \mid 0 < \Delta t \leq \Delta t_0, \ m\Delta t \leq T\}$$

are uniformly bounded; i.e., there exists a constant $M_0 > 0$ such that

$$\|C(\Delta t)^m\|_{V \to V} \leq M_0 \quad \forall m : m\Delta t \leq T, \ \forall \Delta t \leq \Delta t_0.$$

We now come to the central result of the section.

Theorem 5.2.11 (LAX EQUIVALENCE THEOREM) *Suppose the initial value problem (5.2.1) is well-posed. For a consistent difference method, stability is equivalent to convergence.*

Proof. ($\Longrightarrow$) Consider the error

$$C(\Delta t)^m u_0 - u(t)$$
$$= \sum_{j=1}^{m-1} C(\Delta t)^j [C(\Delta t)u((m - 1 - j)\Delta t) - u((m - j)\Delta t)]$$
$$+ u(m\Delta t) - u(t).$$

First assume $u_0 \in V_c$. Then since the method is stable,

$$\|C(\Delta t)^m u_0 - u(t)\| \leq M_0\, m\Delta t \sup_t \left\| \frac{C(\Delta t)u(t) - u(t+\Delta t)}{\Delta t} \right\|$$
$$+ \|u(m\Delta t) - u(t)\|. \tag{5.2.10}$$

By continuity, $\|u(m\Delta t) - u(t)\| \to 0$, and by the consistency,

$$\sup_t \left\| \frac{C(\Delta t)u(t) - u(t+\Delta t)}{\Delta t} \right\| \to 0.$$

So we have the convergence.

Next consider the convergence for the general case where $u_0 \in V$. We have a sequence $\{u_{0,n}\} \subseteq V_0$ such that $u_{0,n} \to u_0$ in V. Writing

$$C(\Delta t)^m u_0 - u(t)$$
$$= C(\Delta t)^m (u_0 - u_{0,n}) + [C(\Delta t)^m - S(t)]u_{0,n} - S(t)\,(u_0 - u_{0,n}),$$

we obtain

$$\|C(\Delta t)^m u_0 - u(t)\| \leq \|C(\Delta t)^m (u_0 - u_{0,n})\|$$
$$+ \|[C(\Delta t)^m - S(t)]u_{0,n}\| + \|S(t)\,(u_0 - u_{0,n})\|.$$

Since the initial value problem (5.2.1) is well-posed and the method is stable,

$$\|C(\Delta t)^m u_0 - u(t)\| \leq c\,\|u_0 - u_{0,n}\| + \|[C(\Delta t)^m - S(t)]u_{0,n}\|.$$

Given any $\varepsilon > 0$, there is an n sufficiently large such that

$$c\,\|u_0 - u_{0,n}\| < \frac{\varepsilon}{2}.$$

For this n, let Δt be sufficiently small,

$$\|[C(\Delta t)^m - S(t)]u_{0,n}\| < \frac{\varepsilon}{2} \quad \forall\, \Delta t \text{ small, } |m\Delta t - t| < \Delta t.$$

Then we obtain the convergence.

($\Longleftarrow$) Suppose the method is not stable. Then there are sequences $\{\Delta t_k\}$ and $\{m_k\}$ such that $m_k \Delta t_k \leq T$ and

$$\lim_{k\to\infty} \|C(\Delta t_k)^{m_k}\| = \infty.$$

Since $\Delta t_k \leq \Delta t_0$, we may assume the sequence $\{\Delta t_k\}$ is convergent. If the sequence $\{m_k\}$ is bounded, then

$$\sup_k \|C(\Delta t_k)^{m_k}\| \leq \sup_k \|C(\Delta t_k)\|^{m_k} < \infty.$$

This is a contradiction. Thus $m_k \to \infty$ and $\Delta t_k \to 0$ as $k \to \infty$.

By the convergence of the method,

$$\sup_k \|C(\Delta t_k)^{m_k} u_0\| < \infty \quad \forall\, u_0 \in V.$$

Applying Theorem 2.4.4, we have

$$\lim_{k \to \infty} \|C(\Delta t_k)^{m_k}\| < \infty,$$

contradicting the assumption that the method is not stable.   ∎

Corollary 5.2.12 (CONVERGENCE ORDER) *Under the assumptions of Theorem 5.2.11, if u is a solution with initial value $u_0 \in V_c$ satisfying*

$$\sup_{0 \le t \le T} \left\| \frac{C(\Delta t)u(t) - u(t + \Delta t)}{\Delta t} \right\| \le c\,(\Delta t)^k \quad \forall \Delta t \in (0, \Delta t_0],$$

then we have the error estimate

$$\|C(\Delta t)^m u_0 - u(t)\| \le c\,(\Delta t)^k$$

where m is a positive integer with $m\Delta t = t$.

Proof. The error estimate follows immediately from (5.2.10).   ∎

Example 5.2.13 (continuation of Example 5.2.8) *Let us apply the Lax equivalence theorem to the forward and backward schemes for the sample problem 5.2.4. For the forward scheme, we assume $r \le 1/2$. Then according to (5.2.8), $\|C(\Delta t)\| \le 1$ and so*

$$\|C(\Delta t)^m\| \le 1, \quad m = 1, 2, \ldots$$

Thus under the condition $r \le 1/2$, the forward scheme is stable. Since the scheme is consistent, we have the convergence

$$\lim_{\Delta t_i \to 0} \|u_{\Delta t}(\cdot, m_i \Delta t_i) - u(\cdot, t)\|_V = 0, \tag{5.2.11}$$

where $\lim_{\Delta t_i \to 0} m_i \Delta t_i = t$.

Actually, it can be shown that

$$\|C(\Delta t)\| = |1 - 2r| + 2r$$

and $r \le 1/2$ is a necessary and sufficient condition for stability and then convergence (cf. Exercise 5.2.3).

By the relation (5.2.9), we see that for the finite difference solution $\{v_j^m\}$ defined in (5.1.8)–(5.1.10) with $f_j^m = 0$, we have the convergence

$$\lim_{h_t \to 0} \max_{0 \le j \le N_x} |v_j^m - u(x_j, t)| = 0,$$

where m depends on h_t and $\lim_{h_t \to 0} m h_t = t$.

Since we need a condition ($r \le 1/2$ in this case) for convergence, the forward scheme is said to be conditionally stable *and* conditionally convergent.

For the backward scheme, for any r, $\|C(\Delta t)\| \le 1$. Then

$$\|C(\Delta t)^m\| \le 1 \quad \forall m.$$

So the backward scheme is unconditionally stable, *which leads to un-conditional convergence of the backward scheme. We skip the detailed presentation of the arguments for the above statement since the arguments are similar to those for the forward scheme.*

We can also apply Corollary 5.2.12 to claim convergence order for the forward and backward schemes (see examples 5.3.3 and 5.3.4 in the next section for some similar arguments).

Exercise 5.2.1 *Show that the subspace V_0 defined in (5.2.5) is dense in V.*

Exercise 5.2.2 *Analyze the Crank-Nicolson scheme for the problem (5.2.4).*

Exercise 5.2.3 *Consider the forward scheme for solving the sample problem 5.2.4. Show that*

$$\|C(\Delta t)\| = |1 - 2r| + 2r$$

and $r \leq 1/2$ is a necessary and sufficient condition for both stability and convergence.

5.3   More on convergence

In the literature, one can find various slightly different variants of the Lax equivalence theorem presented in the preceding section. Here we consider one such variant that is usually more convenient to apply in analyzing convergence of difference schemes for solving initial-boundary value problems.

Consider an initial-boundary value problem of the form

$$Lu = f \quad \text{in } (0, a) \times (0, T), \tag{5.3.1}$$
$$u(0, t) = u(a, t) = 0 \quad t \in [0, T], \tag{5.3.2}$$
$$u(x, 0) = u_0(x) \quad x \in [0, a]. \tag{5.3.3}$$

Here f and u_0 are given data, and L is a linear partial differential operator of first order with respect to the time variable. For the problem (5.1.5)–(5.1.7), $L = \partial_t - \nu \partial_x^2$. We assume for the given data f and u_0, the problem (5.3.1)–(5.3.3) has a unique solution u with certain smoothness that makes the following calculations meaningful (e.g., derivatives of u up to certain order are continuous).

Again denote N_x and N_t positive integers, $h_x = a/N_x$, $h_t = T/N_t$ and we use the other notations introduced in Example 5.1.1. Corresponding to the time level $t = t_m$, we introduce the solution vector

$$\mathbf{v}^m = (v_1^m, \ldots, v_{N_x-1}^m)^T \in \mathbb{R}^{N_x-1},$$

where the norm in the space $\mathbb{R}^{N_x-1}$ is denoted by $\|\cdot\|$; this norm depends on the dimension $N_x - 1$, but we do not indicate this dependence explicitly for notational simplicity. We will be specific about the norm when we consider concrete examples.

Consider a general two-level scheme

$$\mathbf{v}^{m+1} = Q\mathbf{v}^m + h_t \mathbf{g}^m, \quad 0 \le m \le N_t - 1, \qquad (5.3.4)$$
$$\mathbf{v}^0 = \mathbf{u}^0. \qquad (5.3.5)$$

Here the matrix $Q \in \mathbb{R}^{(N_x-1)\times(N_x-1)}$ may depend on h_t and h_x. We use $\|Q\|$ to denote the operator matrix norm induced by the vector norm on $\mathbb{R}^{N_x-1}$. The vector $\mathbf{g}^m$ is usually constructed from values of f at $t = t_m$,

$$\mathbf{u}^0 = (u_0(x_1), \dots, u_0(x_{N_x-1}))^T,$$

and in general

$$\mathbf{u}^m = (u_1^m, \dots, u_{N_x-1}^m)^T, \quad 1 \le m \le N_t,$$

with u the solution of (5.3.1)–(5.3.3).

We now introduce definitions of consistency, stability, and convergence for the scheme (5.3.4)–(5.3.5). For this, we need to define a quantity $\boldsymbol{\tau}^m$ through the relation

$$\mathbf{u}^{m+1} = Q\mathbf{u}^m + h_t \mathbf{g}^m + h_t \boldsymbol{\tau}^m. \qquad (5.3.6)$$

This quantity $\boldsymbol{\tau}^m$ can be called the *local truncation error* of the scheme. As we will see from examples below, for an explicit method, $\boldsymbol{\tau}^m$ defined in (5.3.6) is indeed the local truncation error used in many references. In the case of an implicit method, $\boldsymbol{\tau}^m$ defined here is related to the usual local truncation error by a linear transformation.

Definition 5.3.1 *We say the difference method* (5.3.4)–(5.3.5) *is consistent if*

$$\sup_{m:mh_t \le T} \|\boldsymbol{\tau}^m\| \to 0 \quad \text{as } h_t, h_x \to 0.$$

The method is of order (p_1, p_2) if, when the solution u is sufficiently smooth, there is a constant c such that

$$\sup_{m:mh_t \le T} \|\boldsymbol{\tau}^m\| \le c\,(h_x^{p_1} + h_t^{p_2}). \qquad (5.3.7)$$

The method is said to be stable if for some constant $M_0 < \infty$, which may depend on T, we have

$$\sup_{m:mh_t \le T} \|Q^m\| \le M_0.$$

The method is convergent *if*

$$\sup_{m:mh_t \le T} \|\mathbf{u}^m - \mathbf{v}^m\| \to 0 \quad \text{as } h_t, h_x \to 0.$$

We have the following theorem concerning convergence and convergence order of the difference method.

Theorem 5.3.2 *Assume the scheme* (5.3.4)–(5.3.5) *is consistent and stable. Then the method is convergent.*

Suppose the solution u is sufficiently smooth so that (5.3.7) *holds. Then we have the error estimate*

$$\sup_{m:mh_t \le T} \|\mathbf{u}^m - \mathbf{v}^m\| \le c\,(h_x^{p_1} + h_t^{p_2}).$$

Proof. Introduce the error vectors: $\mathbf{e}^m = \mathbf{u}^m - \mathbf{v}^m$ for $m = 0, 1, \ldots, N_t$. Then $\mathbf{e}^0 = \mathbf{0}$ by the definition of the initial value for the scheme. We have the error recursion relation:

$$\mathbf{e}^{m+1} = Q\,\mathbf{e}^m + \Delta t\,\boldsymbol{\tau}^m.$$

Using this relation repeatedly and remembering $\mathbf{e}^0 = \mathbf{0}$, we find

$$\mathbf{e}^{m+1} = \Delta t \sum_{l=0}^{m} Q^l \boldsymbol{\tau}^{m-l}.$$

Thus

$$\|\mathbf{e}^{m+1}\| \le \Delta t \sum_{l=0}^{m} \|Q^l\|\,\|\boldsymbol{\tau}^{m-l}\|.$$

Apply the stability condition,

$$\|\mathbf{e}^{m+1}\| \le M_0(m+1)\Delta t \sup_{0 \le l \le m} \|\boldsymbol{\tau}^{m-l}\|.$$

Then we have the inequality

$$\sup_{m:mh_t \le T} \|\mathbf{u}^m - \mathbf{v}^m\| \le M_0 T \sup_{m:mh_t \le T} \|\boldsymbol{\tau}^m\|,$$

and the claims of the theorem follow.   ∎

Example 5.3.3 *We give a convergence analysis of the scheme* (5.1.8)–(5.1.10). *Assume* $u_{tt}, u_{xxxx} \in C([0,\pi] \times [0,T])$. *Then* (cf. (5.1.11))

$$u_j^{m+1} = (1 - 2r)\,u_j^m + r\,(u_{j+1}^m + u_{j-1}^m) + h_t f_j^m + h_t \tau_j^m,$$

where, following easily from Taylor expansions,

$$|\tau_j^m| \le c\,(h_x^2 + h_t),$$

with the constant c depending on u_{tt} *and* u_{xxxx}.

The scheme (5.1.8)–(5.1.10) *can be written in the form* (5.3.4)–(5.3.5) *with*

$$Q = \begin{pmatrix} 1 - 2r & r & & & \\ r & 1 - 2r & r & & \\ & & \ddots & \ddots & \ddots & \\ & & & r & 1 - 2r & r \\ & & & & r & 1 - 2r \end{pmatrix}$$

and

$$\mathbf{g}^m = \mathbf{f}^m \equiv (f_1^m, \ldots, f_{N_x-1}^m)^T.$$

Let us assume the condition $r \leq 1/2$. *Then if we choose to use the maximum-norm, we have*

$$\|Q\|_\infty = 1, \quad \|\boldsymbol{\tau}^m\|_\infty \leq c\,(h_x^2 + h_t).$$

Thus the method is stable, and we can apply Theorem 5.3.2 to conclude that under the conditions $u_{tt}, u_{xxxx} \in C([0, \pi] \times [0, T])$ *and* $r \leq 1/2$,

$$\max_{0 \leq m \leq N_x} \|\mathbf{u}^m - \mathbf{v}^m\|_\infty \leq c\,(h_x^2 + h_t).$$

Now suppose we use the discrete weighted 2-norm:

$$\|\mathbf{v}\|_{2,h_x} = \sqrt{h_x}\|\mathbf{v}\|_2.$$

It is easy to see that the induced matrix norm is the usual spectral norm $\|Q\|_2$. *Since* Q *is symmetric and its eigenvalues are (see Exercise 5.3.1)*

$$\lambda_j(Q) = 1 - 4r \sin^2\left(\frac{j\pi}{2N_x}\right), \quad 1 \leq j \leq N_x - 1,$$

we see that under the condition $r \leq 1/2$,

$$\|Q\|_2 = \max_j |\lambda_j(Q)| < 1,$$

; i.e., the method is stable. It is easy to verify that

$$\|\boldsymbol{\tau}^m\|_{2,h_x} \leq c\,(h_x^2 + h_t).$$

Thus by Theorem 5.3.2, we conclude that under the conditions $u_{tt}, u_{xxxx} \in C([0, \pi] \times [0, T])$ *and* $r \leq 1/2$,

$$\max_{0 \leq m \leq N_x} \|\mathbf{u}^m - \mathbf{v}^m\|_{2,h_x} \leq c\,(h_x^2 + h_t).$$

Example 5.3.4 *Now consider the backward scheme* (5.1.12)–(5.1.14). *Assume* $u_{tt}, u_{xxxx} \in C([0, \pi] \times [0, T])$. *Then (cf.* (5.1.15))

$$(1 + 2r)\,u_j^m - r\,(u_{j+1}^m + u_{j-1}^m) = u_j^{m-1} + h_t f_j^m + h_t \bar{\tau}_j^m,$$

where

$$|\bar{\tau}_j^m| \leq c\,(h_x^2 + h_t)$$

with the constant c *depending on* u_{tt} *and* u_{xxxx}. *Define the matrix*

$$Q = Q_1^{-1},$$

where

$$Q_1 = \begin{pmatrix} 1+2r & -r & & & \\ -r & 1+2r & -r & & \\ & \ddots & \ddots & \ddots & \\ & & -r & 1+2r & -r \\ & & & -r & 1+2r \end{pmatrix}.$$

Let $\mathbf{g}^m = Q\mathbf{f}^m$ *and* $\boldsymbol{\tau}^m = Q\bar{\boldsymbol{\tau}}^m$. *Then the scheme* (5.1.12)–(5.1.14) *can be written in the form* (5.3.4)–(5.3.5).

First we consider the convergence in $\|\cdot\|_\infty$. *Let us estimate* $\|Q\|_\infty$. *From the definition of* Q,

$$\mathbf{y} = Q\mathbf{x} \Longleftrightarrow \mathbf{x} = Q_1\mathbf{y} \quad \text{for } \mathbf{x}, \mathbf{y} \in \mathbb{R}^{N_x-1}.$$

Thus

$$y_i = \frac{r}{1+2r}(y_{i-1} + y_{i+1}) + \frac{x_i}{1+2r}, \quad 1 \le i \le N_x - 1.$$

Suppose $\|\mathbf{y}\|_\infty = |y_i|$. *Then*

$$\|\mathbf{y}\|_\infty = |y_i| \le \frac{r}{1+2r}\, 2\, \|\mathbf{y}\|_\infty + \frac{\|\mathbf{x}\|_\infty}{1+2r}.$$

So

$$\|Q\mathbf{x}\|_\infty = \|\mathbf{y}\|_\infty \le \|\mathbf{x}\|_\infty \quad \forall \mathbf{x} \in \mathbb{R}^{N_x-1}.$$

Hence

$$\|Q\|_\infty \le 1,$$

the backward scheme is unconditionally stable and it is easy to see

$$\|\boldsymbol{\tau}^m\|_\infty \le \|\bar{\boldsymbol{\tau}}^m\|_\infty.$$

Applying Theorem 5.3.2, for the backward scheme (5.1.12)–(5.1.14), *we conclude that under the conditions* $u_{tt}, u_{xxxx} \in C([0,\pi] \times [0,T])$,

$$\max_{0 \le m \le N_x} \|\mathbf{u}^m - \mathbf{v}^m\|_\infty \le c\,(h_x^2 + h_t).$$

Now we consider the convergence in $\|\cdot\|_{2,h_x}$. *By Exercise 5.3.1, the eigenvalues of* Q_1 *are*

$$\lambda_j(Q_1) = 1 + 4r \cos^2 \frac{j\pi}{2N_x}, \quad 1 \le j \le N_x - 1.$$

Since $Q = Q_1^{-1}$, *the eigenvalues of* Q *are*

$$\lambda_j(Q) = \lambda_j(Q_1)^{-1} \in (0,1), \quad 1 \le j \le N_x - 1.$$

Now that Q *is symmetric because* Q_1 *is,*

$$\|Q\|_2 = \max_{1 \le j \le N_x - 1} |\lambda_j(Q)| < 1.$$

So the backward scheme is unconditionally stable measured in $\|\cdot\|_{2,h_x}$, and it is also easy to deduce

$$\|\boldsymbol{\tau}^m\|_{2,h_x} \leq \|\overline{\boldsymbol{\tau}}^m\|_{2,h_x}.$$

So for the backward scheme (5.1.12)–(5.1.14), we apply Theorem 5.3.2 to conclude that under the conditions $u_{tt}, u_{xxxx} \in C([0,\pi] \times [0,T])$,

$$\max_{0 \leq m \leq N_x} \|\mathbf{u}^m - \mathbf{v}^m\|_{2,h_x} \leq c\,(h_x^2 + h_t).$$

Exercise 5.3.1 *Show that for the matrix*

$$Q = \begin{pmatrix} a & c & & & \\ b & a & c & & \\ & \ddots & \ddots & \ddots & \\ & & b & a & c \\ & & & b & a \end{pmatrix}$$

of order $N \times N$, the eigenvalues are

$$\lambda_j = a + 2\sqrt{bc}\,\cos\left(\frac{j\pi}{N+1}\right), \quad 1 \leq j \leq N.$$

Hint: *For the nontrivial case $bc \neq 0$, write*

$$Q = a\,I + \sqrt{bc}\,D^{-1}\Lambda D$$

with D is a diagonal matrix with the diagonal elements $\sqrt{c/b}$, $(\sqrt{c/b})^2$, ..., $(\sqrt{c/b})^N$, and Λ is a tridiagonal matrix

$$\Lambda = \begin{pmatrix} 0 & 1 & & & \\ 1 & 0 & 1 & & \\ & \ddots & \ddots & \ddots & \\ & & 1 & 0 & 1 \\ & & & 1 & 0 \end{pmatrix}.$$

Then find the eigenvalues of Λ by following the definition of the eigenvalue problem and solving a difference system for components of associated eigenvectors. An alternative approach is to relate the characteristic equation of Λ through its recurssion formula to Chebyshev polynomials of the second kind (cf. [11, p. 497]).

Exercise 5.3.2 *Give a convergence analysis for the Crank-Nicolson scheme (5.1.16)–(5.1.18) by applying Theorem 5.3.2, as is done for the forward and backward schemes in examples.*

Exercise 5.3.3 *The forward, backward, and Crank-Nicolson schemes are all particular members in a family of difference schemes called generalized mid-point methods. Let $\theta \in [0,1]$ be a parameter. Then a generalized mid-*

point scheme for the initial-boundary value problem (5.1.5)–(5.1.7) *is*

$$\frac{v_j^m - v_j^{m-1}}{h_t} = \nu\,\theta\,\frac{v_{j+1}^m - 2v_j^m + v_{j-1}^m}{h_x^2} + \nu\,(1-\theta)\,\frac{v_{j+1}^{m-1} - 2v_j^{m-1} + v_{j-1}^{m-1}}{h_x^2}$$

$$+\,\theta\,f_j^m + (1-\theta)\,f_j^{m-1}, \quad 1 \le j \le N_x - 1,\ 1 \le m \le N_t,$$

supplemented by the boundary condition (5.1.13) *and the initial condition* (5.1.14). *Show that for* $\theta \in [1/2, 1]$, *the scheme is unconditionally stable in both* $\|\cdot\|_{2,h_x}$ *and* $\|\cdot\|_\infty$ *norms; for* $\theta \in [0, 1/2)$, *the scheme is stable in* $\|\cdot\|_{2,h_x}$ *norm if* $2\,(1-2\,\theta)\,r \le 1$, *and it is stable in* $\|\cdot\|_\infty$ *norm if* $2\,(1-\theta)\,r \le 1$. *Determine the convergence orders of the schemes.*

Suggestion for Further Readings

More details on theoretical analysis of the finite difference method, e.g., treatment of other kind of boundary conditions, general spatial domains for higher spatial dimension problems, approximation of hyperbolic or elliptic problems, can be found in several books on the topic, e.g., [154]. For the finite difference method for parabolic problems, [158] is an in-depth survey. Another popular approach to developing finite difference methods for parabolic problems is the *method of lines*; see [11, p. 414] for an introduction that discusses some of the finite difference methods of this chapter.

For initial-boundary value problems of evolution equations in high spatial dimensions, stability for an explicit scheme usually requires the time stepsize to be prohibitively small. On the other hand, some implicit schemes are unconditionally stable, and stability requirement does not impose restriction on the time stepsize. The disadvantage of an implicit scheme is that at each time level we may need to solve an algebraic system of very large scale. The idea of *operator splitting technique* is to split the computation for each time step into several substeps such that each substep is implicit only in one spatial variable and at the same time good stability property is maintained. The resulting schemes are called *alternating direction methods* or *fractional step methods*. See [113, 169] for detailed discussions.

Many physical phenomena are described by conservation laws (conservation of mass, momentum, and energy). Finite difference methods for conservation laws constitute a large research area. The interested reader can consult [60] and [107] .

Extrapolation methods are efficient means to accelerate the convergence of numerical solutions. For extrapolation methods in the context of the finite difference method, see [114].

6

Sobolev Spaces

In this chapter, we review definitions and properties of Sobolev spaces, which are indispensable for a theoretical analysis of partial differential equations and boundary integral equations, as well as being necessary for the analysis of some numerical methods for solving such equations. Most results are stated without proof; detailed proofs of the results can be found in standard references on Sobolev spaces, e.g., [1].

6.1 Weak derivatives

We need the multi-index notation for partial derivatives introduced in Section 1.4.

Our purpose here is to extend the definition of derivatives. To do this, we start with the classical "integration by parts" formula

$$\int_{\Omega} v(\mathbf{x}) \, D^{\alpha}\phi(\mathbf{x}) \, dx = (-1)^{|\alpha|} \int_{\Omega} D^{\alpha}v(\mathbf{x}) \, \phi(\mathbf{x}) \, dx, \qquad (6.1.1)$$

which holds for $v \in C^m(\Omega)$, $\phi \in C_0^{\infty}(\Omega)$ and $|\alpha| \leq m$. This formula, relating differentiation and integration, is a most important formula in calculus. The *weak derivative* is defined in such a way that, first, if the classical derivative exists then the two derivatives coincide so that the weak derivative is an extension of the classical derivative; second, the integration by parts formula (6.1.1) holds. A more general approach for the extension of the classical derivatives is to first introduce the derivatives in the distributional sense.

A detailed discussion of distributions and the derivatives in the distributional sense can be found in several monographs, e.g., [149]. Here we choose to introduce the concept of the weak derivatives directly, which is sufficient for this text.

As preparation, we first introduce the notion of locally integrable functions.

Definition 6.1.1 *Let $1 \leq p < \infty$. A function $v : \Omega \subseteq \mathbb{R}^d \to \mathbb{R}$ is said to be locally p-integrable, $v \in L_{\text{loc}}^p(\Omega)$, if for every $\mathbf{x} \in \Omega$, there is an open neighborhood Ω' of $\mathbf{x}$ such that $\overline{\Omega'} \subseteq \Omega$ and $v \in L^p(\Omega')$.*

Notice that a locally integrable function can behave arbitrarily badly near the boundary $\partial\Omega$. One such example is the function $e^{d(\mathbf{x})^{-1}} \sin(d(\mathbf{x})^{-1})$, where $d(\mathbf{x}) \equiv \inf_{\mathbf{y} \in \partial\Omega} \|\mathbf{x} - \mathbf{y}\|$ is the distance from $\mathbf{x}$ to $\partial\Omega$.

We have the following useful result which will be used repeatedly ([171, p. 18]).

Lemma 6.1.2 (GENERALIZED VARIATIONAL LEMMA) *Let $v \in L_{\text{loc}}^1(\Omega)$ with Ω a non-empty open set in $\mathbb{R}^d$. If*

$$\int_\Omega v(\mathbf{x}) \, \phi(\mathbf{x}) \, dx = 0 \qquad \forall \phi \in C_0^\infty(\Omega),$$

then $v = 0$ a.e. on Ω.

Now we are ready to introduce the concept of a weak derivative.

Definition 6.1.3 *Let Ω be a non-empty open set in $\mathbb{R}^d$, $v, w \in L_{\text{loc}}^1(\Omega)$. Then w is called a weak α^{th} derivative of v if*

$$\int_\Omega v(\mathbf{x}) \, D^\alpha \phi(\mathbf{x}) \, dx = (-1)^{|\alpha|} \int_\Omega w(\mathbf{x}) \, \phi(\mathbf{x}) \, dx \qquad \forall \phi \in C_0^\infty(\Omega). \quad (6.1.2)$$

Lemma 6.1.4 *A weak derivative, if it exists, is uniquely defined up to a set of measure zero.*

Proof. Suppose $v \in L_{\text{loc}}^1(\Omega)$ has two weak α^{th} derivatives $w_1, w_2 \in L_{\text{loc}}^1(\Omega)$. Then from the definition, we have

$$\int_\Omega (w_1(\mathbf{x}) - w_2(\mathbf{x})) \, \phi(\mathbf{x}) \, dx = 0 \quad \forall \phi \in C_0^\infty(\Omega).$$

Applying Lemma 6.1.2, we conclude $w_1 = w_2$ a.e. on Ω. ∎

From the definition of the weak derivative and Lemma 6.1.4, we immediately see the following result holds.

Lemma 6.1.5 *If $v \in C^m(\Omega)$, then for each α with $|\alpha| \leq m$, the classical partial derivative $D^\alpha v$ is also the weak α^{th} partial derivative of v.*

Because of Lemma 6.1.5, it is natural to use all those notations of the classical derivative also for the weak derivative. For example, $\partial_i v = v_{x_i}$ denote the first-order weak derivative of v with respect to x_i.

The weak derivatives defined here coincide with the extension of the classical derivatives discussed in Section 2.4. Let us return to the situation of Example 2.4.2, where the classical differentiation operator $D : C^1[0,1] \to L^2(0,1)$ is extended to the differentiation operator $\hat{D}$ defined over $H^1(0,1)$, the completion of $C^1[0,1]$ under the norm $\|\cdot\|_{1,2}$. For any $v \in H^1(0,1)$, there exists a sequence $\{v_n\} \subseteq C^1[0,1]$ such that

$$\|v_n - v\|_{1,2} \to 0 \quad \text{as } n \to \infty,$$

which implies, as $n \to \infty$,

$$v_n \to v \quad \text{and} \quad Dv_n \to \hat{D}v \quad \text{in } L^2(0,1).$$

Now by the classical integration by parts formula,

$$\int_0^1 v_n(x)\, D\phi(x)\, dx = -\int_0^1 Dv_n(x)\, \phi(x)\, dx \qquad \forall \phi \in C_0^\infty(0,1).$$

Taking the limit as $n \to \infty$ in the above relation, we obtain

$$\int_0^1 v(x)\, D\phi(x)\, dx = -\int_0^1 \hat{D}v(x)\, \phi(x)\, dx \quad \forall \phi \in C_0^\infty(0,1).$$

Hence, $\hat{D}v$ is also the first-order weak derivative of v.

Now we examine some examples of weakly differentiable functions that are not differentiable in the classical sense, as well as some examples of functions that are not weakly differentiable.

Example 6.1.6 *The absolute value function $v(x) = |x|$ is not differentiable at $x = 0$ in the classical sense. Nevertheless the first-order weak derivative of $|x|$ at $x = 0$ exists. Indeed, it is easy to verify that*

$$w(x) = \begin{cases} 1, & x > 0, \\ -1, & x < 0, \\ c_0, & x = 0, \end{cases}$$

where $c_0 \in \mathbb{R}$ is arbitrary, is a first-order weak derivative of the absolute value function.

Example 6.1.7 *Functions with jump discontinuities are not weakly differentiable. For example, define*

$$v(x) = \begin{cases} -1, & -1 < x < 0, \\ c_0, & x = 0, \\ 1, & 0 < x < 1, \end{cases}$$

where $c_0 \in \mathbb{R}$. Let us show that the function v does not have a weak derivative. We argue by contradiction. Suppose v is weakly differentiable with the derivative $w \in L^1_{\text{loc}}(-1,1)$. By definition, we have the identity

$$\int_{-1}^1 v(x)\, \phi'(x)\, dx = -\int_{-1}^1 w(x)\, \phi(x)\, dx \qquad \forall \phi \in C_0^\infty(-1,1).$$

Figure 6.1. A continuous function that is piecewisely smooth

The left-hand side of the relation can be simplified to $-2\phi(0)$. *Hence we have the identity*

$$\int_{-1}^{1} w(x)\,\phi(x)\,dx = 2\phi(0) \qquad \forall\,\phi \in C_0^{\infty}(-1,1).$$

Taking $\phi \in C_0^{\infty}(0,1)$, *we get*

$$\int_{0}^{1} w(x)\,\phi(x)\,dx = 0 \qquad \forall\,\phi \in C_0^{\infty}(0,1).$$

By Lemma 6.1.2, we conclude that $w(x) = 0$ *a.e. on* $(0,1)$. *Similarly,* $w(x) = 0$ *on* $(-1,0)$. *So* $w(x) = 0$ *a.e. on* $(-1,1)$, *and we arrive at the contradictory relation*

$$0 = 2\phi(0) \qquad \forall\,\phi \in C_0^{\infty}(-1,1).$$

Thus the function v *is not weakly differentiable.*

Example 6.1.8 *More generally, assume* $v \in C[a,b]$ *is piecewisely continuously differentiable (Figure 6.1); i.e., there exists a partition of the interval,* $a = x_0 < x_1 < \cdots < x_n = b$, *such that* $v \in C^1[x_{i-1}, x_i]$, $1 \leq i \leq n$. *Then the first-order weak derivative of* v *is*

$$w(x) = \begin{cases} v'(x), & x \in \cup_{i=1}^{n}(x_{i-1}, x_i), \\ \text{arbitrary}, & x = x_i, \quad 0 \leq i \leq n. \end{cases}$$

This result can be verified directly by applying the definition of the weak derivative. Notice that a second-order weak derivative of v *does not exist.*

Example 6.1.9 *In the finite element analysis for solving differential and integral equations, we frequently deal with piecewise polynomials, or piecewise images of polynomials. Suppose* $\Omega \subseteq \mathbb{R}^2$ *is a polygonal domain and is partitioned into polygonal subdomains:*

$$\overline{\Omega} = \bigcup_{n=1}^{N} \overline{\Omega}_n.$$

Each subdomain Ω_i is usually taken to be a triangle or a quadrilateral. Suppose for some non-negative integer k,

$$v \in C^k(\overline{\Omega}), \quad v|_{\Omega_n} \in C^{k+1}(\overline{\Omega}_n), \ 1 \le n \le N.$$

Then the $(k+1)^{th}$ weak partial derivatives of v exist, and for a multi-index α of length $k+1$, the αth weak derivative of v is given by the formula

$$\begin{cases} D^\alpha v(\mathbf{x}), & \mathbf{x} \in \cup_{i=1}^I \Omega_i, \\ \text{arbitrary, otherwise.} \end{cases}$$

When Ω is a general curved domain, Ω_n may be a curved triangle or quadrilateral. Finite element functions are piecewise polynomials or images of piecewise polynomials. The index k is determined by the order of the PDE problem and the type of the finite elements (conforming or non-conforming). For example, for a second-order elliptic boundary value problem, finite element functions for a conforming method are globally continuous and have first-order weak derivatives. For a non-conforming method, the finite element functions are not globally continuous, and hence do not have first-order weak derivatives. Nevertheless, such functions are smooth in each subdomain (element). For details, see Chapter 9.

Most differentiation rules for classical derivatives can be carried over to weak derivatives. Two such examples are the following results, which can be verified directly from the definition of a weak derivative.

Proposition 6.1.10 *Let α be a multi-index, $c_1, c_2 \in \mathbb{R}$. If $D^\alpha u$ and $D^\alpha v$ exist, then $D^\alpha(c_1 u + c_2 v)$ exists and*

$$D^\alpha(c_1 u + c_2 v) = c_1 D^\alpha u + c_2 D^\alpha v.$$

Proposition 6.1.11 *Let $p, q \in (1, \infty)$ be related by $\frac{1}{p} + \frac{1}{q} = 1$. Assume $u, u_{x_i} \in L^p_{\text{loc}}(\Omega)$ and $v, v_{x_i} \in L^q_{\text{loc}}(\Omega)$. Then $(uv)_{x_i}$ exists and*

$$(uv)_{x_i} = u_{x_i} v + u v_{x_i}.$$

We have the following specialized form of the chain rule.

Proposition 6.1.12 *Assume $f \in C^1(\mathbb{R}, \mathbb{R})$ with f' bounded. Suppose $\Omega \subseteq \mathbb{R}^d$ is open bounded, and for some $p \in (1, \infty)$, $v \in L^p(\Omega)$ and $v_{x_i} \in L^p(\Omega)$, $1 \le i \le d$ (i.e., $v \in W^{1,p}(\Omega)$ using the notation of the Sobolev space $W^{1,p}(\Omega)$ to be introduced in the following section). Then $(f(v))_{x_i} \in L^p(\Omega)$, and $(f(v))_{x_i} = f'(v) v_{x_i}$, $1 \le i \le d$.*

Exercise 6.1.1 *Prove Proposition 6.1.12 using the definition of the weak derivatives.*

Exercise 6.1.2 *Let*

$$f(x) = \begin{cases} 1, & -1 < x < 0, \\ ax + b, & 0 \le x < 1. \end{cases}$$

Find a necessary and sufficient condition on a and b for $f(x)$ to be weakly differentiable on $(-1, 1)$. Calculate the weak derivative $f'(x)$ when it exists.

Exercise 6.1.3 *Show that if $v \in W^{1,p}(\Omega)$, $1 \leq p \leq \infty$, then $|v|, v^+, v^- \in W^{1,p}(\Omega)$, and*

$$Dv^+ = \begin{cases} Dv, & \text{a.e. on } \{\mathbf{x} \in \Omega \mid v(\mathbf{x}) > 0\}, \\ 0, & \text{a.e. on } \{\mathbf{x} \in \Omega \mid v(\mathbf{x}) \leq 0\}, \end{cases}$$

$$Dv^- = \begin{cases} 0, & \text{a.e. on } \{\mathbf{x} \in \Omega \mid v(\mathbf{x}) \geq 0\}, \\ -Dv, & \text{a.e. on } \{\mathbf{x} \in \Omega \mid v(\mathbf{x}) < 0\}, \end{cases}$$

where, $v^+ = (|v| + v)/2$ is the positive part of v, and $v^- = (|v| - v)/2$ is the negative part.

Exercise 6.1.4 *Assume v has the αth weak derivative $w_\alpha = D^\alpha u$ and w_α has the βth weak derivative $w_{\alpha+\beta} = D^\beta w_\alpha$. Show that $w_{\alpha+\beta}$ is the $(\alpha+\beta)^{th}$ weak derivative of v.*

6.2   Sobolev spaces

Some properties of Sobolev spaces require a certain degree of regularity of the boundary $\partial\Omega$ of the domain Ω.

Definition 6.2.1 *Let Ω be open and bounded in $\mathbb{R}^d$, and let V denote a function space on $\mathbb{R}^{d-1}$. We say $\partial\Omega$ is of class V if for each point $\mathbf{x}_0 \in \partial\Omega$, there exist an $r > 0$ and a function $g \in V$ such that upon a transformation of the coordinate system if necessary, we have*

$$\Omega \cap B(\mathbf{x}_0, r) = \{\mathbf{x} \in B(\mathbf{x}_0, r) \mid x_d > g(x_1, \ldots, x_{d-1})\}.$$

Here, $B(\mathbf{x}_0, r)$ denotes the d-dimensional ball centered at $\mathbf{x}_0$ with radius r.

In particular, when V consists of Lipschitz continuous functions, we say Ω is a Lipschitz domain. When V consists of C^k functions, we say Ω is a C^k domain. When V consists of $C^{k,\alpha}$ $(0 < \alpha \leq 1)$ functions, we say $\partial\Omega$ is a Hölder boundary of class $C^{k,\alpha}$. See Figure 6.2.

We remark that in engineering applications, most domains are Lipschitz continuous (Figures 6.3 and 6.4). A well-known non-Lipschitz domain is one with cracks (Figure 6.5).

Since $\partial\Omega$ is a compact set in $\mathbb{R}^d$, we can actually find a finite number of points $\{\mathbf{x}_i\}_{i=1}^I$ on the boundary so that for some positive numbers $\{r_i\}_{i=1}^I$ and functions $\{g_i\}_{i=1}^I \subseteq V$,

$$\Omega \cap B(\mathbf{x}_i, r_i) = \{\mathbf{x} \in B(\mathbf{x}_i, r_i) \mid x_d > g_i(x_1, \ldots, x_{d-1})\}$$

upon a transformation of the coordinate system if necessary, and

$$\partial\Omega \subseteq \bigcup_{i=1}^{I} B(\mathbf{x}_i, r_i).$$

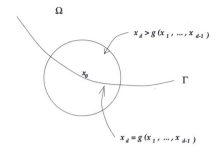

Figure 6.2. Smoothness of the boundary

Figure 6.3. Smooth domains

Figure 6.4. Lipschitz domains

6.2.1   Sobolev spaces of integer order

Definition 6.2.2 *Let k be a non-negative integer, $p \in [1, \infty]$. The Sobolev space $W^{k,p}(\Omega)$ is the set of all the functions $v \in L^1_{loc}(\Omega)$ such that for each multi-index α with $|\alpha| \leq k$, the αth weak derivative $D^\alpha v$ exists and*

Figure 6.5. A crack domain

$D^\alpha v \in L^p(\Omega)$. *The norm in the space* $W^{k,p}(\Omega)$ *is defined as*

$$\|v\|_{W^{k,p}(\Omega)} = \begin{cases} \left(\displaystyle\sum_{|\alpha|\leq k} \|D^\alpha v\|^p_{L^p(\Omega)} \right)^{1/p} , & 1 \leq p < \infty, \\ \displaystyle\max_{|\alpha|\leq k} \|D^\alpha v\|_{L^\infty(\Omega)}, & p = \infty. \end{cases}$$

When $p = 2$, *we write* $H^k(\Omega) \equiv W^{k,2}(\Omega)$.

Usually we replace $\|v\|_{W^{k,p}(\Omega)}$ by the simpler notations $\|v\|_{k,p,\Omega}$, or even $\|v\|_{k,p}$ when no confusion results. The standard semi-norm over the space $W^{k,p}(\Omega)$ is

$$|v|_{W^{k,p}(\Omega)} = \begin{cases} \left(\displaystyle\sum_{|\alpha|= k} \|D^\alpha v\|^p_{L^p(\Omega)} \right)^{1/p} , & 1 \leq p < \infty, \\ \displaystyle\max_{|\alpha|= k} \|D^\alpha v\|_{L^\infty(\Omega)}, & p = \infty. \end{cases}$$

It is not difficult to see that $W^{k,p}(\Omega)$ is a normed space. Moreover, we have the following result.

Theorem 6.2.3 *The Sobolev space* $W^{k,p}(\Omega)$ *is a Banach space.*

Proof. Let $\{v_n\}$ be a Cauchy sequence in $W^{k,p}(\Omega)$. Then for any multi-index α with $|\alpha| \leq k$, $\{D^\alpha v_n\}$ is a Cauchy sequence in $L^p(\Omega)$. Since $L^p(\Omega)$ is complete, there exists a $v_\alpha \in L^p(\Omega)$ such that

$$D^\alpha v_n \to v_\alpha \quad \text{in } L^p(\Omega), \text{ as } n \to \infty.$$

Let us show that $v_\alpha = D^\alpha v$ where v is the limit of the sequence $\{v_n\}$ in $L^p(\Omega)$. For any $\phi \in C_0^\infty(\Omega)$,

$$\int_\Omega v_n(\mathbf{x}) \, D^\alpha \phi(\mathbf{x}) \, dx = (-1)^{|\alpha|} \int_\Omega D^\alpha v_n(\mathbf{x}) \, \phi(\mathbf{x}) \, dx.$$

Letting $n \to \infty$, we obtain

$$\int_\Omega v(\mathbf{x}) \, D^\alpha \phi(\mathbf{x}) \, dx = (-1)^{|\alpha|} \int_\Omega v_\alpha(\mathbf{x}) \, \phi(\mathbf{x}) \, dx \quad \forall \, \phi \in C_0^\infty(\Omega).$$

Therefore, $v_\alpha = D^\alpha v$ and $v_n \to v$ in $W^{k,p}(\Omega)$ as $n \to \infty$. $\blacksquare$

A simple consequence of the theorem is the following result.

Corollary 6.2.4 *The Sobolev space $H^k(\Omega)$ is a Hilbert space with the inner product*

$$(u, v)_k = \int_\Omega \sum_{|\alpha| \le k} D^\alpha u(\mathbf{x}) \, D^\alpha v(\mathbf{x}) \, dx, \qquad u, v \in H^k(\Omega).$$

Like the case for Lebegue spaces $L^p(\Omega)$, it can be shown that the Sobolev space $W^{k,p}(\Omega)$ is reflexive if and only if $p \in (1, \infty)$.

Let us examine some examples of Sobolev functions.

Example 6.2.5 *Assume $\Omega = \{\mathbf{x} \in \mathbb{R}^d \mid |\mathbf{x}| < 1\}$ is the unit ball, and let $v(\mathbf{x}) = |\mathbf{x}|^\lambda$, where λ is real. Let $p \in [1, \infty)$. Notice that*

$$\|v\|_{L^p(\Omega)}^p = \int_\Omega |\mathbf{x}|^{\lambda p} dx = c \int_0^1 r^{\lambda p + d - 1} dr.$$

So

$$v \in L^p(\Omega) \quad \Longleftrightarrow \quad \lambda > -d/p.$$

It can be verified that the first-order weak derivative v_{x_i} is given by the formula, if v_{x_i} exists,

$$v_{x_i}(\mathbf{x}) = \lambda \, |\mathbf{x}|^{\lambda - 2} x_i, \quad \mathbf{x} \ne \mathbf{0}.$$

Thus

$$|\nabla v(\mathbf{x})| = |\lambda| \, |\mathbf{x}|^{\lambda - 1}, \quad \mathbf{x} \ne \mathbf{0}.$$

Now

$$\||\nabla v|\|_{L^p(\Omega)}^p = |\lambda|^p \int_\Omega |\mathbf{x}|^{(\lambda - 1) p} dx = c \int_0^1 r^{(\lambda - 1) p + d - 1} dr.$$

We see that

$$v \in W^{1,p}(\Omega) \quad \Longleftrightarrow \quad \lambda > 1 - \frac{d}{p}.$$

More generally, for a non-negative integer k, we have

$$v \in W^{k,p}(\Omega) \quad \Longleftrightarrow \quad \lambda > k - \frac{d}{p}.$$

Example 6.2.6 *Are elements of $H^1(\Omega)$ continuous? Not necessarily! Consider the example*

$$v(\mathbf{x}) = \log\left(\log\left(\frac{1}{r}\right)\right), \qquad \mathbf{x} \in \mathbb{R}^2, \quad r = |\mathbf{x}|,$$

with $\Omega = B(0, \beta)$, a circle of radius $\beta < 1$ in the plane. Then

$$\int_\Omega |\nabla v(\mathbf{x})|^2 \, dx = \frac{-2\pi}{\log \beta} < \infty$$

and also easily, $\|v\|_{L^2(\Omega)} < \infty$. Thus $v \in H^1(\Omega)$, but $v(\mathbf{x})$ is unbounded as $\mathbf{x} \to \mathbf{0}$. For conditions ensuring continuity of functions from a Sobolev space, see Theorem 6.3.7.

Example 6.2.7 *In the theory of finite elements, we need to analyze the global regularity of a finite element function from its regularity on each element. Let $\Omega \subseteq \mathbb{R}^d$ be a bounded Lipschitz domain, partitioned into Lipschitz subdomains:*

$$\overline{\Omega} = \bigcup_{n=1}^N \overline{\Omega}_n.$$

Suppose for some non-negative integer k and some real $p \in [1, \infty)$ or $p = \infty$,

$$v|_{\Omega_n} \in W^{k+1,p}(\Omega_n), \qquad 1 \le n \le N, \quad v \in C^k(\overline{\Omega}).$$

Let us show that

$$v \in W^{k+1,p}(\Omega).$$

Evidently, it is enough to prove the result for the case $k = 0$. Thus let $v \in C(\overline{\Omega})$ be such that for each $n = 1, \ldots, N$, $v \in W^{1,p}(\Omega_n)$. For each i, $1 \le i \le d$, we need to show that $\partial_i v$ exists as a weak derivative and belongs to $L^p(\Omega)$. An obvious candidate for $\partial_i v$ is

$$w_i(\mathbf{x}) = \begin{cases} \partial_i v(\mathbf{x}), & \mathbf{x} \in \cup_{n=1}^N \Omega_n, \\ \text{arbitrary, otherwise.} \end{cases}$$

Certainly $w_i \in L^p(\Omega)$. So we only need to verify $w_i = \partial_i v$. By definition, we need to prove

$$\int_\Omega w_i \, \phi \, dx = - \int_\Omega v \, \partial_i \phi \, dx \quad \forall \phi \in C_0^\infty(\Omega).$$

Denote the unit outward normal vector on $\partial\Omega_n$ by $\boldsymbol{\nu} = (\nu_1, \ldots, \nu_d)^T$ that exists a.e. since Ω_n has a Lipschitz boundary. We have

$$\int_\Omega w_i \, \phi \, dx = \sum_{n=1}^N \int_{\Omega_n} \partial_i v \, \phi \, dx$$

$$= \sum_{n=1}^N \int_{\partial\Omega_n} v|_{\Omega_n} \, \phi \, \nu_i \, ds - \sum_{n=1}^N \int_{\Omega_n} v \, \partial_i \phi \, dx$$

$$= \sum_{n=1}^N \int_{\partial\Omega_n} v|_{\Omega_n} \, \phi \, \nu_i \, ds - \int_\Omega v \, \partial_i \phi \, dx,$$

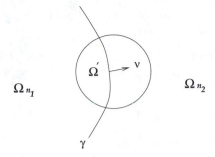

Ω_{n_1}
Ω_{n_2}

Figure 6.6. Two adjacent subdomains

where we apply an integration by parts formula to the integrals on the subdo-mains Ω_n. Integration by parts formulas are valid for functions from certain Sobolev spaces (cf. Section 6.4). Now the sum of the boundary integrals is zero: Either a portion of $\partial\Omega_n$ is a part of $\partial\Omega$ and $\phi = 0$ along this portion, or the contributions from the adjacent subdomains cancel each other. Thus,

$$\int_\Omega w_i\,\phi\,dx = -\int_\Omega v\,\partial_i\phi\,dx \quad \forall\phi \in C_0^\infty(\Omega).$$

By definition, $w_i = \partial_i v$.

Example 6.2.8 *Continuing the preceding example, let us show that if $v \in W^{k+1,p}(\Omega)$ and $v \in C^k(\overline{\Omega}_n)$, $1 \le n \le N$, then $v \in C^k(\overline{\Omega})$. Obviously it is enough to prove the statement for $k = 0$. Let us argue by contradiction. Thus we assume $v \in W^{1,p}(\Omega)$ and $v \in C(\overline{\Omega}_n)$, $1 \le n \le N$, but there are two adjacent subdomains Ω_{n_1} and Ω_{n_2} and a non-empty open set $\Omega' \subseteq \Omega_{n_1} \cup \Omega_{n_2}$ (Figure 6.6) such that*

$$v|_{\Omega_{n_1}} > v|_{\Omega_{n_2}} \qquad on\ \gamma \cap \Omega',$$

where $\gamma = \partial\Omega_{n_1} \cap \partial\Omega_{n_2}$.

By shrinking the set Ω' if necessary, we may assume there is an i between 1 and d, such that $\nu_i > 0$ (or $\nu_i < 0$) on $\gamma \cap \Omega'$. Here ν_i is the ith component of the unit outward normal $\boldsymbol{\nu}$ on γ with respect to Ω_{n_1}. Then we choose $\phi \in C_0^\infty(\Omega') \subseteq C_0^\infty(\Omega)$ with the property $\phi > 0$ on $\gamma \cap \Omega'$. Now

$$\int_\Omega \partial_i v\,\phi\,dx = \sum_{l=1}^2 \int_{\Omega_{n_l}} \partial_i v\,\phi\,dx$$

$$= \sum_{l=1}^2 \int_{\partial\Omega_{n_l}} v|_{\Omega_{n_l}}\,\phi\,\nu_i\,ds - \sum_{l=1}^2 \int_{\Omega_{n_l}} v\,\partial_i\phi\,dx$$

$$= \int_\gamma (v|_{\Omega_{n_1}} - v|_{\Omega_{n_2}})\,\phi\,\nu_i\,ds - \int_\Omega v\,\partial_i\phi\,dx.$$

By the assumptions, the boundary integral is non-zero. This contradicts the definition of the weak derivative.

Combining the results from Examples 6.2.7 and 6.2.8, we see that under the assumption $v|_{\Omega_n} \in C^k(\overline{\Omega}_n) \cap W^{k+1,p}(\Omega_n)$, $1 \leq n \leq N$, we have the conclusion

$$v \in C^k(\overline{\Omega}) \Longleftrightarrow v \in W^{k+1,p}(\Omega).$$

Some important properties of the Sobolev spaces will be discussed in the next section. In general the space $C_0^\infty(\Omega)$ is not dense in $W^{k,p}(\Omega)$. So it makes sense to bring in the following definition.

Definition 6.2.9 *Let* $W_0^{k,p}(\Omega)$ *be the closure of* $C_0^\infty(\Omega)$ *in* $W^{k,p}(\Omega)$. *When* $p = 2$, *we denote* $H_0^k(\Omega) \equiv W_0^{k,2}(\Omega)$.

We interpret $W_0^{k,p}(\Omega)$ to be the space of all the functions v in $W^{k,p}(\Omega)$ with the "property" that

$$D^\alpha v(\mathbf{x}) = 0 \quad \text{on } \partial\Omega, \quad \forall \alpha \text{ with } |\alpha| \leq k - 1.$$

The meaning of this statement is made clear later after the trace theorems are presented.

6.2.2   Sobolev spaces of real order

It is possible to extend the definition of Sobolev spaces with non-negative integer order to any real order. We first introduce Sobolev spaces of positive real order. In this subsection, we assume $p \in [1, \infty)$.

Definition 6.2.10 *Let* $s = k + \sigma$ *with* $k \geq 0$ *an integer and* $\sigma \in (0, 1)$. *Then we define the Sobolev space*

$$W^{s,p}(\Omega) = \left\{ v \in W^{k,p}(\Omega) \mid \frac{|D^\alpha v(\mathbf{x}) - D^\alpha v(\mathbf{y})|}{\|\mathbf{x} - \mathbf{y}\|^{\sigma + d/p}} \in L^p(\Omega \times \Omega) \right.$$
$$\left. \forall \alpha : |\alpha| = k \right\}$$

with the norm

$$\|v\|_{s,p,\Omega} = \left(\|v\|_{k,p,\Omega}^p \right.$$
$$\left. + \sum_{|\alpha|=k} \int_{\Omega \times \Omega} \frac{|D^\alpha v(\mathbf{x}) - D^\alpha v(\mathbf{y})|^p}{\|\mathbf{x} - \mathbf{y}\|^{\sigma p + d}} \, dx \, dy \right)^{1/p}.$$

It can be shown that the space $W^{s,p}(\Omega)$ is a Banach space. It is reflexive if and only if $p \in (1, \infty)$. When $p = 2$, $H^s(\Omega) \equiv W^{s,2}(\Omega)$ is a Hilbert space with the inner product

$$(u, v)_{s,\Omega} = (u, v)_{k,\Omega}$$
$$+ \sum_{|\alpha|=k} \int_{\Omega \times \Omega} \frac{(D^\alpha u(\mathbf{x}) - D^\alpha u(\mathbf{y}))\,(D^\alpha v(\mathbf{x}) - D^\alpha v(\mathbf{y}))}{\|\mathbf{x} - \mathbf{y}\|^{2\sigma + d}}\, dx\, dy.$$

Most properties of Sobolev spaces of integer order, such as density of smooth functions, extension theorem and Sobolev embedding theorems discussed in the next section, carry over to Sobolev spaces of positive real order introduced here. The introduction of the spaces $W^{s,p}(\Omega)$ in this text serves two purposes: as a preparation for the definition of Sobolev spaces over boundaries and for a more precise statement of Sobolev trace theorems. Therefore, we will not give detailed discussions of the properties of the spaces $W^{s,p}(\Omega)$. An interested reader can consult [90, Chap. 4, Part I].

The space $C_0^\infty(\Omega)$ does not need to be dense in $W^{s,p}(\Omega)$. So we introduce the following definition.

Definition 6.2.11 *Let $s \geq 0$. Then we define $W_0^{s,p}(\Omega)$ to be the closure of the space $C_0^\infty(\Omega)$ in $W^{s,p}(\Omega)$. When $p = 2$, we have a Hilbert space $H_0^s(\Omega) \equiv W_0^{s,2}(\Omega)$.*

With the spaces $W_0^{s,p}(\Omega)$, we can then define Sobolev spaces with negative order.

Definition 6.2.12 *Let $s \geq 0$, either an integer or a non-integer. Let $p \in [1, \infty)$ and denote its conjugate exponent p' defined by the relation $1/p + 1/p' = 1$. Then we define $W^{-s,p'}(\Omega)$ to be the dual space of $W_0^{s,p}(\Omega)$. In particular, $H^{-s}(\Omega) \equiv W^{-s,2}(\Omega)$.*

On several occasions later, we need to use in particular the Sobolev space $H^{-1}(\Omega)$, defined as the dual of $H_0^1(\Omega)$. Thus, any $\ell \in H^{-1}(\Omega)$ is a bounded linear functional on $H_0^1(\Omega)$:

$$|\ell(v)| \leq M \|v\| \qquad \forall\, v \in H_0^1(\Omega).$$

The norm of ℓ is

$$\|\ell\|_{H^{-1}(\Omega)} = \sup_{v \in H_0^1(\Omega)} \frac{\ell(v)}{\|v\|_{H_0^1(\Omega)}}.$$

Any function $f \in L^2(\Omega)$ naturally induces a bounded linear functional $f \in H^{-1}(\Omega)$ by the relation

$$\langle f, v \rangle = \int_\Omega f\, v\, dx \qquad \forall\, v \in H_0^1(\Omega).$$

Sometimes even when $f \in H^{-1}(\Omega) \backslash L^2(\Omega)$, we write $\int_\Omega f\, v\, dx$ for the duality pairing $\langle f, v \rangle$ between $H^{-1}(\Omega)$ and $H_0^1(\Omega)$, although integration in this situation does not make sense.

It can be shown that if $\ell \in H^{-1}(\Omega)$, then there exist $L^2(\Omega)$ functions ℓ_0, $\ldots$, ℓ_d, such that

$$\ell(v) = \int_\Omega \left[\ell_0 v + \sum_{i=1}^d \ell_i v_{x_i} \right] dx \qquad \forall\, v \in H_0^1(\Omega).$$

Thus formally (or in the sense of distributions),

$$\ell = \ell_0 - \sum_{i=1}^d \frac{\partial \ell_i}{\partial x_i};$$

i.e., $H^{-1}(\Omega)$ functions can be obtained by differentiating $L^2(\Omega)$ functions.

6.2.3 Sobolev spaces over boundaries

To deal with function spaces over boundaries, we introduce the following Sobolev spaces.

Definition 6.2.13 *Let $k \geq 0$ be an integer, $\alpha \in (0,1]$, $s \in [0, k + \alpha]$ and $p \in [1,\infty)$. Assume a set of local representations of the boundary given by*

$$\partial\Omega \cap B(\mathbf{x}_i, r_i) = \{\mathbf{x} \in B(\mathbf{x}_i, r_i) \mid x_d = g_i(x_1, \ldots, x_{d-1})\}$$

for $i = 1, \ldots, I$, with open $D_i \subseteq \mathbb{R}^{d-1}$ the domain of g_i; and assume every point of $\partial\Omega$ lies in at least one of these local representations. We assume $g_i \in C^{k,\alpha}(D_i)$ for all i. A decomposition of $\partial\Omega$ into a finite number I of such subdomains $g_i(D_i)$ is called a "patch system." For $s \leq k + \alpha$, we define the Sobolev space $W^{s,p}(\partial\Omega)$ as follows:

$$W^{s,p}(\partial\Omega) = \{v \in L^2(\partial\Omega) \mid v \circ g_i \in W^{s,p}(D_i),\ i = 1, \ldots, I\}.$$

The norm in $W^{s,p}(\partial\Omega)$ is defined by

$$\|v\|_{W^{s,p}(\partial\Omega)} = \max_i \|v \circ g_i\|_{W^{s,p}(D_i)}.$$

Other definitions equivalent to this norm are possible. When $p = 2$, we obtain a Hilbert space $H^s(\partial\Omega) \equiv W^{s,2}(\partial\Omega)$.

Exercise 6.2.1 *Show that for non-negative integers k and real $p \in [1,\infty]$, the quantity $\|\cdot\|_{W^{k,p}(\Omega)}$ defines a norm.*

Exercise 6.2.2 *Consider the function*

$$f(x) = \begin{cases} x^2, & 0 \leq x \leq 1, \\ x^3, & -1 \leq x \leq 0. \end{cases}$$

Determine the largest possible integer k for which $f \in H^k(-1,1)$.

Exercise 6.2.3 *Show that $C^k(\overline{\Omega}) \subseteq W^{k,p}(\Omega)$ for any $p \in [1,\infty]$.*

Exercise 6.2.4 *Is it true that $C^\infty(\Omega) \subseteq W^{k,p}(\Omega)$?*

Exercise 6.2.5 *Show that there exists a constant c depending only on k such that*

$$\|av\|_{H^k(\Omega)} \leq c \|a\|_{C^k(\overline{\Omega})} \|v\|_{H^k(\Omega)} \qquad \forall\, a \in C^k(\overline{\Omega}),\ v \in H^k(\Omega).$$

6.3   Properties

We collect some important properties of the Sobolev spaces in this section. Most properties are stated for Sobolev spaces of non-negative integer order, although they can be extended to Sobolev spaces of real order. We refer to Sobolev spaces of real order only when it is necessary to do so, e.g., in presentation of trace theorems. Properties of the Sobolev spaces over boundaries are summarized in [90, Chap. 4, Part I].

6.3.1   Approximation by smooth functions

Inequalities involving Sobolev functions are usually proved for smooth functions first, followed by a density argument. A theoretical basis for this technique is density results of smooth functions in Sobolev spaces.

Theorem 6.3.1 *Assume $v \in W^{k,p}(\Omega)$, $1 \leq p < \infty$. Then there exists a sequence $\{v_n\} \subseteq C^\infty(\Omega) \cap W^{k,p}(\Omega)$ such that*

$$\|v_n - v\|_{k,p} \to 0 \qquad \text{as } n \to \infty.$$

Note that in this theorem the approximation functions v_n are smooth only in the interior of Ω. To have the smoothness up to the boundary of the approximating sequence, we need to make a smoothness assumption on the boundary of Ω.

Theorem 6.3.2 *Assume Ω is a Lipschitz domain, $v \in W^{k,p}(\Omega)$, $1 \leq p < \infty$. Then there exists a sequence $\{v_n\} \subseteq C^\infty(\overline{\Omega})$ such that*

$$\|v_n - v\|_{k,p} \to 0 \quad \text{as } n \to \infty.$$

Proofs of these density theorems can be found, e.g., in [48].

Since $C^\infty(\overline{\Omega}) \subseteq C^k(\overline{\Omega}) \subseteq W^{k,p}(\Omega)$, we see from Theorem 6.3.2 that under the assumption Ω is Lipschitz continuous, the space $W^{k,p}(\Omega)$ is the completion of the space $C^\infty(\overline{\Omega})$ with respect to the norm $\|\cdot\|_{k,p}$.

From the definition of the space $W_0^{k,p}(\Omega)$, we immediately obtain the following density result.

Theorem 6.3.3 *For any $v \in W_0^{k,p}(\Omega)$, there exists a sequence $\{v_n\} \subseteq C_0^\infty(\Omega)$ such that*

$$\|v_n - v\|_{k,p} \to 0 \quad \text{as } n \to \infty.$$

The definitions of the Sobolev spaces over Ω can be extended in a straightforward fashion to those over the whole space $\mathbb{R}^d$ or other unbounded domains. When $\Omega = \mathbb{R}^d$, smooth functions are dense in Sobolev spaces.

Theorem 6.3.4 *Assume* $k \geq 0$, $p \in [1, \infty)$. *Then the space* $C_0^\infty(\mathbb{R}^d)$ *is dense in* $W^{k,p}(\mathbb{R}^d)$.

6.3.2   Extensions

Extension theorems are also useful in proving some relations involving Sobolev functions. A rather general form of extension theorems is the following universal extension theorem, proved in [150, Theorem 5, p. 181].

Theorem 6.3.5 *Assume* Ω *is an open half-space or an open, bounded Lipschitz domain in* $\mathbb{R}^d$. *Then there is an extension operator* E *such that for any non-negative integer* k *and any* $p \in [1, \infty]$, E *is a linear continuous operator from* $W^{k,p}(\Omega)$ *to* $W^{k,p}(\mathbb{R}^d)$; *in other words, for any* $v \in W^{k,p}(\Omega)$, *we have* $Ev \in W^{k,p}(\mathbb{R}^d)$, $Ev = v$ *in* Ω, Ev *is infinitely smooth on* $\mathbb{R}^d \backslash \overline{\Omega}$, *and*

$$\|Ev\|_{W^{k,p}(\mathbb{R}^d)} \leq c\, \|v\|_{W^{k,p}(\Omega)}$$

for some constant c *independent of* v.

Notice that in the above theorem, the extension operator E works for all possible values of k and p. In Exercise 6.3.1, we consider a simple extension operator from $W^{k,p}(\mathbb{R}_+^d)$ to $W^{k,p}(\mathbb{R}^d)$, whose definition depends on the value k.

6.3.3   Sobolev embedding theorems

Sobolev embedding theorems are important, e.g., in analyzing the regularity of a weak solution of a boundary value problem.

Definition 6.3.6 *Let* V *and* W *be two Banach spaces with* $V \subseteq W$. *We say the space* V *is continuously embedded in* W *and write* $V \hookrightarrow W$, *if*

$$\|v\|_W \leq c\, \|v\|_V \quad \forall\, v \in V. \tag{6.3.1}$$

We say the space V *is compactly embedded in* W *and write* $V \hookrightarrow\hookrightarrow W$, *if* (6.3.1) *holds and each bounded sequence in* V *has a convergent subsequence in* W.

If $V \hookrightarrow W$, the functions in V are more smooth than the remaining functions in W. Proofs of most parts of the following two theorems can be found in [48]. The first theorem is on embedding of Sobolev spaces, and the second on compact embedding.

Theorem 6.3.7 *Let* $\Omega \subseteq \mathbb{R}^d$ *be a non-empty open bounded Lipschitz domain. Then the following statements are valid.*

(a) *If* $k < \frac{d}{p}$, *then* $W^{k,p}(\Omega) \hookrightarrow L^q(\Omega)$ *for any* $q \leq p^*$, *where* $\frac{1}{p^*} = \frac{1}{p} - \frac{d}{p}$.

(b) *If* $k = \frac{d}{p}$, *then* $W^{k,p}(\Omega) \hookrightarrow L^q(\Omega)$ *for any* $q < \infty$.

(c) *If* $k > \frac{d}{p}$, *then*

$$W^{k,p}(\Omega) \hookrightarrow C^{k-\left[\frac{d}{p}\right]-1,\beta}(\Omega),$$

where

$$\beta = \begin{cases} \left[\frac{d}{p}\right] + 1 - \frac{d}{p}, & \text{if } \frac{d}{p} \neq \text{integer,} \\ \text{any positive number } < 1, & \text{if } \frac{d}{p} = \text{integer.} \end{cases}$$

Here $[x]$ denotes the integer part of x, i.e., the largest integer less than or equal to x. We remark that in the one-dimensional case, with $\Omega = (a, b)$ a bounded interval, we have

$$W^{k,p}(a, b) \hookrightarrow C[a, b]$$

for any $k \geq 1$, $p \geq 1$.

Theorem 6.3.8 *Let* $\Omega \subseteq \mathbb{R}^d$ *be a non-empty open bounded Lipschitz domain. Then the following statements are valid.*

(a) *If* $k < \frac{d}{p}$, *then* $W^{k,p}(\Omega) \hookrightarrow\hookrightarrow L^q(\Omega)$ *for any* $q < p^*$, *where* $\frac{1}{p^*} = \frac{1}{p} - \frac{d}{p}$.

(b) *If* $k = \frac{d}{p}$, *then* $W^{k,p}(\Omega) \hookrightarrow\hookrightarrow L^q(\Omega)$ *for any* $q < \infty$.

(c) *If* $k > \frac{d}{p}$, *then*

$$W^{k,p}(\Omega) \hookrightarrow\hookrightarrow C^{k-\left[\frac{d}{p}\right]-1,\beta}(\Omega),$$

where $\beta \in [0, \left[\frac{d}{p}\right] + 1 - \frac{d}{p})$.

How to remember these results? We take Theorem 6.3.7 as an example. The larger the product kp, the smoother the functions from the space $W^{k,p}(\Omega)$. There is a critical value d (the dimension of the domain Ω) for this product such that if $kp > d$, then a $W^{k,p}(\Omega)$ function is actually continuous (or more precisely, is equal to a continuous function a.e.). When $kp < d$, a $W^{k,p}(\Omega)$ function belongs to $L^{p^*}(\Omega)$ for an exponent p^* larger than p. To determine the exponent p^*, we start from the condition $kp < d$, which is written as $1/p - d/k > 0$. Then $1/p^*$ is defined to be the difference $1/p - d/k$. When $kp > d$, it is usually useful to know if a $W^{k,p}(\Omega)$ function has continuous derivatives up to certain order. We begin with

$$W^{k,p}(\Omega) \hookrightarrow C(\overline{\Omega}) \qquad \text{if } k > \frac{p}{d}.$$

Then we apply this embedding result to derivatives of Sobolev functions; it is easy to see that

$$W^{k,p}(\Omega) \hookrightarrow C^l(\overline{\Omega}) \qquad \text{if} \quad k - l > \frac{p}{d}.$$

A direct consequence of Theorem 6.3.8 is the following compact embedding result.

Theorem 6.3.9 *Let k and l be non-negative integers, $k > l$, and $p \in [1, \infty]$. Let $\Omega \subseteq \mathbb{R}^d$ be a non-empty open bounded Lipschitz domain. Then $W^{k,p}(\Omega) \hookrightarrow\hookrightarrow W^{l,p}(\Omega)$.*

6.3.4   Traces

Sobolev spaces are defined through $L^p(\Omega)$ spaces. Hence Sobolev functions are uniquely defined only a.e. in Ω. Now that the boundary $\partial\Omega$ has measure zero in $\mathbb{R}^d$, it seems the boundary value of a Sobolev function is not well-defined. Nevertheless it is possible to define the trace of a Sobolev function on the boundary in such a way that for a Sobolev function that is continuous up to the boundary, its trace coincides with its boundary value.

Theorem 6.3.10 *Assume Ω is an open, bounded Lipschitz domain in $\mathbb{R}^d$, $1 \le p < \infty$. Then there exists a continuous linear operator $\gamma : W^{1,p}(\Omega) \to L^p(\partial\Omega)$ such that—*

(a) *$\gamma v = v|_{\partial\Omega}$ if $v \in W^{1,p}(\Omega) \cap C(\overline{\Omega})$.*

(b) *For some constant $c > 0$, $\|\gamma v\|_{L^p(\partial\Omega)} \le c \|v\|_{W^{1,p}(\Omega)} \; \forall\, v \in W^{1,p}(\Omega)$.*

(c) *The mapping $\gamma : W^{1,p}(\Omega) \to L^p(\partial\Omega)$ is compact; i.e., for any bounded sequence $\{v_n\}$ in $W^{1,p}(\Omega)$, there is a subsequence $\{v_{n'}\} \subseteq \{v_n\}$ such that $\{\gamma v_{n'}\}$ is convergent in $L^p(\partial\Omega)$.*

The operator γ is called the *trace operator*, and γv can be called the *generalized boundary value* of v. The trace operator is neither an injection nor a surjection from $W^{1,p}(\Omega)$ to $L^p(\partial\Omega)$. The range $\gamma(W^{1,p}(\Omega))$ is a space smaller than $L^p(\partial\Omega)$, namely, $W^{1-\frac{1}{p},p}(\partial\Omega)$, a positive order Sobolev space over the boundary.

In studying boundary value problems, necessarily we need to be able to impose essential boundary conditions properly in formulations. For second-order boundary value problems, essential boundary conditions involve only function values on the boundary, so Theorem 6.3.10 is enough for the purpose. For higher-order boundary value problems, we need to use the traces of partial derivatives on the boundary. For example, for fourth-order boundary value problems, any boundary conditions involving derivatives of order at most one are treated as essential boundary conditions. Since a tangential derivative of a function on the boundary can be obtained by taking a differentiation of the boundary value of the function, we only need to use traces of a function and its normal derivative.

Let $\boldsymbol{\nu} = (\nu_1, \ldots, \nu_d)^T$ denote the outward unit normal to the boundary Γ of Ω. Recall that if $v \in C^1(\overline{\Omega})$, then its classical normal derivative on the boundary is

$$\frac{\partial v}{\partial \nu} = \sum_{i=1}^{d} \frac{\partial v}{\partial x_i} \nu_i.$$

The following theorem states the fact that for a function from certain Sobolev spaces, it is possible to define a *generalized normal derivative* that is an extension of the classical normal derivative.

Theorem 6.3.11 *Assume Ω is a bounded, open set with a $C^{1,1}$ boundary Γ. Assume that $1 \le p \le \infty$ and $m > 1 + \frac{1}{p}$. Then there exist unique bounded linear and surjective mappings $\gamma_0 : W^{m,p}(\Omega) \to W^{m-\frac{1}{p},p}(\Gamma)$ and $\gamma_1 : W^{m,p}(\Omega) \to W^{m-1-\frac{1}{p},p}(\Gamma)$ such that $\gamma_0 v = v|_\Gamma$ and $\gamma_1 v = (\partial v/\partial \nu)|_\Gamma$ when $v \in W^{m,p}(\Omega) \cap C^1(\overline{\Omega})$.*

6.3.5   Equivalent norms

In the study of weak formulations of boundary value problems, it is convenient to use equivalent norms over Sobolev spaces or Sobolev subspaces. There are some powerful general results, called norm equivalence theorems, for the purpose of generating various equivalent norms on Sobolev spaces. Beore stating the norm equivalence results, we recall the semi-norm defined by

$$|v|_{k,p,\Omega} = \left(\int_\Omega \sum_{|\alpha|=k} |D^\alpha v|^p \, dx \right)^{1/p}$$

over the space $W^{k,p}(\Omega)$ for $p < \infty$. It can be shown that if Ω is connected and $|v|_{k,p,\Omega} = 0$, then v is a polynomial of degree less than or equal to $k-1$.

Theorem 6.3.12 *Let Ω be an open, bounded, connected set in $\mathbb{R}^d$ with a Lipschitz boundary, $k \ge 1$, $1 \le p < \infty$. Assume $f_j : W^{k,p}(\Omega) \to \mathbb{R}$, $1 \le j \le J$, are semi-norms on $W^{k,p}(\Omega)$ satisfying two conditions:*

(H1) $0 \le f_j(v) \le c \|v\|_{k,p,\Omega} \ \forall v \in W^{k,p}(\Omega), 1 \le j \le J.$

(H2) *If v is a polynomial of degree less than or equal to $k-1$ and $f_j(v) = 0$, $1 \le j \le J$, then $v = 0$.*

Then the quantity

$$\|v\| = |v|_{k,p,\Omega} + \sum_{j=1}^{J} f_j(v) \tag{6.3.2}$$

or

$$\|v\| = \left[|v|_{k,p,\Omega}^p + \sum_{j=1}^J f_j(v)^p\right]^{1/p} \tag{6.3.3}$$

defines a norm on $W^{k,p}(\Omega)$, which is equivalent to the norm $\|v\|_{k,p,\Omega}$.

Proof. We prove that the quantity (6.3.2) is a norm on $W^{k,p}(\Omega)$ equivalent to the norm $\|u\|_{k,p,\Omega}$. That the quantity (6.3.3) is also an equivalent norm can be proved similarly.

By the condition (H1), we see that for some constant $c > 0$,

$$\|v\| \le c\|v\|_{k,p,\Omega} \qquad \forall v \in W^{k,p}(\Omega).$$

So we only need to show that there is another constant $c > 0$ such that

$$\|v\|_{k,p,\Omega} \le c\|v\| \qquad \forall v \in W^{k,p}(\Omega).$$

We argue by contradiction. Suppose this inequality is false. Then we can find a sequence $\{v_l\} \subseteq W^{k,p}(\Omega)$ with the properties

$$\|v_l\|_{k,p,\Omega} = 1, \tag{6.3.4}$$

$$\|v_l\| \le \frac{1}{l} \tag{6.3.5}$$

for $l = 1, 2, \dots$. From (6.3.5), we see that as $l \to \infty$,

$$|v_l|_{k,p,\Omega} \to 0 \tag{6.3.6}$$

and

$$f_j(v_l) \to 0, \quad 1 \le j \le J. \tag{6.3.7}$$

Since $\{v_l\}$ is a bounded sequence in $W^{k,p}(\Omega)$ from the property (6.3.4), and since

$$W^{k,p}(\Omega) \hookrightarrow\hookrightarrow W^{k-1,p}(\Omega),$$

there is a subsequence of the sequence $\{v_l\}$, still denoted as $\{v_l\}$, and a function $v \in W^{k-1,p}(\Omega)$ such that

$$v_l \to v \quad \text{in } W^{k-1,p}(\Omega), \text{ as } l \to \infty. \tag{6.3.8}$$

This property and (6.3.6), together with the uniqueness of a limit, imply that

$$v_l \to v \quad \text{in } W^{k,p}(\Omega), \text{ as } l \to \infty$$

and

$$|v|_{k,p,\Omega} = \lim_{l\to\infty} |v_l|_{k,p,\Omega} = 0.$$

We then conclude that v is a polynomial of degree less than or equal to $k-1$. On the other hand, from the continuity of the functionals $\{f_j\}_{1 \le j \le J}$

and (6.3.7), we find that

$$f_j(v) = \lim_{l \to \infty} f_j(v_l) = 0, \quad 1 \le j \le J.$$

Using the condition (H2), we see that $v = 0$, which contradicts the relation that

$$\|v\|_{k,p,\Omega} = \lim_{l \to \infty} \|v_l\|_{k,p,\Omega} = 1.$$

The proof of the result is now completed. ∎

Notice that in Theorem 6.3.12, we need to assume Ω to be connected. This assumption is used to guarantee that from $|v|_{k,p,\Omega} = 0$ we can conclude that v is a (global) polynomial of degree less than or equal to $k-1$. The above proof of Theorem 6.3.12 can be easily modified to yield the next result.

Theorem 6.3.13 *Let Ω be an open, bounded set in $\mathbb{R}^d$ with a Lipschitz boundary, $k \ge 1$, $1 \le p < \infty$. Assume $f_j : W^{k,p}(\Omega) \to \mathbb{R}$, $1 \le j \le J$, are semi-norms on $W^{k,p}(\Omega)$ satisfying two conditions:*

(H1) $0 \le f_j(v) \le c\,\|v\|_{k,p,\Omega} \;\; \forall\, v \in W^{k,p}(\Omega), 1 \le j \le J.$

(H2)′ *If $|v|_{k,p,\Omega} = 0$ and $f_j(v) = 0$, $1 \le j \le J$, then $v = 0$.*

Then the quantities (6.3.2) and (6.3.3) are norms on $W^{k,p}(\Omega)$, equivalent to the norm $\|v\|_{k,p,\Omega}$.

We may also state the norm-equivalence result for the case where Ω is a union of separated open connected sets.

Theorem 6.3.14 *Let Ω be an open, bounded set in $\mathbb{R}^d$, $\Omega = \cup_{\lambda \in \Lambda}\Omega_\lambda$ with each Ω_λ having a Lipschitz boundary, $\Omega_\lambda \cap \Omega_\mu = \emptyset$ for $\lambda \ne \mu$. Let $k \ge 1$, $1 \le p < \infty$. Assume $f_j : W^{k,p}(\Omega) \to \mathbb{R}$, $1 \le j \le J$, are semi-norms on $W^{k,p}(\Omega)$ satisfying two conditions:*

(H1) $0 \le f_j(v) \le c\,\|v\|_{k,p,\Omega} \;\; \forall\, v \in W^{k,p}(\Omega), 1 \le j \le J.$

(H2) *If v is a polynomial of degree less than or equal to $k-1$ on each Ω_λ, $\lambda \in \Lambda$, and $f_j(v) = 0$, $1 \le j \le J$, then $v = 0$.*

Then the quantities (6.3.2) and (6.3.3) are norms on $W^{k,p}(\Omega)$, equivalent to the norm $\|v\|_{k,p,\Omega}$.

Many useful inequalities can be derived as consequences of the previous theorems. We present some examples below.

Example 6.3.15 *Assume Ω is an open, bounded set in $\mathbb{R}^d$ with a Lipschitz boundary. Let us apply Theorem 6.3.13 with $k = 1$, $p = 2$, $J = 1$ and*

$$f_1(v) = \int_{\partial\Omega} |v|\, ds.$$

We can then conclude that there exists a constant $c > 0$, depending only on Ω such that

$$\|v\|_{1,\Omega} \le c\left(|v|_{1,\Omega} + \|v\|_{L^1(\partial\Omega)}\right) \quad \forall\, v \in H^1(\Omega).$$

Therefore, the Poincaré-Friedrichs inequality holds:

$$\|v\|_{1,\Omega} \le c|v|_{1,\Omega} \quad \forall\, v \in H_0^1(\Omega).$$

From this inequality it follows that the semi-norm $|\cdot|_{1,\Omega}$ is a norm on $H_0^1(\Omega)$, equivalent to the usual $H^1(\Omega)$-norm.

Example 6.3.16 Let Ω be an open, bounded, connected set in $\mathbb{R}^d$ with a Lipschitz boundary. Assume Γ_0 is an open, non-empty subset of the boundary $\partial\Omega$, then there is a constant $c > 0$, depending only on Ω, such that

$$\|v\|_{1,\Omega} \le c\left(|v|_{1,\Omega} + \|v\|_{L^1(\Gamma_0)}\right) \quad \forall\, v \in H^1(\Omega).$$

This inequality can be derived by applying Theorem 6.3.12 with $k = 1$, $p = 2$, $J = 1$ and

$$f_1(v) = \int_{\Gamma_0} |v|\, ds.$$

Therefore,

$$\|v\|_{1,\Omega} \le c|v|_{1,\Omega} \quad \forall\, v \in H_{\Gamma_0}^1(\Omega),$$

where

$$H_{\Gamma_0}^1(\Omega) = \{v \in H^1(\Omega) \mid v = 0 \text{ a.e. on } \Gamma_0\}.$$

Some other useful inequalities, however, cannot be derived from the norm equivalence theorem. One example is

$$\|v\|_{2,\Omega} \le c|\Delta v|_{0,\Omega} \quad \forall\, v \in H^2(\Omega) \cap H_0^1(\Omega), \tag{6.3.9}$$

which is valid if Ω is smooth or is convex. This result is proved by using a regularity estimate for the (weak) solution of the boundary value problem (cf. [48])

$$-\Delta u = f \text{ in } \Omega,$$
$$u = 0 \text{ on } \partial\Omega.$$

Another example is Korn's inequality, which is useful in theoretical mechanics. Let Ω be a non-empty, open, bounded, and connected set in $\mathbb{R}^3$ with a Lipschitz boundary. Given a function $\mathbf{u} \in [H^1(\Omega)]^3$, the linearized strain tensor is defined by

$$\varepsilon(\mathbf{u}) = \frac{1}{2}(\nabla\mathbf{u} + (\nabla\mathbf{u})^T);$$

in component form,

$$\varepsilon_{ij}(\mathbf{u}) = \frac{1}{2}\left(\partial_{x_i} u_j + \partial_{x_j} u_i\right), \quad 1 \le i, j \le 3.$$

Let Γ_0 be a measurable subset of $\partial\Omega$ with meas $(\Gamma_0) > 0$, and define

$$[H^1_{\Gamma_0}(\Omega)]^3 = \{\mathbf{v} \in [H^1(\Omega)]^3 \mid \mathbf{v} = \mathbf{0} \text{ a.e. on } \Gamma_0\}.$$

Korn's inequality states that there exists a constant $c > 0$ depending only on Ω such that

$$\|\mathbf{u}\|^2_{[H^1(\Omega)]^3} \leq c \int_{\Omega} |\boldsymbol{\varepsilon}(\mathbf{u})|^2 dx \quad \forall \mathbf{u} \in [H^1_{\Gamma_0}(\Omega)]^3.$$

A proof of Korn's inequality can be found in [95] or [123].

6.3.6 A Sobolev quotient space

Later in error analysis for the finite element method, we need an inequality involving the norm of the Sobolev quotient space

$$V = W^{k+1,p}(\Omega)/\mathcal{P}_k(\Omega)$$
$$= \{[v] \mid [v] = \{v + q \mid q \in \mathcal{P}_k(\Omega)\}, \ v \in W^{k+1,p}(\Omega)\}.$$

Here $k \geq 0$ is an integer and $\mathcal{P}_k(\Omega)$ is the space of polynomials of degree less than or equal to k. Any element $[v]$ of the space V is an equivalence class, the difference between any two elements in the equivalence class being a polynomial in the space $\mathcal{P}_k(\Omega)$. Any $v \in [v]$ is called a representative element of $[v]$. The quotient norm in the space V is defined to be

$$\|[v]\|_V = \inf_{q \in \mathcal{P}_k(\Omega)} \|v + q\|_{k+1,p,\Omega}.$$

Theorem 6.3.17 *Assume $1 \leq p < \infty$. Let $\Omega \subseteq \mathbb{R}^d$ be an open, bounded, connected set with a Lipschitz continuous boundary. Then the quantity $|v|_{k+1,p,\Omega}, \forall v \in [v]$, is a norm on V, equivalent to the quotient norm $\|[v]\|_V$.*

Proof. Obviously, for any $[v] \in V$ and any $v \in [v]$,

$$\|[v]\|_V = \inf_{q \in \mathcal{P}_k(\Omega)} \|v + q\|_{k+1,p,\Omega} \geq |v|_{k+1,p,\Omega}.$$

Thus we only need to prove that there is a constant c, depending only on Ω, such that

$$\inf_{q \in \mathcal{P}_k(\Omega)} \|v + q\|_{k+1,p,\Omega} \leq c |v|_{k+1,p,\Omega} \quad \forall v \in W^{k+1,p}(\Omega). \qquad (6.3.10)$$

Denote $N = \dim(\mathcal{P}_k(\Omega))$. Define N independent linear continuous functionals on $\mathcal{P}_k(\Omega)$, the continuity being with respect to the norm of $W^{k+1,p}(\Omega)$. By the Hahn-Banach theorem, we can extend these functionals to linear continuous functionals over the space $W^{k+1,p}(\Omega)$, denoted by $f_i(\cdot)$, $1 \leq i \leq N$, such that for $q \in \mathcal{P}_k(\Omega)$, $f_i(q) = 0$, $1 \leq i \leq N$, if and only if $q = 0$. Then $|f_i|$, $1 \leq i \leq N$, are semi-norms on $W^{k+1,p}(\Omega)$ satisfying the assumptions of Theorem 6.3.12. Applying Theorem 6.3.12, we have

$$\|v\|_{k+1,p,\Omega} \leq c \left(|v|_{k+1,p,\Omega} + \sum_{i=1}^{N} |f_i(v)|\right) \quad \forall v \in W^{k+1,p}(\Omega).$$

Since f_i, $1 \le i \le N$, are linearly independent on $\mathcal{P}_k(\Omega)$, for any fixed $v \in W^{k+1,p}(\Omega)$, there exists $q \in \mathcal{P}_k(\Omega)$ such that $f_i(v+q) = 0$, $1 \le i \le N$. Thus,

$$\|v+q\|_{k+1,p,\Omega} \le c \left(|v+q|_{k+1,p,\Omega} + \sum_{i=1}^{N} |f_i(v+q)| \right) = c\,|v|_{k+1,p,\Omega},$$

and hence (6.3.10) holds.

It is possible to prove (6.3.10) without using the Hahn-Banach theorem. For this, we apply Theorem 6.3.12 to obtain the inequality

$$\|v\|_{k+1,p,\Omega} \le c \left(|v|_{k+1,p,\Omega} + \sum_{|\alpha| \le k} \left| \int_\Omega D^\alpha v(\mathbf{x})\, dx \right| \right) \qquad \forall\, v \in W^{k+1,p}(\Omega).$$

Replacing v by $v+q$ and noting that $D^\alpha q = 0$ for $|\alpha| = k+1$, we have

$$\|v+q\|_{k+1,p,\Omega} \le c \left(|v|_{k+1,p,\Omega} + \sum_{|\alpha| \le k} \left| \int_\Omega D^\alpha (v+q)\, dx \right| \right) \tag{6.3.11}$$

$$\forall\, v \in W^{k+1,p}(\Omega), \qquad q \in \mathcal{P}_k(\Omega).$$

Now construct a polynomial $\bar{q} \in \mathcal{P}_k(\Omega)$ satisfying

$$\int_\Omega D^\alpha(v+\bar{q})\, dx = 0 \qquad \text{for } |\alpha| \le k. \tag{6.3.12}$$

This can always be done: Set $|\alpha| = k$; then $D^\alpha \bar{p}$ equals $\alpha_1! \cdots \alpha_d!$ times the coefficient of $\mathbf{x}^\alpha \equiv x_1^{\alpha_1} \cdots x_d^{\alpha_d}$, so the coefficient can be computed by using (6.3.12). Having found all the coefficients for terms of degree k, we set $|\alpha| = k-1$, and use (6.3.12) to compute all the coefficients for terms of degree $k-1$. Proceeding in this way, we obtain the polynomial $\bar{q}$ satisfying the condition (6.3.12) for the given function v.

With $q = \bar{q}$ in (6.3.11), we have

$$\inf_{q \in \mathcal{P}_k(\Omega)} \|v+q\|_{k+1,p,\Omega} \le \|v+\bar{q}\|_{k+1,p,\Omega} \le c\,|v|_{k+1,p,\Omega},$$

from which (6.3.10) follows.   ∎

Corollary 6.3.18 *For any open, bounded, connected set $\Omega \subseteq \mathbb{R}^d$ with a Lipschitz continuous boundary, there is a constant c, depending only on Ω, such that*

$$\inf_{p \in \mathcal{P}_k(\Omega)} \|v+p\|_{k+1,\Omega} \le c\,|v|_{k+1,\Omega} \qquad \forall\, v \in H^{k+1}(\Omega). \tag{6.3.13}$$

Exercise 6.3.1 *It is possible to construct a simple extension operator when the domain is a half-space, say, $\mathbb{R}^d_+ = \{x \in \mathbb{R}^d \mid x_d \ge 0\}$. Let $k \ge 1$ be an* *integer, $p \in [1, \infty]$. For any $v \in W^{k,p}(\mathbb{R}^d_+)$, we define*

$$Ev(\mathbf{x}) = \begin{cases} v(\mathbf{x}), & \mathbf{x} \in \mathbb{R}^d_+, \\ \sum_{j=0}^{k-1} c_j v(x_1, \dots, x_{d-1}, -2^j x_d), & \mathbf{x} \in \mathbb{R}^d \backslash \mathbb{R}^d_+, \end{cases}$$

where the coefficients $c_0, \dots, c_{k-1}$ are determined from the system

$$\sum_{j=0}^{k-1} c_j (-2^j)^i = 1, \qquad i = 0, 1, \dots, k-1.$$

Show that $Ev \in W^{k,p}(\mathbb{R}^d)$, and E is a continuous operator from $W^{k,p}(\mathbb{R}^d_+)$ to $W^{k,p}(\mathbb{R}^d)$.

Exercise 6.3.2 *In general, an embedding result (Theorems 6.3.7, 6.3.8) is not easy to prove. On the other hand, it is usually not difficult to prove an embedding result for one-dimensional domains. Let $-\infty < a < b < \infty$, $p > 1$, and let q be the conjugate of p defined by the relation $1/p + 1/q = 1$. Prove the embedding result $W^{1,p}(a,b) \hookrightarrow C^{0,1/q}(a,b)$ with the following steps.*

First, let $v \in C^1[a,b]$. By the Mean Value Theorem in calculus, there exists a $\xi \in [a,b]$ such that

$$\int_a^b v(x)\, dx = (b-a)\, v(\xi).$$

Then we can write

$$v(x) = \frac{1}{b-a} \int_a^b v(s)\, ds + \int_\xi^x v'(s)\, ds,$$

from which, it is easy to find

$$|v(x)| \le c \, \|v\|_{W^{1,p}(a,b)} \qquad \forall\, x \in [a,b].$$

Hence,

$$\|v\|_{C[a,b]} \le c \, \|v\|_{W^{1,p}(a,b)}.$$

Furthermore, for $x \ne y$,

$$|v(x) - v(y)| = \left| \int_y^x v'(s)\, ds \right| \le |x - y|^{1/q} \left(\int_a^b |v'(s)|^p ds \right)^{1/p}.$$

Therefore,

$$\|v\|_{C^{0,1/q}(a,b)} \le c \, \|v\|_{W^{1,p}(a,b)} \qquad \forall\, v \in C^1[a,b].$$

Second, for any $v \in W^{1,p}(a,b)$, using the density of $C^1[a,b]$ in $W^{1,p}(a,b)$, we can find a sequence $\{v_n\} \subseteq C^1[a,b]$, such that

$$\|v_n - v\|_{W^{1,p}(a,b)} \to 0 \qquad \text{as } n \to \infty.$$

Apply the inequality proved in the first step,

$$\|v_n - v\|_{C^{0,1/q}(a,b)} \leq c \|v_n - v\|_{W^{1,p}(a,b)} \to 0 \qquad \text{as } n \to \infty.$$

So $\{v_n\}$ is a Cauchy sequence in $C^{0,1/q}(a,b)$. Since the space $C^{0,1/q}(a,b)$ is complete, the sequence $\{v_n\}$ converges to some $\tilde{v}$ in $C^{0,1/q}(a,b)$. We also have $v_n \to u$ a.e. By the uniqueness of a limit, we conclude $\tilde{v} = u$.

Exercise 6.3.3 *Prove Theorem 6.3.9 by applying Theorem 6.3.8.*

Exercise 6.3.4 *Let $\Omega \subseteq \mathbb{R}^d$ be an open, bounded, connected domain with a Lipschitz continuous boundary $\partial\Omega$. Assume $\Gamma_0 \subseteq \partial\Omega$ is such that $\text{meas}\,\Gamma_0 > 0$. Define*

$$H^2_{\Gamma_0}(\Omega) = \{v \in H^2(\Omega) \mid v = \partial v/\partial\nu = 0 \text{ a.e. on } \Gamma_0\}.$$

Prove the following inequality:

$$\|v\|_{2,\Omega} \leq c |v|_{2,\Omega} \qquad \forall v \in H^2_{\Gamma_0}(\Omega).$$

This result implies that under the stated assumptions, $|v|_{2,\Omega}$ is a norm on $H^2_{\Gamma_0}(\Omega)$, which is equivalent to the norm $\|v\|_{2,\Omega}$.

Exercise 6.3.5 *Apply the norm equivalence theorems to derive the following inequalities, using the previously stated assumptions on Ω:*

$$\|v\|_{1,p,\Omega} \leq c \left(|v|_{1,p,\Omega} + \left|\int_{\Omega_0} v\,dx\right|\right), \qquad \forall v \in W^{1,p}(\Omega),$$
$$\text{if } \Omega_0 \subseteq \Omega, \ \text{meas}(\Omega_0) > 0;$$
$$\|v\|_{1,p,\Omega} \leq c \left(|v|_{1,p,\Omega} + \|v\|_{L^p(\Gamma)}\right), \qquad \forall v \in W^{1,p}(\Omega),$$
$$\text{if } \Gamma \subseteq \partial\Omega, \ \text{meas}_{d-1}(\Gamma) > 0.$$
$$\|v\|_{1,p,\Omega} \leq c |v|_{1,p,\Omega}, \qquad \forall v \in W^{1,p}_0(\Omega).$$

Can you think of some more inequalities of the above kind?

Exercise 6.3.6 *In some applications, it is important to find or estimate the best constant in a Sobolev inequality. For example, let Ω be an open, bounded, connected Lipschitz domain; and let Γ_1 and Γ_2 be two disjoint, nonempty open subsets of the boundary $\partial\Omega$. Then there is a Sobolev inequality*

$$\|v\|_{L^2(\Gamma_1)} \leq c \|\nabla v\|_{L^2(\Omega)} \qquad \forall v \in H^1_{\Gamma_2}(\Omega).$$

By the best constant c_0 of the inequality, we mean that c_0 is the smallest constant such that the inequality holds. The best constant c_0 can be characterized by the expression

$$c_0 = \sup\{\|v\|_{L^2(\Gamma_1)}/\|\nabla v\|_{L^2(\Omega)} \mid v \in H^1_{\Gamma_2}(\Omega)\}.$$

Show that $c_0 = 1/\sqrt{\lambda_1}$ *where* $\lambda_1 > 0$ *is the smallest eigenvalue of the eigenvalue problem*

$$u \in H^1_{\Gamma_2}(\Omega), \quad \int_\Omega \nabla u \cdot \nabla v \, dx = \lambda \int_{\Gamma_1} u \, v \, ds \quad \forall \, v \in H^1_{\Gamma_2}(\Omega).$$

Exercise 6.3.7 *In the preceding exercise, the best constant of an inequality is related to a linear eigenvalue boundary value problem. In some other applications, we need the best constant of an inequality, which can be found or estimated by solving a linear elliptic boundary value problem. Keeping the notations of the previous exercise, we have the Sobolev inequality*

$$\|v\|_{L^1(\Gamma_1)} \le c \, \|\nabla v\|_{L^2(\Omega)} \quad \forall \, v \in H^1_{\Gamma_2}(\Omega).$$

The best constant c_0 *of the inequality can be characterized by the expression*

$$c_0 = \sup\{\|v\|_{L^1(\Gamma_1)}/\|\nabla v\|_{L^2(\Omega)} \mid v \in H^1_{\Gamma_2}(\Omega)\}.$$

Show that

$$c_0 = \|\nabla u\|_{L^2(\Omega)} = \|u\|_{L^1(\Gamma_1)}^{1/2},$$

where u *is the solution of the problem*

$$u \in H^1_{\Gamma_2}(\Omega), \quad \int_\Omega \nabla u \cdot \nabla v \, dx = \int_{\Gamma_1} v \, ds \quad \forall \, v \in H^1_{\Gamma_2}(\Omega).$$

Hint: *Use the result of Exercise 6.1.3 (cf. [68]).*

6.4 Characterization of Sobolev spaces via the Fourier transform

When $\Omega = \mathbb{R}^d$, it is possible to define Sobolev spaces $H^k(\mathbb{R}^d)$ by using the Fourier transform. All the functions in this section are complex-valued. The reader is referred to [149] for a detailed discussion of the Fourier transform and its properties, including proofs of the Theorems 6.4.2 and 6.4.3 below.

Definition 6.4.1 *For* $v \in L^1(\mathbb{R}^d)$, *the Fourier transform is defined by*

$$\mathcal{F}(v)(\mathbf{y}) = \frac{1}{(2\pi)^{d/2}} \int_{\mathbb{R}^d} \exp(-i\,\mathbf{x} \cdot \mathbf{y})\, v(\mathbf{x})\, dx,$$

and the inverse Fourier transform is defined by

$$\mathcal{F}^{-1}(v)(\mathbf{y}) = \frac{1}{(2\pi)^{d/2}} \int_{\mathbb{R}^d} \exp(i\,\mathbf{x} \cdot \mathbf{y})\, v(\mathbf{x})\, dx.$$

An important property of the Fourier transform is the Plancherel's theorem, stated next.

Theorem 6.4.2 *Assume $v \in L^1(\mathbb{R}^d) \cap L^2(\mathbb{R}^d)$. Then $\mathcal{F}v, \mathcal{F}^{-1}v \in L^2(\mathbb{R}^d)$, and*

$$\|\mathcal{F}v\|_{L^2(\mathbb{R}^d)} = \|\mathcal{F}^{-1}v\|_{L^2(\mathbb{R}^d)} = \|v\|_{L^2(\mathbb{R}^d)}. \tag{6.4.1}$$

Using (6.4.1) and the density of $L^1(\mathbb{R}^d) \cap L^2(\mathbb{R}^d)$ in $L^2(\mathbb{R}^d)$, one can extend the definitions of the Fourier transform and its inverse to $L^2(\mathbb{R}^d)$ functions. Some useful properties of the transforms are recorded in the next theorem.

Theorem 6.4.3 *Assume $u, v \in L^2(\mathbb{R}^d)$. Then*

(a)

$$\int_{\mathbb{R}^d} \mathcal{F}u(\mathbf{y}) \, \overline{\mathcal{F}v}(\mathbf{y}) \, dy = \int_{\mathbb{R}^d} u(\mathbf{x}) \, \overline{v}(\mathbf{x}) \, dx;$$

(b)

$$\mathcal{F}(D^\alpha v)(\mathbf{y}) = (i\,\mathbf{y})^\alpha \mathcal{F}v(\mathbf{y}) \quad \text{if} \quad D^\alpha v \in L^2(\mathbb{R}^d);$$

(c)

$$\mathcal{F}u = v \qquad \Longleftrightarrow \qquad u = \mathcal{F}^{-1}v.$$

It is then straightforward to show the next result.

Theorem 6.4.4 *A function $v \in L^2(\mathbb{R}^d)$ belongs to $H^k(\mathbb{R}^d)$ if and only if $(1 + |\mathbf{y}|^k)\,\mathcal{F}v \in L^2(\mathbb{R}^d)$. Moreover, there exist $c_1, c_2 > 0$ such that*

$$c_1 \|v\|_{H^k(\mathbb{R}^d)} \le \|(1 + |\mathbf{y}|^k)\,\mathcal{F}v\|_{L^2(\mathbb{R}^d)} \le c_2 \|v\|_{H^k(\mathbb{R}^d)} \quad \forall\, v \in H^k(\mathbb{R}^d). \tag{6.4.2}$$

Thus we see that $\|(1 + |\mathbf{y}|^k)\,\mathcal{F}v\|_{L^2(\mathbb{R}^d)}$ is a norm on $H^k(\mathbb{R}^d)$, which is equivalent to the canonical norm $\|v\|_{H^k(\mathbb{R}^d)}$. It is equally good to define the space $H^k(\mathbb{R}^d)$ by

$$H^k(\mathbb{R}^d) = \{v \in L^2(\mathbb{R}^d) \mid (1 + |\mathbf{y}|^k)\,\mathcal{F}v \in L^2(\mathbb{R}^d)\}.$$

We notice that in this equivalent definition, there is no need to assume k to be an integer. It is natural to define Sobolev spaces of any (positive) order $s \ge 0$:

$$H^s(\mathbb{R}^d) = \{v \in L^2(\mathbb{R}^d) \mid (1 + |\mathbf{y}|^s)\,\mathcal{F}v \in L^2(\mathbb{R}^d)\} \tag{6.4.3}$$

with the norm

$$\|v\|_{H^s(\mathbb{R}^d)} = \|(1 + |\mathbf{y}|^s)\,\mathcal{F}v\|_{L^2(\mathbb{R}^d)}$$

and the inner product

$$(u, v)_{H^s(\mathbb{R}^d)} = \int_{\mathbb{R}^d} (1 + |\mathbf{y}|^s)^2 \, \mathcal{F}u(\mathbf{y}) \, \overline{\mathcal{F}v(\mathbf{y})} \, dy.$$

In particular, when $s = 0$, we recover the $L^2(\mathbb{R}^d)$ space: $H^0(\mathbb{R}^d) = L^2(\mathbb{R}^d)$.

Actually, the definition of the Fourier transform can be extended to distributions of slow growth that are continuous linear functionals on smooth functions decaying sufficiently rapidly at infinity (for detail, cf. [149]). Then we can define the Sobolev space $H^s(\mathbb{R}^d)$ for negative index s to be the set of the distributions v of slow growth such that

$$\|v\|_{H^s(\mathbb{R}^d)} = \|(1 + |\mathbf{y}|^s)\, \mathcal{F}v\|_{L^2(\mathbb{R}^d)} < \infty.$$

We can combine the extension theorem, the approximation theorems and the Fourier transform characterization of Sobolev spaces to prove some properties of Sobolev spaces over bounded Lipschitz domains.

Example 6.4.5 *Let* $\Omega \subseteq \mathbb{R}^d$ *be a bounded domain with a Lipschitz boundary. Assume* $k > d/2$*. Let us prove* $H^k(\Omega) \hookrightarrow C(\overline{\Omega})$*; i.e.,*

$$\|v\|_{C(\overline{\Omega})} \le c\, \|v\|_{H^k(\Omega)} \qquad \forall\, v \in H^k(\Omega). \tag{6.4.4}$$

Proof. STEP 1. We prove (6.4.4) for $\Omega = \mathbb{R}^d$ and $v \in C_0^\infty(\mathbb{R}^d)$. We have

$$v(\mathbf{x}) = \frac{1}{(2\pi)^{d/2}} \int_{\mathbb{R}^d} \exp(i\,\mathbf{x}\cdot\mathbf{y}) \mathcal{F}v(\mathbf{y})\, d\mathbf{y}.$$

Thus

$$|v(\mathbf{x})| \le c \int_{\mathbb{R}^d} |\mathcal{F}v(\mathbf{y})|\, d\mathbf{y}$$

$$= c \int_{\mathbb{R}^d} \frac{(1 + |\mathbf{y}|^k)\, |\mathcal{F}v(\mathbf{y})|}{1 + |\mathbf{y}|^k}\, d\mathbf{y}$$

$$\le c \left(\int_{\mathbb{R}^d} (1 + |\mathbf{y}|^k)^{-2} d\mathbf{y} \right)^{\frac{1}{2}} \left(\int_{\mathbb{R}^d} (1 + |\mathbf{y}|^k)^2 |\mathcal{F}v(\mathbf{y})|^2 d\mathbf{y} \right)^{\frac{1}{2}}$$

$$= c \left(\int_{\mathbb{R}^d} (1 + |\mathbf{y}|^k)^2 |\mathcal{F}v(\mathbf{y})|^2 d\mathbf{y} \right)^{\frac{1}{2}},$$

where we used the fact that

$$\int_{\mathbb{R}^d} (1 + |\mathbf{y}|^k)^{-2} d\mathbf{y} < \infty \qquad \text{if } k > d/2.$$

Hence,

$$\|v\|_{C(\mathbb{R}^d)} \le c\, \|v\|_{H^k(\mathbb{R}^d)} \qquad \forall\, v \in C_0^\infty(\mathbb{R}^d).$$

STEP 2. Since $C_0^\infty(\mathbb{R}^d)$ is dense in $H^k(\mathbb{R}^d)$, the relation (6.4.4) holds for any $v \in H^k(\mathbb{R}^d)$.

STEP 3. We now use the extension theorem. For any $v \in H^k(\Omega)$, we can extend it to $Ev \in H^k(\mathbb{R}^d)$ with

$$\|Ev\|_{H^k(\mathbb{R}^d)} \le c\, \|v\|_{H^k(\Omega)}.$$

Therefore,

$$\|v\|_{C(\overline{\Omega})} \le \|Ev\|_{C(\mathbb{R}^d)} \le c\, \|Ev\|_{H^k(\mathbb{R}^d)} \le c\, \|v\|_{H^k(\Omega)}.$$

Thus we have proved (6.4.4).   ■

Example 6.4.6 *Let* $\Omega \subseteq \mathbb{R}^d$ *be a bounded domain with a Lipschitz boundary. Then*

$$\|v\|_{C(\overline{\Omega})} \le c \|v\|_{H^d(\Omega)}^{\frac{1}{2}} \|v\|_{L^2(\Omega)}^{\frac{1}{2}} \qquad \forall v \in H^d(\Omega). \qquad (6.4.5)$$

Proof. As in the previous example, it is enough to show (6.4.5) for the case $\Omega = \mathbb{R}^d$ and $v \in C_0^\infty(\Omega)$. For any $\lambda > 0$,

$$|v(\mathbf{x})|^2 \le c \left(\int_{\mathbb{R}^d} |\mathcal{F}v(\mathbf{y})| \, dy \right)^2$$

$$= c \left(\int_{\mathbb{R}^d} \frac{(1 + \lambda |\mathbf{y}|^d) |\mathcal{F}v(\mathbf{y})|}{1 + \lambda |\mathbf{y}|^d} \, dy \right)^2$$

$$\le c \int_{\mathbb{R}^d} (1 + \lambda |\mathbf{y}|^d)^2 |\mathcal{F}v(\mathbf{y})|^2 \, dy \int_{\mathbb{R}^d} (1 + \lambda |\mathbf{y}|^d)^{-2} dy$$

$$\le c \frac{1}{\lambda} \int_{\mathbb{R}^d} \left(|\mathcal{F}v(\mathbf{y})|^2 + \lambda^2 (1 + |\mathbf{y}|^d)^2 |\mathcal{F}v(\mathbf{y})|^2 \right) dy$$

$$= c \left(\lambda^{-1} \|v\|_{L^2(\Omega)}^2 + \lambda \|v\|_{H^d(\Omega)}^2 \right).$$

Taking $\lambda = \|v\|_{L^2(\Omega)}/\|v\|_{H^d(\Omega)}$, we then get the required inequality.   ■

Exercise 6.4.1 *Let* $v(x)$ *be a step function defined by:* $v(x) = 1$ *for* $x \in [0,1]$, *and* $v(x) = 0$ *for* $x \notin [0,1]$. *Find the range of s for which* $v \in H^s(\mathbb{R})$.

Exercise 6.4.2 *Provide a detailed argument for Step 2 in the proof of Example 6.4.5.*

6.5   Periodic Sobolev spaces

When working with an equation defined over the boundary of a bounded and simply connected region in the plane, the functions being discussed are periodic. Therefore, it is useful to consider Sobolev spaces of such functions. Since periodic functions are often discussed with reference to their Fourier series expansion, we use this expansion to discuss Sobolev spaces of such functions.

From Example 1.3.11, we write the Fourier series of $\varphi \in L^2(0, 2\pi)$ as

$$\varphi(x) = \sum_{m=-\infty}^{\infty} a_m \psi_m(x)$$

with

$$\psi_m(x) = \frac{1}{\sqrt{2\pi}} e^{imx}, \qquad m = 0, \pm 1, \pm 2, \dots$$

forming an orthonormal basis of $L^2(0, 2\pi)$. The Fourier coefficients are given by

$$a_m = (\varphi, \psi_m) = \frac{1}{\sqrt{2\pi}} \int_0^{2\pi} \varphi(x) \, e^{-imx} dx.$$

The convergence of the Fourier series was discussed earlier in Sections 3.3 and 3.5. Also, for a non-negative integer k, recall that $C_p^k(2\pi)$ denotes the set of all periodic functions on $(-\infty, \infty)$ with period 2π that are also k-times continuously differentiable.

Definition 6.5.1 *For an integer $k \geq 0$, $H^k(2\pi)$ is defined to be the closure of $C_p^k(2\pi)$ under the inner product norm*

$$\|\varphi\|_{H^k} \equiv \left[\sum_{j=0}^{k} \left\| \varphi^{(j)} \right\|_{L^2}^2 \right]^{\frac{1}{2}}.$$

For arbitrary real $s \geq 0$, $H^s(2\pi)$ can be obtained as in earlier sections, with the formulas of Section 6.4 being closest to the discussion given below, especially (6.4.3).

The following can be shown without too much difficulty; e.g., see [100, Chap. 8].

Theorem 6.5.2 *For $s \in \mathbb{R}$, $H^s(2\pi)$ is the set of all series*

$$\varphi(x) = \sum_{m=-\infty}^{\infty} a_m \psi_m(x) \tag{6.5.1}$$

for which

$$\|\varphi\|_{*,s}^2 \equiv |a_0|^2 + \sum_{|m|>0} |m|^{2s} |a_m|^2 < \infty. \tag{6.5.2}$$

Moreover, the norm $\|\varphi\|_{*,s}$ is equivalent to the standard Sobolev norm $\|\varphi\|_{H^s}$ for $\varphi \in H^s(2\pi)$.

The norm $\| \cdot \|_{*,s}$ is based on the inner product defined by

$$(\varphi, \rho)_{*,s} \equiv a_0 \bar{b}_0 + \sum_{|m|>0} |m|^{2s} a_m \bar{b}_m,$$

where $\varphi = \sum a_m \psi_m$ and $\rho = \sum b_m \psi_m$.

For $s < 0$, the space $H^s(2\pi)$ contains series that are divergent according to most usual definitions of convergence. These new "functions" (6.5.1) are referred to as both *generalized functions* and *distributions*. One way of giving meaning to these new functions is to introduce the concept of *distributional derivative*, which generalizes the derivative of ordinary sense and the weak derivative introduced in Section 6.1. With the ordinary differentiation operator $\mathcal{D} \equiv d/dt$, we have

$$\mathcal{D}\varphi(t) \equiv \frac{d\varphi(t)}{dt} = i \sum_{m=-\infty}^{\infty} m a_m \psi_m(t) \qquad (6.5.3)$$

and $\mathcal{D} : H^s(2\pi) \to H^{s-1}(2\pi)$, $s \geq 1$. The distributional derivative gives meaning to differentiation of periodic functions in $L^2(0, 2\pi)$, and also to repeated differentiation of generalized functions. To prove that there exists a unique such extension of the definition of $\mathcal{D}$, proceed as follows.

Introduce the space $\mathcal{T}$ of all trigonometric polynomials,

$$\mathcal{T} = \left\{ \varphi \equiv \sum_{m=-n}^{n} a_m \psi_m \;\middle|\; a_m \in \mathbb{C}, \; |m| \leq n, \; n = 0, 1, \dots \right\}. \qquad (6.5.4)$$

It is straightforward to show this is a dense subspace of $H^s(2\pi)$ for arbitrary s, meaning that when using the norm (6.5.2), the closure of $\mathcal{T}$ equals $H^s(2\pi)$. Considering $\mathcal{T}$ as a subspace of $H^s(2\pi)$, define $\mathcal{D} : \mathcal{T} \to H^{s-1}(2\pi)$ by

$$\mathcal{D}\varphi = \varphi', \qquad \varphi \in \mathcal{T}. \qquad (6.5.5)$$

This is a bounded operator; and using the representation of φ in (6.5.4), it is straightforward that

$$\|\mathcal{D}\| = 1.$$

Since $\mathcal{T}$ is dense in $H^s(2\pi)$, and since $\mathcal{D} : \mathcal{T} \subseteq H^s(2\pi) \to H^{s-1}(2\pi)$ is bounded, we have that there is a unique bounded extension of $\mathcal{D}$ to all of $H^s(2\pi)$; see Theorem 2.4.1. We will retain the notation $\mathcal{D}$ for the extension. Combining the representation of $\varphi \in \mathcal{T}$ in (6.5.4) with the definition (6.5.5), and using the continuity of the extension $\mathcal{D}$, the formula (6.5.5) remains valid for any $\varphi \in H^s(2\pi)$ for all s.

Example 6.5.3 *Define*

$$\varphi(t) = \begin{cases} 0, & (2k-1)\pi < t < 2k\pi, \\ 1, & 2k\pi < t < (2k+1)\pi, \end{cases}$$

for all integers k, a so-called "square wave." The Fourier series of this function is given by

$$\varphi(t) = \frac{1}{2} - \frac{i}{\pi} \sum_{k=0}^{\infty} \frac{1}{2k+1} \left[e^{(2k+1)it} - e^{-(2k+1)it} \right], \qquad -\infty < t < \infty,$$

which converges almost everywhere. Regarding this series as a function defined on $\mathbb{R}$, the distributional derivative of $\varphi(t)$ is

$$\varphi'(t) = \frac{1}{\pi} \sum_{k=0}^{\infty} \left[e^{(2k+1)it} + e^{-(2k+1)it} \right] = \sum_{j=-\infty}^{\infty} (-1)^j \delta(t - \pi j).$$

The function $\delta(t)$ is the Dirac delta function, and it is a well-studied linear functional on elements of $H^s(2\pi)$ for $s > \frac{1}{2}$:

$$\delta[\varphi] \equiv \langle \varphi, \delta \rangle = \varphi(0), \qquad \varphi \in H^s(2\pi), \quad s > \frac{1}{2}$$

with $\delta[\varphi]$ denoting the action of δ on φ.

6.5.1 The dual space

The last example suggests another interpretation of $H^s(2\pi)$ for negative s, that of a dual space. Let ℓ be a bounded linear functional on $H^t(2\pi)$ for some $t \geq 0$, bounded with respect to the norm $\| \cdot \|_t$. Then using the *Riesz representation theorem* (Theorem 2.5.7), it can be shown that there is a unique element $\eta_\ell \in H^{-t}(2\pi)$, $\eta_\ell = \sum b_m \psi_m$, with

$$\ell[\varphi] \equiv \langle \varphi, \eta_\ell \rangle$$
$$= \sum_{m=-\infty}^{\infty} a_m \bar{b}_m, \qquad \text{for } \varphi = \sum_{m=-\infty}^{\infty} a_m \psi_m \in H^t(2\pi). \tag{6.5.6}$$

It is also straightforward to show that when given two such linear functionals, say ℓ_1 and ℓ_2, we have

$$\eta_{c_1 \ell_1 + c_2 \ell_2} = \bar{c}_1 \eta_{\ell_1} + \bar{c}_2 \eta_{\ell_2}$$

for all scalars c_1, c_2. Moreover,

$$|\langle \varphi, \eta_\ell \rangle| \leq \|\varphi\|_{*,t} \|\eta_\ell\|_{*,-t}$$

and

$$\|\ell\| = \|\eta_\ell\|_{*,-t} .$$

The space $H^{-t}(2\pi)$ can be used to represent the space of bounded linear functionals on $H^t(2\pi)$, and it is usually called the *dual space* for $H^t(2\pi)$. In this framework, we are regarding $H^0(2\pi) \equiv L^2(0, 2\pi)$ as self-dual. The evaluation of linear functionals on $H^t(2\pi)$, as in (6.5.6), can be considered as a bilinear function defined on $H^t(2\pi) \times H^{-t}(2\pi)$; and in that case,

$$|\langle \varphi, \eta \rangle| \leq \|\varphi\|_{*,t} \|\eta\|_{*,-t} , \qquad \varphi \in H^t(2\pi), \quad \eta \in H^{-t}(2\pi). \tag{6.5.7}$$

Define $b : H^t(2\pi) \times L^2(0, 2\pi) \to \mathbb{C}$ by

$$b(\varphi, \psi) = (\varphi, \psi), \qquad \varphi \in H^t(2\pi), \quad \psi \in L^2(0, 2\pi) \tag{6.5.8}$$

using the usual inner product of $L^2(0, 2\pi)$. Then the bilinear *duality pairing* $\langle \cdot, \cdot \rangle$ is the unique bounded extension of $b(\cdot, \cdot)$ to $H^t(2\pi) \times H^{-t}(2\pi)$, when regarding $L^2(0, 2\pi)$ as a dense subspace of $H^{-t}(2\pi)$. For a more extensive discussion of this topic with much greater generality, see Aubin [17, Chapter 3].

We have considered $\langle \cdot, \cdot \rangle$ as defined on $H^t(2\pi) \times H^{-t}(2\pi)$ with $t \geq 0$. But we can easily extend this definition to allow $t < 0$. Using (6.5.6), we define

$$\langle \varphi, \eta \rangle = \sum_{m=-\infty}^{\infty} a_m \overline{b}_m \tag{6.5.9}$$

for $\varphi = \sum a_m \psi_m$ in $H^t(2\pi)$ and $\eta = \sum b_m \psi_m$ in $H^{-t}(2\pi)$, for any real number t. The bound (6.5.7) is also still valid. This extension of $\langle \cdot, \cdot \rangle$ is merely a statement that the dual space for $H^t(2\pi)$ with $t < 0$ is just $H^{-t}(2\pi)$.

6.5.2 Embedding results

We give another variant of the Sobolev embedding theorem.

Proposition 6.5.4 *Let $s > k + \frac{1}{2}$ for some integer $k \geq 0$, and let $\varphi \in H^s(2\pi)$. Then $\varphi \in C_p^k(2\pi)$.*

Proof. We give a proof for only the case $k = 0$, as the general case is quite similar. We show the Fourier series (6.5.1) for φ is absolutely and uniformly convergent on $[0, 2\pi]$; and it then follows by standard arguments that φ is continuous and periodic. From the definition (6.5.1), and by using the Cauchy-Schwarz inequality,

$$|\varphi(s)| \leq \sum_{m=-\infty}^{\infty} |a_m|$$

$$= |a_0| + \sum_{|m|>0} |m|^{-s} |m|^s |a_m|$$

$$\leq |a_0| + \sqrt{\sum_{|m|>0} |m|^{-2s}} \sqrt{\sum_{|m|>0} |m|^{2s} |a_m|^2}.$$

Denoting $\zeta(r)$ the *zeta* function

$$\zeta(r) = \sum_{m=1}^{\infty} \frac{1}{m^r}, \qquad r > 1,$$

we then have

$$|\varphi(s)| \leq |a_0| + \sqrt{2\zeta(2s)} \, \|\varphi\|_s. \tag{6.5.10}$$

By standard arguments on the convergence of infinite series, (6.5.10) implies that the Fourier series (6.5.1) for φ is absolutely and uniformly convergent. In addition,

$$\|\varphi\|_\infty \leq [1 + \sqrt{2\zeta(2s)}] \, \|\varphi\|_{*,s}, \qquad \varphi \in H^s(2\pi). \tag{6.5.11}$$

Thus the identity mapping from $H^s(2\pi)$ into $C_p(2\pi)$ is bounded. ∎
  The proof of the following result is left as Exercise 6.5.1.

Proposition 6.5.5 *Let $s > t$. Then $H^s(2\pi)$ is dense in $H^t(2\pi)$, and the identity mapping*

$$I : H^s(2\pi) \to H^t(2\pi), \qquad I(\varphi) \equiv \varphi \quad \text{for } \varphi \in H^s(2\pi)$$

is a compact operator.

6.5.3 *Approximation results*

When integrals of periodic functions are approximated numerically, the trapezoidal rule is the method of choice in most cases. Let us explain why. Suppose the integral to be evaluated is

$$I(\varphi) = \int_0^{2\pi} \varphi(x)\, dx$$

with $\varphi \in H^s(2\pi)$ and $s > \frac{1}{2}$. The latter assumption guarantees φ is continuous so that evaluation of $\varphi(x)$ makes sense for all x. For an integer $k \geq 1$, let $h = \frac{2\pi}{k}$ and write the rule as

$$T_k(\varphi) = h \sum_{j=1}^{k} \varphi(jh). \qquad (6.5.12)$$

We give a nonstandard error bound, but one that shows the rapid rate of convergence for smooth periodic functions.

Proposition 6.5.6 *Assume $s > \frac{1}{2}$, and let $\varphi \in H^s(2\pi)$. Then*

$$|I(\varphi) - T_k(\varphi)| \leq \frac{\sqrt{4\pi\zeta(2s)}}{k^s} \|\varphi\|_s, \qquad k \geq 1. \qquad (6.5.13)$$

Proof. We begin with the following result, on the application of the trapezoidal rule to e^{imx}.

$$T_k(e^{imx}) = \begin{cases} 2\pi, \, m = jk, \quad j = 0, \pm 1, \pm 2, \ldots, \\ 0, \ \text{otherwise.} \end{cases} \qquad (6.5.14)$$

Using it, and applying T_k to the Fourier series representation (6.5.1) of $\varphi(s)$, we have

$$I(\varphi) - T_k(\varphi) = -\sqrt{2\pi} \sum_{|m|>0} a_{km} = -\sqrt{2\pi} \sum_{|m|>0} a_{km}(km)^s (km)^{-s}.$$

Applying the Cauchy-Schwarz inequality to the last sum,

$$|I(\varphi) - T_k(\varphi)| \leq \sqrt{2\pi} \left[\sum_{|m|>0} |a_{km}|^2 (km)^{2s} \right]^{\frac{1}{2}} \left[\sum_{|m|>0} (km)^{-2s} \right]^{\frac{1}{2}}$$

$$\leq \sqrt{2\pi} \|\varphi\|_s \, k^{-s} \sqrt{2\zeta(2s)}.$$

This completes the proof. ■

Recall the discussion of the trigonometric interpolation polynomial $\mathcal{I}_n\varphi$ in Chapter 3, and in particular the theoretical error bound of Theorem 3.6.2. We give another error bound for this interpolation. A complete derivation of it can be found in [102].

Theorem 6.5.7 *Let* $s > \frac{1}{2}$, *and let* $\varphi \in H^s(2\pi)$. *Then for* $0 \le r \le s$,

$$\|\varphi - \mathcal{I}_n\varphi\|_r \le \frac{c}{n^{s-r}} \|\varphi\|_s, \qquad n \ge 1. \tag{6.5.15}$$

The constant c *depends on* s *and* r *only.*

Proof. The proof is based on using the Fourier series representation (6.5.1) of φ to obtain a Fourier series for $\mathcal{I}_n\varphi$. This is then subtracted from that for φ, and the remaining terms are bounded to give the result (6.5.15). For details, see [102]. This is only a marginally better result than Theorem 3.6.2, but it is an important tool when doing error analyses of numerical methods in the Sobolev spaces $H^s(2\pi)$. ∎

6.5.4   An illustrative example of an operator

To illustrate the usefulness of the Sobolev spaces $H^s(2\pi)$, we consider the following important integral operator:

$$\mathcal{A}\varphi(x) = -\frac{1}{\pi}\int_0^{2\pi} \varphi(y)\log\left|2e^{-\frac{1}{2}}\sin\left(\frac{x-y}{2}\right)\right|\,dy, \qquad -\infty < x < \infty \tag{6.5.16}$$

for $\varphi \in L^2(0, 2\pi)$. It plays a critical role in the study of boundary integral equation reformulations of Laplace's equation in the plane. Using results from the theory of functions of a complex variable, it can be shown that

$$\mathcal{A}\varphi(t) = a_0\psi_0(t) + \sum_{|m|>0} \frac{a_m}{|m|}\psi_m(t) \tag{6.5.17}$$

where $\varphi = \sum_{m=-\infty}^{\infty} a_m\psi_m$. In turn, this implies that

$$\mathcal{A} : H^s(2\pi) \underset{onto}{\overset{1-1}{\to}} H^{s+1}(2\pi), \qquad s \ge 0 \tag{6.5.18}$$

and

$$\|\mathcal{A}\| = 1.$$

The definition of $\mathcal{A}$ as an integral operator in (6.5.16) requires that the function φ be a function to which integration can be applied. However, the formula (6.5.17) permits us to extend the domain for $\mathcal{A}$ to any Sobolev space $H^s(2\pi)$ with $s < 0$. This is important in that one important approach to the numerical analysis of the equation $\mathcal{A}\varphi = f$ requires $\mathcal{A}$ to be regarded as an operator from $H^{-\frac{1}{2}}(2\pi)$ to $H^{\frac{1}{2}}(2\pi)$.

We will return to the study and application of $\mathcal{A}$ in Chapter 12; and it is discussed at length in [13, Chap. 7]; [148].

6.5.5 Spherical polynomials and spherical harmonics

The Fourier series can be regarded as an expansion of functions defined on the unit circle in the plane. For functions defined on the unit sphere U in $\mathbb{R}^3$, the analogue of the Fourier series is the *Laplace expansion*, and it uses *spherical harmonics* as the generalizations of the trigonometric functions. We begin by first considering *spherical polynomials*, and then we introduce the spherical harmonics and Laplace expansion. Following the tradition in the literature on the topic, we use (x, y, z) for a generic point in $\mathbb{R}^3$ in this subsection.

Definition 6.5.8 *Consider an arbitrary polynomial in* (x, y, z) *of degree* N, *say,*

$$p(x, y, z) = \sum_{\substack{i,j,k \geq 0 \\ i+j+k \leq N}} a_{i,j,k} x^i y^j z^k, \qquad (6.5.19)$$

and restrict (x, y, z) *to lie on the unit sphere* U. *The resulting function is called a* spherical polynomial *of degree* $\leq N$.

Spherical polynomials are the analogues of the trigonometric polynomials, which can be obtained by replacing (x, y) in

$$\sum_{\substack{i,j \geq 0 \\ i+j \leq N}} a_{i,j} x^i y^j$$

with $(\cos\theta, \sin\theta)$. Note that a polynomial $p(x, y, z)$ may reduce to an expression of lower degree. For example, $p(x, y, z) = x^2 + y^2 + z^2$ reduces to 1 when $(x, y, z) \in U$. Denote by $\mathcal{S}_N$ the set of all such spherical polynomials of degree $\leq N$.

An alternative way of obtaining polynomials on U is to begin with *homogeneous harmonic polynomials*.

Definition 6.5.9 *Let* $p = p(x, y, z)$ *be a polynomial of degree* n *which satisfies Laplace's equation,*

$$\Delta p(x, y, z) \equiv \frac{\partial^2 p}{\partial x^2} + \frac{\partial^2 p}{\partial y^2} + \frac{\partial^2 p}{\partial z^2} = 0, \qquad (x, y, z) \in \mathbb{R}^3,$$

and further, let p *be homogeneous of degree* n:

$$p(tx, ty, tz) = t^n p(x, y, z), \qquad -\infty < t < \infty, \quad (x, y, z) \in \mathbb{R}^3.$$

Restrict all such polynomials to U. *Such functions are called* spherical harmonics *of degree* n.

As examples of spherical harmonics, we have the following.

1. $n = 0$

$$p(x, y, z) = 1,$$

2. $n = 1$

$$p(x, y, z) = x, \ y, \ z,$$

3. $n = 2$

$$p(x, y, z) = xy, \ xz, \ yz, \ x^2 + y^2 - 2z^2, \ x^2 + z^2 - 2y^2,$$

where in all cases, we use $(x, y, z) = (\cos \phi \sin \theta, \sin \phi \sin \theta, \cos \theta)$. Nontrivial linear combinations of spherical harmonics of a given degree are again spherical harmonics of that same degree. For example,

$$p(x, y, z) = x + y + z$$

is also a spherical harmonic of degree 1. The number of linearly independent spherical harmonics of degree n is $2n + 1$; and thus the above sets are maximal independent sets for each of the given degrees $n = 0, 1, 2$.

Define $\widehat{\mathcal{S}}_N$ to be the smallest vector space to contain all of the spherical harmonics of degree $n \leq N$. Alternatively, $\widehat{\mathcal{S}}_N$ is the set of all finite linear combinations of spherical harmonics of all possible degrees $n \leq N$. Then it can be shown that

$$\mathcal{S}_N = \widehat{\mathcal{S}}_N \tag{6.5.20}$$

and

$$\dim \mathcal{S}_N = (N + 1)^2. \tag{6.5.21}$$

Below, we give a basis for $\mathcal{S}_N$. See MacRobert [111, Chap. 7] for a proof of these results.

There are well-known formulas for spherical harmonics, and we will make use of some of them in working with spherical polynomials. The subject of spherical harmonics is quite large one, and we can only touch on a few small parts of it. For further study, see the classical book [111] by T. MacRobert. The spherical harmonics of degree n are the analogues of the trigonometric functions $\cos n\theta$ and $\sin n\theta$, which are restrictions to the unit circle of the homogeneous harmonic polynomials

$$r^n \cos(n\theta), \quad r^n \sin(n\theta)$$

written in polar coordinates form.

The standard basis for spherical harmonics of degree n is

$$\begin{aligned}
S_n^1(x, y, z) &= c_n L_n(\cos \theta), \\
S_n^{2m}(x, y, z) &= c_{n,m} L_n^m(\cos \theta) \cos(m\phi), \\
S_n^{2m+1}(x, y, z) &= c_{n,m} L_n^m(\cos \theta) \sin(m\phi), \quad m = 1, \ldots, n,
\end{aligned} \tag{6.5.22}$$

with $(x, y, z) = (\cos \phi \sin \theta, \sin \phi \sin \theta, \cos \theta)$. In this formula, $L_n(t)$ is a *Legendre polynomial* of degree n,

$$L_n(t) = \frac{1}{2^n n!} \frac{d^n}{dt^n} \left[(t^2 - 1)^n \right] \tag{6.5.23}$$

and $L_n^m(t)$ is an *associated Legendre function*,

$$L_n^m(t) = (-1)^m (1-t^2)^{\frac{m}{2}} \frac{d^m}{dt^m} L_n(t), \qquad 1 \le m \le n. \qquad (6.5.24)$$

The constants in (6.5.22) are given by

$$c_n = \sqrt{\frac{2n+1}{4\pi}},$$

$$c_{n,m} = \sqrt{\frac{2n+1}{2\pi} \frac{(n-m)!}{(n+m)!}}.$$

We will occasionally denote the Legendre polynomial L_n by L_n^0, to simplify referring to these Legendre functions.

The standard inner product on $L^2(U)$ is given by

$$(f,g) = \int_U f(Q)g(Q)\, dS_Q.$$

Using this definition, we can verify that the functions of (6.5.22) satisfy

$$(S_n^k, S_q^p) = \delta_{n,q}\delta_{k,p}$$

for $n,q = 0,1,\ldots$ and $1 \le k \le 2n+1$, $1 \le p \le 2q+1$. The set of functions

$$\{S_n^k \mid 1 \le k \le 2n+1,\ 0 \le n \le N\}$$

is an orthonormal basis for $\mathcal{S}_N$. To avoid some double summations, we will sometimes write this basis for $\mathcal{S}_N$ as

$$\{\Psi_1, \ldots, \Psi_{d_N}\} \qquad (6.5.25)$$

with $d_N = (N+1)^2$ the dimension of the subspace.

The set $\{S_n^k \mid 1 \le k \le 2n+1,\ 0 \le n < \infty\}$ of spherical harmonics is an orthonormal basis for $L^2(U)$, and it leads to the expansion formula

$$g(Q) = \sum_{n=0}^{\infty} \sum_{k=1}^{2n+1} (g, S_n^k)\, S_n^k(Q), \qquad g \in L^2(U). \qquad (6.5.26)$$

This is called the *Laplace expansion* of the function g, and it is the generalization to $L^2(U)$ of the Fourier series on the unit circle in the plane. The function $g \in L^2(U)$ if and only if

$$\|g\|_{L^2}^2 = \sum_{n=0}^{\infty} \sum_{k=1}^{2n+1} \left| (g, S_n^k) \right|^2 < \infty. \qquad (6.5.27)$$

In analogy with the use of the Fourier series to define the Sobolev spaces $H^s(2\pi)$, we can characterize the Sobolev spaces $H^s(U)$ by using the Laplace expansion.

Of particular interest is the truncation of the series (6.5.26) to terms of degree at most N, to obtain

$$P_N g(Q) = \sum_{n=0}^{N} \sum_{k=1}^{2n+1} (g, S_n^k) \, S_n^k(Q). \tag{6.5.28}$$

This defines the orthogonal projection of $L^2(U)$ onto $\mathcal{S}_N$; and of course, $P_N g \to g$ as $N \to \infty$. Since it is an orthogonal projection on $L^2(U)$, we have $\|\mathcal{P}_N\| = 1$ as an operator from $L^2(U)$ in $L^2(U)$. However, we can also regard $\mathcal{S}_N$ as a subset of $C(S)$; and then regarding P_N as a projection from $C(S)$ to $\mathcal{S}_N$, we have

$$\|P_N\| = \left(\sqrt{\frac{8}{\pi}} + \delta_N \right) \sqrt{N} \tag{6.5.29}$$

with $\delta_N \to 0$ as $N \to \infty$. A proof of this is quite involved, and we refer the reader to Gronwall [66] and Ragozin [136]. In a later chapter, we use the projection P_N to define a Galerkin method for solving integral equations defined on U, with $\mathcal{S}_N$ as the approximating subspace.

Best approximations

Given $g \in C(U)$, define

$$\rho_N(g) = \inf_{p \in \mathcal{S}_N} \|g - p\|_\infty. \tag{6.5.30}$$

This is called the *minimax error* for the approximation of g by spherical polynomials of degree $\leq N$. With the Stone-Weierstraß theorem (e.g., see [117]), it can be shown that $\rho_N(g) \to 0$ as $N \to \infty$. In the error analysis of numerical methods that use spherical polynomials, it is important to have bounds on the rate at which $\rho_N(g)$ converges to zero. An initial partial result was given by Gronwall [66]; and a much more complete theory was given many years later by Ragozin [135], a special case of which we give here. We first introduce some notation.

For given positive integer k, let $D^k g$ denote an arbitrary k^{th} order derivative of g on U, formed with respect to local surface coordinates on U. (One should consider a set of local patch coordinate systems over U, as in Definition 6.2.13, and Sobolev spaces based on these patches. But what is intended is clear and the present notation is simpler.) Let γ be a real number, $0 < \gamma \leq 1$. Define $C^{k,\gamma}(U)$ to be the set of all functions $g \in C(U)$ for which all of its derivatives $D^k g \in C(U)$, with each of these derivatives satisfying a Hölder condition with exponent γ:

$$\left| D^k g(P) - D^k g(Q) \right| \leq H_{k,\gamma}(g) \, |P - Q|^\gamma, \qquad P, Q \in U.$$

Here $|P - Q|$ denotes the usual distance between the two points P and Q. The Hölder constant $H_{k,\gamma}(g)$ is to be uniform over all k^{th}-order derivatives of g.

Theorem 6.5.10 *Let $g \in C^{k,\gamma}(U)$. Then there is a sequence of spherical polynomials $\{p_N\}$ for which $\|g - p_N\|_\infty = \rho_N(g)$ and*

$$\rho_N(g) \leq \frac{c_k H_{k,\gamma}(g)}{N^{k+\gamma}}, \qquad N \geq 1. \tag{6.5.31}$$

The constant c_k is dependent on only k.

For the case $k = 0$, see Gronwall [66] for a proof; and for the general case, a proof can be found in Ragozin [135, Theorem 3.3].

This result leads immediately to results on the rate of convergence of the Laplace series expansion of a function $g \in C^{k,\gamma}(U)$, given in (6.5.26). Using the norm of $L^2(U)$, and using the definition of the orthogonal projection $\mathcal{P}_N g$ being the best approximation in the inner product norm, we have

$$\begin{aligned}
\|g - \mathcal{P}_N g\| &\leq \|g - p_N\| \\
&\leq 4\pi \|g - p_N\|_\infty \\
&\leq \frac{4\pi c_{k,\gamma}(g)}{N^{k+\gamma}}, \qquad N \geq 1.
\end{aligned} \tag{6.5.32}$$

We can also consider the uniform convergence of the Laplace series. Write

$$\begin{aligned}
\|g - \mathcal{P}_N g\|_\infty &= \|g - p_N - P_N(g - p_N)\|_\infty \\
&\leq (1 + \|P_N\|) \|g - p_N\|_\infty \\
&\leq c N^{-(k+\gamma-\frac{1}{2})}
\end{aligned} \tag{6.5.33}$$

with the last step using (6.5.29). In particular, if $g \in C^{0,\gamma}(U)$ with $\frac{1}{2} < \gamma \leq 1$, we have uniform convergence of $\mathcal{P}_N g$ to g on U. From (6.5.31), the constant c is a multiple of $H_{k,\gamma}(g)$.

No way is known to interpolate with spherical polynomials in a manner that generalizes trigonometric interpolation. For a more complete discussion of this and other problems in working with spherical polynomial approximations, see [13, Section 5.5].

Sobolev spaces on the unit sphere

The function spaces $L^2(U)$ and $C(U)$ are the most widely used function spaces over U, but we need to also introduce the Sobolev spaces $H^r(U)$. There are several equivalent ways to define $H^r(U)$. The standard way is to proceed as in Definition 6.2.13, using local coordinate systems based on a local set of patches covering U and then using Sobolev spaces based on these patches.

Another approach, used less often but possibly more intuitive, is based on the Laplace expansion of a function g defined on the unit sphere U. Recalling the Laplace expansion (6.5.26), define the Sobolev space $H^r(U)$ to be the set of functions whose Laplace expansion satisfies

$$\|g\|_{*,r} \equiv \left[\sum_{n=0}^{\infty} (2n+1)^{2r} \sum_{k=1}^{2n+1} |(g, S_n^k)|^2 \right]^{\frac{1}{2}} < \infty. \tag{6.5.34}$$

This definition can be used for any real number $r \geq 0$. For r a positive integer, the norm $\|g\|_{*,r}$ can be shown to be equivalent to a standard Sobolev norm based on a set of local patch coordinate systems for U.

Exercise 6.5.1 *Prove Proposition 6.5.5.*

Exercise 6.5.2 *Prove that the space $\mathcal{T}$ defined in (6.5.4) is dense in $H^s(2\pi)$ for $-\infty < s < \infty$.*

Exercise 6.5.3 *Let φ be a continuous periodic function with period 2π. Write Simpson's rule as*

$$I(\varphi) = \int_0^{2\pi} \varphi(x)\,dx \approx \frac{h}{3}\left[4\sum_{j=1}^k \varphi(x_{2j-1}) + 2\sum_{j=1}^k \varphi(x_{2j})\right] \equiv S_{2k}(\varphi),$$

where $h = 2\pi/(2k) = \pi/k$. Show that if $\varphi \in H^s(2\pi)$, $s > 1/2$, then

$$|I(\varphi) - S_{2k}(\varphi)| \leq c\,k^{-s}\|\varphi\|_s.$$

Exercise 6.5.4 *For $t > 0$, demonstrate that elements of $H^{-t}(2\pi)$ are indeed bounded linear functionals on $H^t(2\pi)$ when using (6.5.6) to define the linear functional.*

6.6   Integration by parts formulas

We comment on the validity of integration by parts formulas. Assume Ω is a bounded domain in $\mathbb{R}^d$ with a Lipschitz continuous boundary Γ. Denote $\boldsymbol{\nu} = (\nu_1, \ldots, \nu_d)^T$ the unit outward normal vector on Γ, which is defined almost everywhere giving that Γ is Lipschitz continuous. It is a well-known classical result that

$$\int_\Omega u_{x_i} v\,dx = \int_\Gamma u\,v\,\nu_i\,ds - \int_\Omega u\,v_{x_i}dx \qquad \forall\,u,v \in C^1(\overline{\Omega}).$$

This is often called *Gauss's formula* or the *divergence theorem*; and for planar regions Ω, it is also called *Green's formula*. This formula can be extended to functions from certain Sobolev spaces so that the smoothness of the functions is exactly enough for the integrals to be well defined in the sense of Lebesgue.

Proposition 6.6.1 *Assume $\Omega \subset \mathbb{R}^d$ is a bounded domain with a Lipschitz continuous boundary Γ. Then*

$$\int_\Omega u_{x_i} v\,dx = \int_\Gamma u\,v\,\nu_i\,ds - \int_\Omega u\,v_{x_i}dx \qquad \forall\,u,v \in H^1(\Omega). \qquad (6.6.1)$$

Proof. Since $C^1(\overline{\Omega})$ is dense in $H^1(\Omega)$, we have sequences $\{u_n\}, \{v_n\} \subseteq C^1(\overline{\Omega})$, such that

$$\|u_n - u\|_{H^1(\Omega)} \to 0 \qquad \text{as } n \to \infty,$$
$$\|v_n - v\|_{H^1(\Omega)} \to 0 \qquad \text{as } n \to \infty.$$

Since $u_n, v_n \in C^1(\overline{\Omega})$, we have

$$\int_\Omega (u_n)_{x_i} v_n \, dx = \int_\Gamma u_n v_n \nu_i \, ds - \int_\Omega u_n (v_n)_{x_i} dx. \qquad (6.6.2)$$

We take the limit as $n \to \infty$ in (6.6.2). Let us estimate

$$\left| \int_\Omega u_{x_i} v \, dx - \int_\Omega (u_n)_{x_i} v_n \, dx \right|$$
$$\leq \int_\Omega |(u - u_n)_{x_i}| \, |v| \, dx + \int_\Omega |(u_n)_{x_i}| \, |v - v_n| \, dx$$
$$\leq \|u - u_n\|_{H^1(\Omega)} \|v\|_{L^2(\Omega)} + \|u_n\|_{H^1(\Omega)} \|v - v_n\|_{L^2(\Omega)}.$$

Since the sequences $\{u_n\}$ and $\{v_n\}$ are convergent in $H^1(\Omega)$, the quantities $\|u_n\|_{H^1(\Omega)}$ are uniformly bounded. Hence,

$$\int_\Omega u_{x_i} v \, dx - \int_\Omega (u_n)_{x_i} v_n \, dx \to 0 \qquad \text{as } n \to \infty.$$

Similarly,

$$\int_\Omega u \, v_{x_i} dx - \int_\Omega u_n (v_n)_{x_i} dx \to 0 \qquad \text{as } n \to \infty.$$

With regard to the boundary integral terms, we need to use the trace theorem, $H^1(\Omega) \hookrightarrow L^2(\Gamma)$. From this, we see that

$$\|u_n - u\|_{L^2(\Gamma)} \leq c \|u_n - u\|_{H^1(\Omega)} \to 0 \qquad \text{as } n \to \infty,$$

and similarly,

$$\|v_n - v\|_{L^2(\Gamma)} \to 0 \qquad \text{as } n \to \infty.$$

Then use the argument technique above,

$$\int_\Gamma u v \nu_i \, ds - \int_\Gamma u_n v_n \nu_i \, ds \to 0 \qquad \text{as } n \to \infty.$$

Taking the limit $n \to \infty$ in (6.6.2) we obtain (6.6.1). ∎

The above proof technique is called a "density argument." Roughly speaking, a classical integral relation can often be extended to functions from certain Sobolev spaces, as long as all the expressions in the integral relation make sense. The formula (6.6.1) suffices in studying linear second-order boundary value problems. For analyzing nonlinear problems, it is beneficial to extend the formula (6.6.1) even further. Indeed we have

$$\int_\Omega u_{x_i} v \, dx = \int_\Gamma u \, v \, \nu_i \, ds - \int_\Omega u \, v_{x_i} dx \quad \forall\, u \in W^{1,p}(\Omega),\ v \in W^{1,p^*}(\Omega),$$

(6.6.3)

where $p \in (1,\infty)$, and $p^* \in (1,\infty)$ is the conjugate exponent defined through the relation $1/p + 1/p^* = 1$.

Various other useful formulas can be derived from (6.6.2). One such formula is

$$\int_\Omega \Delta u \, v \, dx = \int_\Gamma \frac{\partial u}{\partial \nu} v \, ds - \int_\Omega \nabla u \cdot \nabla v \, dx \quad \forall\, u \in H^2(\Omega),\ v \in H^1(\Omega).$$

(6.6.4)

Here,

$$\Delta : u \mapsto \Delta u = \sum_{i=1}^d \frac{\partial^2 u}{\partial x_i^2}$$

is the Laplacian operator;

$$\nabla u = (u_{x_1}, \dots, u_{x_d})^T$$

is the gradient of u; and

$$\frac{\partial u}{\partial \nu} = \nabla u \cdot \boldsymbol{\nu}$$

is the outward normal derivative.

Another useful formula derived from (6.6.2) is

$$\int_\Omega (\operatorname{div} \mathbf{u}) \, v \, dx = \int_\Gamma u_\nu v \, ds - \int_\Omega \mathbf{u} \cdot \nabla v \, dx \quad \forall\, \mathbf{u} \in H^1(\Omega)^d,\ v \in H^1(\Omega).$$

(6.6.5)

Here $\mathbf{u} = (u_1, \dots, u_d)^T$ is a vector-valued function;

$$\operatorname{div} \mathbf{u} = \sum_{i=1}^d \frac{\partial u_i}{\partial x_i}$$

is the divergence of $\mathbf{u}$; and $u_\nu = \mathbf{u} \cdot \boldsymbol{\nu}$ is the normal component of $\mathbf{u}$ on Γ.

Exercise 6.6.1 *Use the density argument to prove the formula (6.6.3).*

Exercise 6.6.2 *Prove the formulas (6.6.4) and (6.6.5) by using (6.6.1).*

Suggestion for Further Readings

ADAMS [1] and LIONS AND MAGENES [109] provides a comprehensive treatment of basic aspects of Sobolev spaces, including proofs of various results. Many references on modern PDEs contain an introduction to the theory of Sobolev spaces, e.g., EVANS [48], MCOWEN [116], WLOKA [167].

Sobolev spaces of any real order (i.e., $W^{s,p}(\Omega)$ with $s \in \mathbb{R}$) can be studied in the framework of interpolation spaces. Several methods are possible to develop a theory of interpolation spaces; see TRIEBEL [161]. A relatively easily accessible reference on the topic is BERGH AND LÖFSTRÖM [22].

7

Variational Formulations of Elliptic Boundary Value Problems

In this chapter, we formulate variational (or weak) forms of some elliptic boundary value problems and study the well-posedness of the variational problems. We begin with a derivation of the weak formulation of the homogeneous Dirichlet boundary value problem for the Poisson equation. In the abstract form, a weak formulation can be viewed as an operator equation. In the second section, we provide some general results on existence and uniqueness for linear operator equations. In the third section, we present and discuss the well-known Lax-Milgram lemma, which is applied, in the section following, to the study of well-posedness of variational formulations for various linear elliptic boundary value problems. We also apply the Lax-Milgram lemma in studying a boundary value problem in linearized elasticity. The framework in the Lax-Milgram lemma is suitable for the development of the Galerkin method for numerically solving linear elliptic boundary value problems. In Section 7.6, we provide a brief discussion of two different weak formulations: the mixed formulation and the dual formulation. For the development of Petrov-Galerkin method, where the trial function space and the test function space are different, we discuss a generalization of Lax-Milgram lemma in Section 7.7. Most of the chapter is concerned with boundary value problems with linear differential operators. In the last section, we analyze a nonlinear elliptic boundary value problem.

Recall that we use Ω to denote an open bounded set in $\mathbb{R}^d$, and we assume the boundary $\Gamma = \partial\Omega$ is Lipschitz continuous. Occasionally, we need to further assume Ω to be connected, and we state this assumption explicitly when it is needed.

7.1   A model boundary value problem

To begin, we use the following model boundary value problem as an illustrative example:

$$\begin{cases} -\Delta u = f \text{ in } \Omega, \\ u = 0 \quad\;\; \text{on } \Gamma. \end{cases} \tag{7.1.1}$$

The differential equation in (7.1.1) is called the Poisson equation. The Poisson equation can be used to describe many physical processes, e.g., steady-state heat conduction, electrostatics, deformation of a thin elastic membrane (cf. [133]). We discuss a weak solution of the problem and its relation to a classical solution of the problem.

A classical solution of the problem (7.1.1) is a smooth function $u \in C^2(\Omega) \cap C(\overline{\Omega})$ that satisfies the differential equation $(7.1.1)_1$ and the boundary condition $(7.1.1)_2$ pointwise. Necessarily we have to assume $f \in C(\Omega)$, but this condition, or even the stronger condition $f \in C(\overline{\Omega})$, does not guarantee the existence of a classical solution of the problem (cf. [64]). A purpose of the introduction of the weak formulation is to remove the high smoothness requirement on the solution and as a result it is easier to have the existence of a (weak) solution.

To derive the weak formulation corresponding to (7.1.1), we temporarily assume it has a classical solution $u \in C^2(\Omega) \cap C(\overline{\Omega})$. We multiply the differential equation $(7.1.1)_1$ by an arbitrary function $v \in C_0^\infty(\Omega)$ (so-called smooth test functions), and integrate the relation on Ω,

$$-\int_\Omega \Delta u\, v\, dx = \int_\Omega f\, v\, dx.$$

Now an integration by parts yields (recall that $v = 0$ on Γ)

$$\int_\Omega \nabla u \cdot \nabla v\, dx = \int_\Omega f\, v\, dx. \tag{7.1.2}$$

This relation was proved under the assumptions $u \in C^2(\Omega) \cap C(\overline{\Omega})$ and $v \in C_0^\infty(\Omega)$. However, for each term in the relation (7.1.2) to make sense, we only need to require the following regularities of u and v: $u, v \in H^1(\Omega)$, assuming $f \in L^2(\Omega)$. Recalling the homogeneous Dirichlet boundary condition $(7.1.1)_2$, we thus seek a solution $u \in H_0^1(\Omega)$ satisfying the relation (7.1.2) for any $v \in C_0^\infty(\Omega)$. Since $C_0^\infty(\Omega)$ is dense in $H_0^1(\Omega)$, the relation (7.1.2) is valid for any $v \in H_0^1(\Omega)$. Therefore, the weak formulation of the boundary value problem (7.1.1) is

$$u \in H_0^1(\Omega), \quad \int_\Omega \nabla u \cdot \nabla v\, dx = \int_\Omega f\, v\, dx \quad \forall\, v \in H_0^1(\Omega). \tag{7.1.3}$$

Actually, we can even weaken the assumption $f \in L^2(\Omega)$. It is enough for us to assume $f \in H^{-1}(\Omega) = (H_0^1(\Omega))'$, as long as we interpret the integral $\int_\Omega f\, v\, dx$ as the duality pairing $\langle f, v \rangle$ between $H^{-1}(\Omega)$ and $H_0^1(\Omega)$. We adopt the convention of using $\int_\Omega f v \, dx$ for $\langle f, v \rangle$ when $f \in H^{-1}(\Omega)$ and $v \in H_0^1(\Omega)$.

We have shown that if u is a classical solution of (7.1.1), then it is also a solution of the weak formulation (7.1.3). Conversely, suppose u is a weak solution with the additional regularity $u \in C^2(\Omega) \cap C(\overline{\Omega})$. Then for any $v \in C_0^\infty(\Omega) \subseteq H_0^1(\Omega)$, from (7.1.3) we obtain

$$\int_\Omega (-\Delta u - f) v \, dx = 0.$$

Then we must have $-\Delta u = f$ in Ω; i.e., the differential equation $(7.1.1)_1$ is satisfied. Also u satisfies the homogeneous Dirichlet boundary condition pointwisely.

Thus we have shown that the boundary value problem (7.1.1) and the variational problem (7.1.3) are formally equivalent. In case the weak solution u does not have the regularity $u \in C^2(\Omega) \cap C(\overline{\Omega})$, we will say u formally solves the boundary value problem (7.1.1).

We set $V = H_0^1(\Omega)$, and let $a(\cdot, \cdot) : V \times V \to \mathbb{R}$ be the bilinear form defined by

$$a(u, v) = \int_\Omega \nabla u \cdot \nabla v \, dx \quad \text{for } u, v \in V,$$

and $\ell : V \to \mathbb{R}$ the linear functional defined by

$$\ell(v) = \int_\Omega f v \, dx \quad \text{for } v \in V.$$

Then the weak formulation of the problem is to find $u \in V$ such that

$$a(u, v) = \ell(v) \quad \forall v \in V. \tag{7.1.4}$$

We define a differential operator A associated with the boundary value problem (7.1.1) by

$$A : H_0^1(\Omega) \to H^{-1}(\Omega), \quad \langle Au, v \rangle = a(u, v) \quad \forall u, v \in H_0^1(\Omega).$$

Here, $\langle \cdot, \cdot \rangle$ denotes the duality pairing between $H^{-1}(\Omega)$ and $H_0^1(\Omega)$. Then the problems (7.1.4) can be viewed as a linear operator equation

$$Au = \ell \quad \text{in } H^{-1}(\Omega).$$

A formulation of the type (7.1.1) in the form of a partial differential equation and a set of boundary conditions is referred to as a classical formulation of a boundary value problem, while a formulation of the type (7.1.4) is known as a weak formulation. One advantage of weak formulations over classical formulations is that questions related to existence and uniqueness of solutions can be answered more satisfactorily. Another advantage is that weak formulations naturally lead to the development of Galerkin-type numerical methods.

7.2   Some general results on existence and uniqueness

We first present some general ideas and results on existence and uniqueness for a linear operator equation of the form

$$u \in V, \quad Lu = f, \tag{7.2.1}$$

where $L : \mathcal{D}(L) \subseteq V \to W$, V and W are Hilbert spaces, and $f \in W$. Notice that the solvability of the equation is equivalent to the condition $\mathcal{R}(L) = W$, while the uniqueness of a solution is equivalent to the condition $\mathcal{N}(L) = \{0\}$.

A very basic existence result is the following theorem.

Theorem 7.2.1 *Let V and W be Hilbert spaces, $L : \mathcal{D}(L) \subseteq V \to W$ a linear operator. Then $\mathcal{R}(L) = W$ if and only if $\mathcal{R}(L)$ is closed and $\mathcal{R}(L)^{\perp} = \{0\}$.*

Proof. If $\mathcal{R}(L) = W$, then obviously $\mathcal{R}(L)$ is closed and $\mathcal{R}(L)^{\perp} = \{0\}$.

Now assume $\mathcal{R}(L)$ is closed and $\mathcal{R}(L)^{\perp} = \{0\}$, but $\mathcal{R}(L) \neq W$. Then $\mathcal{R}(L)$ is a closed subspace of W. Let $w \in W \backslash \mathcal{R}(L)$. By the Hahn-Banach theorem, the compact set $\{w\}$ and the closed convex set $\mathcal{R}(L)$ can be strictly separated by a closed hyperplane; i.e., there exists a $w^* \in W'$ such that $\langle w^*, w \rangle > 0$ and $\langle w^*, Lv \rangle \leq 0$ for all $v \in \mathcal{D}(L)$. Since L is a linear operator, $\mathcal{D}(L)$ is a subspace of V. Hence, $\langle w^*, Lv \rangle = 0$ for all $v \in \mathcal{D}(L)$. Therefore, $0 \neq w^* \in \mathcal{R}(L)^{\perp}$. This is a contradiction. ∎

Let us see under what conditions $\mathcal{R}(L)$ is closed. We first introduce an important generalization of the notion of continuity.

Definition 7.2.2 *An operator $T : \mathcal{D}(T) \subseteq V \to W$, where V and W are Banach spaces, is said to be a closed operator if for any sequence $\{v_n\} \subseteq \mathcal{D}(T)$, $v_n \to v$ and $T(v_n) \to w$ imply $v \in \mathcal{D}(T)$ and $w = T(v)$.*

We notice that a continuous operator is closed. The next example shows that a closed operator is not necessarily continuous.

Example 7.2.3 *Let us consider a linear differential operator, $Lv = -\Delta v$. This operator is not continuous from $L^2(\Omega)$ to $L^2(\Omega)$. Nevertheless, L is a closed operator on $L^2(\Omega)$. To see this, let $\{v_n\}$ be a sequence converging to v in $L^2(\Omega)$, such that the sequence $\{-\Delta v_n\}$ converges to w in $L^2(\Omega)$. In the relation*

$$\int_{\Omega} -\Delta v_n \, \phi \, dx = - \int_{\Omega} v_n \Delta \phi \, dx \quad \forall \phi \in C_0^{\infty}(\Omega)$$

we take the limit $n \to \infty$ to obtain

$$\int_{\Omega} w \, \phi \, dx = - \int_{\Omega} v \, \Delta \phi \, dx \quad \forall \phi \in C_0^{\infty}(\Omega).$$

Therefore $w = -\Delta v$, and the operator L is closed.

Theorem 7.2.4 *Let V and W be Hilbert spaces, $L : \mathcal{D}(L) \subseteq V \to W$ a linear closed operator. Assume for some constant $c > 0$, the following a priori estimate holds:*

$$\|Lv\|_W \geq c\,\|v\|_V \quad \forall\, v \in \mathcal{D}(L), \tag{7.2.2}$$

which is usually called a stability estimate. Also assume $\mathcal{R}(L)^\perp = \{0\}$. Then for each $f \in W$, the equation (7.2.1) has a unique solution.

Proof. Let us verify that $\mathcal{R}(L)$ is closed. Let $\{f_n\}$ be a sequence in $\mathcal{R}(L)$, converging to f. Then there is a sequence $\{v_n\} \subseteq \mathcal{D}(L)$ with $f_n = Lv_n$. By (7.2.2),

$$\|v_n - v_m\| \leq c\,\|f_n - f_m\|.$$

Thus $\{v_n\}$ is a Cauchy sequence in V. Since V is a Hilbert space, the sequence $\{v_n\}$ converges: $v_n \to v \in V$. Now L is assumed to be closed, we conclude that $v \in \mathcal{D}(L)$ and $f = Lv \in \mathcal{R}(L)$. So we can invoke Theorem 7.2.1 to obtain the existence of a solution. The uniqueness of the solution follows from the stability estimate (7.2.2). ∎

Noticing that a continuous operator is closed, we can replace the closedness of the operator by the continuity of the operator in the above Theorem 7.2.4.

Example 7.2.5 *Let V be a Hilbert space, $L \in \mathcal{L}(V, V')$ be strongly monotone, i.e., for some constant $c > 0$,*

$$\langle Lv, v \rangle \geq c\,\|v\|_V^2 \quad \forall\, v \in V.$$

Then (7.2.2) holds because, from the monotonicity,

$$\|Lv\|_{V'}\,\|v\|_V \geq c\,\|v\|_V^2,$$

which implies

$$\|Lv\|_{V'} \geq c\,\|v\|_V.$$

Also $\mathcal{R}(L)^\perp = \{0\}$, since from $v \perp \mathcal{R}(L)$ we have

$$c\,\|v\|_V^2 \leq \langle Lv, v \rangle = 0,$$

and hence $v = 0$. Therefore from Theorem 7.2.4, under the stated assumptions, for any $f \in V'$, there is a unique solution $u \in V$ to the equation $Lu = f$ in V'.

Example 7.2.6 *As a concrete example, we consider the weak formulation of the model elliptic boundary value problem (7.1.1). Here, $\Omega \subseteq \mathbb{R}^d$ is a bounded domain with Lipschitz boundary $\partial\Omega$, $V = H_0^1(\Omega)$ with the norm $\|v\|_V = |v|_{H^1(\Omega)}$, and $V' = H^{-1}(\Omega)$. Given $f \in H^{-1}(\Omega)$, consider the problem*

$$\begin{cases} -\Delta u = f \text{ in } \Omega, \\ \quad\;\; u = 0 \text{ on } \partial\Omega. \end{cases} \tag{7.2.3}$$

We define the operator $L : V \to V'$ by

$$\langle Lu, v \rangle = \int_\Omega \nabla u \cdot \nabla v \, dx, \quad u, v \in V.$$

Then L is linear, continuous, and strongly monotone; indeed, we have

$$\|L\| = 1$$

and

$$\langle Lv, v \rangle = \|v\|_V^2 \quad \forall v \in V.$$

Thus from Example 7.2.5, for any $f \in H^{-1}(\Omega)$, there is a unique $u \in H_0^1(\Omega)$ such that

$$\int_\Omega \nabla u \cdot \nabla v \, dx = \langle f, v \rangle \quad \forall v \in V;$$

i.e., the boundary value problem (7.2.3) has a unique weak solution.

It is possible to extend the existence results presented in Theorems 7.2.1 and 7.2.4 to linear operator equations on Banach spaces. Let V and W be Banach spaces, with duals V' and W'. Let $L : \mathcal{D}(L) \subseteq V \to W$ be a densely defined linear operator; i.e., L is a linear operator and $\mathcal{D}(L)$ is a dense subspace of V. Because $\mathcal{D}(L)$ is dense in V, one can define the dual operator $L^* : \mathcal{D}(L^*) \subseteq W' \to V'$ by

$$\langle L^* w^*, v \rangle = \langle w^*, Lv \rangle \quad \forall v \in \mathcal{D}(L), \ w^* \in W'.$$

We then define

$$\mathcal{N}(L)^\perp = \{ v^* \in V' \mid \langle v^*, v \rangle = 0 \ \forall v \in \mathcal{D}(L) \},$$
$$\mathcal{N}(L^*)^\perp = \{ w \in W \mid \langle w^*, w \rangle = 0 \ \forall w^* \in \mathcal{D}(L^*) \}.$$

The most important theorem on dual operators in Banach spaces is the following closed range theorem of Banach (cf., e.g., [175, p. 210]).

Theorem 7.2.7 *Assume V and W are Banach spaces, $L : \mathcal{D}(L) \subseteq V \to W$ is a densely defined linear closed operator. Then the following four statements are equivalent.*
(a) *$\mathcal{R}(L)$ is closed in W.*
(b) *$\mathcal{R}(L) = \mathcal{N}(L^*)^\perp$.*
(c) *$\mathcal{R}(L^*)$ is closed in V'.*
(d) *$\mathcal{R}(L^*) = \mathcal{N}(L)^\perp$.*

In particular, this theorem implies the abstract Fredholm alternative result: If $\mathcal{R}(L)$ is closed, then $\mathcal{R}(L) = \mathcal{N}(L^*)^\perp$; i.e., the equation $Lu = f$ has a solution $u \in \mathcal{D}(L)$ if and only if $\langle w^*, f \rangle = 0$ for any $w^* \in W'$ with $L^* w^* = 0$. The closedness of $\mathcal{R}(L)$ follows from the stability estimate

$$\|Lv\| \geq c \, \|v\| \quad \forall v \in \mathcal{D}(L),$$

as we have seen in the proof of Theorem 7.2.4.

Now we consider the issue of uniqueness of a solution to a nonlinear operator equation. We have the following general result.

Theorem 7.2.8 *Assume V and W are Banach spaces, $T : \mathcal{D}(T) \subseteq V \to W$. Then for any $f \in W$, there exists at most one solution $u \in V$ of the equation $T(u) = f$, if one of the following conditions is satisfied.*
(a) *Stability: for some constant $c > 0$,*

$$\|T(u) - T(v)\| \geq c \|u - v\| \quad \forall u, v \in \mathcal{D}(T).$$

(b) *Contractivity of $T - I$:*

$$\|(T(u) - u) - (T(v) - v)\| < \|u - v\| \quad \forall u, v \in \mathcal{D}(T), \ u \neq v.$$

Proof. (a) Assume both u_1 and u_2 are solutions. Then $T(u_1) = T(u_2) = f$. Apply the stability condition,

$$c \|u_1 - u_2\| \leq \|T(u_1) - T(u_2)\| = 0.$$

Therefore, $u_1 = u_2$.

(b) Suppose there are two solutions $u_1 \neq u_2$. Then from the contractivity condition, we have

$$\|u_1 - u_2\| > \|(T(u_1) - u_1) - (T(u_2) - u_2)\| = \|u_1 - u_2\|.$$

This is a contradiction. ∎

We remark that the result of Theorem 7.2.8 certainly holds in the special case of Hilbert spaces V and W, and when $T = L : \mathcal{D}(L) \subseteq V \to W$ is a linear operator. In the case of a linear operator, the stability condition reduces to the estimate (7.2.2).

Exercise 7.2.1 *Consider a linear system on $\mathbb{R}^d$: $A\mathbf{x} = \mathbf{b}$, where $A \in \mathbb{R}^{d \times d}$, and $\mathbf{b} \in \mathbb{R}^d$. Recall the well-known result that for such a linear system, existence and uniqueness are equivalent. Apply Theorem 7.2.8 to find sufficient conditions on A that guarantee the unique solvability of the linear system for any given $\mathbf{b} \in \mathbb{R}^d$.*

7.3 The Lax-Milgram Lemma

The Lax-Milgram lemma is employed frequently in the study of linear elliptic boundary value problems of the form (7.1.4). For a real Banach space V, let us first explore the relation between a linear operator $A : V \to V'$ and a bilinear form $a : V \times V \to \mathbb{R}$ related by

$$\langle Au, v \rangle = a(u, v) \quad \forall u, v \in V. \tag{7.3.1}$$

Theorem 7.3.1 *There exists a one-to-one correspondence between linear continuous operators $A : V \to V'$ and continuous bilinear forms $a : V \times V \to \mathbb{R}$, given by the formula (7.3.1).*

Proof. If $A \in \mathcal{L}(V, V')$, then $a : V \times V \to \mathbb{R}$ defined in (7.3.1) is bilinear and bounded:

$$|a(u, v)| \leq \|Au\| \, \|v\| \leq \|A\| \, \|u\| \, \|v\| \quad \forall u, v \in V.$$

Conversely, let $a(\cdot, \cdot)$ be given as a continuous bilinear form on V. For any fixed $u \in V$, the map $v \mapsto a(u, v)$ defines a linear continuous operator on V. Thus, there is an element $Au \in V'$ such that (7.3.1) holds. From the bilinearity of $a(\cdot, \cdot)$, we obtain the linearity of A. From the boundedness of $a(\cdot, \cdot)$, we obtain the boundedness of A. ∎

With a linear operator A and a bilinear form a related through (7.3.1), many properties of the linear operator A can be defined through those of the bilinear form a, or vice versa. Some examples are (assuming V is a real Hilbert space):

- a is bounded ($a(u, v) \leq M \|u\| \, \|v\| \; \forall u, v \in V$) if and only if A is bounded ($\|Av\| \leq M \|v\| \; \forall v \in V$).

- a is positive ($a(v, v) \geq 0 \; \forall v \in V$) if and only if A is positive ($\langle Av, v \rangle \geq 0 \; \forall v \in V$).

- a is strictly positive ($a(v, v) > 0 \; \forall 0 \neq v \in V$) if and only if A is strictly positive ($\langle Av, v \rangle > 0 \; \forall 0 \neq v \in V$).

- a is strongly positive or V-elliptic ($a(v, v) \geq \alpha \|v\|^2 \; \forall v \in V$) if and only if A is strongly positive ($\langle Av, v \rangle \geq \alpha \|v\|^2 \; \forall v \in V$).

- a is symmetric ($a(u, v) = a(v, u) \; \forall u, v \in V$) if and only if A is symmetric ($\langle Au, v \rangle = \langle Av, u \rangle \; \forall u, v \in V$).

We now recall the following minimization principle from Chapter 3.

Theorem 7.3.2 *Assume K is a non-empty, closed, convex subset of the Hilbert space V, $\ell \in V'$. Let*

$$E(v) = \frac{1}{2} \|v\|^2 - \ell(v), \quad v \in V.$$

Then there exists a unique $u \in K$ such that

$$E(u) = \inf_{v \in K} E(v).$$

The minimizer u is uniquely characterized by the inequality

$$u \in K, \quad (u, v - u) \geq \ell(v - u) \quad \forall v \in K.$$

If additionally, K is a subspace of V, then u is equivalently defined by

$$u \in K, \quad (u, v) = \ell(v) \quad \forall v \in K.$$

Let us apply this result to get the Lax-Milgram lemma in case the bilinear form is symmetric.

Theorem 7.3.3 *Assume K is a non-empty, closed, convex subset of the Hilbert space V, $a(\cdot,\cdot) : V \times V \to \mathbb{R}$ bilinear, symmetric, bounded, and V-elliptic, $\ell \in V'$. Let*

$$E(v) = \frac{1}{2}a(v,v) - \ell(v), \quad v \in V.$$

Then there exists a unique $u \in K$ such that

$$E(u) = \inf_{v \in K} E(v), \tag{7.3.2}$$

which is also the unique solution of the variational inequality

$$u \in K, \quad a(u, v - u) \geq \ell(v - u) \quad \forall v \in K, \tag{7.3.3}$$

or

$$u \in K, \quad a(u, v) = \ell(v) \quad \forall v \in K \tag{7.3.4}$$

in the special case K is a subspace.

Proof. By the assumptions,

$$(u, v)_a = a(u, v), \quad u, v \in V$$

defines an inner product on V with the induced norm

$$\|v\|_a = \sqrt{a(v, v)}$$

which is equivalent to the original norm,

$$c_1 \|v\| \leq \|v\|_a \leq c_2 \|v\| \quad \forall v \in V,$$

for some constants $0 < c_1 \leq c_2 < \infty$. Also notice that ℓ is continuous with respect to the original norm if and only if it is continuous with respect to the norm $\|\cdot\|_a$. Now

$$E(v) = \frac{1}{2}\|v\|_a^2 - \ell(v)$$

and we can apply the results of Theorem 7.3.2.   ∎

In case the bilinear form $a(\cdot,\cdot)$ is not symmetric, there is no longer an associated minimization problem, yet we can still discuss the solvability of the variational equation (7.3.4) (the next theorem) or the variational inequality (7.3.3) (see Chapter 10).

Theorem 7.3.4 (LAX-MILGRAM LEMMA) *Assume V is a Hilbert space, $a(\cdot,\cdot)$ is a bounded, V-elliptic bilinear form on V, $\ell \in V'$. Then there is a unique solution of the problem*

$$u \in V, \quad a(u, v) = \ell(v) \quad \forall v \in V. \tag{7.3.5}$$

Before proving the result, let us consider the simple real linear equation

$$x \in \mathbb{R}, \quad a\,x = \ell.$$

Its weak formulation is

$$x \in \mathbb{R}, \quad a\,x\,y = \ell\,y \quad \forall\, y \in \mathbb{R}.$$

We observe that the real linear equation has a solution if and only if $0 < a < \infty$ (we multiply the equation by (-1) to make a positive, if necessary) and $|\ell| < \infty$, i.e., if and only if the bilinear form $a(x, y) \equiv a\,x\,y$ is continuous and $\mathbb{R}$-elliptic, and the linear form $\ell(y) \equiv \ell\,y$ is bounded. Thus the assumptions made in Theorem 7.3.4 are quite natural.

Several different proofs are possible for this important result. Here we present two of them.

Proof. [#1] For any $\theta > 0$, the problem (7.3.5) is equivalent to

$$(u, v) = (u, v) - \theta\,[a(u, v) - \ell(v)] \quad \forall\, v \in V,$$

i.e., the fixed-point problem,

$$u = P_\theta(u),$$

where $P_\theta(u) \in V$ is defined through the relation

$$(P_\theta(u), v) = (u, v) - \theta\,[a(u, v) - \ell(v)], \quad v \in V.$$

We will apply the Banach fixed-point theorem with a proper choice of θ. Let $A : V \to V'$ be the linear operator associated with the bilinear form $a(\cdot, \cdot)$, cf. (7.3.1). Then A is bounded and strongly positive: $\forall\, v \in V$,

$$\|Av\| \le M\,\|v\|,$$
$$\langle Av, v \rangle \ge \alpha\,\|v\|^2.$$

Denote $\mathcal{J} : V' \to V$ the isometric dual mapping from the Riesz representation theorem. Then

$$a(u, v) = \langle Au, v \rangle = (\mathcal{J}Au, v) \quad \forall\, u, v \in V,$$

and

$$\|\mathcal{J}Au\| = \|Au\| \quad \forall\, u \in V.$$

For any $u \in V$, by Theorem 7.3.3, the problem

$$(w, v) = (u, v) - \theta\,[a(u, v) - \ell(v)] \quad \forall\, v \in V$$

has a unique solution $w = P_\theta(u)$. Let us show that for $\theta \in (0, 2\,\alpha/M^2)$, the operator P_θ is a contraction. Indeed let $u_1, u_2 \in V$, and denote $w_1 = P_\theta(u_1)$, $w_2 = P_\theta(u_2)$. Then

$$(w_1 - w_2, v) = (u_1 - u_2, v) - \theta\,a(u_1 - u_2, v) = ((I - \theta\,\mathcal{J}A)(u_1 - u_2), v),$$

i.e., $w_1 - w_2 = (I - \theta\,\mathcal{J}A)(u_1 - u_2)$. We then have

$$\|w_1 - w_2\|^2 = \|u_1 - u_2\|^2 - 2\,\theta\,(\mathcal{J}A(u_1 - u_2), u_1 - u_2)$$
$$+ \theta^2\|\mathcal{J}A(u_1 - u_2)\|^2$$
$$\le (1 - 2\,\theta\,\alpha + \theta^2 M^2)\,\|u_1 - u_2\|^2.$$

Since $\theta \in (0, 2\,\alpha/M^2)$, we have

$$1 - 2\,\theta\,\alpha + \theta^2 M^2 < 1$$

and the mapping P_θ is a contraction. By the Banach fixed-point theorem, P_θ has a unique fixed-point $u \in V$, which is the solution of the problem (7.3.5).

[#2] The uniqueness of a solution follows from the V-ellipticity of the bilinear form. We prove the existence by applying Theorem 7.2.1. We will use the linear operator $L = \mathcal{J}A : V \to V$ constructed in the first proof. We recall that $\mathcal{R}(L) = V$ if and only if $\mathcal{R}(L)$ is closed and $\mathcal{R}(L)^\perp = \{0\}$.

To show $\mathcal{R}(L)$ is closed, we let $\{u_n\} \subseteq \mathcal{R}(L)$ be a sequence converging to u. Then $u_n = \mathcal{J}Aw_n$ for some $w_n \in V$. We have

$$\|u_n - u_m\| = \|\mathcal{J}A(w_n - w_m)\| = \|A(w_n - w_m)\| \geq \alpha\,\|w_n - w_m\|.$$

Hence $\{w_n\}$ is a Cauchy sequence and so has a limit $w \in V$. Then

$$\|u_n - \mathcal{J}Aw\| = \|\mathcal{J}A(w_n - w)\| = \|A(w_n - w)\| \leq M\,\|w_n - w\| \to 0.$$

Hence, $u = \mathcal{J}Aw \in \mathcal{R}(L)$ and $\mathcal{R}(L)$ is closed.

Now suppose $u \in \mathcal{R}(L)^\perp$. Then for any $v \in V$,

$$0 = (\mathcal{J}Av, u) = a(v, u).$$

Taking $v = u$ above, we have $a(u, u) = 0$. By the V-ellipticity of $a(\cdot, \cdot)$, we conclude $u = 0$. ∎

Example 7.3.5 *Applying the Lax-Milgram lemma, we see that the boundary value problem (7.2.3) has a unique weak solution $u \in H_0^1(\Omega)$.*

7.4   Weak formulations of linear elliptic boundary value problems

In this section, we formulate and analyze weak formulations of some linear elliptic boundary value problems. To present the ideas clearly, we will frequently use boundary value problems associated with the Poisson equation,

$$-\Delta u = f,$$

and the Helmholtz equation,

$$-\Delta u + u = f,$$

as examples.

7.4.1 Problems with homogeneous Dirichlet boundary conditions

So far, we have studied the model elliptic boundary value problem corresponding to the Poisson equation with the homogeneous Dirichlet boundary condition

$$-\Delta u = f \quad \text{in } \Omega, \tag{7.4.1}$$
$$u = 0 \quad \text{in } \Gamma, \tag{7.4.2}$$

where $f \in L^2(\Omega)$. The weak formulation of the problem is

$$u \in V, \quad a(u, v) = \ell(v) \quad \forall\, v \in V. \tag{7.4.3}$$

Here

$$V = H_0^1(\Omega),$$
$$a(u, v) = \int_\Omega \nabla u \cdot \nabla v \, dx \quad \text{for } u, v \in V,$$
$$\ell(v) = \int_\Omega f v \, dx \quad \text{for } v \in V.$$

The problem (7.4.3) has a unique solution $u \in V$ by the Lax-Milgram lemma.

Dirichlet boundary conditions are also called *essential boundary conditions* since they are explicitly required by the weak formulations.

7.4.2 Problems with non-homogeneous Dirichlet boundary conditions

Suppose that instead of (7.4.2) the boundary condition is

$$u = g \quad \text{on } \Gamma. \tag{7.4.4}$$

To derive a weak formulation, we proceed similarly as in Section 7.1. We first assume the boundary value problem (7.4.1)–(7.4.4) has a classical solution $u \in C^2(\Omega) \cap C(\overline{\Omega})$. Multiplying the equation (7.4.1) by a test function v with certain smoothness which validates the following calculations, and integrating over Ω, we have

$$\int_\Omega -\Delta u \, v \, dx = \int_\Omega f v \, dx.$$

Integrate by parts,

$$-\int_\Gamma \frac{\partial u}{\partial \nu} v \, ds + \int_\Omega \nabla u \cdot \nabla v \, dx = \int_\Omega f v \, dx.$$

We now assume $v = 0$ on Γ so that the boundary integral term vanishes; the boundary integral term would otherwise be difficult to deal with under the expected regularity condition $u \in H^1(\Omega)$ on the weak solution. Thus we arrive at the relation

$$\int_\Omega \nabla u \cdot \nabla v \, dx = \int_\Omega f v \, dx$$

if v is smooth and $v = 0$ on Γ. For each term in the above relation to make sense, we assume $f \in L^2(\Omega)$, and let $u \in H^1(\Omega)$ and $v \in H_0^1(\Omega)$. Recall that the solution u should satisfy the boundary condition $u = g$ on Γ. We observe that it is necessary to assume $g \in H^{1/2}(\Gamma)$. Finally, we obtain the weak formulation for the boundary value problem (7.4.1)–(7.4.4):

$$u \in H^1(\Omega), \ u = g \text{ on } \Gamma, \quad \int_\Omega \nabla u \cdot \nabla v \, dx = \int_\Omega f v \, dx \quad \forall v \in H_0^1(\Omega).$$

$$(7.4.5)$$

For the weak formulation (7.4.5), though, we cannot apply Lax-Milgram lemma directly, since the trial function u and the test function v do not lie in the same space. There is a standard way to get rid of this problem. Since $g \in H^{1/2}(\Gamma)$ and $\gamma(H^1(\Omega)) = H^{1/2}(\Gamma)$, we have the existence of a function $G \in H^1(\Omega)$ such that $\gamma G = g$. We remark that finding the function G in practice may be nontrivial. Thus, setting

$$u = w + G,$$

the problem may be transformed into one of seeking w such that

$$w \in H_0^1(\Omega), \quad \int_\Omega \nabla w \cdot \nabla v \, dx = \int_\Omega (f v - \nabla G \cdot \nabla v) \, dx \quad \forall v \in H_0^1(\Omega).$$

$$(7.4.6)$$

The classical form of the boundary value problem for w is

$$-\Delta w = f + \Delta G \quad \text{in } \Omega,$$
$$w = 0 \quad \text{on } \Gamma.$$

Applying the Lax-Milgram lemma, we have a unique solution $w \in H_0^1(\Omega)$ of the problem (7.4.6). Then we set $u = w + G$ to get a solution u of the problem (7.4.5). Notice that the choice of the function G is not unique, so the uniqueness of the solution u of the problem (7.4.5) does not follow from the above argument. Nevertheless, we can show the uniqueness of u by a standard approach. Assume both u_1 and u_2 are solution of the problem (7.4.5). Then the difference $u_1 - u_2$ satisfies

$$u_1 - u_2 \in H_0^1(\Omega), \quad \int_\Omega \nabla(u_1 - u_2) \cdot \nabla v \, dx = 0 \quad \forall v \in H_0^1(\Omega).$$

Taking $v = u_1 - u_2$, we obtain

$$\int_\Omega |\nabla(u_1 - u_2)|^2 dx = 0.$$

Thus, $\nabla(u_1 - u_2) = 0$ a.e. in Ω, and hence $u_1 - u_2 = c$ a.e. in Ω. Using the boundary condition $u_1 - u_2 = 0$ a.e. on Γ, we see that $u_1 = u_2$ a.e. in Ω.

Since non-homogeneous Dirichlet boundary conditions can be rendered homogeneous in the way described above, for convenience only problems with homogeneous Dirichlet conditions will be considered later.

7.4.3 Problems with Neumann boundary conditions

Consider next the *Neumann* problem of determining u that satisfies

$$\begin{cases} -\Delta u + u = f & \text{in } \Omega, \\ \partial u/\partial \nu = g & \text{on } \Gamma. \end{cases} \tag{7.4.7}$$

Again we first derive a weak formulation. Assume $u \in C^2(\Omega) \cap C^1(\overline{\Omega})$ is a classical solution of the problem (7.4.7). Multiplying $(7.4.7)_1$ by an arbitrary test function v with certain smoothness for the following calculations to make sense, integrating over Ω and using Green's theorem, we obtain

$$\int_\Omega (\nabla u \cdot \nabla v + uv)\, dx = \int_\Omega fv\, dx + \int_\Gamma \frac{\partial u}{\partial \nu} v\, ds.$$

Then, substitution of the Neumann boundary condition $(7.4.7)_2$ in the boundary term leads to the relation

$$\int_\Omega (\nabla u \cdot \nabla v + uv)\, dx = \int_\Omega fv\, dx + \int_\Gamma gv\, ds.$$

Assume $f \in L^2(\Omega)$, $g \in L^2(\Gamma)$. For each term in the above relation to make sense, it is natural to choose the space $H^1(\Omega)$ for both the trial function u and the test function v. Thus, the weak formulation of the boundary value problem (7.4.7) is

$$u \in H^1(\Omega), \quad \int_\Omega (\nabla u \cdot \nabla v + uv)\, dx = \int_\Omega fv\, dx + \int_\Gamma gv\, ds \quad \forall\, v \in H^1(\Omega). \tag{7.4.8}$$

This problem has the form (7.4.3), where $V = H^1(\Omega)$, $a(\cdot, \cdot)$ and $\ell(\cdot)$ are defined by

$$a(u, v) = \int_\Omega (\nabla u \cdot \nabla v + uv)\, dx,$$

$$\ell(v) = \int_\Omega fv\, dx + \int_\Gamma gv\, ds,$$

respectively. Applying the Lax-Milgram lemma, it is straightforward to show that the weak formulation (7.4.8) has a unique solution $u \in H^1(\Omega)$.

Above we have shown that a classical solution $u \in C^2(\Omega) \cap C^1(\overline{\Omega})$ of the boundary value problem (7.4.7) is also the solution $u \in H^1(\Omega)$ of the weak formulation (7.4.8). Conversely, reversing the arguments leading to (7.4.8) to (7.4.7), it is readily seen that a weak solution of the problem (7.4.8) with sufficient smoothness is also the classical solution of the problem (7.4.7).

Neumann boundary conditions are also called natural boundary conditions since they are naturally incorporated in the weak formulations of the boundary value problems, as can be seen from (7.4.8).

It is more delicate to study the Neumann problem for the Poisson equation

$$\begin{cases} -\Delta u = f & \text{in } \Omega, \\ \partial u/\partial \nu = g & \text{on } \Gamma, \end{cases} \tag{7.4.9}$$

where $f \in L^2(\Omega)$ and $g \in L^2(\Gamma)$ are given. In general, the problem (7.4.9) does not have a solution, and when the problem has a solution u, any function of the form $u+c$, $c \in \mathbb{R}$, is a solution. Formally, the corresponding weak formulation is

$$u \in H^1(\Omega), \quad \int_\Omega \nabla u \cdot \nabla v \, dx = \int_\Omega fv \, dx + \int_\Gamma gv \, ds \quad \forall v \in H^1(\Omega).$$
$$\tag{7.4.10}$$

A necessary condition for (7.4.10) to have a solution is

$$\int_\Omega f \, dx + \int_\Gamma g \, ds = 0 \tag{7.4.11}$$

which is derived from (7.4.10) by taking the test function $v = 1$. Assume $\Omega \subseteq \mathbb{R}^d$ is an open, bounded, connected set with a Lipschitz boundary. Let us show that the condition (7.4.11) is also a sufficient condition for the problem (7.4.10) to have a solution. Indeed, the problem (7.4.10) is most conveniently studied in the quotient space $V = H^1(\Omega)/\mathbb{R}$ (cf. Exercise 1.2.14 for the definition of a quotient space), where each element $[v] \in V$ is an equivalence class $[v] = \{v + \alpha \mid \alpha \in \mathbb{R}\}$, and any $v \in [v]$ is called a representative element. The following result is a special case of Theorem 6.3.17.

Lemma 7.4.1 *Assume $\Omega \subseteq \mathbb{R}^d$ is an open, bounded, connected set with a Lipschitz boundary. Then over the space $V = H^1(\Omega)/\mathbb{R}$, the quotient norm*

$$\|[v]\|_V \equiv \inf_{v \in [v]} \|v\|_1 = \inf_{\alpha \in \mathbb{R}} \|v + \alpha\|_1$$

is equivalent to the $H^1(\Omega)$ semi-norm $|v|_1$ for any $v \in [v]$.

It is now easy to see that

$$\bar{a}([u], [v]) = \int_\Omega \nabla u \cdot \nabla v \, dx, \quad u \in [u], \ v \in [v]$$

defines a bilinear form on V, which is continuous and V-elliptic. Because of the condition (7.4.11),

$$\bar{\ell}([v]) = \int_\Omega f \, v \, dx + \int_\Gamma g \, v \, ds$$

is a well-defined linear continuous form on V. Hence, we can apply the Lax-Milgram lemma to conclude that the problem

$$[u] \in V, \quad \bar{a}([u], [v]) = \bar{\ell}([v]) \quad \forall [v] \in V$$

has a unique solution $[u]$. It is easy to see that any $u \in [u]$ is a solution of (7.4.10).

Another approach to studying the Neumann boundary value problem (7.4.9) is to add a side condition, such as

$$\int_\Omega u \, dx = 0.$$

Then we introduce the space

$$V = \left\{ v \in H^1(\Omega) \ \middle| \ \int_\Omega v \, dx = 0 \right\}.$$

An application of Theorem 6.3.12 shows that over the space V, $|\cdot|_1$ is a norm equivalent to the norm $\|\cdot\|_1$. The bilinear form $a(u, v) = \int_\Omega \nabla u \cdot \nabla v \, dx$ is both continuous and V-elliptic. So there is a unique solution to the problem

$$u \in V, \quad \int_\Omega \nabla u \cdot \nabla v \, dx = \int_\Omega f v \, dx + \int_\Gamma g v \, ds \quad \forall v \in V.$$

7.4.4   Problems with mixed boundary conditions

It is also possible to specify different kind of boundary conditions on different portions of the boundary. One such example is

$$\begin{cases} -\Delta u + u = f & \text{in } \Omega, \\ u = 0 & \text{on } \Gamma_D, \\ \partial u / \partial \nu = g & \text{on } \Gamma_N, \end{cases} \tag{7.4.12}$$

where Γ_D and Γ_N form a non-overlapping decomposition of the boundary $\partial\Omega$: Γ_D and Γ_N are relatively open, $\partial\Omega = \bar{\Gamma}_D \cup \bar{\Gamma}_N$, and $\Gamma_D \cap \Gamma_N = \emptyset$. Assume Ω is connected. The appropriate space in which to pose this problem in weak form is now

$$V = H^1_{\Gamma_D}(\Omega) = \{ v \in H^1(\Omega) \mid v = 0 \text{ on } \Gamma_D \}.$$

Then the weak problem is again of the form (7.4.7) with

$$a(u, v) = \int_\Omega (\nabla u \cdot \nabla v + u \, v) \, dx$$

and

$$\ell(v) = \int_\Omega f v \, dx + \int_{\Gamma_N} g v \, ds.$$

Under suitable assumptions, say $f \in L^2(\Omega)$ and $g \in L^2(\Gamma_N)$, we can again apply the Lax-Milgram lemma to conclude that the weak problem has a unique solution.

7.4.5 A general linear second-order elliptic boundary value problem

The issue of existence and uniqueness of solutions to the problems just discussed may be treated in the more general framework of arbitrary linear elliptic PDEs of second order. Let $\Omega \subseteq \mathbb{R}^d$ be an open, bounded, connected set with a Lipschitz continuous boundary $\partial\Omega$. Let $\partial\Omega = \overline{\Gamma}_D \cup \overline{\Gamma}_N$ with $\Gamma_D \cap \Gamma_N = \emptyset$, Γ_D and Γ_N being open subsets of $\partial\Omega$. Consider the boundary value problem

$$\begin{cases} -\partial_j(a_{ij}\partial_i u) + b_i\partial_i u + cu = f & \text{in } \Omega, \\ \qquad\qquad\qquad\quad u = 0 & \text{on } \Gamma_D, \\ \qquad\qquad\quad a_{ij}\partial_i u\, \nu_j = g & \text{on } \Gamma_N. \end{cases} \tag{7.4.13}$$

Here $\boldsymbol{\nu} = (\nu_1, \ldots, \nu_d)^T$ is the unit outward normal on Γ_N.

The given functions a_{ij}, b_i, c, f, and g are assumed to satisfy the following conditions:

$$a_{ij}, b_i, c \in L^\infty(\Omega); \tag{7.4.14}$$

there exists a constant $\theta > 0$ such that

$$a_{ij}\xi_i\xi_j \geq \theta\,|\boldsymbol{\xi}|^2 \quad \forall\boldsymbol{\xi} = (\xi_i) \in \mathbb{R}^d, \text{ a.e. in } \Omega; \tag{7.4.15}$$

$$f \in L^2(\Omega); \tag{7.4.16}$$

$$g \in L^2(\Gamma_N). \tag{7.4.17}$$

The weak formulation of the problem (7.4.13) is obtained again in the usual way by multiplying the differential equation in (7.4.13) by an arbitrary test function v that vanishes on Γ_D, integrating over Ω, performing an integration by parts, and applying the specified boundary conditions. As a result, we get the weak formulation (7.4.3) with

$$V = H^1_{\Gamma_D}(\Omega),$$

$$a(u,v) = \int_\Omega (a_{ij}\partial_i u\partial_j u + b_i\,(\partial_i u)\,v + c\,u\,v)\,dx, \tag{7.4.18}$$

$$\ell(v) = \int_\Omega f\,v\,dx + \int_{\Gamma_N} g\,v\,ds.$$

We can again apply Lax-Milgram lemma to study the well-posedness of the boundary value problem. The space $V = H^1_{\Gamma_D}(\Omega)$ is a Hilbert space, with the standard H^1-norm. The assumptions (7.4.14)–(7.4.17) ensure that the bilinear form is bounded on V, and the linear form is bounded on V. What remains to be established is the V-ellipticity of the bilinear form.

Some sufficient conditions for the V-ellipticity of the bilinear form of the left hand side of (7.4.18) are discussed in Exercise 7.4.3.

Exercise 7.4.1 *The boundary value problem*

$$\begin{cases} -\Delta u + c\,u = f & \text{in } \Omega, \\ u = \text{constant} & \text{on } \Gamma, \\ \displaystyle\int_\Gamma \frac{\partial u}{\partial \nu}\,ds = \int_\Gamma g\,ds \end{cases}$$

is called an Adler problem. Derive a weak formulation, and show that the weak formulation and the boundary value problem are formally equivalent. Assume $c > 0$, $f \in L^2(\Omega)$, and $g \in L^2(\Gamma)$. Prove that the weak formulation has a unique solution.

Exercise 7.4.2 *A boundary condition can involve both the unknown function and its normal derivative; such a boundary condition is called the* third boundary condition *or* Robin boundary condition *for second-order differential equations. Consider the boundary value problem*

$$-\Delta u = f \quad \text{in } \Omega,$$
$$\frac{\partial u}{\partial \nu} + a\,u = g \quad \text{on } \Gamma.$$

Derive a weak formulation of the Robin boundary value problem for the Poisson equation. Find conditions on the given data for the existence and uniqueness of a solution to the weak formulation; prove your assertion.

Exercise 7.4.3 *Assume $\Omega \subseteq \mathbb{R}^d$ is an open, bounded, connected Lipschitz domain. Show that the bilinear form defined in (7.4.18) is V-elliptic with $V = H^1_{\Gamma_D}(\Omega)$, if (7.4.14)–(7.4.17) hold and one of the following three conditions is satisfied, with $\mathbf{b} = (b_1, \ldots, b_d)^T$ and θ the ellipticity constant in (7.4.15):*

$$c \geq c_0 > 0, \quad |\mathbf{b}| \leq B \text{ a.e. in } \Omega, \text{ and } B^2 < 4\,\theta\,c_0,$$

or

$$\mathbf{b} \cdot \boldsymbol{\nu} \geq 0 \text{ a.e. on } \Gamma_N, \text{ and } c - \frac{1}{2}\,\mathrm{div}\,\mathbf{b} \geq c_0 > 0 \text{ a.e. in } \Omega,$$

or

$$\mathrm{meas}(\Gamma_D) > 0, \quad \mathbf{b} = \mathbf{0}, \text{ and } \inf_\Omega c > -\theta/\bar{c},$$

where $\bar{c}$ is the best constant in the Poincaré inequality

$$\int_\Omega v^2 dx \leq \bar{c} \int_\Omega |\nabla v|^2 dx \quad \forall\, v \in H^1_{\Gamma_D}(\Omega).$$

This best constant can be computed by solving a linear elliptic eigenvalue problem: $\bar{c} = 1/\lambda_1$, with $\lambda_1 > 0$ the smallest eigenvalue of the eigenvalue *problem*

$$\begin{cases} -\Delta u = \lambda u \text{ in } \Omega, \\ \quad u = 0 \quad \text{ on } \Gamma_D, \\ \dfrac{\partial u}{\partial \nu} = 0 \quad \text{ on } \Gamma_N. \end{cases}$$

A special and important case is that corresponding to $b_i = 0$; in this case the bilinear form is symmetric, and V-ellipticity is assured if

$$c \geq c_0 > 0.$$

Exercise 7.4.4 *It is not always necessary to assume Ω to be connected. Let $\Omega \subseteq \mathbb{R}^d$ be open, bounded with a Lipschitz boundary, and let us consider the boundary value problem 7.4.13 with $\Gamma_D = \partial\Omega$ and $\Gamma_N = \emptyset$ (i.e., a pure Dirichlet boundary value problem). Keep the assumptions (7.4.14)–(7.4.16). Show that the boundary value problem has a unique solution if one of the following three conditions is satisfied:*

$$c \geq c_0 > 0, \ |\mathbf{b}| \leq B \text{ a.e. in } \Omega, \text{ and } B^2 < 4\,\theta\,c_0,$$

or

$$c - \frac{1}{2}\operatorname{div}\mathbf{b} \geq c_0 > 0 \text{ a.e. in } \Omega,$$

or

$$\mathbf{b} = \mathbf{0} \text{ and } \inf_{\Omega} c > -\theta/\bar{c}$$

where $\bar{c}$ is the best constant in the Poincaré inequality

$$\int_{\Omega} v^2 dx \leq \bar{c} \int_{\Omega} |\nabla v|^2 dx \quad \forall\, v \in H_0^1(\Omega).$$

This best constant can be computed by solving a linear elliptic eigenvalue problem: $\bar{c} = 1/\lambda_1$, with $\lambda_1 > 0$ the smallest eigenvalue of the eigenvalue problem

$$\begin{cases} -\Delta u = \lambda u \text{ in } \Omega, \\ u = 0 \qquad \text{ on } \Gamma. \end{cases}$$

Exercise 7.4.5 *The biharmonic equation*

$$\Delta^2 u = f \quad \text{in } \Omega$$

arises in fluid mechanics as well as thin elastic plate problems. Let us consider the biharmonic equation together with the homogeneous Dirichlet boundary conditions

$$u = \frac{\partial u}{\partial \nu} = 0 \quad \text{on } \Gamma.$$

Note that the differential equation is of fourth-order, so boundary conditions involving the unknown function and first-order derivatives are treated as Dirichlet (or essential) boundary conditions, while Neumann (or natural) boundary conditions refer to those involving second- and third-order derivatives of the unknown function. Give a weak formulation of the homogeneous Dirichlet boundary value problem for the biharmonic equation and demonstrate its unique solvability.

7.5 A boundary value problem of linearized elasticity

We study a boundary value problem of linearized elasticity in this section. The quantities describing the mechanical behavior of the deformation of an elastic material are the displacement $\mathbf{u}$, the strain tensor ε, and the stress tensor σ. A reader with little background on elasticity may simply view $\mathbf{u}$ as a d-dimensional vector-valued function, and ε and σ as $d \times d$ symmetric matrix-valued functions. Here d is the dimension of the material; in applications of the linearized elasticity theory, $d \leq 3$.

We consider the problem of the deformation of a linearly elastic body occupying a bounded, connected domain $\Omega \subseteq \mathbb{R}^d$. The boundary Γ of the domain is assumed Lipschitz continuous so that the unit outward normal ν exists almost everywhere on Γ. We divide the boundary Γ into two complementary parts $\overline{\Gamma}_u$ and $\overline{\Gamma}_g$, where Γ_u and Γ_g are open, $\Gamma_u \cap \Gamma_g = \emptyset$ and $\Gamma_u \neq \emptyset$. The body is subject to the action of a body force of the density $\mathbf{f}$ and the surface traction of density $\mathbf{g}$ on Γ_g. We assume the body is fixed along $\overline{\Gamma}_u$. As a result of the applications of the external forces, the body experiences some deformation and reaches an equilibrium state. A material point $\mathbf{x} \in \Omega$ in the undeformed body will be moved to the location $\mathbf{x} + \mathbf{u}$ after the deformation. The quantity $\mathbf{u} = \mathbf{u}(\mathbf{x})$ is the *displacement* of the point $\mathbf{x}$.

Mathematical relations in a mechanical problem can be divided into two kinds: one of them consists of material-independent relations, and the other material-dependent relations, or *constitutive laws*. The material-independent relations include the strain-displacement relation, the equation of equilibrium, and boundary conditions. The equation of equilibrium takes the form

$$-\operatorname{div}\sigma = \mathbf{f} \quad \text{in } \Omega. \tag{7.5.1}$$

Here div σ is a d-dimensional vector-valued function whose i^{th} component equals $\sum_{j=1}^{d} \sigma_{ij,j}$. We assume the deformation is small (i.e., both the displacement and its gradient are small in size), and use the linearized strain tensor

$$\varepsilon(\mathbf{u}) = \frac{1}{2}\left(\nabla\mathbf{u} + (\nabla\mathbf{u})^T\right) \quad \text{in } \Omega. \tag{7.5.2}$$

The specified boundary conditions take the form

$$\mathbf{u} = \mathbf{0} \quad \text{on } \Gamma_u, \tag{7.5.3}$$

$$\boldsymbol{\sigma}\boldsymbol{\nu} = \mathbf{g} \quad \text{on } \Gamma_g. \tag{7.5.4}$$

Here $\boldsymbol{\sigma}\boldsymbol{\nu}$ is the action of the stress tensor $\boldsymbol{\sigma}$ on the unit outward normal $\boldsymbol{\nu}$. It can be viewed as a matrix-vector multiplication. The result is a d-dimensional vector-valued function whose i^{th} component is $\sum_{j=1}^{d} \sigma_{ij}\nu_j$.

The above relations are supplemented by a constitutive relation, which describes the mechanical response of the material to the external forces. The simplest constitutive relation is provided by that of linearized elasticity,

$$\boldsymbol{\sigma} = \mathbf{C}\boldsymbol{\varepsilon}(\mathbf{u}). \tag{7.5.5}$$

The elasticity tensor $\mathbf{C}$ is of fourth order, and can be viewed as a linear mapping from the space of symmetric second-order tensors to itself. With respect to the Cartesian coordinate system, the tensor $\mathbf{C}$ has the components C_{ijkl}, $1 \le i, j, k, l \le d$. The expression $\mathbf{C}\boldsymbol{\varepsilon}$ stands for a second-order tensor whose $(i, j)^{th}$ component is $\sum_{k,l=1}^{d} C_{ijkl}\varepsilon_{kl}$. In component form, the constitutive relation (7.5.5) is rewritten as

$$\sigma_{ij} = \sum_{k,l=1}^{d} C_{ijkl}\varepsilon_{kl}(\mathbf{u}), \quad 1 \le i, j \le d.$$

We assume the elasticity tensor $\mathbf{C}$ is bounded

$$C_{ijkl} \in L^{\infty}(\Omega), \tag{7.5.6}$$

symmetric

$$C_{ijkl} = C_{jikl} = C_{klij}, \tag{7.5.7}$$

and pointwise stable

$$\boldsymbol{\varepsilon} : \mathbf{C}\boldsymbol{\varepsilon} \ge \alpha \, |\boldsymbol{\varepsilon}|^2 \quad \text{for all symmetric second-order tensors } \boldsymbol{\varepsilon} \tag{7.5.8}$$

with a constant $\alpha > 0$. Here for two second-order tensors (or matrices) $\boldsymbol{\sigma}$ and $\boldsymbol{\varepsilon}$, we define their inner product by

$$\boldsymbol{\sigma} : \boldsymbol{\varepsilon} = \sum_{i,j=1}^{d} \sigma_{ij}\varepsilon_{ij}.$$

In the special case of an isotropic, homogenous linearly elastic material, we have

$$C_{ijkl} = \lambda \, \delta_{ij}\delta_{kl} + \mu \, (\delta_{ik}\delta_{jl} + \delta_{il}\delta_{jk}), \tag{7.5.9}$$

where $\lambda, \mu > 0$ are called Lamé moduli; then the constitutive relation is reduced to

$$\boldsymbol{\sigma} = \lambda \, (\text{tr}\,\boldsymbol{\varepsilon})\,\mathbf{I} + 2\,\mu\,\boldsymbol{\varepsilon}. \tag{7.5.10}$$

Here we use $\mathbf{I}$ to denote the unit tensor of the second order (think of it as the unit matrix of order d), and $\operatorname{tr}\varepsilon$ is the trace of the tensor (matrix) ε. The Lamé moduli λ and μ are related to Young modulus (modulus of elasticity) E and Poisson ratio (the contraction ratio) ν by

$$\lambda = \frac{E\nu}{(1+\nu)(1-2\nu)}, \quad \mu = \frac{E}{1+\nu}.$$

The classical formulation of the boundary value problem for the linearized elasticity then consists of the equations (7.5.1)–(7.5.5). We now derive the corresponding weak formulation with regard to the unknown variable $\mathbf{u}$. We assume that

$$\mathbf{f} \in [L^2(\Omega)]^d, \quad \mathbf{g} \in [L^2(\Gamma_g)]^d. \tag{7.5.11}$$

Combining the equations (7.5.1) and (7.5.2), we see that the differential equation is of second order for $\mathbf{u}$. Keeping in mind the Dirichlet boundary condition (7.5.3), we are led to the function space

$$V = \{\mathbf{v} \in (H^1(\Omega))^d \mid \mathbf{v} = \mathbf{0} \text{ a.e. on } \Gamma_u\}.$$

We now multiply the equation (7.5.1) by an arbitrary test function $\mathbf{v} \in V$, integrate over Ω,

$$-\int_\Omega \operatorname{div}\boldsymbol{\sigma} \cdot \mathbf{v}\, dx = \int_\Omega \mathbf{f} \cdot \mathbf{v}\, dx.$$

We transform the left-hand side by integration by parts to obtain

$$-\int_\Gamma (\boldsymbol{\sigma}\nu) \cdot \mathbf{v}\, ds + \int_\Omega \boldsymbol{\sigma} : \nabla\mathbf{v}\, dx = \int_\Omega \mathbf{f} \cdot \mathbf{v}\, dx.$$

Upon the use of the boundary conditions, the boundary integral term can be written as

$$-\int_\Gamma (\boldsymbol{\sigma}\nu) \cdot \mathbf{v}\, ds = -\int_{\Gamma_g} \mathbf{g} \cdot \mathbf{v}\, ds.$$

Since $\boldsymbol{\sigma}$ is symmetric, we have $\boldsymbol{\sigma} : \nabla\mathbf{v} = \boldsymbol{\sigma} : \varepsilon(\mathbf{v})$. Therefore, we obtain the relation

$$\int_\Omega \boldsymbol{\sigma} : \varepsilon(\mathbf{v})\, dx = \int_\Omega \mathbf{f} \cdot \mathbf{v}\, dx + \int_{\Gamma_g} \mathbf{g} \cdot \mathbf{v}\, ds.$$

Recalling the constitutive law (7.5.5), we have thus derived the following weak formulation for the displacement variable,

$$\mathbf{u} \in V, \quad \int_\Omega (\mathbf{C}\varepsilon(\mathbf{u})) : \varepsilon(\mathbf{v})\, dx = \int_\Omega \mathbf{f} \cdot \mathbf{v}\, dx + \int_{\Gamma_g} \mathbf{g} \cdot \mathbf{v}\, ds \quad \forall \mathbf{v} \in V. \tag{7.5.12}$$

We can apply the Lax-Milgram lemma to conclude the existence and uniqueness of the problem (7.5.12).

Theorem 7.5.1 *Assume* (7.5.6)–(7.5.8), (7.5.11), *and* meas $(\Gamma_u) > 0$. *Then there is a unique solution to the problem* (7.5.12). *The problem* (7.5.12) *is equivalent to the minimization problem*

$$\mathbf{u} \in V, \quad E(\mathbf{u}) = \inf\{E(\mathbf{v}) \mid \mathbf{v} \in V\}, \tag{7.5.13}$$

where the energy functional is defined by

$$E(\mathbf{v}) = \frac{1}{2} \int_\Omega (C\varepsilon(\mathbf{v})) : \varepsilon(\mathbf{v}) \, dx - \int_\Omega \mathbf{f} \cdot \mathbf{v} \, dx - \int_{\Gamma_g} \mathbf{g} \cdot \mathbf{v} \, ds. \tag{7.5.14}$$

A proof of this result is left as Exercise 7.5.2. In verifying the V-ellipticity of the bilinear form

$$a(\mathbf{u}, \mathbf{v}) = \int_\Omega (C\varepsilon(\mathbf{u})) : \varepsilon(\mathbf{v}) \, dx,$$

we need to apply Korn's inequality: There exists a constant $c > 0$ depending only on Ω such that

$$\|\mathbf{v}\|_{[H^1(\Omega)]^d}^2 \leq c \int_\Omega |\varepsilon(\mathbf{v})|^2 dx \quad \forall \mathbf{v} \in V. \tag{7.5.15}$$

Exercise 7.5.1 *In the case of an isotropic, homogeneous linearly elastic material* (7.5.10), *show that the classical formulation of the equilibrium equation written in terms of the displacement is*

$$-\mu \, \Delta \mathbf{u} - (\lambda + \mu) \, \nabla \mathrm{div} \, \mathbf{u} = \mathbf{f} \quad \text{in } \Omega.$$

Give a derivation of the weak formulation of the problem: Find $\mathbf{u} \in V$ *such that*

$$a(\mathbf{u}, \mathbf{v}) = \ell(\mathbf{v}) \quad \forall \mathbf{v} \in V,$$

where

$$V = \{\mathbf{v} \in (H^1(\Omega))^d \mid \mathbf{v} = \mathbf{0} \text{ on } \Gamma_u\},$$

$$a(\mathbf{u}, \mathbf{v}) = \int_\Omega [\lambda \, \mathrm{div} \, \mathbf{u} \, \mathrm{div} \, \mathbf{v} + 2 \, \mu \, \varepsilon(\mathbf{u}) : \varepsilon(\mathbf{v})] \, dx,$$

$$\ell(\mathbf{v}) = \int_\Omega \mathbf{f} \cdot \mathbf{v} \, dx + \int_{\Gamma_g} \mathbf{g} \cdot \mathbf{v} \, ds.$$

Prove that the weak formulation has a unique solution.

Exercise 7.5.2 *Apply the Lax-Milgram lemma to prove Theorem 7.5.1.*

7.6   Mixed and dual formulations

This section is intended as a brief introduction to two different weak formulations for boundary value problems, namely, the mixed formulation and the dual formulation. We use the model problem

$$-\Delta u = f \quad \text{in } \Omega, \tag{7.6.1}$$
$$u = 0 \quad \text{on } \partial\Omega \tag{7.6.2}$$

for a description of the new weak formulations. Assume $f \in L^2(\Omega)$. We have seen that the weak formulation discussed in previous sections is

$$u \in H_0^1(\Omega), \quad \int_\Omega \nabla u \cdot \nabla v \, dx = \int_\Omega f v \, dx \quad \forall\, v \in H_0^1(\Omega). \tag{7.6.3}$$

This weak formulation is called the *primal formulation*, since the unknown variable is u. Here the bilinear form is symmetric, so the weak formulation (7.6.3) is equivalent to the minimization problem

$$u \in H_0^1(\Omega), \quad J(u) = \inf_{v \in H_0^1(\Omega)} J(v) \tag{7.6.4}$$

with

$$J(v) = \frac{1}{2} \int_\Omega |\nabla v|^2 dx - \int_\Omega f v \, dx. \tag{7.6.5}$$

In the context of a heat conduction problem, u is the temperature and ∇u has the physical meaning of the heat flux. In many situations, the heat flux is a more important quantity than the temperature variable. It is then desirable to develop equivalent weak formulations of the boundary value problem (7.6.1)–(7.6.2) that involves $\mathbf{p} = \nabla u$ as an unknown. For this purpose, let $\mathbf{q} = \nabla v$. Then noticing that

$$\frac{1}{2} \int_\Omega |\nabla v|^2 dx = \sup_{\mathbf{q} \in (L^2(\Omega))^d} \int_\Omega \left(\mathbf{q} \cdot \nabla v - \frac{1}{2} |\mathbf{q}|^2 \right) dx,$$

we can replace the minimization problem (7.6.4) by

$$\inf_{v \in H_0^1(\Omega)} \sup_{\mathbf{q} \in (L^2(\Omega))^d} L(\mathbf{q}, v), \tag{7.6.6}$$

where

$$L(\mathbf{q}, v) = \int_\Omega \left(\mathbf{q} \cdot \nabla v - \frac{1}{2} |\mathbf{q}|^2 - f v \right) dx. \tag{7.6.7}$$

This is called a *saddle point* problem as its solution $(u, \mathbf{p}) \in H_0^1(\Omega) \times (L^2(\Omega))^d$, if it exists, satisfies the inequalities

$$L(\mathbf{q}, u) \le L(\mathbf{q}, v) \le L(\mathbf{p}, v) \quad \forall\, \mathbf{q} \in (L^2(\Omega))^d, \ v \in H_0^1(\Omega). \tag{7.6.8}$$

It is left as an exercise to show that the inequalities (7.6.8) are equivalent to $(u, \mathbf{p}) \in H_0^1(\Omega) \times (L^2(\Omega))^d$ satisfying

$$\int_\Omega \mathbf{p} \cdot \mathbf{q} \, dx - \int_\Omega \mathbf{q} \cdot \nabla u \, dx = 0 \qquad \forall\, \mathbf{q} \in (L^2(\Omega))^d, \tag{7.6.9}$$

$$-\int_\Omega \mathbf{p} \cdot \nabla v \, dx = -\int_\Omega f v \, dx \qquad \forall\, v \in H_0^1(\Omega). \tag{7.6.10}$$

Upon an integration by parts, another weak formulation is: Find $(u, \mathbf{p}) \in L^2(\Omega) \times H(\mathrm{div}; \Omega)$ such that

$$\int_\Omega \mathbf{p} \cdot \mathbf{q}\, dx + \int_\Omega \mathrm{div}\, \mathbf{q}\, u\, dx = 0 \qquad \forall \mathbf{q} \in H(\mathrm{div}; \Omega), \qquad (7.6.11)$$

$$\int_\Omega \mathrm{div}\, \mathbf{p}\, v\, dx = -\int_\Omega f\, v\, dx \qquad \forall v \in L^2(\Omega). \qquad (7.6.12)$$

Here

$$H(\mathrm{div}; \Omega) = \{\mathbf{q} \in (L^2(\Omega))^d \mid \mathrm{div}\, \mathbf{q} \in L^2(\Omega)\}.$$

Formulations (7.6.9)–(7.6.10) and (7.6.11)–(7.6.12) are examples of *mixed formulations* and they fall in the following abstract framework:

Let V and Q be two Hilbert spaces. Assume $a(u, v)$ is a continuous bilinear form on $V \times V$, $b(v, q)$ is a continuous bilinear form on $V \times Q$. Given $f \in V'$ and $g \in Q'$, find $u \in V$ and $p \in Q$ such that

$$a(u, v) + b(v, p) = \langle f, v \rangle_{V' \times V} \quad \forall v \in V, \qquad (7.6.13)$$

$$b(u, q) = \langle g, q \rangle_{Q' \times Q} \quad \forall q \in Q. \qquad (7.6.14)$$

In addition to the need of bringing in quantities of physical importance into play in a weak formulation (e.g., $\mathbf{p} = \nabla u$ for the model problem) in the hope of achieving more accurate numerical approximations for them, we frequently arrive at mixed formulation of the type (7.6.13)–(7.6.14) in dealing with constraints such as the incompressibility in fluids or in certain solids (cf. Exercise 7.6.3). Finite element approximations based on mixed formulations are called *mixed finite element methods*, and they have been extensively analyzed and applied in solving mechanical problems. In Chapter 9, we only discuss the finite element method based on the primal weak formulation. The interested reader can consult [29] for a detailed discussion of the mixed formulation (7.6.13)–(7.6.14) and mixed finite element methods.

Back to the model problem, we can eliminate from the mixed formulation (7.6.9)–(7.6.10) the variable u and find a problem with $\mathbf{p}$ as the only unknown. Let us interchange inf and sup in (7.6.6) to get another problem

$$\sup_{\mathbf{q} \in (L^2(\Omega))^d} \inf_{v \in H_0^1(\Omega)} L(\mathbf{q}, v). \qquad (7.6.15)$$

We have

$$\inf_{v \in H_0^1(\Omega)} L(\mathbf{q}, v) = \begin{cases} -\dfrac{1}{2} \displaystyle\int_\Omega |\mathbf{q}|^2 dx & \text{if } \mathbf{q} \in Q_f, \\ -\infty & \text{if } \mathbf{q} \notin Q_f, \end{cases}$$

where

$$Q_f = \left\{ \mathbf{q} \in (L^2(\Omega))^d \mid \int_\Omega \mathbf{q} \cdot \nabla v\, dx = \int_\Omega f\, v\, dx\ \forall v \in H_0^1(\Omega) \right\}.$$

The constraint in Q_f is the weak form of the relation $-\mathrm{div}\, \mathbf{q} = f$.

Thus the problem (7.6.15) is equivalent to

$$\sup_{\mathbf{q}\in Q_f}\left[-\frac{1}{2}\int_\Omega |\mathbf{q}|^2 dx\right]$$

or

$$\inf_{\mathbf{q}\in Q_f}\frac{1}{2}\int_\Omega |\mathbf{q}|^2 dx. \tag{7.6.16}$$

This is called the *dual formulation*. The functional

$$\frac{1}{2}\int_\Omega |\mathbf{q}|^2 dx$$

is the complementary (or dual) energy, and (7.6.16) is the classical complementary energy principle.

Exercise 7.6.1 *Show the equivalence between (7.6.8) and (7.6.9)–(7.6.10).*

Exercise 7.6.2 *Show that (7.6.9) and (7.6.10) are the weak formulation of the equations*

$$\mathbf{p} = \nabla u,$$
$$-\operatorname{div}\mathbf{p} = f.$$

Exercise 7.6.3 *As an example of a mixed formulation for the treatment of a constraint, we consider the following boundary value problem for the Stokes equation: Given a force density* $\mathbf{f}$*, find a velocity* $\mathbf{u}$ *and a pressure* p *such that*

$$-\Delta\mathbf{u} + \nabla p = \mathbf{f} \quad \text{in } \Omega,$$
$$\operatorname{div}\mathbf{u} = 0 \quad \text{in } \Omega,$$
$$\mathbf{u} = \mathbf{0} \quad \text{on } \partial\Omega.$$

To uniquely determine p*, we impose the condition*

$$\int_\Omega p\,dx = 0.$$

Assume $\mathbf{f} \in (L^2(\Omega))^d$. *Let*

$$V = (H_0^1(\Omega))^d,$$
$$Q = \left\{q \in L^2(\Omega) \;\middle|\; \int_\Omega q\,dx = 0\right\}.$$

Show that a mixed weak formulation of the boundary value problem is: Find $\mathbf{u} = (u_1,\ldots,u_d)^T \in V$ *and* $p \in Q$ *such that*

$$\sum_{i=1}^d \int_\Omega \nabla u_i \cdot \nabla v_i\,dx - \int_\Omega p\operatorname{div}\mathbf{v}\,dx = \int_\Omega \mathbf{f}\cdot\mathbf{v}\,dx \quad \forall\mathbf{v}\in V,$$

$$\int_\Omega q\operatorname{div}\mathbf{u}\,dx = 0 \quad \forall q \in Q.$$

7.7   Generalized Lax-Milgram Lemma

The following result extends the Lax-Milgram lemma, and is due to Nečas [122].

Theorem 7.7.1 *Let U and V be real Hilbert spaces, $a : U \times V \to \mathbb{R}$ a bilinear form, and $\ell \in V'$. Assume there are constants $M > 0$ and $\alpha > 0$ such that*

$$|a(u, v)| \leq M \|u\|_U \|v\|_V \quad \forall u \in U,\ v \in V; \tag{7.7.1}$$

$$\sup_{0 \neq v \in V} \frac{a(u, v)}{\|v\|_V} \geq \alpha \|u\|_U \quad \forall u \in U; \tag{7.7.2}$$

$$\sup_{u \in U} a(u, v) > 0 \quad \forall v \in V,\ v \neq 0. \tag{7.7.3}$$

Then there exists a unique solution u of the problem

$$u \in U, \quad a(u, v) = \ell(v) \quad \forall v \in V. \tag{7.7.4}$$

Moreover,

$$\|u\|_U \leq \frac{\|\ell\|_{V'}}{\alpha}. \tag{7.7.5}$$

Proof. The proof is similar to the second proof of Lax-Milgram lemma, and we apply Theorem 7.2.1.

Again, let $A : U \to V$ be the linear continuous operator defined by the relation

$$a(u, v) = (Au, v)_V \quad \forall u \in U,\ v \in V.$$

Using the condition (7.7.1), we have

$$\|Au\|_V \leq M \|u\|_U \quad \forall u \in U.$$

Then the problem (7.7.4) can be rewritten as

$$u \in U, \quad Au = \mathcal{J}\ell, \tag{7.7.6}$$

where $\mathcal{J} : V' \to V$ is the Riesz isometric operator.

From the condition (7.7.2) and the definition of A, it follows immediately that A is injective; i.e., $Au = 0$ for some $u \in U$ implies $u = 0$.

To show that the range $\mathcal{R}(A)$ is closed, let $\{u_n\} \subseteq U$ be a sequence such that $\{Au_n\}$ converges in V, the limit being denoted by $v \in V$. Using the condition (7.7.2), we have

$$\|u_m - u_n\|_U \leq \frac{1}{\alpha} \sup_{0 \neq v \in V} \frac{(A(u_m - u_n), v)_V}{\|v\|_V} \leq \frac{1}{\alpha} \|Au_m - Au_n\|_V.$$

Hence, $\{u_n\}$ is a Cauchy sequence in U, and hence has a limit $u \in U$. Moreover, by the continuity condition (7.7.1), $Au_n \to Au = v$ in V. Thus, the range $\mathcal{R}(A)$ is closed.

Now if $v \in \mathcal{R}(A)^{\perp}$, then

$$(Au, v)_V = a(u, v) = 0 \quad \forall u \in U.$$

Applying the condition (7.7.3), we conclude $v = 0$. So $\mathcal{R}(A)^{\perp} = \{0\}$.

Therefore, the equation (7.7.6) and hence also the problem (7.7.4) has a unique solution.

The estimate (7.7.5) follows easily from another application of the condition (7.7.2). ■

Exercise 7.7.1 *Show that Theorem 7.7.1 is a generalization of the Lax-Milgram lemma.*

Exercise 7.7.2 *As an application of Theorem 7.7.1, we consider the model boundary value problem*

$$-\Delta u = f \quad \text{in } \Omega, \tag{7.7.7}$$

$$u = 0 \quad \text{on } \partial\Omega. \tag{7.7.8}$$

The "standard" weak formulation of the problem is

$$u \in H_0^1(\Omega), \quad \int_{\Omega} \nabla u \cdot \nabla v \, dx = \langle f, v \rangle_{H^{-1}(\Omega) \times H_0^1(\Omega)} \quad \forall v \in H_0^1(\Omega). \tag{7.7.9}$$

This formulation makes sense as long as $f \in H^{-1}(\Omega)$ (e.g., if $f \in L^2(\Omega)$). Performing an integration by part on the bilinear form, we are led to a new weak formulation:

$$u \in L^2(\Omega), \quad -\int_{\Omega} \Delta u \, v \, dx = \langle f, v \rangle_{(H^2(\Omega))' \times H^2(\Omega)} \quad \forall v \in H^2(\Omega) \cap H_0^1(\Omega). \tag{7.7.10}$$

This formulation makes sense even when $f \notin H^{-1}(\Omega)$ as long as $f \in (H^2(\Omega))'$. One example is the point load

$$f(\mathbf{x}) = c_0 \delta(\mathbf{x} - \mathbf{x}_0)$$

for some $c_0 \in \mathbb{R}$ and $\mathbf{x}_0 \in \Omega$. In this case, we interpret $\langle f, v \rangle_{(H^2(\Omega))' \times H^2(\Omega)}$ as $c_0 v(\mathbf{x}_0)$, which is well-defined if $d \leq 3$ since $H^2(\Omega)$ is embedded in $C(\overline{\Omega})$.

Assume $f \in (H^2(\Omega))'$ and $\Omega \subseteq \mathbb{R}^d$ is smooth or convex. Apply Theorem 7.7.1 to show that there is a unique "weaker" solution $u \in L^2(\Omega)$ to the problem (7.7.10). In verifying the condition (7.7.2), you need the estimate (6.3.9).

7.8 A nonlinear problem

A number of physical applications lead to partial differential equations of the type (see [172]):

$$-\text{div} \left[\alpha(|\nabla u|) \nabla u \right] = f.$$

In this section, we consider one such nonlinear equation. Specifically, we study the boundary value problem

$$-\mathrm{div}\left[(1+|\nabla u|^2)^{p/2-1}\nabla u\right] = f \quad \text{in } \Omega, \tag{7.8.1}$$

$$u = 0 \quad \text{on } \partial\Omega, \tag{7.8.2}$$

where $p \geq 2$. We use p^* to denote the conjugate exponent defined through the relation

$$\frac{1}{p} + \frac{1}{p^*} = 1.$$

When $p = 2$, (7.8.1)–(7.8.2) reduces to a linear problem: the homogeneous Dirichlet boundary value problem for the Poisson equation, which was studied in Section 7.1.

Let us first formally derive a weak formulation for the problem (7.8.1)–(7.8.2). For this, we assume the problem has a solution, sufficiently smooth so that all the following calculations leading to the weak formulation are meaningful. Multiplying the equation (7.8.1) with an arbitrary test function $v \in C_0^\infty(\Omega)$ and integrating the relation over Ω, we have

$$-\int_\Omega \mathrm{div}\left[(1+|\nabla u|^2)^{p/2-1}\nabla u\right] v\,dx = \int_\Omega f v\,dx.$$

Then perform an integration by parts to obtain

$$\int_\Omega (1+|\nabla u|^2)^{p/2-1}\nabla u \cdot \nabla v\,dx = \int_\Omega f v\,dx. \tag{7.8.3}$$

Let us introduce the space

$$V = W_0^{1,p}(\Omega), \tag{7.8.4}$$

and define the norm

$$\|v\|_V = \left(\int_\Omega |\nabla v|^p dx\right)^{1/p}. \tag{7.8.5}$$

It can be verified that $\|\cdot\|_V$ defined in (7.8.5) is a norm over the space V, which is equivalent to the standard norm $\|\cdot\|_{W^{1,p}(\Omega)}$ (see Exercise 7.8.1). Since $p \in [2, \infty)$, the space V is a reflexive Banach space. The dual space of V is

$$V' = W^{-1,p^*}(\Omega)$$

and we assume throughout this section

$$f \in V'.$$

Notice that the dual space V' is pretty large, e.g., $L^{p^*}(\Omega) \subseteq V'$.

It can be shown that the left side of (7.8.3) makes sense as long as $u, v \in V$ (see Exercise 7.8.2). Additionally, the right side of (7.8.3) is well defined for $f \in V'$ and $v \in V$ when we interpret the right side of (7.8.3) as the duality pair between V^* and V.

Now we are ready to introduce the weak formulation for the boundary value problem (7.8.1)–(7.8.2):

$$u \in V : \quad a(u; u, v) = \ell(v) \quad \forall \, v \in V. \tag{7.8.6}$$

Here

$$a(w; u, v) = \int_\Omega (1 + |\nabla w|^2)^{p/2-1} \nabla u \cdot \nabla v \, dx, \tag{7.8.7}$$

$$\ell(v) = \int_\Omega f \, v \, dx. \tag{7.8.8}$$

Related to the weak formulation (7.8.6), we introduce a minimization problem

$$u \in V : \quad E(u) = \inf_{v \in V} E(v), \tag{7.8.9}$$

where the "energy functional" $E(\cdot)$ is

$$E(v) = \frac{1}{p} \int_\Omega (1 + |\nabla v|^2)^{p/2} dx - \int_\Omega f \, v \, dx. \tag{7.8.10}$$

We first explore some properties of the energy functional.

Lemma 7.8.1 *The energy functional $E(\cdot)$ is coercive, i.e.,*

$$E(v) \to \infty \quad \text{as } \|v\|_V \to \infty.$$

Proof. It is easy to see that

$$E(v) \geq \frac{1}{p} \|v\|_V^p - \|f\|_{V'} \|v\|_V.$$

Since $p > 1$, we have $E(v) \to \infty$ as $\|v\|_V \to \infty$. ∎

Lemma 7.8.2 *The energy functional $E(\cdot)$ is Fréchet differentiable and*

$$\langle E'(u), v \rangle = \int_\Omega (1 + |\nabla u|^2)^{p/2-1} \nabla u \cdot \nabla v \, dx - \int_\Omega f \, v \, dx, \quad u, v \in V.$$

Proof. We only need to prove that the functional

$$E_1(v) = \frac{1}{p} \int_\Omega (1 + |\nabla v|^2)^{p/2} dx \tag{7.8.11}$$

is Fréchet differentiable over V and

$$\langle E_1'(u), v \rangle = \int_\Omega (1 + |\nabla u|^2)^{p/2-1} \nabla u \cdot \nabla v \, dx, \quad u, v \in V. \tag{7.8.12}$$

For any $\boldsymbol{\xi}, \boldsymbol{\eta} \in \mathbb{R}^d$, we define a real function

$$\phi(t) = \frac{1}{p} \left(1 + |\boldsymbol{\xi} + t\,(\boldsymbol{\eta} - \boldsymbol{\xi})|^2 \right)^{p/2}, \quad t \in \mathbb{R}.$$

We have

$$\phi'(t) = \left(1 + |\boldsymbol{\xi} + t(\boldsymbol{\eta} - \boldsymbol{\xi})|^2\right)^{p/2-1} (\boldsymbol{\xi} + t(\boldsymbol{\eta} - \boldsymbol{\xi})) \cdot (\boldsymbol{\eta} - \boldsymbol{\xi}).$$

Hence

$$\phi'(0) = (1 + |\boldsymbol{\xi}|^2)^{p/2-1} \boldsymbol{\xi} \cdot (\boldsymbol{\eta} - \boldsymbol{\xi}).$$

For the second derivative,

$$\phi''(t) = (p-2)\left(1 + |\boldsymbol{\xi} + t(\boldsymbol{\eta} - \boldsymbol{\xi})|^2\right)^{p/2-2} [(\boldsymbol{\xi} + t(\boldsymbol{\eta} - \boldsymbol{\xi})) \cdot (\boldsymbol{\eta} - \boldsymbol{\xi})]^2$$
$$+ \left(1 + |\boldsymbol{\xi} + t(\boldsymbol{\eta} - \boldsymbol{\xi})|^2\right)^{p/2-1} |\boldsymbol{\eta} - \boldsymbol{\xi}|^2.$$

Apply Taylor's theorem,

$$\phi(1) - \phi(0) = \phi'(0) + \phi''(\theta) \quad \text{for some } \theta \in (0, 1);$$

i.e.,

$$\frac{1}{p}(1 + |\boldsymbol{\eta}|^2)^{p/2} - \frac{1}{p}(1 + |\boldsymbol{\xi}|^2)^{p/2} - (1 + |\boldsymbol{\xi}|^2)^{p/2-1} \boldsymbol{\xi} \cdot (\boldsymbol{\eta} - \boldsymbol{\xi}) = \phi''(\theta).$$

$$(7.8.13)$$

It can be verified that for some constant c,

$$|\phi''(\theta)| \le c(1 + |\boldsymbol{\xi}|^{p-2} + |\boldsymbol{\eta}|^{p-2}) |\boldsymbol{\eta} - \boldsymbol{\xi}|^2.$$

We take $\boldsymbol{\eta} = \nabla v$ and $\boldsymbol{\xi} = \nabla u$ in (7.8.13), and then integrate the relation over Ω to obtain

$$E_1(v) - E_1(u) - \int_\Omega (1 + |\nabla u|^2)^{p/2-1} \nabla u \cdot (\nabla v - \nabla u) \, dx = R(u, v)$$

with

$$|R(u, v)| \le c \int_\Omega (1 + |\nabla u|^{p-2} + |\nabla v|^{p-2}) |\nabla(v - u)|^2 dx.$$

Using Hölder's inequality (Lemma 1.5.4), we have

$$|R(u, v)| \le c\left(1 + \|u\|_V^{1-2/p} + \|v\|_V^{1-2/p}\right) \|v - u\|_V^2.$$

By Definition 4.3.1, we see that $E_1(v)$ is Fréchet differentiable over V and we have the formula (7.8.12). ∎

Since Fréchet differentiability implies continuity, we immediately obtain the next property.

Corollary 7.8.3 *The energy functional $E(\cdot)$ is continuous over V.*

Lemma 7.8.4 *The energy functional $E(\cdot)$ is strictly convex.*

Proof. We only need to show that $E_1(\cdot)$ defined in (7.8.11) is strictly convex. However, this follows immediately from the strict convexity of the real-valued function $\boldsymbol{\xi} \mapsto \frac{1}{p}(1 + |\boldsymbol{\xi}|^2)^{p/2}$ (cf. Exercise 4.3.7). ∎

We can now state the main result concerning the existence and uniqueness for the weak formulation (7.8.6) and the minimization problem (7.8.9).

Theorem 7.8.5 *Assume $f \in V'$ and $p \in [2, \infty)$. Then the weak formulation (7.8.6) and the minimization problem (7.8.9) are equivalent, and both admits a unique solution.*

Proof. Since V is reflexive and $E : V \to \mathbb{R}$ is coercive, continuous, and strictly convex, by Theorem 3.2.12, we conclude the minimization problem (7.8.9) has a unique minimizer $u \in V$. By Theorem 4.3.17, we know that the weak formulation (7.8.6) and the minimization problem (7.8.9) are equivalent. ∎

Exercise 7.8.1 *Use Theorem 6.3.13 to show that (7.8.5) defines a norm over the space V, which is equivalent to the standard norm $\| \cdot \|_{W^{1,p}(\Omega)}$.*

Exercise 7.8.2 *Apply Hölder's inequality (Lemma 1.5.4) to show that the left side of (7.8.3) makes sense for $u, v \in W^{1,p}(\Omega)$.*

Exercise 7.8.3 *Show that the minimization problem (7.8.9) has a unique minimizer for the case $p \in (1, 2)$.*

Suggestion for Further Readings

Many books can be consulted on detailed treatment of PDEs, for both steady and evolution equations, e.g., EVANS [48], LIONS AND MAGENES [109], MCOWEN [116], WLOKA [167].

8

The Galerkin Method and Its Variants

In this chapter, we briefly discuss some numerical methods for solving boundary value problems. These are the Galerkin method and its variants: the Petrov-Galerkin method and the generalized Galerkin method.

8.1 The Galerkin method

The Galerkin method provides a general framework for approximation of operator equations, which includes the finite element method as a special case. In this section, we discuss the Galerkin method for a linear operator equation in a form directly applicable to the study of the finite element method.

Let V be a Hilbert space, $a(\cdot, \cdot) : V \times V \to \mathbb{R}$ be a bilinear form, and $\ell \in V'$. We consider the problem

$$u \in V, \quad a(u, v) = \ell(v) \quad \forall\, v \in V. \tag{8.1.1}$$

Throughout this section, we assume $a(\cdot, \cdot)$ is bounded,

$$|a(u, v)| \leq M \, \|u\|_V \|v\|_V \quad \forall\, u, v \in V, \tag{8.1.2}$$

and V-elliptic,

$$a(v, v) \geq c_0 \|v\|_V^2 \quad \forall\, v \in V. \tag{8.1.3}$$

Then according to the Lax-Milgram lemma, the variational problem (8.1.1) has a unique solution.

In general, it is impossible to find the exact solution of the problem (8.1.1) because the space V is infinite dimensional. A natural approach to constructing an approximate solution is to solve a finite-dimensional analog of the problem (8.1.1). Thus, let $V_N \subseteq V$ be an N-dimensional subspace. We project the problem (8.1.1) onto V_N,

$$u_N \in V_N, \quad a(u_N, v) = \ell(v) \quad \forall v \in V_N. \tag{8.1.4}$$

Under the assumptions that the bilinear form $a(\cdot, \cdot)$ is bounded and V-elliptic, and $\ell \in V'$, we can again apply the Lax-Milgram lemma and conclude that the problem (8.1.4) has a unique solution u_N.

We can express the problem (8.1.4) in the form of a linear system. Indeed, let $\{\phi_i\}_{i=1}^N$ be a basis of the finite-dimensional space V_N. We write

$$u_N = \sum_{j=1}^N \xi_j \phi_j$$

and take $v \in V_N$ in (8.1.4) each of the basis functions ϕ_i. As a result, (8.1.4) is equivalent to a linear system

$$A\boldsymbol{\xi} = \mathbf{b}. \tag{8.1.5}$$

Here, $\boldsymbol{\xi} = (\xi_j) \in \mathbb{R}^N$ is the unknown vector, $A = (a(\phi_j, \phi_i)) \in \mathbb{R}^{N \times N}$ is called the *stiffness matrix*, $\mathbf{b} = (\ell(\phi_i)) \in \mathbb{R}^N$ is the *load vector*. So the solution of the problem (8.1.4) can be found by solving a linear system.

The approximate solution u_N is, in general, different from the exact solution u. To increase the accuracy, it is natural to seek the approximate solution u_N in a larger subspace V_N. Thus, for a sequence of subspaces $V_{N_1} \subseteq V_{N_2} \subseteq \cdots \subseteq V$, we compute a corresponding sequence of approximate solutions $u_{N_i} \in V_{N_i}$, $i = 1, 2, \ldots$. This solution procedure is called the *Galerkin method*.

In the special case when the bilinear form $a(\cdot, \cdot)$ is also symmetric,

$$a(u, v) = a(v, u) \quad \forall u, v \in V,$$

the original problem (8.1.1) is equivalent to a minimization problem (see Chapter 3)

$$u \in V, \quad E(u) = \inf_{v \in V} E(v), \tag{8.1.6}$$

where the energy functional

$$E(v) = \frac{1}{2} a(v, v) - \ell(v). \tag{8.1.7}$$

Now with a finite-dimensional subspace $V_N \subseteq V$ chosen, it is equally natural to develop a numerical method by minimizing the energy functional over the finite-dimensional space V_N,

$$u_N \in V_N, \quad E(u_N) = \inf_{v \in V_N} E(v). \tag{8.1.8}$$

It is easy to verify that the two approximate problems (8.1.4) and (8.1.8) are equivalent. The method based on minimizing the energy functional over finite dimensional subspaces is called the *Ritz method*. From the above discussion, we see that the Galerkin method is more general than the Ritz method, while when both methods are applicable, they are equivalent. Because of this, the Galerkin method is also called the *Ritz-Galerkin method*.

Example 8.1.1 *We examine a concrete example of the Galerkin method. Consider the boundary value problem*

$$\begin{cases} -u'' = f & \text{in } (0,1), \\ u(0) = u(1) = 0. \end{cases} \tag{8.1.9}$$

The weak formulation of the problem is

$$u \in V, \quad \int_0^1 u'v'\,dx = \int_0^1 fv\,dx \quad \forall v \in V,$$

where $V = H_0^1(0,1)$. Applying the Lax-Milgram lemma, we see that the weak problem has a unique solution. To develop a Galerkin method, we need to choose a finite-dimensional subspace of V. Notice that a function in V must vanish at both $x = 0$ and $x = 1$. Thus a natural choice is

$$V_N = \text{span} \{x^i(1-x), \ i = 1, \ldots, N\}.$$

We write

$$u_N(x) = \sum_{j=1}^N \xi_j x^j (1-x);$$

the coefficients $\{\xi_j\}_{j=1}^N$ are determined by the Galerkin equations

$$\int_0^1 u_N' v'\,dx = \int_0^1 fv\,dx \quad \forall v \in V_N.$$

Taking v to be each of the basis functions $x^i(1-x)$, $1 \leq i \leq N$, we derive a linear system for the coefficients:

$$A\xi = b.$$

Here $\xi = (\xi_1, \ldots, \xi_N)^T$ is the vector of unknowns, $b \in \mathbb{R}^N$ is a vector whose i^{th} component is $\int_0^1 f(x)\,x^i(1-x)\,dx$. The coefficient matrix is A, whose (i,j)-th entry is

$$\int_0^1 [x^j(1-x)]'[x^i(1-x)]'\,dx = \frac{(i+1)(j+1)}{i+j+1} + \frac{(i+2)(j+2)}{i+j+3}$$
$$- \frac{(i+1)(j+2)+(i+2)(j+1)}{i+j+2}.$$

The coefficient matrix is rather ill-conditioned, indicating it is difficult to solve the above Galerkin system numerically. The following table shows how *rapidly the condition number of the matrix (measured in 2-norm) increases with the order N. We conclude that the seemingly natural choice of the basis functions $\{x^i(1-x)\}$ is not suitable for solving the problem (8.1.9).*

N	Cond (A)
3	8.92E+02
4	2.42E+04
5	6.56E+05
6	1.79E+07
7	4.95E+08
8	1.39E+10
9	3.93E+11
10	1.14E+13

Example 8.1.2 *Let us consider the problem (8.1.9) again. This time, the finite-dimensional subspace is chosen to be*

$$V_N = \text{span}\{\sin(i\pi x), \ i = 1, \dots, N\}.$$

The basis functions are orthogonal with respect to the inner product defined by the bilinear form:

$$\int_0^1 (\sin j\pi x)'(\sin i\pi x)'dx = ij\pi^2 \int_0^1 \cos j\pi x \, \cos i\pi x \, dx = \frac{ij\pi^2}{2}\delta_{ij}.$$

Writing

$$u_N(x) = \sum_{j=1}^N \xi_j \sin j\pi x,$$

we see that the coefficients $\{\xi_j\}_{j=1}^N$ are determined by the linear system

$$\sum_{j=1}^N \xi_j \int_0^1 (\sin j\pi x)'(\sin i\pi x)'dx = \int_0^1 f(x)\sin i\pi x \, dx, \quad i = 1, \dots, N,$$

which is a diagonal system and we find the solution immediately:

$$\xi_i = \frac{2}{\pi^2 i^2} \int_0^1 f(x)\sin i\pi x \, dx, \quad i = 1, \dots, N.$$

It is worth noticing that the Galerkin solution can be written in the form of a kernel approximation:

$$u_N(x) = \int_0^1 f(t) K_N(x, t) \, dt,$$

where the kernel function

$$K_N(x, t) = \frac{2}{\pi^2} \sum_{j=1}^{N} \frac{\sin j\pi x \, \sin j\pi t}{j^2}.$$

From the above two examples, we see that in applying the Galerkin method it is very important to choose appropriate basis functions for finite-dimensional subspaces. Before the invention of computers, the Galerkin method was applied mainly with the use of global polynomials or global trignometric polynomials. For the simple model problem (8.1.9) we see that the seemingly natural choice of the polynomial basis functions $\{x^i(1-x)\}$ leads to a severely ill-conditioned linear system. For the same model problem, the trigonometric polynomial basis functions $\{\sin(i\pi x)\}$ is ideal in the sense that it leads to a diagonal linear system so that its conditioning is best possible. We need to be aware, though, that trigonometric polynomial basis functions can lead to severely ill-conditioned linear systems in different but equally simple model problems. The idea of the finite element method (see Chapter 9) is to use basis functions with small supports so that, among various advantages of the method, the conditioning of the resulting linear system can be moderately maintained (see Exercise 9.3.4 for an estimate on the growth of the condition number of stiffness matrices as the mesh is refined).

Now we consider the important issue of convergence and error estimation for the Galerkin method. A key result is the following Céa's inequality.

Proposition 8.1.3 *Assume V is a Hilbert space, $V_N \subseteq V$ is a subspace, $a(\cdot, \cdot)$ is a bounded, V-elliptic bilinear on V, and $\ell \in V'$. Let $u \in V$ be the solution of the problem (8.1.1), and $u_N \in V_N$ be the Galerkin approximation defined in (8.1.4). Then there is a constant c such that*

$$\|u - u_N\|_V \leq c \inf_{v \in V_N} \|u - v\|_V. \tag{8.1.10}$$

Proof. Subtracting (8.1.4) from (8.1.1) with $v \in V_N$, we find an error relation

$$a(u - u_N, v) = 0 \quad \forall v \in V_N. \tag{8.1.11}$$

Using the V-ellipticity of $a(\cdot, \cdot)$, the error relation and the boundedness of $a(\cdot, \cdot)$, we have, for any $v \in V_N$,

$$\begin{aligned}
c_0 \|u - u_N\|_V^2 &\leq a(u - u_N, u - u_N) \\
&= a(u - u_N, u - v) \\
&\leq M \|u - u_N\|_V \|u - v\|_V.
\end{aligned}$$

Thus

$$\|u - u_N\|_V \leq c \|u - v\|_V,$$

where we may take $c = M/c_0$. Since v is arbitrary in V_N, we have the inequality (8.1.10). ∎

The inequality (8.1.10) is known as Céa's lemma in the literature. Such an inequality was first proved by Céa [32] for the case when the bilinear form is symmetric and was extended to the nonsymmetric case in [25]. The inequality (8.1.10) states that to estimate the error of the Galerkin solution, it suffices to estimate the approximation error $\inf_{v \in V_N} \|u - v\|$.

In the special case when $a(\cdot, \cdot)$ is symmetric, we may assign a geometrical interpretation of the error relation (8.1.11). Indeed, in this special case, the bilinear form $a(\cdot, \cdot)$ defines an inner product over the space V and its induced norm $\|v\|_a = \sqrt{a(v, v)}$, called the energy norm, is equivalent to the norm $\|v\|_V$. With respect to this new inner product, the Galerkin solution error $u - u_N$ is orthogonal to the subspace V_N, or in other words, the Galerkin solution u_N is the orthogonal projection of the exact solution u to the subspace V_N. Also in this special case, Céa's inequality (8.1.10) can be replaced by

$$\|u - u_N\|_a = \inf_{v \in V_N} \|u - v\|_a;$$

i.e., measured in the energy norm, u_N is the optimal approximation of u from the subspace V_N.

Céa's inequality is a basis for convergence analysis and error estimations. As a simple consequence, we have the next convergence result.

Corollary 8.1.4 *We make the assumptions stated in Proposition 8.1.3. Assume $V_{N_1} \subseteq V_{N_2} \subseteq \cdots$ is a sequence of subspaces of V with the property*

$$\overline{\bigcup_{i \geq 1} V_{N_i}} = V. \tag{8.1.12}$$

Then the Galerkin method converges:

$$\|u - u_{N_i}\|_V \to 0 \quad \text{as } i \to \infty. \tag{8.1.13}$$

Proof. By the density assumption (8.1.12), we can find a sequence $v_i \in V_{N_i}$, $i \geq 1$, such that

$$\|u - v_i\|_V \to 0 \quad \text{as } i \to \infty.$$

Applying Céa's inequality (8.1.10), we have

$$\|u - u_{N_i}\|_V \leq c \|u - v_i\|_V.$$

Therefore, we have the convergence statement (8.1.13). ∎

When the finite-dimensional space V_N is constructed from piecewise (images of) polynomials, the Galerkin method leads to a finite element method, which will be discussed in some detail in the chapter following. We will see in the context of the finite element method that Céa's inequality also serves as a basis for error estimates.

Exercise 8.1.1 *Show that the discrete problems (8.1.4) and (8.1.8) are equivalent.*

Exercise 8.1.2 *Show that if the bilinear form $a(\cdot, \cdot)$ is symmetric, then the stiffness matrix A is symmetric; if $a(\cdot, \cdot)$ is V-elliptic, then A is positive definite.*

8.2   The Petrov-Galerkin method

The Petrov-Galerkin method for a linear boundary value problem can be developed based on the framework of the generalized Lax-Milgram lemma presented in Section 6.5. Let U and V be two real Hilbert spaces, $a : U \times V \to \mathbb{R}$ a bilinear form, and $\ell \in V'$. The problem to be solved is

$$u \in U, \quad a(u, v) = \ell(v) \quad \forall v \in V. \tag{8.2.1}$$

From the generalized Lax-Milgram lemma, we know that the problem (8.2.1) has a unique solution $u \in U$, if the following conditions are satisfied: there exist constants $M > 0$ and $\alpha > 0$, such that

$$|a(u, v)| \leq M \|u\|_U \|v\|_V \quad \forall u \in U, \ v \in V; \tag{8.2.2}$$

$$\sup_{0 \neq v \in V} \frac{a(u, v)}{\|v\|_V} \geq \alpha \|u\|_U \quad \forall u \in U; \tag{8.2.3}$$

$$\sup_{u \in U} a(u, v) > 0 \quad \forall v \in V, \ v \neq 0. \tag{8.2.4}$$

Now let $U_N \subseteq U$ and $V_N \subseteq V$ be finite-dimensional subspaces of U and V with $\dim(U_N) = \dim(V_N) = N$. Then a Petrov-Galerkin method to solve the problem (8.2.1) is given by

$$u_N \in U_N, \quad a(u_N, v_N) = \ell(v_N) \quad \forall v_N \in V_N. \tag{8.2.5}$$

Well-posedness and error analysis for the method (8.2.5) are discussed in the next result (see [19]).

Theorem 8.2.1 *We keep the above assumptions on the spaces U, V, U_N and V_N, and the forms $a(\cdot, \cdot)$ and $\ell(\cdot)$. Assume further that there exists a constant $\alpha_N > 0$, such that*

$$\sup_{0 \neq v_N \in V_N} \frac{a(u_N, v_N)}{\|v_N\|_V} \geq \alpha_N \|u_N\|_U \quad \forall u_N \in U_N. \tag{8.2.6}$$

Then the discrete problem (8.2.5) has a unique solution u_N, and we have the error estimate

$$\|u - u_N\|_U \leq \left(1 + \frac{M}{\alpha_N}\right) \inf_{w_N \in U_N} \|u - w_N\|_U. \tag{8.2.7}$$

Proof. From the generalized Lax-Milgram lemma, we conclude immediately that under the stated assumptions, the problem (8.2.5) has a unique solution u_N. Subtracting (8.2.5) from (8.2.1) with $v = v_N \in V_N$, we obtain the error relation

$$a(u - u_N, v_N) = 0 \quad \forall\, v_N \in V_N. \tag{8.2.8}$$

Now for any $w_N \in V_N$, we write

$$\|u - u_N\|_U \leq \|u - w_N\|_U + \|u_N - w_N\|_U. \tag{8.2.9}$$

Using the condition (8.2.6), we have

$$\alpha_N \|u_N - w_N\|_U \leq \sup_{0 \neq v_N \in V_N} \frac{a(u_N - w_N, v_N)}{\|v_N\|_V}.$$

Using the error relation (8.2.8), we then obtain

$$\alpha_N \|u_N - w_N\|_U \leq \sup_{0 \neq v_N \in V_N} \frac{a(u - w_N, v_N)}{\|v_N\|_V}.$$

The right-hand side can be bounded by $M \|u - w_N\|_U$. Therefore,

$$\|u_N - w_N\|_U \leq \frac{M}{\alpha_N} \|u - w_N\|_U.$$

This inequality and (8.2.9) imply the estimate (8.2.7).   ∎

As in Corollary 8.1.4, we have a convergence result based on the estimate (8.2.7).

Corollary 8.2.2 *We make the assumptions stated in Theorem 8.2.1. Furthermore, we assume that there is a constant $\alpha_0 > 0$ such that*

$$\alpha_N \geq \alpha_0 \quad \forall\, N. \tag{8.2.10}$$

Assume the sequence of subspaces $U_{N_1} \subseteq U_{N_2} \subseteq \cdots \subseteq U$ has the property

$$\overline{\bigcup_{i \geq 1} U_{N_i}} = U. \tag{8.2.11}$$

Then the Petrov-Galerkin method (8.2.5) converges:

$$\|u - u_{N_i}\|_U \to 0 \quad \text{as } i \to \infty.$$

We remark that to achieve convergence of the method, we can allow α_N to approach 0 under certain rule, as long as

$$\max\{1, \alpha_N^{-1}\} \inf_{w_N \in U_N} \|u - w_N\|_U \to 0$$

as is seen from the estimate (8.2.7). Nevertheless, the condition (8.2.10) is crucial in obtaining optimal order error estimates. This condition is usually written as

$$\sup_{0 \neq v_N \in V_N} \frac{a(u_N, v_N)}{\|v_N\|_V} \geq \alpha_0 \|u_N\|_U \quad \forall\, u_N \in U_N, \ \forall\, N, \tag{8.2.12}$$

or equivalently,

$$\inf_{0 \neq u_N \in U_N} \sup_{0 \neq v_N \in V_N} \frac{a(u_N, v_N)}{\|u_N\|_U \|v_N\|_V} \geq \alpha_0 \quad \forall N. \tag{8.2.13}$$

In the literature, this condition is called the inf-sup condition or Babuška-Brezzi condition. This condition states that the two finite-dimensional spaces must be compatible in order to yield convergent numerical solutions. The condition is most important in the context of the study of mixed finite element methods (cf. [29]).

8.3   Generalized Galerkin method

In the Galerkin method discussed in Section 7.1, the finite-dimensional space V_N is assumed to be a subspace of V. The resulting numerical method is called an internal approximation method. For certain problems, we will need to relax this assumption and to allow the variational "crime" $V_N \not\subset V$. This, for instance, is the case for non-conforming method (see Example 9.1.3 for an example of a non-conforming finite element method). There are situations where considerations of other variational "crimes" are needed. Two such situations are when a general curved domain is approximated by a polygonal domain then functions are integrated over a slightly different domain, and when numerical quadratures are used to compute the integrals defining the bilinear form and the linear form. These considerations lead to the following framework of a generalized Galerkin method for the approximate solution of the problem (8.1.1)

$$u_N \in V_N, \quad a_N(u_N, v_N) = \ell_N(v_N) \quad \forall v_N \in V_N. \tag{8.3.1}$$

Here, V_N is a finite-dimensional space, but it is no longer assumed to be a subspace of V; the bilinear form $a_N(\cdot, \cdot)$ and the linear form $\ell_N(\cdot)$ are suitable approximations of $a(\cdot, \cdot)$ and $\ell_N(\cdot)$.

We have the following result related to the approximation method (8.3.1).

Theorem 8.3.1 *Assume a discretization-dependent norm $\| \cdot \|_N$, the approximation bilinear form $a_N(\cdot, \cdot)$, and the linear form $\ell_N(\cdot)$ are defined on the space*

$$V + V_N = \{w \mid w = v + v_N, \ v \in V, \ v_N \in V_N\}.$$

Assume there exist constants $M, \alpha_0, c_0 > 0$, independent of N, such that

$$|a_N(w, v_N)| \leq M \|w\|_N \|v_N\|_N \quad \forall w \in V + V_N, \ \forall v_N \in V_N,$$
$$a_N(v_N, v_N) \geq \alpha_0 \|v_N\|_N^2 \quad \forall v_N \in V_N,$$
$$|\ell_N(v_N)| \leq c_0 \|v_N\|_N \quad \forall v_N \in V_N.$$

Then the problem (8.3.1) *has a unique solution* $u_N \in V_N$, *and we have the error estimate*

$$\|u - u_N\|_N \le \left(1 + \frac{M}{\alpha_0}\right) \inf_{w_N \in V_N} \|u - w_N\|_N$$

$$+ \frac{1}{\alpha_0} \sup_{v_N \in V_N} \frac{|a_N(u, v_N) - \ell_N(v_N)|}{\|v_N\|_N}. \qquad (8.3.2)$$

Proof. The unique solvability of the problem (8.3.1) follows from an application of the Lax-Milgram lemma. Let us derive the error estimate (8.3.2). For any $w_N \in V_N$, we write

$$\|u - u_N\|_N \le \|u - w_N\|_N + \|w_N - u_N\|_N.$$

Using the assumptions on the approximate bilinear form and the definition of the approximate solution u_N, we have

$$\alpha_0 \|w_N - u_N\|_N^2$$
$$\le a_N(w_N - u_N, w_N - u_N)$$
$$= a_N(w_N - u, w_N - u_N) + a_N(u, w_N - u_N) - \ell_N(w_N - u_N)$$
$$\le M \|w_N - u\|_N \|w_N - u_N\|_N + |a_N(u, w_N - u_N) - \ell_N(w_N - u_N)|.$$

Thus

$$\alpha_0 \|w_N - u_N\|_N \le M \|w_N - u\|_N + \frac{|a_N(u, w_N - u_N) - \ell_N(w_N - u_N)|}{\|w_N - u_N\|_N}.$$

We replace $w_N - u_N$ by v_N and take the supremum of the second term of the right-hand side with respect to $v_N \in V_N$ to obtain (8.3.2). ∎

The estimate (8.3.2) is a Strang-type estimate for the effect of the variational "crimes" on the numerical solution. We notice that in the bound of the estimate (8.3.2), the first term is on the approximation property of the solution u by functions from the finite-dimensional space V_N, while the second term describes the extent to which the exact solution u satisifes the approximate problem.

Example 8.3.2 *In Subsection 9.1.3, we use a "non-conforming finite element method" to solve the fourth-order elliptic boundary value problem*

$$\begin{cases} u^{(4)} = f & in\ (0, 1), \\ u(0) = u'(0) = u(1) = u'(1) = 0. \end{cases}$$

Here $f \in L^2(0, 1)$ *is given. The weak formulation is the problem* (8.1.1) *with the choice*

$$V = H_0^2(0, 1),$$

$$a(u, v) = \int_0^1 u'' v'' dx,$$

$$\ell(v) = \int_0^1 f v\, dx.$$

Let $0 = x_0 < x_1 < \cdots < x_N = 1$ be a partition of the domain $[0, 1]$, and denote $I_i = [x_{i-1}, x_i]$ for $1 \leq i \leq N$, and $h = \max_{1 \leq i \leq N}(x_i - x_{i-1})$ for the meshsize. We choose the finite-dimensional space V_N to be the finite element space

$$V_h = \{v_h \in C(\overline{I}) \mid v_h|_{I_i} \in P_2(I_i), \ 1 \leq i \leq N,$$
$$v_h(x) = v_h'(x) = 0 \text{ at } x = 0, 1\}.$$

Notice that $V_h \not\subseteq V$. The corresponding non-conforming finite element method is (8.3.1) with

$$a_N(u_h, v_h) \equiv a_h(u_h, v_h) = \sum_{i=1}^{N} \int_{I_i} u_h''(x)\, v_h''(x)\, dx,$$

$$\ell_N(v_h) \equiv \ell_h(v_h) = \ell(v_h).$$

We use the norm

$$\|v\|_N \equiv \|v\|_h = \left(\sum_{i=1}^{N} |v|_{2, I_i}^2 \right)^{\frac{1}{2}}.$$

Then all the assumptions of Theorem 8.3.1 are satisfied with $M = \alpha_0 = 1$, and so we have the estimate

$$\|u - u_h\|_h \leq 2 \inf_{w_h \in V_h} \|u - w_h\|_h + \sup_{v_h \in V_h} \frac{|a_h(u, v_h) - \ell(v_h)|}{\|v_h\|_h}.$$

It is then possible to find an order error estimate, which will not be done here; an interested reader can consult [35, 36].

Exercise 8.3.1 *Show that in the case of a conforming method (i.e., $V_N \subseteq V$, $a_N(\cdot, \cdot) \equiv a(\cdot, \cdot)$ and $\ell_N(\cdot) \equiv \ell(\cdot)$), the error estimate (8.3.2) reduces to Céa's inequality (8.1.10).*

Suggestion for Further Readings

Based on any weak formulation, we can develop a particular Galerkin-type numerical method. Mixed formulations are the basis for mixed Galerkin finite element methods. We refer the reader to [29] for an extensive treatment of the mixed methods.

Many numerical methods exist for solving differential equations. In this text, we do not touch upon some other popular methods,— e.g., the collocation method, the spectral method, the finite volume method, etc., and various combinations of these methods such as the spectral collocation method. The well-written book [132] can be consulted for discussions of many of the existing methods for numerical solution of partial differential equations and for a rather comprehensive list of related references.

9
Finite Element Analysis

The finite element method is the most popular numerical method for solving elliptic boundary value problems. In this chapter, we introduce the concept of the finite element method, the finite element interpolation theory and its application in order error estimates of finite element solutions of elliptic boundary value problems. The boundary value problems considered in this chapter are linear.

Detailed mathematical analysis of the finite element method can be found in numerous monographs and textbooks, e.g., [19, 27, 28, 35, 36, 84, 126, 153].

From the discussion in the previous chapter, we see that the Galerkin method for a linear boundary value problem reduces to the solution of a linear system. In solving the linear system, properties of the coefficient matrix A play an essential role. For example, if the condition number of A is too big, then from a practical perspective, it is impossible to find directly an accurate solution of the system (see [11]). Another important issue is the sparsity of the matrix A. The matrix A is said to be *sparse*, if most of its entries are zero; otherwise the matrix is said to be *dense*. Sparseness of the matrix can be utilized for two purposes. First, the stiffness matrix is less costly to form (observing that the computation of each entry of the matrix involves a domain integration and sometimes a boundary integration as well). Second, if the coefficient matrix is sparse, then the linear system can usually be solved more efficiently (often using an iterative method to solve the linear system). We have seen from Example 8.1.1 that the Galerkin method usually produces a dense stiffness matrix. In order to get a sparse matrix, we have to be careful in choosing the finite-dimensional approximation spaces and their basis functions. More precisely, the requirements are that, firstly, the support of a basis function should be as small as possible, and secondly, the number of basis functions whose supports intersect with the interior of the support of an arbitrary basis function should be as small as possible. This consideration gives rise to the idea of the finite element method, where we use piecewise (images of) smooth functions (usually polynomials) for approximations.

Loosely speaking, the finite element method is a Galerkin method with the use of piecewise (images of) polynomials.

For a linear elliptic boundary value problem defined on a Lipschitz domain Ω, the weak formulation is of the form

$$u \in V, \quad a(u, v) = \ell(v) \quad \forall v \in V, \tag{9.0.1}$$

where V is a Sobolev space on Ω, $a(\cdot, \cdot) : V \times V \to \mathbb{R}$ is a bilinear form, and $\ell \in V'$. For a second-order differential equation problem, $V = H^1(\Omega)$ if the given boundary condition is natural (i.e., if the condition involves first order derivatives), and $V = H_0^1(\Omega)$ if the homogeneous Dirichlet boundary condition is specified over the whole boundary. As discussed in Chapter 7, a problem with a non-homogeneous Dirichlet boundary condition on a part of the boundary, $\Gamma_D \subseteq \partial\Omega$, can be reduced to one with the homogeneous Dirichlet boundary condition on Γ_D after a change of dependent variables. In this case, then, the space V equals $H^1_{\Gamma_D}(\Omega)$. The form $a(\cdot, \cdot)$ is assumed to be bilinear, continuous, and V-elliptic, while ℓ is a given linear continuous form on V.

When the subspace V_N consists of piecewise polynomials (or more precisely, piecewise images of polynomials) associated with a partition (or mesh) of the domain Ω, the Galerkin method discussed in Section 8.1 becomes the celebrated finite element method. Convergence of the finite element method may be achieved by progressively refining the mesh, or by increasing the polynomial degree, or by doing both simultaneously. Then we get the h-version, p-version, or h-p-version of the finite element method. It is customary to use h as the parameter for the meshsize and p as the parameter for the polynomial degree. Efficient selection among the three versions of the method depends on the *a priori* knowledge on the regularity of the exact solution of the problem. Roughly speaking, over a region where the solution is smooth, high-degree polynomials with large elements are more efficient, while in a region where the solution has singularities, low-order elements together with a locally refined mesh should be used. Here, we will focus on the h-version finite element method. Detailed discussion of the p-version method can be found in the reference [155]. Conventionally, for the h-version finite element method, we write V_h instead of V_N to denote the finite element space. Thus, with a finite element space V_h chosen, the finite element method is

$$u_h \in V_h, \quad a(u_h, v_h) = \ell(v_h) \quad \forall v_h \in V_h. \tag{9.0.2}$$

All the discussions made on the Galerkin method in Section 8.1 are valid for the finite element method. In particular, we still have Céa's inequality, and the problem of estimating the finite element solution error is reduced to one of estimating the approximation error

$$\|u - u_h\|_V \le c\,\|u - \Pi_h u\|_V,$$

where $\Pi_h u$ is a finite element interpolant of u. We will study in some detail affine families of finite elements and derive some order error estimates for finite element interpolants.

9.1   One-dimensional examples

To have some idea of the finite element method, in this section we examine some examples on solving one-dimensional boundary value problems. These examples exhibit various aspects of the finite element method in the simple context of one-dimensional problems.

9.1.1   Linear elements for a second-order problem

Let us consider a finite element method to solve the boundary value problem

$$\begin{cases} -u'' + u = f & \text{in } (0,1), \\ u(0) = 0, \ u'(1) = b, \end{cases} \tag{9.1.1}$$

where $f \in L^2(0,1)$ and $b \in \mathbb{R}$ are given. Let

$$V = H_{(0}(0,1) = \{v \in H^1(0,1) \mid v(0) = 0\},$$

a subspace of $H^1(0,1)$. The weak formulation of the problem is

$$u \in V, \quad \int_0^1 (u'v' + u\,v)\,dx = \int_0^1 f\,v\,dx + b\,v(1) \quad \forall\,v \in V. \tag{9.1.2}$$

Applying the Lax-Milgram lemma, we see that the problem (9.1.2) has a unique solution.

Let us develop a finite element method for the problem. For a natural number N, we partition the domain $I = [0,1]$ into N parts: $0 = x_0 < x_1 < \cdots < x_N = 1$. The points x_i, $0 \le i \le N$, are called the nodes, and the subintervals $I_i = [x_{i-1}, x_i]$, $1 \le i \le N$, are called the elements. In this example, we have a Dirichlet condition at the node x_0. Denote $h_i = x_i - x_{i-1}$, and $h = \max_{1 \le i \le N} h_i$. The value h is called the meshsize or mesh parameter. We use piecewise linear functions for the approximation; i.e., we choose

$$V_h = \{v_h \in V \mid v_h|_{I_i} \in \mathcal{P}_1(I_i),\ 1 \le i \le N\}.$$

From the discussion in Chapter 6, we know that for a piecewisely smooth function v_h, $v_h \in H^1(I)$ if and only if $v_h \in C(\overline{I})$. Thus a more transparent

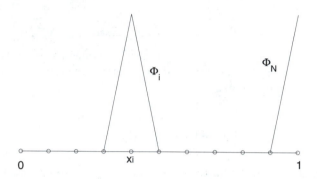

Figure 9.1. Piecewise linear basis functions

yet equivalent definition of the finite element space is

$$V_h = \{v_h \in C(\overline{I}) \mid v_h|_{I_i} \in \mathcal{P}_1(I_i),\ 1 \le i \le N,\ v_h(0) = 0\}.$$

For the basis functions, we introduce hat functions associated with the nodes $x_1, \ldots, x_N$. For $i = 1, \ldots, N - 1$, let

$$\phi_i(x) = \begin{cases} (x - x_{i-1})/h_i, & x_{i-1} \le x \le x_i, \\ (x_{i+1} - x)/h_{i+1}, & x_i \le x \le x_{i+1}, \\ 0, & \text{otherwise}, \end{cases}$$

and for $i = N$,

$$\phi_N(x) = \begin{cases} (x - x_{N-1})/h_N, & x_{N-1} \le x \le x_N, \\ 0, & \text{otherwise}. \end{cases}$$

These functions are continuous and piecewise linear (cf. Figure 9.1). It is easy to see they are linearly independent. The first-order weak derivatives of the basis functions are piecewise constants. Indeed, for $i = 1, \ldots, N - 1$,

$$\phi_i'(x) = \begin{cases} 1/h_i, & x_{i-1} \le x \le x_i, \\ -1/h_{i+1}, & x_i \le x \le x_{i+1}, \\ 0, & \text{otherwise}, \end{cases}$$

and for $i = N$,

$$\phi_N'(x) = \begin{cases} 1/h_N, & x_{N-1} \le x \le x_N, \\ 0, & \text{otherwise}. \end{cases}$$

Notice that the weak derivatives are defined only almost everywhere.

Then we see that the finite element space is

$$V_h = \operatorname{span} \{\phi_i \mid 1 \le i \le N\};$$

i.e., any function in V_h is a linear combination of the hat functions $\{\phi_i\}_{i=1}^N$. The corresponding finite element method is

$$u_h \in V_h, \quad \int_0^1 (u_h' v_h' + u_h\, v_h)\, dx = \int_0^1 f\, v_h\, dx + b\, v_h(1) \quad \forall\, v_h \in V_h,$$

$$(9.1.3)$$

which admits a unique solution by another application of the Lax-Milgram lemma. Writing

$$u_h = \sum_{j=1}^N u_j \phi_j,$$

we see that the finite element method (9.1.3) is equivalent to the following linear system for the unknowns $u_1, \dots, u_N$,

$$\sum_{j=1}^N u_j \int_0^1 (\phi_i' \phi_j' + \phi_i \phi_j)\, dx = \int_0^1 f\, \phi_i\, dx + b\, \phi_i(1), \quad 1 \le i \le N. \quad (9.1.4)$$

Let us find the coefficient matrix of the system (9.1.4) in the case of a uniform partition, i.e., $h_1 = \cdots = h_N = h$. The following formulas are useful for this purpose.

$$\int_0^1 \phi_i' \phi_{i-1}' dx = -\frac{1}{h}, \quad 2 \le i \le N,$$

$$\int_0^1 (\phi_i')^2 dx = \frac{2}{h}, \quad 1 \le i \le N-1,$$

$$\int_0^1 \phi_i \phi_{i-1} dx = \frac{h}{6}, \quad 2 \le i \le N,$$

$$\int_0^1 (\phi_i)^2 dx = \frac{2\,h}{3}, \quad 1 \le i \le N-1,$$

$$\int_0^1 (\phi_N')^2 dx = \frac{1}{h},$$

$$\int_0^1 (\phi_N)^2 dx = \frac{h}{3}.$$

We see that in matrix/vector notation, in the case of a uniform partition, the finite element system (9.1.4) can be written as

$$A\,\mathbf{u} = \mathbf{b},$$

where

$$\mathbf{u} = (u_1, \dots, u_N)^T$$

is the unknown vector,

$$A = \begin{pmatrix} \dfrac{2h}{3} + \dfrac{2}{h} & \dfrac{h}{6} - \dfrac{1}{h} & & & & \\[2ex] \dfrac{h}{6} - \dfrac{1}{h} & \dfrac{2h}{3} + \dfrac{2}{h} & \dfrac{h}{6} - \dfrac{1}{h} & & & \\[2ex] & \ddots & \ddots & \ddots & & \\[2ex] & & \dfrac{h}{6} - \dfrac{1}{h} & \dfrac{2h}{3} + \dfrac{2}{h} & \dfrac{h}{6} - \dfrac{1}{h} \\[2ex] & & & \dfrac{h}{6} - \dfrac{1}{h} & \dfrac{h}{3} + \dfrac{1}{h} \end{pmatrix}$$

is the stiffness matrix, and

$$\mathbf{b} = \left(\int_0^1 f\,\phi_1 dx, \ldots, \int_0^1 f\,\phi_{N-1} dx, \int_0^1 f\,\phi_N dx + b \right)^T$$

is the load vector. The matrix A is sparse, owing to the fact that the supports of the basis functions are small. The finite element system can be solved more efficiently with a sparse coefficient matrix. One distinguished feature of the finite element method is that the basis functions are constructed in such a way that their supports are as small as possible, so that the corresponding stiffness matrix is as sparse as possible.

9.1.2   High-order elements and the condensation technique

We still consider the finite element method for solving the boundary value problem (9.1.1). This time we use piecewise quadratic functions. So the finite element space is

$$V_h = \{v_h \in V \mid v_h|_{I_i} \text{ is quadratic}\}.$$

Equivalently,

$$V_h = \{v_h \in C(\overline{I}) \mid v_h|_{I_i} \text{ is quadratic, } v_h(0) = 0\}.$$

Let us introduce a basis for the space V_h. We denote the mid-points of the subintervals by $x_{i-1/2} = (x_{i-1} + x_i)/2$, $1 \le i \le N$. Associated with each node x_i, $1 \le i \le N - 1$, we define

$$\phi_i(x) = \begin{cases} 2(x - x_{i-1})(x - x_{i-1/2})/h_i^2, & x \in [x_{i-1}, x_i], \\ 2(x_{i+1} - x)(x_{i+1/2} - x)/h_{i+1}^2, & x \in [x_i, x_{i+1}], \\ 0, & \text{otherwise.} \end{cases}$$

Associated with x_N, we define

$$\phi_N(x) = \begin{cases} 2(x - x_{N-1})(x - x_{N-1/2})/h_N^2, & x \in [x_{N-1}, x_N], \\ 0, & \text{otherwise.} \end{cases}$$

We also need basis functions associated with the mid-points $x_{i-1/2}$, $1 \leq i \leq N$,

$$\psi_{i-1/2}(x) = \begin{cases} 4(x_i - x)(x - x_{i-1})/h_i^2, & x \in [x_{i-1}, x_i], \\ 0, & \text{otherwise.} \end{cases}$$

We notice that a mid-point basis function is non-zero only in one element. Now the finite element space can be represented as

$$V_h = \text{span} \{\phi_i, \psi_{i-1/2} \mid 1 \leq i \leq N\},$$

and we write

$$u_h = \sum_{j=1}^{N} u_j \phi_j + \sum_{j=1}^{N} u_{j-1/2} \psi_{j-1/2}.$$

The finite element system

$$\begin{cases} a(u_h, \phi_i) = \ell(\phi_i), & 1 \leq i \leq N, \\ a(u_h, \psi_{i-1/2}) = \ell(\psi_{i-1/2}), & 1 \leq i \leq N \end{cases}$$

can be written as, in the matrix/vector notation,

$$M_{11}\mathbf{u} + M_{12}\tilde{\mathbf{u}} = \mathbf{b}_1, \qquad (9.1.5)$$

$$M_{21}\mathbf{u} + D_{22}\tilde{\mathbf{u}} = \mathbf{b}_2. \qquad (9.1.6)$$

Here, $\mathbf{u} = (u_1, \ldots, u_N)^T$, $\tilde{\mathbf{u}} = (u_{1/2}, \ldots, u_{N-1/2})^T$, $M_{11} = (a(\phi_j, \phi_i))_{N \times N}$ is a tridiagonal matrix, $M_{12} = (a(\psi_{j-1/2}, \phi_i))_{N \times N}$ is a matrix with two diagonals, $M_{21} = M_{12}^T$, and $D_{22} = (a(\psi_{j-1/2}, \psi_{i-1/2}))_{N \times N}$ is a diagonal matrix with positive diagonal elements. We can eliminate $\tilde{\mathbf{u}}$ from the system (9.1.5)–(9.1.6) easily (both theoretically and practically). From (9.1.6), we have

$$\tilde{\mathbf{u}} = D_{22}^{-1}(\mathbf{b}_2 - M_{21}\mathbf{u}).$$

This relation is substituted into (9.1.5),

$$M\mathbf{u} = \mathbf{b}, \qquad (9.1.7)$$

where $M = M_{11} - M_{12}D_{22}^{-1}M_{21}$ is a tridiagonal matrix, $\mathbf{b} = \mathbf{b}_1 - M_{12}D_{22}^{-1}\mathbf{b}_2$. It can be shown that M is positive definite.

As a result we see that for the finite element solution with quadratic elements, we only need to solve a tridiagonal system of order N, just as in the case of using linear elements in Subsection 9.1.1. The procedure of eliminating $\tilde{\mathbf{u}}$ from (9.1.5)–(9.1.6) to form a smaller system (9.1.7) is called *condensation*. The key for the success of the condensation technique is that the supports of some basis functions are limited to a single element.

This condensation technique is especially useful in using high-order elements to solve higher-dimensional problems.

9.1.3   Reference element technique, non-conforming method

Here we introduce the *reference element technique* in a natural way. At the same time, we will comment on the use of the *non-conforming* finite element method.

Consider a clamped beam, initially occupying the region $[0, 1]$, which is subject to the action of a transversal force of density f. Denote u as the deflection of the beam. Then the boundary value problem is

$$\begin{cases} u^{(4)} = f & \text{in } (0,1), \\ u(0) = u'(0) = u(1) = u'(1) = 0. \end{cases} \tag{9.1.8}$$

The weak formulation of the problem is

$$u \in V, \quad \int_0^1 u'' v'' dx = \int_0^1 f v \, dx \quad \forall v \in V, \tag{9.1.9}$$

where $V = H_0^2(0,1)$. If we choose the finite element space V_h to be a subspace of V, then any function in V_h must be C^1 continuous. Suppose V_h consists of piecewise polynomials of degree less than or equal to p. The requirement that a finite element function be C^1 is equivalent to the C^1 continuity of the function across the interior nodal points $\{x_i\}_{i=1}^{N-1}$, which places $2(N-1)$ constraints. Additionally, the Dirichlet boundary conditions impose 4 constraints. Hence,

$$\dim (V_h) = (p+1) N - 2 (N-1) - 4 = (p-1) N - 2.$$

Now it is evident that the polynomial degree p must be at least 2. However, with $p = 2$, we cannot construct basis functions with small supports. Thus we should choose p to be at least 3. For $p = 3$, our finite element space is taken to be

$$V_h = \{v_h \in C^1(\bar{I}) \mid v_h|_{I_i} \in \mathcal{P}_3(I_i),\ 1 \le i \le N,$$
$$v_h(x) = v'_h(x) = 0 \text{ at } x = 0, 1\}.$$

It is then possible to construct basis functions with small supports using interpolation conditions of the function and its first derivative at the interior nodes $\{x_i\}_{i=1}^{N-1}$. More precisely, associated with each interior node x_i, there are two basis functions ϕ_i and ψ_i satisfying the interpolation conditions

$$\phi_i(x_j) = \delta_{ij},\ \phi'_i(x_j) = 0,$$
$$\psi_i(x_j) = 0,\ \psi'_i(x_j) = \delta_{ij}.$$

A more plausible approach to constructing the basis functions is to use the reference element technique. To this end, let us choose $I_0 = [0, 1]$ as the reference element. Then the mapping

$$F_i : I_0 \to I_i,\ F_i(\xi) = x_{i-1} + h_i \xi$$

is a bijection between I_0 and I_i. Over the reference element I_0, we construct cubic functions Φ_0, Φ_1, Ψ_0, and Ψ_1 satisfying the interpolation conditions

$$\Phi_0(0) = 1, \ \Phi_0(1) = 0, \ \Phi_0'(0) = 0, \ \Phi_0'(1) = 0,$$
$$\Phi_1(0) = 0, \ \Phi_1(1) = 1, \ \Phi_1'(0) = 0, \ \Phi_1'(1) = 0,$$
$$\Psi_0(0) = 0, \ \Psi_0(1) = 0, \ \Psi_0'(0) = 1, \ \Psi_0'(1) = 0,$$
$$\Psi_1(0) = 0, \ \Psi_1(1) = 0, \ \Psi_1'(0) = 0, \ \Psi_1'(1) = 1.$$

It is not difficult to find these functions,

$$\Phi_0(\xi) = (1 + 2\,\xi)\,(1 - \xi)^2,$$
$$\Phi_1(\xi) = (3 - 2\,\xi)\,\xi^2,$$
$$\Psi_0(\xi) = \xi\,(1 - \xi)^2,$$
$$\Psi_1(\xi) = -(1 - \xi)\,\xi^2.$$

These functions, defined on the reference element, are called *shape functions*. With the shape functions, it is an easy matter to construct the basis functions with the aid of the mapping functions $\{F_i\}_{i=1}^{N-1}$. We have

$$\phi_i(x) = \begin{cases} \Phi_1(F_i^{-1}(x)), \ x \in I_i, \\ \Phi_0(F_{i+1}^{-1}(x)), \ x \in I_{i+1}, \\ 0, \qquad\qquad \text{otherwise}, \end{cases}$$

and

$$\psi_i(x) = \begin{cases} h_i \Psi_1(F_i^{-1}(x)), \quad x \in I_i, \\ h_{i+1} \Psi_0(F_{i+1}^{-1}(x)), \ x \in I_{i+1}, \\ 0, \qquad\qquad\quad \text{otherwise}. \end{cases}$$

Once the basis functions are available, it is a routine work to form the finite element system. We emphasize that the computations of the stiffness matrix and the load are done on the reference element. For example, by definition,

$$a_{i-1,i} = \int_0^1 (\phi_{i-1})''(\phi_i)'' dx = \int_{I_i} (\phi_{i-1})''(\phi_i)'' dx;$$

using the mapping function F_i and the definition of the basis functions, we have

$$a_{i-1,i} = \int_{I_0} (\Phi_0)'' h_i^{-2} (\Phi_1)'' h_i^{-2}\, h_i d\xi$$
$$= \frac{1}{h_i^3} \int_{I_0} 6\,(2\,\xi - 1)\,6\,(1 - 2\,\xi)\, d\xi$$
$$= -\frac{12}{h_i^3}.$$

For higher-dimensional problems, the use of the reference element technique is essential for both theoretical error analysis and practical implementation of the finite element method. The computations of the stiffness matrix and the load vector involve a large number of integrals that usually cannot be computed analytically. With the reference element technique, all the integrals are done on a single region (the reference element), and therefore only numerical quadratures on the reference element are needed. Later, we will see how the reference element technique is used to derive error estimates for the finite element interpolations.

The discussion of the selection of the finite element space V_h also alerts us that locally high-degree polynomials are needed in order to have C^1 continuity for a piecewise polynomial. The polynomial degrees are even higher for C^1 elements in higher-dimensional problems. An impact of this is that the basis functions are more difficult to construct, and less efficient to use. An alternative approach overcoming this difficulty is to use *non-conforming elements*. For the sample problem (9.1.8), we may try to use continuous piecewise polynomials to approximate the space $H_0^2(0,1)$. As an example, we take

$$V_h = \{v_h \in C(\overline{I}) \mid v_h|_{I_i} \in \mathcal{P}_2(I_i),\ 1 \le i \le N,$$
$$v_h(x) = v_h'(x) = 0 \text{ at } x = 0, 1\}.$$

Notice that $V_h \not\subseteq V$. The finite element method is then defined as

$$u_h \in V_h, \quad \sum_{i=1}^{N} \int_{I_i} u_h'' v_h'' dx = \int_I f\, v_h\, dx, \quad \forall\, v_h \in V_h.$$

It is still possible to discuss the convergence of the non-conforming finite element solution. Some preliminary discussion of this topic was made in Example 8.3.2. See [35] for some detail.

In this chapter, we only discuss the finite element method for solving second-order boundary value problems, and so we will focus on conforming finite element methods.

Exercise 9.1.1 *In Subsection 9.1.1, we computed the stiffness matrix for the case of a uniform partition. Find the stiffness matrix when the partition is non-uniform.*

Exercise 9.1.2 *Use the fact that the coefficient matrix of the system (9.1.5) and (9.1.6) is symmetric, positive definite to show that the coefficient matrix of the system (9.1.7) is symmetric, positive definite.*

Exercise 9.1.3 *Show that in solving (9.1.8) with a conforming finite element method with piecewise polynomials of degree less than or equal to 2, it is impossible to construct basis functions with small supports.*

9.2   Basics of the finite element method

We have seen from the one-dimensional examples in the preceding section that there are some typical steps in a finite element solution of a boundary value problem. First we need a weak formulation of the boundary value problem; this topic was discussed in Chapter 7. Then we need a partition (or triangulation) of the domain into subdomains called elements. Associated with the partition, we define a finite element space. Further, we choose basis functions for the finite element space. The basis functions should have small supports so that the resulting stiffness matrix is sparse. With the basis functions defined, the finite element system can be formed. The reference element technique is used from time to time in this process. Once the finite element system is formed, we then need to solve the system; see some discussions in Section 4.2. More details can be found in ATKINSON [11, Chap. 8], GOLUB AND VAN LOAN [61], and STEWART [152].

9.2.1   Triangulation

A triangulation or a mesh is a partition $\mathcal{T}_h = \{K\}$ of the domain $\overline{\Omega}$ into a finite number of subsets K, called *elements*, with the following properties:

1. $\overline{\Omega} = \cup_{K \in \mathcal{T}_h} K$;

2. each K is closed with a nonempty interior $\overset{\circ}{K}$ and a Lipschitz continuous boundary;

3. for distinct $K_1, K_2 \in \mathcal{T}_h$, $\overset{\circ}{K}_1 \cap \overset{\circ}{K}_2 = \emptyset$.

From now on, we restrict our discussion to two-dimensional problems; most of the discussion can be extended to higher-dimensional problems straightforwardly. We will assume the domain Ω is a polygon so that it can be partitioned into straight-sided triangles and quadrilaterals. When Ω is a general domain with a curved boundary, it cannot be partitioned into straight-sided triangles and quadrilaterals, and usually curved-sided elements need to be used. The reader is referred to [35, 36] for some detailed discussion on the use of the finite element method in this case. We will emphasize the use of the reference element technique to estimate the error; for this reason, we need a particular structure on the finite elements, namely, we will consider only affine families of finite elements. In general, bilinear functions are needed for a bijective mapping between a four-sided reference element and a general quadrilateral. So a further restriction is we will mostly use triangular elements, for any triangle is affine equivalent to a fixed triangle (the reference triangle).

For a triangulation of a polygon into triangles (and quadrilaterals), we usually impose one more condition, called the *regularity condition*.

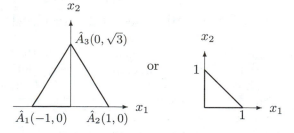

Figure 9.2. Reference triangular elements in $\mathbb{R}^2$

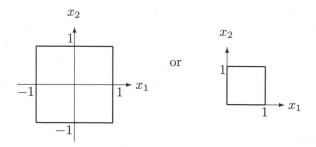

Figure 9.3. Reference rectangular elements in $\mathbb{R}^2$

4. For distinct $K_1, K_2 \in \mathcal{T}_h$, $K_1 \cap K_2$ is either empty, or a common vertex, or a common side of K_1 and K_2.

For a triangulation of a three-dimensional domain into tetrahedral, hexahedral, or pentahedral elements, the regularity condition requires that the intersection of two distinct elements is either empty, or a common vertex, or a common face of the two elements.

For convenience in practical implementation as well as in theoretical analysis, it is assumed that there exist a finite number of fixed Lipschitz domains, ambiguously represented by one symbol $\hat{K}$, such that for each element K, there is a smooth mapping function F_K with $K = F_K(\hat{K})$.

Example 9.2.1 *We will use finite elements over two-dimensional domains as examples. The reference element can be taken to be either an equilateral or right isosceles triangle for triangular elements, and squares with side length 1 or 2 for quadrilateral elements. Correspondingly, the mapping function F_K is linear if K is a triangle, and bilinear if K is a quadrilateral. The triangular and rectangular reference elements are shown in Figures 9.2 and 9.3 respectively. We will use the equilateral triangle and the square $[-1,1]^2$ as the reference elements.* ∎

For an arbitrary element K, we denote

$$h_K = \mathrm{diam}\,(K) = \max\left\{\|\mathbf{x} - \mathbf{y}\| \mid \mathbf{x}, \mathbf{y} \in K\right\}$$

and

$$\rho_K = \text{diameter of the largest sphere } S_K \text{ inscribed in } K.$$

When dealing with the reference element $\hat{K}$ we denote the corresponding quantities by $\hat{h}$ and $\hat{\rho}$. The quantity h_K describes the size of K, while the ratio h_K/ρ_K is an indication whether the element is flat.

9.2.2 Polynomial spaces on the reference elements

Function spaces over a general element will be constructed from those on the reference element. Thus we will first introduce a polynomial space $\hat{X}$ on $\hat{K}$. Although it is possible to choose any finite-dimensional function space as $\hat{X}$, the overwhelming choice for $\hat{X}$ for practical use is a polynomial space.

Example 9.2.2 *For definiteness, we choose the equilateral triangle $\hat{K}$ to be the reference triangle with the vertices $\hat{A}_1(-1,0)$, $\hat{A}_2(1,0)$, and $\hat{A}_3(0,\sqrt{3})$. We introduce three functions,*

$$\hat{\lambda}_1(\hat{\mathbf{x}}) = \frac{1}{2}\left(1 - \hat{x}_1 - \frac{\hat{x}_2}{\sqrt{3}}\right),$$

$$\hat{\lambda}_2(\hat{\mathbf{x}}) = \frac{1}{2}\left(1 + \hat{x}_1 - \frac{\hat{x}_2}{\sqrt{3}}\right),$$

$$\hat{\lambda}_3(\hat{\mathbf{x}}) = \frac{\hat{x}_2}{\sqrt{3}}.$$

These functions are linear and satisfy the relations

$$\hat{\lambda}_i(\hat{A}_j) = \delta_{ij}.$$

They are called the barycentric coordinates associated with the triangle $\hat{K}$. It is convenient to use the barycentric coordinates to represent polynomials. For example, any linear function $\hat{v} \in \mathcal{P}_1(\hat{K})$ is uniquely determined by its values at the vertices $\{\hat{A}_i\}_{i=1}^3$ and we have the representation

$$\hat{v}(\hat{\mathbf{x}}) = \sum_{i=1}^{3} \hat{v}(\hat{A}_i)\, \hat{\lambda}_i(\hat{\mathbf{x}}).$$

In this case, the vertices $\{\hat{A}_i\}_{i=1}^3$ are called the nodes, the function values $\{\hat{v}(\hat{A}_i)\}_{i=1}^3$ are called the parameters (used to determine the linear function), and $\{\hat{\lambda}_i\}_{i=1}^3$ are the basis functions on $\hat{K}$, called the shape functions.

A quadratic function has six coefficients and we need six interpolation conditions to determine it. For this, we introduce the side mid-points,

$$\hat{A}_{ij} = \frac{1}{2}(\hat{A}_i + \hat{A}_j), \quad 1 \le i < j \le 3.$$

Then any quadratic function $\hat{v} \in \mathcal{P}_2(\hat{K})$ is uniquely determined by its values at the vertices $\{\hat{A}_i\}_{i=1}^{3}$ and the side mid-points $\{\hat{A}_{ij}\}_{1 \leq i < j \leq 3}$. Indeed we have the representation formula

$$\hat{v}(\hat{\mathbf{x}}) = \sum_{i=1}^{3} \hat{v}(\hat{A}_i) \, \hat{\lambda}_i(\hat{\mathbf{x}}) \, (2 \, \hat{\lambda}_i(\hat{\mathbf{x}}) - 1) + \sum_{1 \leq i < j \leq 3} 4 \, \hat{v}(\hat{A}_{ij}) \, \hat{\lambda}_i(\hat{\mathbf{x}}) \, \hat{\lambda}_j(\hat{\mathbf{x}})$$

for $\hat{v} \in \mathcal{P}_2(\hat{K})$. This formula is derived from the observations that
(1) for each k, $1 \leq k \leq 3$, $\hat{\lambda}_k(\hat{\mathbf{x}}) \, (2 \, \hat{\lambda}_k(\hat{\mathbf{x}}) - 1)$ is a quadratic function, takes on the value 1 at $\hat{A}_k$, and the value 0 at the other vertices and the side mid-points;
(2) for $1 \leq i < j \leq 3$, $4 \, \hat{\lambda}_i(\hat{\mathbf{x}}) \, \hat{\lambda}_j(\hat{\mathbf{x}})$ is a quadratic function, takes on the value 1 at $\hat{A}_{ij}$, and the value 0 at the other side mid-points and the vertices.

In this case, the vertices and the side mid-points are called the nodes, the function values at the nodes are called the parameters (used to determine the quadratic function).

In the above example, the parameters are function values at the nodes. The corresponding finite elements to be constructed later are called Lagrange finite elements. Lagrange finite elements are the natural choice in solving second-order boundary value problems, where weak formulations only need the use of weak derivatives of the first order, and hence only the continuity of finite element functions across interelement boundaries is requires, as we will see later. It is also possible to use other types of parameters to determine polynomials. For example, we may choose some parameters to be derivatives of the function at some nodes; in this case, we will get Hermite finite elements. We can construct Hermite finite elements that are globally continuously differentiable; such Hermite finite elements can be used to solve fourth-order boundary value problems, as was discussed in Example 9.1.3 in the context of a one-dimensional problem. For some applications, it may be advantageous to use average values of the function along the sides. Different selections of the parameters lead to different basis functions, and thus lead to different finite element system. Here we will focus on the discussion of Lagrange finite elements. The discussion of the above example can be extended to the case of higher degree polynomials and to domains of any (finite) dimension. A reader interested in a more complete discussion of general finite elements can consult the references [35, 36].

In general, let $\hat{X}$ be a polynomial space over $\hat{K}$, $\dim \hat{X} = I$. We choose a set of nodal points $\{\hat{\mathbf{x}}_i\}_{i=1}^{I}$ in $\hat{K}$, such that any function $\hat{v} \in \hat{X}$ is uniquely determined by its values at the nodes $\{\hat{\mathbf{x}}_i\}_{i=1}^{I}$, and we have the following formula

$$\hat{v}(\hat{\mathbf{x}}) = \sum_{i=1}^{I} \hat{v}(\hat{\mathbf{x}}_i) \, \hat{\phi}_i(\hat{\mathbf{x}}).$$

The functions $\{\hat{\phi}_i\}_{i=1}^I$ form a basis for the space $\hat{X}$ with the property

$$\hat{\phi}_i(\hat{\mathbf{x}}_j) = \delta_{ij}.$$

9.2.3   Affine-equivalent finite elements

We will then define function spaces over a general element. The domain $\overline{\Omega}$, being polygonal, is partitioned into straight-sided triangles and quadrilaterals. We will only consider *affine-equivalent* families of finite elements. In other words, there exists one or several reference elements, $\hat{K}$, such that each element K is the image of $\hat{K}$ under an invertible affine mapping $F_K : \hat{K} \to K$ of the form

$$F_K(\hat{\mathbf{x}}) = \mathbf{T}_K \hat{\mathbf{x}} + \mathbf{b}_K. \tag{9.2.1}$$

The mapping F_K is a bijection between $\hat{K}$ and K, $\mathbf{T}_K$ is an invertible 2×2 matrix and $\mathbf{b}_K$ is a translation vector.

Over each element K, we define a finite-dimensional function space X_K by the formula

$$X_K = \hat{X} \circ F_K^{-1} \equiv \{v \mid v = \hat{v} \circ F_K^{-1}, \ \hat{v} \in \hat{X}\}. \tag{9.2.2}$$

An immediate consequence of this definition is that if $\hat{X}$ is a polynomial space of certain degree, then X_K is a polynomial space of the same degree. In the future, for any function v defined on K, we will use $\hat{v}$ to denote the corresponding function defined on $\hat{K}$ through $\hat{v} = v \circ F_K$. Conversely, for any function $\hat{v}$ on $\hat{K}$, we let v be the function on K defined by $v = \hat{v} \circ F_K^{-1}$. Thus we have the relation

$$v(\mathbf{x}) = \hat{v}(\hat{\mathbf{x}}) \quad \forall \mathbf{x} \in K, \ \hat{\mathbf{x}} \in \hat{K}, \text{ with } \mathbf{x} = F_K(\hat{\mathbf{x}}).$$

Using the nodal points $\hat{\mathbf{x}}_i$, $1 \le i \le I$, of $\hat{K}$, we introduce the nodal points $\mathbf{x}_i^K$, $1 \le i \le I$, of K defined by

$$\mathbf{x}_i^K = F_K(\hat{\mathbf{x}}_i), \quad i = 1, \dots, I. \tag{9.2.3}$$

Recall that $\{\hat{\phi}_i\}_{i=1}^I$ are the basis functions of the space $\hat{X}$ associated with the nodal points $\{\hat{\mathbf{x}}_i\}_{i=1}^I$ with the property that

$$\hat{\phi}_i(\hat{\mathbf{x}}_j) = \delta_{ij}.$$

We define

$$\phi_i^K = \hat{\phi}_i \circ F_K^{-1}, \quad i = 1, \dots, I.$$

Then the functions $\{\phi_i^K\}_{i=1}^I$ have the property that

$$\phi_i^K(\mathbf{x}_j^K) = \delta_{ij}.$$

Hence, $\{\phi_i^K\}_{i=1}^I$ form a set of local polynomial basis functions on K.

We now present a result on the affine transformation (9.2.1), which will be used in estimating finite element interpolation errors. The matrix norm is the spectral norm, i.e., the operator matrix norm induced by the Euclidean vector norm.

Lemma 9.2.3 *For the affine map* $F_K : \hat{K} \to K$ *defined by* (9.2.1), *we have the estimates*

$$\|\mathbf{T}_K\| \le \frac{h_K}{\hat{\rho}} \quad and \quad \|\mathbf{T}_K^{-1}\| \le \frac{\hat{h}}{\rho_K}.$$

Proof. By definition,

$$\|\mathbf{T}_K\| = \sup \left\{ \frac{\|\mathbf{T}_K \hat{\mathbf{x}}\|}{\|\hat{\mathbf{x}}\|} \mid \hat{\mathbf{x}} \ne \mathbf{0} \right\}.$$

Let us rewrite it in the equivalent form

$$\|\mathbf{T}_K\| = \hat{\rho}^{-1} \sup \{ \|\mathbf{T}_K \hat{\mathbf{z}}\| \mid \|\hat{\mathbf{z}}\| = \hat{\rho} \}$$

by taking $\hat{\mathbf{z}} = \hat{\rho}\hat{\mathbf{x}}/\|\hat{\mathbf{x}}\|$. Now for any $\hat{\mathbf{z}}$ with $\|\hat{\mathbf{z}}\| = \hat{\rho}$, pick up any two vectors $\hat{\mathbf{x}}$ and $\hat{\mathbf{y}}$ that lie on the largest sphere $\hat{S}$ of diameter $\hat{\rho}$, which is inscribed in $\hat{K}$, such that $\hat{\mathbf{z}} = \hat{\mathbf{x}} - \hat{\mathbf{y}}$. Then

$$\begin{aligned}
\|\mathbf{T}_K\| &= \hat{\rho}^{-1} \sup \left\{ \|\mathbf{T}_K(\hat{\mathbf{x}} - \hat{\mathbf{y}})\| \mid \hat{\mathbf{x}}, \hat{\mathbf{y}} \in \hat{S} \right\} \\
&= \hat{\rho}^{-1} \sup \left\{ \|(\mathbf{T}_K \hat{\mathbf{x}} + \mathbf{b}_K) - (\mathbf{T}_K \hat{\mathbf{y}} + \mathbf{b}_K)\| \mid \hat{\mathbf{x}}, \hat{\mathbf{y}} \in \hat{S} \right\} \\
&\le \hat{\rho}^{-1} \sup \{ \|\mathbf{x} - \mathbf{y}\| \mid \mathbf{x}, \mathbf{y} \in K \} \\
&\le h_K/\hat{\rho}.
\end{aligned}$$

The second inequality follows from the first one by interchanging the roles played by K and $\hat{K}$. ∎

9.2.4 Finite element spaces

A global finite element function v_h is defined piecewise by the formula

$$v_h|_K \in X_K \quad \forall K \in \mathcal{T}_h.$$

A natural question is whether such finite element functions can be used to solve a boundary value problem. For a linear second-order elliptic boundary value problem, the space V is a subspace of $H^1(\Omega)$. Thus we need to check whether $v_h \in H^1(\Omega)$ holds. Since the restriction of v_h on each element K is a smooth function, a necessary and sufficient condition for $v_h \in H^1(\Omega)$ is $v_h \in C(\overline{\Omega})$ (cf. Chapter 6). We then define a finite element space corresponding to the triangulation $\mathcal{T}_h$,

$$X_h = \{ v_h \in C(\overline{\Omega}) \mid v_h|_K \in X_K \ \forall K \in \mathcal{T}_h \}.$$

We observe that if $\hat{X}$ consists of polynomials, then a function from the space X_h is a piecewise image of polynomials. In our special case of an affine family of finite elements, F_K is an affine mapping, and $v_h|_K$ is a polynomial.

For more general mapping functions F_K (e.g., bilinear mapping functions used for quadrilateral elements), $v_h|_K$ is in general not a polynomial.

Then for a second-order boundary value problem with Neumann boundary condition, we use $V_h = X_h$ as the finite element space; and for a second-order boundary value problem with the homogeneous Dirichlet condition, the finite element space is chosen to be

$$V_h = \{v_h \in X_h \mid v_h = 0 \text{ on } \Gamma\},$$

which is a subspace of $H_0^1(\Omega)$.

We remark that the condition $v_h \in C(\overline{\Omega})$ is guaranteed if v_h is continuous across any interelement boundary; this requirement is satisfied if for any element K and any of its side γ, the restriction of $v_h|_K$ on γ is completely determined by its values at the nodes on γ (cf. Example 9.2.4 below).

For an affine family of finite elements, the mesh is completely described by the reference element $\hat{K}$ and the family of affine mappings $\{F_K \mid K \in \mathcal{T}_h\}$. The finite element space X_h is completely described by the space $\hat{X}$ on $\hat{K}$ and the family of affine mappings.

Example 9.2.4 *Let $\Omega \subseteq \mathbb{R}^2$ be a polygon, and let $\mathcal{T}_h$ be a partition of Ω into triangles. For a general triangle $K \in \mathcal{T}_h$, we let $\mathbf{a}_j = (a_{1j}, a_{2j})^T$, $1 \leq j \leq 3$, denote its three vertices. We define the barycentric coordinates $\lambda_j(\mathbf{x})$, $1 \leq j \leq 3$, associated with the vertices $\{\mathbf{a}_j\}_{j=1}^3$ to be the affine functions satisfying the relations*

$$\lambda_j(\mathbf{a}_i) = \delta_{ij}, \quad 1 \leq i, j \leq 3. \tag{9.2.4}$$

Then by the uniqueness of linear interpolation, we have

$$\mathbf{x} = \sum_{i=1}^3 \lambda_i(\mathbf{x}) \, \mathbf{a}_i. \tag{9.2.5}$$

Also, since $\sum_{i=1}^3 \lambda_i(\mathbf{a}_j) = 1$ for $1 \leq j \leq 3$, we have

$$\sum_{i=1}^3 \lambda_i(\mathbf{x}) = 1. \tag{9.2.6}$$

The equations (9.2.5) and (9.2.6) constitute three equations for the three unknowns $\{\lambda_i\}_{i=1}^3$,

$$\begin{pmatrix} a_{11} & a_{12} & a_{13} \\ a_{21} & a_{22} & a_{23} \\ 1 & 1 & 1 \end{pmatrix} \begin{pmatrix} \lambda_1 \\ \lambda_2 \\ \lambda_3 \end{pmatrix} = \begin{pmatrix} x_1 \\ x_2 \\ 1 \end{pmatrix}. \tag{9.2.7}$$

Since the triangle K is non-degenerate (i.e., the interior of K is non-empty), the system (9.2.7) has a nonsingular coefficient matrix and hence uniquely determines the barycentric coordinates $\{\lambda_i\}_{i=1}^3$. From the *uniqueness of the polynomial interpolation, it is easy to see that*

$$\lambda_i = \hat{\lambda}_i \circ F_K^{-1}, \quad 1 \le i \le 3.$$

Here $\{\hat{\lambda}_i\}_{i=1}^3$ are the barycentric coordinates on $\hat{K}$.
  Now we define a linear element space,

$$X_h = \{v_h \in H^1(\Omega) \mid v_h|_K \text{ is linear, } K \in \mathcal{T}_h\}.$$

The space can be equivalently defined as

$$X_h = \{v_h \in C(\overline{\Omega}) \mid v_h|_K \text{ is linear, } K \in \mathcal{T}_h\}.$$

On each element K, $v_h|_K$ is a linear function. If we use the function values at the vertices as the defining parameters, then we have the expression

$$v_h|_K = \sum_{i=1}^3 v_h(\mathbf{a}_i)\,\lambda_i, \quad K \in \mathcal{T}_h. \tag{9.2.8}$$

The local finite element function $v_h|_K$ can be obtained from a linear function defined on $\hat{K}$. For this, we let

$$\hat{v}(\hat{\mathbf{x}}) = \sum_{i=1}^3 v_h(\mathbf{a}_i)\,\hat{\lambda}_i(\hat{\mathbf{x}}).$$

Then it is easily seen that $v_h|_K = \hat{v} \circ F_K^{-1}$.
  The piecewise linear function v_h defined by (9.2.8) is globally continuous. To see this we only need to prove the continuity of v_h across $\gamma = K_1 \cap K_2$, the common side of two neighboring elements K_1 and K_2. Let us denote $v_1 = v_h|_{K_1}$ and $v_2 = v_h|_{K_2}$, and consider the difference $w = v_1 - v_2$. On γ, w is a linear function of one variable (the tangential variable along γ), and vanishes at the two end points of γ. Hence, $w = 0$, i.e. $v_1 = v_2$ along γ.

  Summarizing, if we use the function values at the vertices to define the function on each element K, then X_h is an affine-equivalent finite element space consisting of continuous piecewise linear functions. In this case, the vertices are the nodes of the finite element space, the nodal function values are called the parameters. On each element, we have the representation formula (9.2.8).

9.2.5   Interpolation

We first introduce an interpolation operator $\hat{\Pi}$ for continuous functions on $\hat{K}$. Recall that $\{\hat{\mathbf{x}}_i\}_{i=1}^I$ are the nodal points while $\{\hat{\phi}_i\}_{i=1}^I$ are the associated basis functions of the polynomial space $\hat{X}$ in the sense that $\hat{\phi}_i(\hat{\mathbf{x}}_j) = \delta_{ij}$ is valid. We define

$$\hat{\Pi} : C(\hat{K}) \to \hat{X}, \quad \hat{\Pi}\hat{v} = \sum_{i=1}^I \hat{v}(\hat{\mathbf{x}}_i)\hat{\phi}_i. \tag{9.2.9}$$

Evidently, $\hat{\Pi}\hat{v} \in \hat{X}$ is uniquely determined by the interpolation conditions

$$\hat{\Pi}\hat{v}(\hat{\mathbf{x}}_i) = \hat{v}(\hat{\mathbf{x}}_i), \quad i = 1, \dots, I.$$

On any element K, we define similarly the interpolation operator Π_K by

$$\Pi_K : C(K) \to X_K, \quad \Pi_K v = \sum_{i=1}^{I} v(\mathbf{x}_i^K)\phi_i^K. \tag{9.2.10}$$

We see that $\Pi_K v \in X_K$ is uniquely determined by the interpolation conditions

$$\Pi_K v(\mathbf{x}_i^K) = v(\mathbf{x}_i^K), \quad i = 1, \dots, I.$$

The following result explores the relation between the two interpolation operators. The result is of fundamental importance in error analysis of finite element interpolation.

Theorem 9.2.5 *For the two interpolation operators $\hat{\Pi}$ and Π_K introduced above, we have $\hat{\Pi}(\hat{v}) = (\Pi_K v) \circ F_K^{-1}$, i.e., $\hat{\Pi}\hat{v} = \widehat{\Pi_K v}$.*

Proof. From the definition (9.2.10), we have

$$\Pi_K v = \sum_{i=1}^{I} v(\mathbf{x}_i^K)\phi_i^K = \sum_{i=1}^{I} \hat{v}(\hat{\mathbf{x}}_i)\phi_i^K.$$

Since $\phi_i^K \circ F_K^{-1} = \hat{\phi}_i$, we obtain

$$(\Pi_K v) \circ F_K^{-1} = \sum_{i=1}^{I} \hat{v}(\hat{\mathbf{x}}_i)\hat{\phi}_i = \hat{\Pi}\hat{v}.$$

∎

Example 9.2.6 *Let us consider a planar polygonal domain $\overline{\Omega}$ partitioned into triangles $\{K\}$. For a generic element K, again use $\mathbf{a}_1$, $\mathbf{a}_2$, and $\mathbf{a}_3$ to denote its vertices. Then with linear elements, for a continuous function v defined on K, its linear interpolant is*

$$\Pi_K v(\mathbf{x}) = \sum_{i=1}^{3} v(\mathbf{a}_i)\,\lambda_i(\mathbf{x}), \quad \mathbf{x} \in K.$$

For the function $\hat{v} = v \circ F_K^{-1}$ defined on the reference element $\hat{K}$, its linear interpolant is

$$\hat{\Pi}\hat{v}(\hat{\mathbf{x}}) = \sum_{i=1}^{3} \hat{v}(\hat{\mathbf{a}}_i)\,\hat{\lambda}_i(\hat{\mathbf{x}}), \quad \hat{\mathbf{x}} \in \hat{K}.$$

Here, $\hat{\mathbf{a}}_i = F_K^{-1}(\mathbf{a}_i)$, $1 \le i \le 3$, are the vertices of $\hat{K}$. Since $v(\mathbf{a}_i) = \hat{v}(\hat{\mathbf{a}}_i)$ by definition and $\lambda_i(\mathbf{x}) = \hat{\lambda}_i(\hat{\mathbf{x}})$ for $\mathbf{x} = F_K(\hat{\mathbf{x}})$, obviously the relation $\hat{\Pi}\hat{v} = \widehat{\Pi_K v}$ holds.

By using the mid-points of the sides, we can give a similar discussion of quadratic elements.

Exercise 9.2.1 *Over the reference triangle $\hat{K}$, define a set of nodes and parameters for cubic functions. Represent a cubic function in terms of the parameters.*

Exercise 9.2.2 *Assume $\overline{\Omega}$ is the union of some rectangles whose sides parallel the coordinate axes. We partition $\overline{\Omega}$ into rectangular elements with sides parallel to the coordinate axes. In this case, the reference element is a square $\hat{K}$, say, the square centered at the origin with sides of length 2, as shown in Figure 9.3. The polynomial space over $\hat{K}$ is usually taken to be*

$$\mathcal{Q}_{k,l}(\hat{K}) = \left\{ v(\hat{\mathbf{x}}) = \sum_{i \leq k} \sum_{j \leq l} a_{ij} \hat{x}_1^i \hat{x}_2^j : a_{ij} \in \mathbb{R}, \ \hat{\mathbf{x}} \in \hat{K} \right\}$$

for non-negative integers k and l. For $k = l = 1$, we get the bilinear functions. Define a set of nodes and parameters over $\mathcal{Q}_{1,1}(\hat{K})$. Represent a bilinear function in terms of the parameters.

9.3   Error estimates of finite element interpolations

In this section, we present some estimates for the finite element interpolation error, which will be used in the next section to bound the error of the finite element solution for a linear elliptic boundary value problem, through the application of Céa's inequality. The interpolation error estimates are derived through the use of the reference element technique; i.e., error estimates are first derived on the reference element, which are then translated to a general finite element. The results discussed in this section can be extended to the case of a general d-dimensional domain. Definitions of triangulation and finite elements in d-dimensional case are similar to those for the two-dimensional case, cf., e.g., [35, 36].

9.3.1   Interpolation error estimates on the reference element

We first derive interpolation error estimates over the reference element.

Theorem 9.3.1 *Let k and m be non-negative integers with $k > 0$, $k+1 \geq m$, and $\mathcal{P}_k(\hat{K}) \subseteq \hat{X}$. Let $\hat{\Pi}$ be the operators defined in (9.2.9). Then there exists a constant c such that*

$$|\hat{v} - \hat{\Pi}\hat{v}|_{m,\hat{K}} \leq c \, |\hat{v}|_{k+1,\hat{K}} \quad \forall \hat{v} \in H^{k+1}(\hat{K}). \tag{9.3.1}$$

Proof. Notice that $k > 0$ implies $H^{k+1}(\hat{K}) \hookrightarrow C(\hat{K})$, so $\hat{v} \in H^{k+1}(\hat{K})$ is continuous and $\hat{\Pi}\hat{v}$ is well-defined. From

$$\|\hat{\Pi}\hat{v}\|_{m,\hat{K}} \le \sum_{i=1}^{I} |\hat{v}(\hat{\mathbf{x}}_i)| \, \|\hat{\phi}_i\|_{m,\hat{K}} \le c \, \|\hat{v}\|_{C(\hat{K})} \le c \, \|\hat{v}\|_{k+1,\hat{K}},$$

we see that $\hat{\Pi}$ is a bounded operator from $H^{k+1}(\hat{K})$ to $H^m(\hat{K})$. By the assumption on the space $\hat{X}$, we have

$$\hat{\Pi}\hat{v} = \hat{v} \quad \forall \, \hat{v} \in \mathcal{P}_k(\hat{K}). \tag{9.3.2}$$

Using (9.3.2), we then have, for all $\hat{v} \in H^{k+1}(\hat{K})$ and all $\hat{p} \in \mathcal{P}_k(\hat{K})$,

$$\begin{aligned}
|\hat{v} - \hat{\Pi}\hat{v}|_{m,\hat{K}} &\le \|\hat{v} - \hat{\Pi}\hat{v}\|_{m,\hat{K}} = \|\hat{v} - \hat{\Pi}\hat{v} + \hat{p} - \hat{\Pi}\hat{p}\|_{m,\hat{K}} \\
&\le \|(\hat{v} + \hat{p}) - \hat{\Pi}(\hat{v} + \hat{p})\|_{m,\hat{K}} \\
&\le \|\hat{v} + \hat{p}\|_{m,\hat{K}} + \|\hat{\Pi}(\hat{v} + \hat{p})\|_{m,\hat{K}} \\
&\le c \, \|\hat{v} + \hat{p}\|_{k+1,\hat{K}}.
\end{aligned}$$

Since $\hat{p} \in \mathcal{P}_k(\hat{K})$ is arbitrary, we have

$$|\hat{v} - \hat{\Pi}\hat{v}|_{m,\hat{K}} \le c \inf_{\hat{p} \in \mathcal{P}_k(\hat{K})} \|\hat{v} + \hat{p}\|_{k+1,\hat{K}}.$$

By an application of Corollary 6.3.18, we get the estimate (9.3.1). ∎

In Theorem 9.3.1, the assumption $k > 0$ is made to warrant the continuity of an $H^{k+1}(\hat{K})$ function. In the d-dimensional case, this assumption is replaced by $k + 1 > d/2$. The property (9.3.2) is called a *polynomial invariance property* of the finite element interpolation operator.

9.3.2   *Local interpolation error estimates*

We now consider the finite element interpolation error over each element K. As in Theorem 9.3.1, we assume $k > 0$; this assumption ensures the property $H^{k+1}(K) \hookrightarrow C(K)$, and so for $v \in H^{k+1}(K)$, pointwise values $v(\mathbf{x})$ are meaningful. Let the projection operator $\Pi_K : H^{k+1}(K) \to X_K \subseteq H^m(K)$ be defined by (9.2.10).

To translate the result of Theorem 9.3.1 from the reference element $\hat{K}$ to the element K, we need to discuss the relations between Sobolev norms over the reference element and a general element.

Theorem 9.3.2 *Assume* $\mathbf{x} = T_K \hat{\mathbf{x}} + \mathbf{b}_K$ *is a bijection from* $\hat{K}$ *to* K. *Then* $v \in H^m(K)$ *if and only if* $\hat{v} \in H^m(\hat{K})$. *Furthermore, for some constant c independent of* K *and* $\hat{K}$, *the estimates*

$$|\hat{v}|_{m,\hat{K}} \le c \, \|\mathbf{T}_K\|^m |\det \mathbf{T}_K|^{-1/2} |v|_{m,K} \tag{9.3.3}$$

and

$$|v|_{m,K} \le c \|\mathbf{T}_K^{-1}\|^m |\det \mathbf{T}_K|^{1/2} |\hat{v}|_{m,\hat{K}} \qquad (9.3.4)$$

hold.

Proof. We only need to prove the inequality (9.3.3); the inequality (9.3.4) follows from (9.3.3) by interchanging the roles played by $\mathbf{x}$ and $\hat{\mathbf{x}}$. Recall the multi-index notation: for $\alpha = (\alpha_1, \alpha_2)$,

$$D_{\hat{\mathbf{x}}}^\alpha = \frac{\partial^{|\alpha|}}{\partial \hat{x}_1^{\alpha_1} \partial \hat{x}_2^{\alpha_2}}, \quad D_{\mathbf{x}}^\alpha = \frac{\partial^{|\alpha|}}{\partial x_1^{\alpha_1} \partial x_2^{\alpha_2}}.$$

By a change of variables, we have

$$
\begin{aligned}
|\hat{v}|_{m,\hat{K}}^2 &= \sum_{|\alpha|=m} \int_{\hat{K}} (D_{\hat{\mathbf{x}}}^\alpha \hat{v}(\hat{\mathbf{x}}))^2 \, d\hat{x} \\
&= \sum_{|\alpha|=m} \int_K \left(D_{\hat{\mathbf{x}}}^\alpha \hat{v}(F_K^{-1}(\mathbf{x})) \right)^2 |\det \mathbf{T}_K|^{-1} \, dx.
\end{aligned}
$$

Since the mapping function is affine, for any multi-index α with $|\alpha| = m$, we have

$$D_{\hat{\mathbf{x}}}^\alpha \hat{v} = \sum_{|\beta|=m} c_{\alpha,\beta}(\mathbf{T}_K) \, D_{\mathbf{x}}^\beta v,$$

where each $c_{\alpha,\beta}(\mathbf{T}_K)$ is a product of m entries of the matrix $\mathbf{T}_K$. Thus

$$\sum_{|\alpha|=m} |D_{\hat{\mathbf{x}}}^\alpha \hat{v}(F_K^{-1}(\mathbf{x}))|^2 \le c \|\mathbf{T}_K\|^{2m} \sum_{|\alpha|=m} |D_{\mathbf{x}}^\alpha v(\mathbf{x})|^2,$$

and so,

$$
\begin{aligned}
|\hat{v}|_{m,\hat{K}}^2 &\le c \sum_{|\alpha|=m} \int_K (D_{\mathbf{x}}^\alpha v(\mathbf{x}))^2 \|\mathbf{T}_K\|^{2m} (\det \mathbf{T}_K)^{-1} \, dx \\
&= c \|\mathbf{T}_K\|^{2m} (\det \mathbf{T}_K)^{-1} |v|_{m,K}^2,
\end{aligned}
$$

from which the inequality (9.3.3) follows. ∎

We now combine the preceding theorems to obtain an estimate for the interpolation error in the semi-norm $|v - \Pi_K v|_{m,K}$.

Theorem 9.3.3 *Let k and m be non-negative integers with $k > 0$, $k+1 \ge m$, and $\mathcal{P}_k(\hat{K}) \subseteq \hat{X}$. Let Π_K be the operators defined in (9.2.10). Then there is a constant c depending only on $\hat{K}$ and $\hat{\Pi}$ such that*

$$|v - \Pi_K v|_{m,K} \le c \frac{h_K^{k+1}}{\rho_K^m} |v|_{k+1,K} \quad \forall v \in H^{k+1}(K). \qquad (9.3.5)$$

Proof. From Theorem 9.2.5 we have $\hat{v} - \hat{\Pi}\hat{v} = (v - \Pi_K v) \circ F_K$. Consequently, using (9.3.4) we obtain

$$|v - \Pi_K v|_{m,K} \le c \|\mathbf{T}_K^{-1}\|^m |\det \mathbf{T}_K|^{1/2} |\hat{v} - \hat{\Pi}\hat{v}|_{m,\hat{K}}.$$

Using the estimate (9.3.1), we have

$$|v - \Pi_K v|_{m,K} \le c \, \|\mathbf{T}_K^{-1}\|^m |\det \mathbf{T}_K|^{1/2} |\hat{v}|_{k+1,\hat{K}}. \qquad (9.3.6)$$

The inequality (9.3.3) with $m = k + 1$ is

$$|\hat{v}|_{k+1,\hat{K}} \le c \, \|\mathbf{T}_K\|^{k+1} |\det \mathbf{T}_K|^{-1/2} |v|_{k+1,K}.$$

So from (9.3.6), we obtain

$$|v - \Pi_K v|_{m,K} \le c \, \|\mathbf{T}_K^{-1}\|^m \|\mathbf{T}_K\|^{k+1} |v|_{k+1,K}.$$

The estimate (9.3.5) now follows from an application of Lemma 9.2.3. ∎

The error estimate (9.3.5) is proved through the use of the reference element $\hat{K}$. The proof method can be termed the *reference element technique*. We notice that in the proof we only use the polynomial invariance property (9.3.2) of the finite element interpolation on the reference element, and we do not need to use a corresponding polynomial invariance property on the real finite element. This feature is important when we analyze finite element spaces that are not based on affine-equivalent elements. For example, suppose the domain is partitioned into quadrilateral elements $\{K \mid K \in \mathcal{T}_h\}$. Then a reference element can be taken to be the unit square $\hat{K} = [0,1]^2$. For each element K, the mapping function F_K is bilinear, and maps each vertex of the reference element $\hat{K}$ to a corresponding vertex of K. The first degree finite element space for approximating $V = H^1(\Omega)$ is

$$V_h = \{v_h \in C(\overline{\Omega}) \mid v_h \circ F_K \in \mathcal{Q}_{1,1}(\hat{K}), \ K \in \mathcal{T}_h\},$$

where

$$\mathcal{Q}_{1,1}(\hat{K}) = \{\hat{v} \mid \hat{v}(\hat{\mathbf{x}}) = a + b \, \hat{x}_1 + c \, \hat{x}_2 + d \, \hat{x}_1 \hat{x}_2, \ a,b,c,d \in \mathbb{R}\}$$

is the space of bilinear functions. We see that for $v_h \in V_h$, on each element K, $v_h|_K$ is not a polynomial (as a function of the variable $\mathbf{x}$), but rather the image of a polynomial on the reference element. Obviously, we do not have the polynomial invariance property for the interpolation operator Π_K, nevertheless (9.3.2) is still valid. For such a finite element space, the proof of Theorem 9.3.3 still goes through.

The error bound in (9.3.5) depends on two parameters h_K and ρ_K. It will be convenient to use the parameter h_K only in an interpolation error bound. For this purpose we introduce the notion of a *regular* family of finite elements. For a triangulation $\mathcal{T}_h$, we denote

$$h = \max_{K \in \mathcal{T}_h} h_K, \qquad (9.3.7)$$

often called the *mesh parameter*. The quantity h is a measure of how refined the mesh is. The smaller h is, the finer the mesh.

Definition 9.3.4 *A family $\mathcal{T}_h = \{K\}$ of finite element partitions is said to be* regular *if*
(a) *there exists a constant σ such that $h_K/\rho_K \leq \sigma$ for all elements K;*
(b) *the mesh parameter h approaches zero.*

A necessary and sufficient condition for the fulfillment of the condition (a) in Definition 9.3.4 is that the minimal angles of all the elements are bounded below away from 0; a proof of this result is left as an exercise (cf. Exercise 9.3.2).

In the case of a regular family of affine finite elements, we can deduce the following error estimate from Theorem 9.3.3.

Corollary 9.3.5 *We keep the assumptions stated in Theorem 9.3.3. Furthermore, assume $\{K \mid K \in \mathcal{T}_h\}$ is a regular family of finite elements. Then there is a constant c such that*

$$\|v - \Pi_K v\|_{m,K} \leq c\, h_K^{k+1-m}|v|_{k+1,K} \quad \forall\, v \in H^{k+1}(K),\ \forall\, K \in \mathcal{T}_h. \quad (9.3.8)$$

Example 9.3.6 *Let K be a triangle in a regular family of affine finite elements. We take the three vertices of K to be the nodal points. The local function space X_K is $\mathcal{P}_1(K)$. Assume $v \in H^2(K)$. Applying the estimate (9.3.8) with $k = 1$, we have*

$$\|v - \Pi_K v\|_{m,K} \leq c\, h_K^{2-m}|v|_{2,K} \quad \forall\, v \in H^2(K). \quad (9.3.9)$$

This estimate holds for $m = 0, 1$.

9.3.3 Global interpolation error estimates

We now estimate the finite element interpolation error of a continuous function over the entire domain Ω. For a function $v \in C(\overline{\Omega})$, we construct its global interpolant $\Pi_h v$ in the finite element space X_h by the formula

$$\Pi_h v|_K = \Pi_K v \quad \forall\, K \in \mathcal{T}_h.$$

Let $\{\mathbf{x}_i\}_{i=1}^{N_h} \subseteq \overline{\Omega}$ be the set of the nodes collected from the nodes of all the elements $K \in \mathcal{T}_h$. We have the representation formula

$$\Pi_h v = \sum_{i=1}^{N_h} v(\mathbf{x}_i)\phi_i \quad (9.3.10)$$

for the global finite element interpolant. Here ϕ_i, $i = 1, \ldots, N_h$, are the global basis functions that span X_h. The basis function ϕ_i is associated with the node $\mathbf{x}_i$—i.e., ϕ_i is a piecewise polynomial of degree less than or equal to k, and $\phi_i(\mathbf{x}_j) = \delta_{ij}$. If the node $\mathbf{x}_i$ is a vertex $\mathbf{x}_l^K$ of the element K, then $\phi_i|_K = \phi_l^K$. If $\mathbf{x}_i$ is not a node of K, then $\phi_i|_K = 0$. Thus the functions ϕ_i are constructed from local basis functions ϕ_i^K.

Example 9.3.7 *We examine an example of linear elements. Assume $\Omega \subseteq \mathbb{R}^2$ is a polygonal domain, which is triangulated into triangles K, $K \in$*

$\mathcal{T}_h$. Denote $\{\mathbf{x}_i\}_{i=1}^{N_{\text{int}}}$ the set of the interior nodes, i.e., the vertices of the triangulation that lie in Ω; and $\{\mathbf{x}_i\}_{i=N_{\text{int}}+1}^{N_h}$, the set of the boundary nodes, i.e., the vertices of the triangulation that lie on $\partial\Omega$. From each vertex $\mathbf{x}_i$, denote $\tilde{K}_i$ the patch of the elements K that contain $\mathbf{x}_i$ as a vertex. The basis function ϕ_i associated with the node $\mathbf{x}_i$ is a continuous function on $\overline{\Omega}$, which is linear on each K and is non-zero only on $\tilde{K}_i$. The corresponding piecewise linear function space is then

$$X_h = \text{span}\{\phi_i, \ 1 \le i \le N_h\}.$$

Suppose we need to solve a linear elliptic boundary value problem with Neumann boundary condition. Then the function space is $V = H^1(\Omega)$, and we choose the linear element space to be $V_h = X_h$. Now suppose the boundary condition is homogeneous Dirichlet. Then the function space is $V = H_0^1(\Omega)$, while the linear element space is

$$V_h = \text{span}\{\phi_i, \ 1 \le i \le N_{\text{int}}\}.$$

In other words, we only use those basis functions associated with the interior nodes, so that again the finite element space is a subspace of the space for the boundary value problem. In case of a mixed boundary value problem, where we have a homogeneous Dirichlet boundary condition on part of the boundary Γ_1 and a Neumann boundary condition on the other part of the boundary Γ_2, then the triangulation should be compatible with the splitting $\partial\Omega = \Gamma_1 \cup \Gamma_2$; i.e., if an element K has one side on the boundary, then that side must belong entirely to either Γ_1 or Γ_2. The corresponding linear element space is then constructed from the basis functions associated with the interior nodes together with those boundary nodes located on Γ_1.

In the context of the finite element approximation of a linear second-order elliptic boundary value problem, Céa's inequality holds:

$$\|u - u_h\|_{1,\Omega} \le c \inf_{v_h \in V_h} \|u - v_h\|_{1,\Omega}.$$

Then

$$\|u - u_h\|_{1,\Omega} \le c\,\|u - \Pi_h u\|_{1,\Omega},$$

and we need to find an estimate of the interpolation error $\|u - \Pi_h u\|_{1,\Omega}$.

Theorem 9.3.8 *Assume that all the conditions of Corollary 9.3.5 hold. Then there exists a constant c independent of h such that*

$$\|v - \Pi_h v\|_{m,\Omega} \le c\,h^{k+1-m}|v|_{k+1,\Omega} \quad \forall v \in H^{k+1}(\Omega), \ m = 0, 1. \quad (9.3.11)$$

Proof. Since the finite element interpolant $\Pi_h u$ is defined piecewisely by $\Pi_h u|_K = \Pi_K u$, we can apply Corollary 9.3.5 with $m = 0$ and 1 to find

$$\|u - \Pi_h u\|_{m,\Omega}^2 = \sum_{K \in \mathcal{T}_h} \|u - \Pi_K u\|_{m,K}^2$$

$$\leq \sum_{K \in \mathcal{T}_h} c\, h_K^{2(k+1-m)} |u|_{k+1,K}^2$$

$$\leq c\, h^{2(k+1-m)} |u|_{k+1,\Omega}^2.$$

Taking the square root of the above relation, we obtain the error estimates (9.3.11). ∎

We make a remark on finite element interpolation of possibly discontinuous functions. The finite element interpolation error analysis discussed above assumes that the function to be interpolated is continuous, so that it is meaningful to talk about its finite element interpolation (9.3.10). In the case of a general Sobolev function v, not necessarily continuous, we can define $\Pi_h v$ by local L^2 projections in such a way that the interpolation error estimates stated in Theorem 9.3.8 are still valid. For detail, see [37]. We will use the same symbol $\Pi_h v$ to denote the "regular" finite element interpolant (9.3.10) when v is continuous, and in case v is discontinuous, $\Pi_h v$ is defined through local L^2 projections. In either case, we have the error estimates (9.3.11).

Exercise 9.3.1 *Let $\mathcal{T}_h$ be a regular partition of the domain Ω, and K any element obtained from the reference element $\hat{K}$ through the affine mapping (9.2.1). Show that there exists a constant c independent of K and h, such that*

$$\int_{\partial K} |v|^2 ds \leq c\,(h_K^{-1} \|v\|_{0,K}^2 + h_K\, |v|_{1,K}^2) \quad \forall v \in H^1(K).$$

Exercise 9.3.2 *Show that a necessary and sufficient condition for requirement (a) in Definition 9.3.4 is that the minimal angles of all the elements are bounded below from 0.*

Exercise 9.3.3 *A family of triangulations $\{\mathcal{T}_h\}$ is said to be* quasiuniform *if each triangulation $\mathcal{T}_h$ is regular, and there is a constant $c_0 > 0$ such that*

$$\min_{K \in \mathcal{T}_h} h_K / \max_{K \in \mathcal{T}_h} h_K \geq c_0.$$

(Hence, all the elements in $\mathcal{T}_h$ are of comparable size.)
Suppose $\{\mathcal{T}_h\}$ is a family of uniform triangulations of the domain $\Omega \subseteq \mathbb{R}^2$, $\{X_h\}$ a corresponding family of affine-equivalent finite elements. Denote $N_h = \dim X_h$, and denote the nodes (of the basis functions) by $\mathbf{x}_i$, $1 \leq i \leq N_h$. Show that there are constants $c_1, c_2 > 0$ independent of h such

that

$$c_1\|v\|^2_{L^2(\Omega)} \le h^2 \sum_{i=1}^{N_h} |v(\mathbf{x}_i)|^2 \le c_2\|v\|^2_{L^2(\Omega)} \quad \forall v \in X_h.$$

Exercise 9.3.4 *In this exercise, we employ the technique of the reference element to estimate the condition number of the stiffness matrix. The boundary value problem considered is a symmetric elliptic second-order problem*

$$u \in V, \quad a(u,v) = \ell(v) \quad \forall v \in V,$$

where $V \subseteq H^1(\Omega)$ is a Hilbert space, $a(\cdot,\cdot) : V \times V \to \mathbb{R}$ is bilinear, symmetric, continuous, and V-elliptic, $\ell \in V'$. Let $V_h \subseteq V$ be an affine family of finite elements of piecewiese polynomials of degree less than or equal to r. The finite element solution $u_h \in V_h$ is defined by

$$u_h \in V_h, \quad a(u_h, v_h) = \ell(v_h) \quad \forall v_h \in V_h.$$

Let $\{\phi_i\}$ be a basis of the space V_h. If we express the finite element solution in terms of the basis, $u_h = \sum_i \xi_i \phi_i$, then the unknown coefficients are determined from a linear system

$$A\boldsymbol{\xi} = \mathbf{b},$$

where the stiffness matrix A has the entries $a(\phi_j, \phi_i)$. Then A is a symmetric positive definite matrix. Let us find an upper bound for the spectral condition number

$$\mathrm{Cond}_2(A) = \|A\|_2 \|A^{-1}\|_2.$$

We assume the finite element spaces $\{V_h\}$ are constructed based on a family of quasiuniform triangulations $\{\mathcal{T}_h\}$.
(1) Show that there exist constants $c_1, c_2 > 0$, such that

$$c_1 h^2 |\boldsymbol{\eta}|^2 \le \|v_h\|_0^2 \le c_2 h^2 |\boldsymbol{\eta}|^2 \quad \forall v_h \in V_h, \ v_h = \sum_i \eta_i \phi_i.$$

(2) Show that there exists a constant $c_3 > 0$ such that

$$\|\nabla v_h\|_0^2 \le c_3\, h^{-2} \|v_h\|_0^2 \quad \forall v_h \in V_h.$$

This result is an example of an inverse inequality for finite element functions.
(3) Show that $\mathrm{Cond}_2(A) = O(h^{-2})$.
(Hint: A is symmetric, positive definite, so $\|A\|_2 = \sup\{(A\boldsymbol{\eta}, \boldsymbol{\eta})/|\boldsymbol{\eta}|^2\}$.)
   In the general d-dimensional case, it can be shown that these results are valid with the h^2 terms in (1) being replaced by h^d; in particular, we notice that the result (3) does not depend on the dimension of the domain Ω.

Exercise 9.3.5 *As an example of the application of the reference element technique in forming the stiffness matrix and the load vector, let us consider* *the computation of the integral* $I = \int_K v(x_1, x_2)\, dx$, *where* K *is a triangle with vertices* $a^i = (a^i_1, a^i_2)$, $1 \leq i \leq 3$. *Let* K_0 *be the unit triangle,*

$$K_0 = \{\hat{\mathbf{x}} \mid \hat{x}_1, \hat{x}_2 \geq 0,\ \hat{x}_1 + \hat{x}_2 \leq 1\}.$$

Show that the mapping

$$\phi_K : \hat{\mathbf{x}} \mapsto \mathbf{a}^1 + \hat{x}_1(\mathbf{a}^2 - \mathbf{a}^1) + \hat{x}_2(\mathbf{a}^3 - \mathbf{a}^1)$$

is a bijection from K_0 *to* K, *and*

$$\int_K v(\hat{\mathbf{x}})\, d\hat{x} = |(a^2_1 - a^1_1)(a^3_2 - a^1_2) - (a^2_2 - a^1_2)(a^3_1 - a^1_1)| \int_{K_0} v(\phi_K(\hat{\mathbf{x}}))\, d\hat{x}.$$

The significance of this result is that we can use quadrature formulas on the (fixed) reference element to compute integrals over finite elements.

9.4 Convergence and error estimates

As an example, we consider the convergence and error estimates for finite element approximations of a linear second-order elliptic problem over a polygonal domain. The function space V is a subspace of $H^1(\Omega)$; e.g., $V = H^1_0(\Omega)$ if the homogeneous Dirichlet condition is specified over the whole boundary, while $V = H^1(\Omega)$ if a Neumann condition is specified over the boundary. Let the weak formulation of the problem be

$$u \in V, \quad a(u, v) = \ell(v) \quad \forall v \in V. \tag{9.4.1}$$

We assume all the assumptions required by the Lax-Milgram lemma; then the problem (9.3.11) has a unique solution u. Let $V_h \subseteq V$ be a finite element space. Then the discrete problem

$$u_h \in V_h, \quad a(u_h, v_h) = \ell(v_h) \quad \forall v_h \in V_h \tag{9.4.2}$$

also has a unique solution $u_h \in V_h$ and Céa's inequality holds:

$$\|u - u_h\|_V \leq c \inf_{v_h \in V_h} \|u - v_h\|_V. \tag{9.4.3}$$

This inequality is a basis for convergence and error analysis.

Theorem 9.4.1 *We keep the assumptions mentioned above. Let* $\{V_h\} \subseteq V$ *be a regular family of affine-equivalent finite element spaces of piecewise polynomials of degree less than or equal to* k. *Then the finite element method converges,*

$$\|u - u_h\|_V \to 0 \quad \text{as } h \to 0.$$

Assume $u \in H^{k+1}(\Omega)$ *for some integer* $k > 0$. *Then there exists a constant* c *such that the following error estimate holds*

$$\|u - u_h\|_{1,\Omega} \leq c\, h^k |u|_{k+1,\Omega}. \tag{9.4.4}$$

Proof. We take $v_h = \Pi_h u$ in Céa's inequality (9.4.3),

$$\|u - u_h\|_{1,\Omega} \le c \|u - \Pi_h u\|_{1,\Omega}.$$

Using the estimate (9.3.11) with $m = 1$, we obtain the error estimate (9.4.4).

The convergence of the finite element solution under the basic solution regularity $u \in V$ follows from the facts that smooth functions are dense in the space V and for a smooth function, its finite element interpolants converge (with a convergence rate k). ∎

Example 9.4.2 *Consider the problem*

$$-\Delta u = f \quad \text{in } \Omega,$$
$$u = 0 \quad \text{on } \Gamma.$$

The corresponding variational formulation is; Find $u \in H_0^1(\Omega)$ such that

$$\int_\Omega \nabla u \cdot \nabla v \, dx = \int_\Omega fv \, dx \quad \forall v \in H_0^1(\Omega),$$

and this problem has a unique solution. Similarly, the discrete problem of finding $u_h \in V_h$ such that

$$\int_\Omega \nabla u_h \cdot \nabla v_h \, dx = \int_\Omega fv_h \, dx \quad \forall v_h \in V_h$$

has a unique solution. Here V_h is a finite element subspace of $H_0^1(\Omega)$, consisting of piecewise polynomials of degree less than or equal to k, corresponding to a regular triangulation of the domain Ω. If $u \in H^{k+2}(\Omega)$, then the error is estimated by

$$\|u - u_h\|_{1,\Omega} \le c\, h^k \|u\|_{k+2,\Omega}.$$

We should notice that in error estimation for finite element solutions, we need to assume certain degree of solution regularity. However, such a regularity condition is not always satisfied (see Exercise 9.4.3). When the solution exhibits singularities, there are two popular approaches to recover the optimal convergence order corresponding to a smooth solution. One approach is by means of *singular elements*: i.e., some singular functions are included in the finite element space. The advantage of this approach is its efficiency, while the weakness is that the form of the singular functions must be known *a priori*. The second approach is by using *mesh refinement* around the singularities. This approach does not need the knowledge on the forms of the singular functions, and is more popular in practical use.

Exercise 9.4.1 *Let us use linear elements to solve the boundary value problem:*

$$\begin{cases} -u'' = f \quad \text{in } (0, 1), \\ u(0) = u(1) = 0, \end{cases}$$

where $f \in L^2(0, 1)$. Divide the domain $\overline{I} = [0, 1]$ with the nodes $0 = x_0 < x_1 < \cdots < x_N = 1$, and denote the elements $I_i = [x_{i-1}, x_i]$, $1 \leq i \leq N$. Then the finite element space is

$$V_h = \{v_h \in H_0^1(0, 1) \mid v_h|_{I_i} \in P_1(I_i), \ 1 \leq i \leq N\}.$$

Let $u_h \in V_h$ denote the corresponding finite element solution of the boundary value problem. Prove that $u_h(x_i) = u(x_i)$, $0 \leq i \leq N$; in other words, the linear finite element solution for the boundary value problem is infinitely accurate at the nodes.

Hint: Show that the finite element interpolant of u is the finite element solution.

Exercise 9.4.2 Show that in $\mathbb{R}^2$, in terms of the polar coordinates

$$x_1 = r \cos \theta, \quad x_2 = r \sin \theta,$$

the Laplacian operator takes the form

$$\Delta = \frac{\partial^2}{\partial r^2} + \frac{1}{r} \frac{\partial}{\partial r} + \frac{1}{r^2} \frac{\partial^2}{\partial \theta^2};$$

and in $\mathbb{R}^3$, in terms of the spherical coordinates

$$x_1 = r \cos \theta \sin \phi, \quad x_2 = r \sin \theta \sin \phi, \quad x_3 = r \cos \phi,$$

the Laplacian operator takes the form

$$\Delta = \frac{\partial^2}{\partial r^2} + \frac{2}{r} \frac{\partial}{\partial r} + \frac{1}{r^2} \left[\frac{1}{\sin^2 \phi} \frac{\partial^2}{\partial \theta^2} + \cot \phi \frac{\partial}{\partial \phi} + \frac{\partial^2}{\partial \phi^2} \right].$$

Exercise 9.4.3 Show that $u = r^{2/3} \sin(2\theta/3)$ is the solution of the boundary value problem

$$\begin{cases} -\Delta u = 0 & \text{in } \Omega, \\ u = 0 & \text{on } \Gamma_1, \\ u = \sin\left(\frac{2\theta}{3}\right) & \text{on } \Gamma_2, \end{cases}$$

where $\Omega = \{(r, \theta) \mid 0 < r < 1, 0 < \theta < 3\pi/2\}$ is an L-shape domain, its boundary consisting of two legs about the corner, $\Gamma_1 = \{(r, \theta) \mid 0 \leq r \leq 1, \theta = 0 \text{ or } 3\pi/2\}$, and a circular curve, $\Gamma_2 = \{(r, \theta) \mid r = 1, 0 \leq \theta \leq 3\pi/2\}$. Verify that $u \in H^1(\Omega)$, but $u \notin H^2(\Omega)$.

Exercise 9.4.4 In Exercise 7.5.1, we studied the weak formulation of an elasticity problem for an isotropic, homogeneous linearly elastic material. Let $d = 2$ and assume Ω is a polygonal domain. Introduce a regular family of finite element partitions $\mathcal{T}_h = \{K\}$ in such a way that each K is a triangle and if $K \cap \Gamma \neq \emptyset$, then either $K \cap \Gamma \subseteq \overline{\Gamma}_u$ or $K \cap \Gamma \subseteq \overline{\Gamma}_g$. Let $V_h \subseteq V$ be the corresponding finite element subspace of continuous piecewise linear functions. Give the formulation of the finite element method and show that there is a unique finite element solution $u_h \in V_h$.

Assume $\mathbf{u} \in (H^2(\Omega))^2$. *Derive error estimates for* $\mathbf{u} - \mathbf{u}_h$ *and* $\boldsymbol{\sigma} - \boldsymbol{\sigma}_h$, *where*

$$\boldsymbol{\sigma}_h = \lambda \operatorname{tr} \boldsymbol{\varepsilon}(\mathbf{u}_h) \, \mathbf{I} + 2 \, \mu \, \boldsymbol{\varepsilon}(\mathbf{u}_h)$$

is a discrete stress field.

Exercise 9.4.5 *Consider a finite element approximation of the nonlinear elliptic boundary value problem studied in Section 7.8. Let us use all the notation introduced there. Let V_h be a finite element space consisting of continuous, piecewise polynomials of certain degree such that the functions vanish on the boundary. Then from Example 6.2.7, $V_h \subseteq V$. Show that the finite element method*

$$u_h \in V_h: \quad a(u_h; u_h, v_h) = \ell(v_h) \quad \forall \, v_h \in V_h$$

has a unique solution. Also show that u_h is the unique minimizer of the energy functional $E(\cdot)$ over the finite element space V_h.

Error estimate for the finite element solution defined above can be derived following that in [35, Section 5.3], where finite element approximation of the homogeneous Dirichlet problem for the nonlinear differential equation

$$-\operatorname{div}\left(|\nabla u|^{p-2}\nabla u\right) = f$$

is considered. However, the error estimate is not of optimal order. The optimal order error estimate for the linear element solution is derived in [20].

Suggestion for Further Readings

Standard references on mathematical analysis of the finite element method include BABUŠKA AND AZIZ [19], BRENNER AND SCOTT [28], BRAESS [27], CIARLET [35, 36], JOHNSON [84], ODEN AND REDDY [126], and STRANG AND FIX [153].

Detailed discussion of the p-version finite element method can be found in SZABÓ AND BABUŠKA [155]. Mathematical theory of the p-version and h-p-version finite element methods with applications in solid and fluid mechanics can be found in SCHWAB [146].

For the theory of mixed and hybrid finite element methods, see BREZZI AND FORTIN [29] and ROBERTS AND THOMAS [139].

For the numerical solution of Navier-Stokes equations by the finite element method, see GIRAULT AND RAVIART [57].

Theory of the finite element method for solving parabolic problems can be found in THOMÉE [157] and more recently [159].

Singularities of solutions to boundary value problems on non-smooth domains are analyzed in detail in GRISVARD [64]. See also KOZLOV, MAZ'YA AND ROSSMANN [97]. To improve the convergence rate of the finite element method when the solution exhibits singularities, one can employ the so-called singular element method where the finite element space contains the singular functions, or the mesh refinement method where the mesh is locally refined around the singular region of the solution. One can find a discussion of the singular element method in STRANG AND FIX [153], and the mesh refinement method in SZABÓ AND BABUŠKA [155].

10

Elliptic Variational Inequalities and Their Numerical Approximations

Variational inequalities form an important family of nonlinear problems. Some of the more complex physical processes are described by variational inequalities. Several comprehensive monographs can be consulted for the theory and numerical solution of variational inequalities, e.g., [45, 54, 58, 59, 95, 96, 128]. We study standard elliptic variational inequalities (EVIs) in this chapter. We give a brief introduction to some well-known results on the existence, uniqueness and stability of solutions to elliptic variational inequalities. We also discuss numerical approximations of EVIs and their error analysis.

10.1 Introductory examples

We first recall the general framework used in the Lax-Milgram lemma. Let V be a real Hilbert space, $a(\cdot, \cdot) : V \times V \rightarrow \mathbb{R}$ a continuous, V-elliptic bilinear form, and $\ell \in V'$. Then there is a unique solution of the problem

$$u \in V, \quad a(u, v) = \ell(v) \quad \forall v \in V. \tag{10.1.1}$$

In the special case where $a(\cdot, \cdot)$ is symmetric,

$$a(u, v) = a(v, u) \quad \forall u, v \in V, \tag{10.1.2}$$

we can introduce an energy functional

$$E(v) = \frac{1}{2} a(v, v) - \ell(v), \tag{10.1.3}$$

and it is easy to verify that the weak formulation (10.1.1) is equivalent to
the minimization problem

$$u \in V, \quad E(u) = \inf_{v \in V} E(v). \tag{10.1.4}$$

We observe that the minimization problem (10.1.4) is a *linear* problem,
owing to the assumptions that $E(\cdot)$ is a quadratic functional, and the set
over which the infimum is sought is a linear space. The problem (10.1.4) becomes *nonlinear* if either the energy functional $J(v)$ is no longer quadratic,
or the energy functional is minimized over a general set instead of a linear space. Indeed, if a quadratic energy functional of the form (10.1.3)
is minimized over a closed convex subset of V, then we obtain an *elliptic variational inequality of the first kind*. When the energy functional is
the summation of a quadratic functional of the form (10.1.3) and a non-
negative non-differentiable term, then the minimization problem (10.1.4)
is equivalent to an *elliptic variational inequality of the second kind*.

We remark that not every inequality problem is derived from a minimization principle. The feature of a variational inequality arising from a
quadratic minimization problem is that the bilinear form of the inequality
is symmetric.

Now let us examine two concrete examples.

Example 10.1.1 (THE OBSTACLE PROBLEM) *In an obstacle problem, we
need to determine the equilibrium position of an elastic membrane which
(1) passes through a closed curve Γ, the boundary of a planar domain Ω;
(2) lies above an obstacle of height ψ; and (3) is subject to the action of a
vertical force of density τf, here τ is the elastic tension of the membrane,
and f is a given function.*

*The unknown of interest for this problem is the vertical displacement u
of the membrane. Since the membrane is fixed along the boundary Γ, we
have the boundary condition $u = 0$ on Γ. To make the problem meaningful,
we assume the obstacle function satisfies the condition $\psi \leq 0$ on Γ. In
the following, we assume $\psi \in H^1(\Omega)$ and $f \in H^{-1}(\Omega)$. Thus the set of
admissible displacements is*

$$K = \{v \in H_0^1(\Omega) \mid v \geq \psi \text{ a.e. in } \Omega\}.$$

*The principle of minimal energy from mechanics asserts that the displace-
ment u is a minimizer of the total energy,*

$$u \in K : \quad E(u) = \inf\{E(v) \mid v \in K\}, \tag{10.1.5}$$

where the energy functional is defined as

$$E(v) = \int_\Omega \left(\frac{1}{2} |\nabla v|^2 - f v \right) dx.$$

*The set K is non-empty, because the function $\max\{0, \psi\}$ belongs to K. It
is easy to verify that the set K is closed and convex, the energy functional*

$E(\cdots)$ *is strictly convex, coercive, and continuous on K. Hence from the theory presented in Section 3.2, the minimization problem (10.1.5) has a unique solution $u \in K$. An argument similar to the proof of Lemma 3.3.1 shows that the solution is also characterized by the variational inequality (see Exercise 10.1.1):*

$$u \in K, \quad \int_\Omega \nabla u \cdot \nabla(v - u)\, dx \geq \int_\Omega f(v - u)\, dx \qquad \forall\, v \in K. \quad (10.1.6)$$

It is possible to derive the corresponding boundary value problem for the variational inequality (10.1.6). For this, we assume $f \in C(\Omega)$, $\psi \in C(\Omega)$, and the solution u of (10.1.6) satisfies $u \in C^2(\Omega) \cap C(\overline{\Omega})$. An integration by parts in (10.1.6) yields

$$\int_\Omega (-\Delta u - f)\,(v - u)\, dx \geq 0 \qquad \forall\, v \in K. \quad (10.1.7)$$

We take $v = u + \phi$ in (10.1.7), with $\phi \in C_0^\infty(\Omega)$ and $\phi \geq 0$, to obtain

$$\int_\Omega (-\Delta u - f)\,\phi\, dx \geq 0 \qquad \forall\, \phi \in C_0^\infty(\Omega), \ \phi \geq 0.$$

We see then that u must satisfy the differential inequality

$$-\Delta u - f \geq 0 \qquad \text{in } \Omega.$$

Now suppose for some $\mathbf{x}_0 \in \Omega$, $u(\mathbf{x}_0) > \psi(\mathbf{x}_0)$. Then there exist a neighborhood $U(\mathbf{x}_0) \subseteq \Omega$ of $\mathbf{x}_0$ and a number $\delta > 0$ such that $u(\mathbf{x}) > \psi(\mathbf{x}) + \delta$ for $\mathbf{x} \in U(\mathbf{x}_0)$. In (10.1.7) we choose $v = u \pm \delta\,\phi$ with any $\phi \in C_0^\infty(U(\mathbf{x}_0))$ satisfying $\|\phi\|_\infty \leq 1$ and obtain the relation

$$\pm \int_\Omega (-\Delta u - f)\,\phi\, dx \geq 0 \qquad \forall\, \phi \in C_0^\infty(U(\mathbf{x}_0)), \ \|\phi\|_\infty \leq 1.$$

Therefore,

$$\int_\Omega (-\Delta u - f)\,\phi\, dx = 0 \qquad \forall\, \phi \in C_0^\infty(U(\mathbf{x}_0)), \ \|\phi\|_\infty \leq 1$$

and then

$$\int_\Omega (-\Delta u - f)\,\phi\, dx = 0 \quad \forall\, \phi \in C_0^\infty(U(\mathbf{x}_0)).$$

Hence, if $u(\mathbf{x}_0) > \psi(\mathbf{x}_0)$ and $\mathbf{x}_0 \in \Omega$, then

$$(-\Delta u - f)(\mathbf{x}_0) = 0.$$

Summarizing, if the solution of the problem (10.1.6) has the smoothness $u \in C^2(\Omega) \cap C(\overline{\Omega})$, then the following relations hold

$$u - \psi \geq 0, \quad -\Delta u - f \geq 0, \quad (u - \psi)(-\Delta u - f) = 0 \quad \text{in } \Omega. \quad (10.1.8)$$

Consequently, the domain Ω is decomposed into two parts. On the first part, denoted by Ω_1, we have

$$u > \psi \quad \text{and} \quad -\Delta u - f = 0 \quad \text{in } \Omega_1,$$

and the membrane has no contact with the obstacle. On the second part, denoted by Ω_2, we have

$$u = \psi \quad \text{and} \quad -\Delta u - f > 0 \quad \text{in } \Omega_2,$$

and there is contact between the membrane and the obstacle. Notice that the region of contact, $\{x \in \Omega : u(x) = \psi(x)\}$, is an unknown a priori.

Next, let us show that if $u \in C^2(\Omega) \cap C(\overline{\Omega})$, $u = 0$ on Γ, satisfies the relations (10.1.8), then u must be a solution of the variational inequality (10.1.6). First we have

$$\int_\Omega (-\Delta u - f)\,(v - \psi)\,dx \geq 0 \quad \forall\, v \in K.$$

Since

$$\int_\Omega (-\Delta u - f)\,(u - \psi)\,dx = 0,$$

we obtain

$$\int_\Omega (-\Delta u - f)\,(v - u)\,dx \geq 0 \quad \forall\, v \in K.$$

Integrating by parts, we get

$$\int_\Omega \nabla u \cdot \nabla (v - u)\,dx \geq \int_\Omega f\,(v - u)\,dx \quad \forall\, v \in K.$$

Thus, u is a solution of the variational inequality (10.1.6).

In this example, we obtain a variational inequality from minimizing a quadratic energy functional over a convex set.

Example 10.1.2 (A FRICTIONAL CONTACT PROBLEM) *Consider a frictional contact problem between a linearly elastic body occupying a bounded connected domain Ω and a rigid foundation. The boundary Γ is assumed to be Lipschitz continuous and is partitioned into three non-overlapping regions Γ_u, Γ_g, and Γ_C. The body is fixed along Γ_u, and is subject to the action of a body force of the density $\mathbf{f}$ and the surface traction of density $\mathbf{g}$ on Γ_g. Over Γ_C, the body is in frictional contact with a rigid foundation (Figure 10.1). The body is assumed to be in equilibrium.*

We begin with the specification of the differential equations and boundary conditions. We use the notations introduced in Section 7.5. First we have the equilibrium equation

$$-\operatorname{div} \boldsymbol{\sigma} = \mathbf{f} \quad \text{in } \Omega, \tag{10.1.9}$$

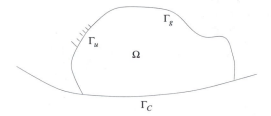

Figure 10.1. A body in frictional contact with a rigid foundation

where $\boldsymbol{\sigma} = (\sigma_{ij})_{d \times d}$ is the stress variable, $\boldsymbol{\sigma} = \boldsymbol{\sigma}^T$. The material is assumed to be linearly elastic with the constitutive relation

$$\boldsymbol{\sigma} = \mathbf{C}\boldsymbol{\varepsilon} \quad \text{in } \Omega, \tag{10.1.10}$$

where $\boldsymbol{\varepsilon} = \boldsymbol{\varepsilon}(\mathbf{u}) = (\varepsilon_{ij}(\mathbf{u}))_{d \times d}$ is the linearized strain tensor

$$\boldsymbol{\varepsilon}(\mathbf{u}) = \frac{1}{2}\left(\nabla\mathbf{u} + (\nabla\mathbf{u})^T\right), \tag{10.1.11}$$

while $\mathbf{C}$ is the elasticity tensor, satisfying the assumptions (7.5.6)–(7.5.8); i.e., $\mathbf{C}$ is bounded, symmetric, and pointwise stable.

The boundary conditions on Γ_u and Γ_g are

$$\mathbf{u} = \mathbf{0} \quad \text{on } \Gamma_u, \tag{10.1.12}$$

$$\boldsymbol{\sigma}\boldsymbol{\nu} = \mathbf{g} \quad \text{on } \Gamma_g. \tag{10.1.13}$$

To describe the boundary condition on Γ_C, we need some more notations. Given a vector field $\mathbf{u}$ on the boundary Γ, we define its normal displacement to be $u_\nu = \mathbf{u} \cdot \boldsymbol{\nu}$ and its tangential displacement by the formula $\mathbf{u}_t = \mathbf{u} - u_\nu \boldsymbol{\nu}$. Then we have the decomposition for the displacement:

$$\mathbf{u} = u_\nu \boldsymbol{\nu} + \mathbf{u}_t.$$

Similarly, given a stress tensor on the boundary, we have the stress vector $\boldsymbol{\sigma}\boldsymbol{\nu}$. We define the normal stress by $\sigma_\nu = (\boldsymbol{\sigma}\boldsymbol{\nu}) \cdot \boldsymbol{\nu}$ and the tangential stress vector by $\boldsymbol{\sigma}_t = \boldsymbol{\sigma}\boldsymbol{\nu} - \sigma_\nu \boldsymbol{\nu}$. In this way, we have the decomposition for the stress vector,

$$\boldsymbol{\sigma}\boldsymbol{\nu} = \sigma_\nu \boldsymbol{\nu} + \boldsymbol{\sigma}_t.$$

On Γ_C, we impose a simplified frictional contact condition (cf. [95, p. 272]):

$$\begin{aligned} &\sigma_\nu = -G, \\ &|\boldsymbol{\sigma}_t| \leq \mu_F G \quad \text{and} \\ &|\boldsymbol{\sigma}_t| < \mu_F G \Longrightarrow \mathbf{u}_t = \mathbf{0}, \\ &|\boldsymbol{\sigma}_t| = \mu_F G \Longrightarrow \mathbf{u}_t = -\lambda \boldsymbol{\sigma}_t \text{ for some } \lambda \geq 0. \end{aligned} \tag{10.1.14}$$

Here, $G > 0$ and the friction coefficient $\mu_F > 0$ are prescribed functions, $G, \mu_F \in L^\infty(\Gamma_C)$. It is easy to derive from the last two relations of (10.1.14) that

$$\boldsymbol{\sigma}_t \cdot \mathbf{u}_t = -\mu_F G \,|\mathbf{u}_t| \quad \text{on } \Gamma_C. \tag{10.1.15}$$

The mechanical problem for the frictional contact consists of the relations (10.1.9)–(10.1.14). Let us derive the corresponding weak formulation. The function space for this problem is chosen to be

$$V = \{\mathbf{v} \in (H^1(\Omega))^d \mid \mathbf{v} = \mathbf{0} \text{ a.e. on } \Gamma_u\}.$$

We assume the function $\mathbf{u}$ is sufficiently smooth so that all the calculations next are valid. We multiply the differential equation (10.1.9) by $\mathbf{v} - \mathbf{u}$ with an arbitrary $\mathbf{v} \in V$,

$$-\int_\Omega \operatorname{div}\boldsymbol{\sigma} \cdot (\mathbf{v} - \mathbf{u}) \, dx = \int_\Omega \mathbf{f} \cdot (\mathbf{v} - \mathbf{u}) \, dx.$$

Performing an integration by parts, using the boundary conditions (10.1.12) and (10.1.13) and the constitutive relation (10.1.10), we obtain

$$-\int_\Omega \operatorname{div}\boldsymbol{\sigma} \cdot (\mathbf{v} - \mathbf{u}) \, dx$$
$$= -\int_\Gamma \boldsymbol{\sigma}\boldsymbol{\nu} \cdot (\mathbf{v} - \mathbf{u}) \, ds + \int_\Omega \boldsymbol{\sigma} : \boldsymbol{\varepsilon}(\mathbf{v} - \mathbf{u}) \, dx$$
$$= -\int_{\Gamma_g} \mathbf{g} \cdot (\mathbf{v} - \mathbf{u}) \, ds - \int_{\Gamma_C} \boldsymbol{\sigma}\boldsymbol{\nu} \cdot (\mathbf{v} - \mathbf{u}) \, ds + \int_\Omega \mathbf{C}\boldsymbol{\varepsilon}(\mathbf{u}) : \boldsymbol{\varepsilon}(\mathbf{v} - \mathbf{u}) \, dx.$$

With the normal and tangential components decompositions of $\boldsymbol{\sigma}\boldsymbol{\nu}$ and $\mathbf{v}-\mathbf{u}$ on Γ_g, we have

$$-\int_{\Gamma_C} \boldsymbol{\sigma}\boldsymbol{\nu} \cdot (\mathbf{v} - \mathbf{u}) \, ds$$
$$= -\int_{\Gamma_C} (\sigma_\nu(v_\nu - u_\nu) + \boldsymbol{\sigma}_t \cdot (\mathbf{v}_t - \mathbf{u}_t)) \, ds$$
$$= \int_{\Gamma_C} G \,(v_\nu - u_\nu) \, ds + \int_{\Gamma_C} (-\boldsymbol{\sigma}_t \cdot \mathbf{v}_t - \mu_F G \,|\mathbf{u}_t|) \, ds$$
$$\leq \int_{\Gamma_C} G \,(v_\nu - u_\nu) \, ds + \int_{\Gamma_C} \mu_F G \,(|\mathbf{v}_t| - |\mathbf{u}_t|) \, ds,$$

where, the boundary condition (10.1.14) (and its consequence (10.1.15)) is used.

Summarizing, the variational inequality of the problem is to find the displacement field $\mathbf{u} \in V$ *such that*

$$\int_{\Omega} \mathbf{C}\varepsilon(\mathbf{u}) : \varepsilon(\mathbf{v} - \mathbf{u}) \, dx + \int_{\Gamma_C} \mu_F G \, |\mathbf{v}_t| \, ds - \int_{\Gamma_C} \mu_F G \, |\mathbf{u}_t| \, ds$$

$$\geq \int_{\Omega} \mathbf{f} \cdot (\mathbf{v} - \mathbf{u}) \, dx + \int_{\Gamma_g} \mathbf{g} \cdot (\mathbf{v} - \mathbf{u}) \, ds - \int_{\Gamma_C} G \, (v_\nu - u_\nu) \, ds \quad \forall \mathbf{v} \in V.$$

$$(10.1.16)$$

We assume that $\mathbf{f} \in [L^2(\Omega)]^d$, $\mathbf{g} \in [L^2(\Gamma_g)]^d$. *Then each term in the variational inequality* (10.1.16) *makes sense.*

The corresponding minimization problem is

$$\mathbf{u} \in V, \qquad E(\mathbf{u}) = \inf\{E(\mathbf{v}) \mid \mathbf{v} \in V\}, \qquad (10.1.17)$$

where the energy functional

$$E(\mathbf{v}) = \frac{1}{2} \int_{\Omega} \mathbf{C}\varepsilon(\mathbf{v}) : \varepsilon(\mathbf{v}) \, dx - \int_{\Omega} \mathbf{f} \cdot \mathbf{v} \, dx$$

$$- \int_{\Gamma_g} \mathbf{g} \cdot \mathbf{v} \, ds + \int_{\Gamma_C} G \, (v_\nu + \mu_F |\mathbf{v}_t|) \, ds.$$

$$(10.1.18)$$

This energy functional is non-differentiable. The non-differentiable term takes the frictional effect into account. The equivalence between the variational inequality (10.1.16) *and the minimization problem* (10.1.17) *is left as Exercise 10.1.2.*

Exercise 10.1.1 *Show that u is a solution of the constraint minimization problem* (10.1.5) *if and only if it satisfies the variational inequality* (10.1.6).

Exercise 10.1.2 *Prove that the variational inequality* (10.1.16) *and the minimization problem* (10.1.17) *are mutually equivalent.*

10.2 Elliptic variational inequalities of the first kind

The obstacle problem (Example 10.1.1), is a representative example of a class of inequality problems known as *elliptic variational inequalities (EVIs) of the first kind*. In the general framework of EVIs of the first kind, though, the bilinear form does not need to be symmetric (and hence there may exist no equivalent minimization principle).

The abstract form of EVIs of the first kind can be described as the following. Let V be a real Hilbert space with inner product $(\cdot, \cdot)$ and associated norm $\| \cdot \|$. Let $a : V \times V \to \mathbb{R}$ be a continuous, V-elliptic bilinear form on V, and K a subset of V. Given a linear functional $\ell : V \to \mathbb{R}$, it is required to find $u \in K$ satisfying

$$a(u, v - u) \geq \ell(v - u) \qquad \forall \, v \in K. \tag{10.2.1}$$

Variational inequalities of the first kind may be characterized by the fact that they are posed on convex subsets. When the set K is in fact a subspace of V, the variational inequality becomes a variational equation (cf. Exercise 10.2.2).

We have the following result for the unique solvability of the variational inequality (10.2.1).

Theorem 10.2.1 *Let V be a real Hilbert space, $a : V \times V \to \mathbb{R}$ a continuous, V-elliptic bilinear form, $\ell : V \to \mathbb{R}$ a bounded linear functional, and $K \subseteq V$ a non-empty, closed, and convex set. Then the EVI of the first kind (10.2.1) has a unique solution $u \in K$.*

Proof. A general principle is that it is easier to work with equations than inequalities. So we rewrite the inequality (10.2.1) in the form of an equivalent fixed-point problem. To do this, we first apply the Riesz representation theorem (cf. Theorem 2.5.7 of Chapter 2) to claim that there exists a unique member $L \in V$ such that $\|L\| = \|\ell\|$ and

$$\ell(v) = (L, v) \qquad \forall \, v \in V.$$

For any fixed $u \in V$, the mapping $v \mapsto a(u, v)$ defines a linear, continuous form on V. Thus applying the Riesz representation theorem again, we have a mapping $A : V \to V$ such that

$$a(u, v) = (Au, v) \qquad \forall \, v \in V.$$

Since $a(\cdot, \cdot)$ is bilinear and continuous, it is easy to verify that A is linear and bounded, with

$$\|A\| \leq M.$$

For any $\theta > 0$, the problem (10.2.1) is therefore equivalent to one of finding $u \in K$ such that

$$((u - \theta \, (Au - L)) - u, v - u) \leq 0 \qquad \forall \, v \in K. \tag{10.2.2}$$

If P_K denotes the orthogonal projection onto K, then (10.2.2) may be written in the form

$$u = P_K \, (u - \theta \, (Au - L)). \tag{10.2.3}$$

We show that by choosing $\theta > 0$ sufficiently small, the operator defined by the right hand side of (10.2.3) is a contraction. Indeed, for any $v_1, v_2 \in V$, we have

$$
\begin{aligned}
\| P_K \left(v_1 - \theta \left(A v_1 - L \right) \right) &- P_K \left(v_2 - \theta \left(A v_2 - L \right) \right) \|^2 \\
&\leq \| \left(v_1 - \theta \left(A v_1 - L \right) \right) - \left(v_2 - \theta \left(A v_2 - L \right) \right) \|^2 \\
&= \| (v_1 - v_2) - \theta\, A(v_1 - v_2) \|^2 \\
&= \| v_1 - v_2 \|^2 - 2\,\theta\, a(v_1 - v_2, v_1 - v_2) + \theta^2 \| A(v_1 - v_2) \|^2 \\
&\leq \left(1 - 2\,\theta\,\alpha + \theta^2 M^2 \right) \| v_1 - v_2 \|^2 .
\end{aligned}
$$

Here $\alpha > 0$ is the V-ellipticity constant for the bilinear form $a(\cdot, \cdot)$.

Thus if we choose $\theta \in (0, 2\,\alpha/M^2)$, then P_K is a contraction and, by the Banach fixed-point theorem (cf. Theorem 4.1.3 in Chapter 4), the problem (10.2.2), and consequently the problem (10.2.1), has a unique solution. $\blacksquare$

Theorem 10.2.1 is a generalization of Lax-Milgram lemma for the well-posedness of a linear elliptic boundary value problem (cf. Exercise 10.2.3).

We also remark that in the case when the bilinear form is symmetric, the variational inequality (10.2.1) is equivalent to the constrained minimization problem

$$
\inf_{v \in K} \left\{ \frac{1}{2}\, a(v, v) - \ell(v) \right\} .
$$

This problem has a unique solution, following the general results discussed in Chapter 3. Indeed in this case we have a useful characterization of the solution of the variational inequality (10.2.1).

Proposition 10.2.2 *Let the assumptions of Theorem 10.2.1 hold. Additionally assume $a(\cdot, \cdot)$ is symmetric. Then the solution $u \in K$ of the variational inequality (10.2.1) is the unique best approximation in K of $w \in V$, in the sense of the inner product defined by $a(\cdot, \cdot)$:*

$$
\| w - u \|_a = \inf_{v \in K} \| w - v \|_a .
$$

Here $w \in V$ is the unique solution of the linear elliptic boundary value problem

$$
w \in V, \quad a(w, v) = \ell(v) \quad \forall\, v \in V.
$$

Proof. We have, for any $v \in K$,

$$
a(u, v - u) \geq \ell(v - u) = a(w, v - u);
$$

i.e.,

$$
a(w - u, v - u) \leq 0 \qquad \forall\, v \in V.
$$

Hence, $u \in K$ is the projection of w with respect to the inner product $a(\cdot, \cdot)$ onto K. $\blacksquare$

Proposition 10.2.2 suggests a possible approach to solve the variational inequality (10.2.1). In the first step, we solve a corresponding linear boundary value problem to get a solution w. In the second step, we compute the projection of w onto K, in the inner product $a(\cdot, \cdot)$.

Example 10.2.3 *The obstacle problem* (10.1.6) *is clearly an elliptic variational inequality of the first kind. All the conditions stated in Theorem 10.2.1 are satisfied, and hence the obstacle problem* (10.1.6) *has a unique solution.*

Exercise 10.2.1 *A subset* $K \subseteq V$ *is said to be a cone in* V *if for any* $v \in K$ *and any* $\alpha > 0$, *we have* $\alpha v \in K$. *Show that if* K *is a closed convex cone of* V, *then the variational inequality* (10.2.1) *is equivalent to the relations*

$$a(u, v) \geq \ell(v) \quad \forall v \in K,$$
$$a(u, u) = \ell(u).$$

Exercise 10.2.2 *Show that if* K *is a subspace of* V, *then the variational inequality* (10.2.1) *reduces to a variational equation.*

Exercise 10.2.3 *Show that Theorem 10.2.1 is a generalization of the Lax-Milgram lemma.*

Exercise 10.2.4 *As another example of EVI of the first kind, we consider the elasto-plastic torsion problem. Let* $\Omega \subseteq \mathbb{R}^2$ *be a bounded domain with a Lipschitz continuous boundary* $\partial\Omega$. *Let*

$$V = H_0^1(\Omega),$$
$$a(u, v) = \int_\Omega \nabla u \cdot \nabla v \, dx,$$
$$\ell(v) = \int_\Omega f v \, dx,$$
$$K = \{v \in V \mid |\nabla v| \leq 1 \text{ a.e. in } \Omega\}.$$

Then the problem is

$$u \in K, \quad a(u, v - u) \geq \ell(v - u) \qquad \forall v \in K.$$

Use Theorem 10.2.1 to show that the elasto-plastic torsion problem has a unique solution.

In general one cannot expect high regularity for the solution of a variational inequality. It can be shown that if Ω is convex or smooth and $f \in L^p(\Omega)$, $1 < p < \infty$, then the solution of the elasto-plastic torsion problem $u \in W^{2,p}(\Omega)$. The following exact solution shows that $u \notin H^3(\Omega)$ even if Ω and f are smooth ([58, Chap. 2]).

Consider the special situation where Ω is the circle centered at O with the radius R, and the external force density $f = c_0 > 0$ is a constant. Denote $r = \|\mathbf{x}\|_2$. Verify that if $c_0 \leq 2/R$, then the solution is given by

$$u(\mathbf{x}) = (c_0/4)(R^2 - r^2);$$

while if $c_0 > 2/R$, then

$$u(\mathbf{x}) = \begin{cases} (c_0/4)\,[R^2 - r^2 - (R - 2/c_0)^2], & \text{if } 0 \le r \le 2/c_0, \\ R - r, & \text{if } 2/c_0 \le r \le R. \end{cases}$$

Exercise 10.2.5 *A simplified version of the well-known Signorini problem can be described as an EVI of the first kind with the following data:*

$$V = H^1(\Omega),$$
$$K = \{v \in V \mid v \ge 0 \text{ a.e. on } \Gamma\},$$
$$a(u, v) = \int_\Omega (\nabla u \cdot \nabla v + u\,v)\,dx,$$
$$\ell(v) = \int_\Omega f\,v\,dx + \int_\Gamma g\,v\,ds,$$

where for simplicity, we assume $f \in L^2(\Omega)$, $g \in L^2(\Gamma)$. Show that the corresponding variational inequality problem

$$u \in K, \quad a(u, v - u) \ge \ell(v - u) \qquad \forall v \in K$$

has a unique solution u. Show that formally, u solves the boundary value problem

$$-\Delta u + u = f \qquad \text{a.e. in } \Omega,$$
$$u \ge 0, \quad \frac{\partial u}{\partial \nu} \ge g, \quad u\left(\frac{\partial u}{\partial \nu} - g\right) = 0 \qquad \text{a.e. on } \Gamma.$$

10.3   Approximation of EVIs of the first kind

For convenience, we recall the general framework for elliptic variational inequalities of the first kind. Let V be a real Hilbert space, $K \subseteq V$ be non-empty, convex, and closed. Assume $a(\cdot, \cdot)$ is a V-elliptic and bounded bilinear form on V, ℓ a continuous linear functional on V. Then according to Theorem 10.2.1, the following elliptic variational inequality of the first kind

$$u \in K, \quad a(u, v - u) \ge \ell(v - u) \qquad \forall v \in K \qquad (10.3.1)$$

has a unique solution.

A general framework for numerical solution of the variational inequality (10.3.1) can be described as follows. Let $V_h \subseteq V$ be a finite element space, and let $K_h \subseteq V_h$ be non-empty, convex, and closed. Then the finite element approximation of the problem (10.3.1) is

$$u_h \in K_h, \quad a(u_h, v_h - u_h) \ge \ell(v_h - u_h) \qquad \forall v_h \in K_h. \qquad (10.3.2)$$

Another application of Theorem 10.2.1 shows that the discrete problem (10.3.2) has a unique solution under the stated assumptions on the given data.

A general convergence result of the finite element method can be found in [58] (cf. Exercise 10.3.1). Here we derive an error estimate for the finite element solution u_h. We follow [49] and first give an abstract error analysis.

Theorem 10.3.1 *There is a constant $c > 0$ independent of h and u, such that*

$$\|u - u_h\| \le c \Big\{ \inf_{v_h \in K_h} \Big[\|u - v_h\| + |a(u, v_h - u) - \ell(v_h - u)|^{\frac{1}{2}} \Big]$$

$$+ \inf_{v \in K} |a(u, v - u_h) - \ell(v - u_h)|^{\frac{1}{2}} \Big\}. \tag{10.3.3}$$

Proof. From (10.3.1) and (10.3.2), we find that

$$a(u, u) \le a(u, v) - \ell(v - u) \quad \forall v \in K,$$
$$a(u_h, u_h) \le a(u_h, v_h) - \ell(v_h - u_h) \quad \forall v_h \in K_h.$$

Using these relations, together with the V-ellipticity and boundedness of the bilinear form $a(\cdot, \cdot)$, we have for any $v \in K$ and $v_h \in K_h$,

$$\begin{aligned}
\alpha \|u - u_h\|^2 &\le a(u - u_h, u - u_h) \\
&= a(u, u) + a(u_h, u_h) - a(u, u_h) - a(u_h, u) \\
&\le a(u, v - u_h) - \ell(v - u_h) + a(u, v_h - u) - \ell(v_h - u) \\
&\quad + a(u_h - u, v_h - u) \\
&\le a(u, v - u_h) - \ell(v - u_h) + a(u, v_h - u) - \ell(v_h - u) \\
&\quad + \frac{1}{2}\alpha \|u - u_h\|^2 + c \|v_h - u\|^2.
\end{aligned}$$

Thus the inequality (10.3.3) holds. ∎

The inequality (10.3.3) is a generalization of Céa's lemma to the finite element approximation of elliptic variational inequalities of the first kind. It is easy to see that the inequality (10.3.3) reduces to Céa's lemma in the case of finite element approximation of a variational equation problem.

Remark 10.3.2 *In the case $K_h \subseteq K$, we have the so-called* internal approximation *of the elliptic variational inequality of the first kind. Since now $u_h \in K$, the second term on the right-hand side of (10.3.3) vanishes, and the error inequality (10.3.3) reduces to*

$$\|u - u_h\| \le c \inf_{v_h \in K_h} \Big[\|u - v_h\| + |a(u, v_h - u) - \ell(v_h - u)|^{1/2} \Big].$$

Example 10.3.3 *Let us apply the inequality (10.3.3) to derive an order error estimate for the approximation of the obstacle problem. Such an estimate was first proved in [49]. We assume $u, \psi \in H^2(\Omega)$ and Ω is a polygon, and use linear elements on a regular mesh of triangles for the approximation. Then the discrete admissible set is*

$$K_h = \{v_h \in H_0^1(\Omega) \mid v_h \text{ is piecewise linear,}$$
$$v_h(\mathbf{x}) \ge \psi(\mathbf{x}) \text{ for any node } \mathbf{x}\}.$$

We see that any function in K_h is a continuous piecewise linear function, vanishing on the boundary and dominating the obstacle function at the interior nodes of the mesh. In general, $K_h \not\subset K$. For the obstacle problem, for any $u \in H^2(\Omega)$ and $v \in H_0^1(\Omega)$ we have

$$a(u, v) - \ell(v) = \int_\Omega (\nabla u \cdot \nabla v - f\, v)\, dx = \int_\Omega (-\Delta u - f)\, v\, dx.$$

Thus from the inequality (10.3.3), we have the following error estimate

$$\|u - u_h\|_1 \leq c \left\{ \inf_{v_h \in K_h} \left[\|u - v_h\|_1 + \| -\Delta u - f\|_0^{1/2} \|u - v_h\|_0^{1/2} \right] \right.$$

$$\left. + \| -\Delta u - f\|_0^{1/2} \inf_{v \in K} \|v - u_h\|_0^{1/2} \right\} \tag{10.3.4}$$

Let $\Pi_h u$ be the piecewise linear interpolant of u. It is not difficult to verify that $\Pi_h u \in K_h$. Then

$$\inf_{v_h \in K_h} \left[\|u - v_h\|_1 + \| -\Delta u - f\|_0^{1/2} \|u - v_h\|_0^{1/2} \right]$$

$$\leq \|u - \Pi_h u\|_1 + \| -\Delta u - f\|_0^{1/2} \|u - \Pi_h u\|_0^{1/2}$$

$$\leq c \left[|u|_2 + \| -\Delta u - f\|_0^{1/2} |u|_2^{1/2} \right] h.$$

To evaluate the term $\inf_{v \in K} \|v - u_h\|_0$, we define

$$u_h^* = \max\{u_h, \psi\}.$$

Since $u_h, \psi \in H^1(\Omega)$, we have $u_h^* \in H^1(\Omega)$. By the definition, certainly $u_h^* \geq \psi$. Finally, since $\psi \leq 0$ on Γ, we have $u_h^* = 0$ on Γ. Hence, $u_h^* \in K$. Let

$$\Omega^* = \{\mathbf{x} \in \Omega \mid u_h(\mathbf{x}) < \psi(\mathbf{x})\}.$$

Then over $\Omega \backslash \Omega^*$, $u_h^* = u_h$, and so

$$\inf_{v \in K} \|v - u_h\|_0^2 \leq \|u_h^* - u_h\|_0^2 = \int_{\Omega^*} |u_h - \psi|^2 dx.$$

Let $\Pi_h \psi$ be the piecewise linear interpolant of ψ. Since at any node, $u_h \geq \psi = \Pi_h \psi$, we have $u_h \geq \Pi_h \psi$ in Ω. Therefore, over Ω^*,

$$0 < |u_h - \psi| = \psi - u_h \leq \psi - \Pi_h \psi = |\psi - \Pi_h \psi|.$$

Thus,

$$\int_{\Omega^*} |u_h - \psi|^2 dx \leq \int_{\Omega^*} |\psi - \Pi_h \psi|^2 dx \leq \int_\Omega |\psi - \Pi_h \psi|^2 dx \leq c\, |\psi|_2^2 h^4,$$

and then

$$\inf_{v \in K} \|v - u_h\|_0^{1/2} \leq c\, |\psi|_2^{1/2} h.$$

From the inequality (10.3.4), we finally get the optimal order error estimate

$$\|u - u_h\|_{H^1(\Omega)} \leq c\,h$$

for some constant $c > 0$ depending only on $|u|_2$, $\|f\|_0$ and $|\psi|_2$.

Exercise 10.3.1 *Here we consider the convergence of the scheme (10.3.2). We first introduce a definition on the closeness between the sets K and K_h. We say the sets K_h converge to K and write $K_h \to K$ if (a) K_h is a non-empty closed convex set in V; (b) for any $v \in K$, there exist $v_h \in K_h$ such that $v_h \to v$ in V; and (c) if $u_h \in K_h$ and $u_h \rightharpoonup u$, then $u \in K$.*

Show the convergence of the scheme (10.3.2) via the following steps, assuming $\{h\}$ is a sequence of mesh parameters converging to 0.

First, prove the boundedness of the sequence $\{u_h\}$ in V.

Second, since V is reflexive, there is a subsequence $\{u_{h'}\}$ converging weakly to w. Use the weak l.s.c. of $a(\cdot, \cdot)$ to conclude that w is the solution u of the problem (10.3.1).

Finally, prove $a(u_h - u, u_h - u) \to 0$ as $h \to 0$.

Exercise 10.3.2 *The elasto-plastic torsion problem was introduced in Exercise 10.2.4. Let us consider its one-dimensional analogue and a linear finite element approximation. Let the domain be the unit interval $[0, 1]$, and Δ_h be a partition of the interval with the meshsize h. Then the admissible set K_h consists of the continuous piecewise linear functions, vanishing at the ends, and the magnitude of the first derivative being bounded by 1. Show that under suitable solution regularity assumptions, there is an optimal order error estimate $\|u - u_h\|_1 \leq c\,h$.*

10.4   Elliptic variational inequalities of the second kind

The frictional contact problem (10.1.16) is an example of an *elliptic variational inequality (EVI) of the second kind*. To give the general framework for this class of problems, in addition to the bilinear form $a(\cdot, \cdot)$ and the linear functional ℓ, we introduce a proper, convex and lower semicontinuous (l.s.c.) functional $j : V \to \overline{\mathbb{R}} \equiv \mathbb{R} \cup \{\pm\infty\}$. The functional j is not assumed to be differentiable. Then the problem of finding $u \in V$ that satisfies

$$a(u, v - u) + j(v) - j(u) \geq \ell(v - u) \qquad \forall v \in V \tag{10.4.1}$$

is referred to as an EVI of the second kind. We observe that we have an inequality owing to the presence of the nondifferentiable term $j(\cdot)$.

Theorem 10.4.1 *Let V be a real Hilbert space, $a : V \times V \to \mathbb{R}$ a continuous, V-elliptic bilinear form, $\ell : V \to \mathbb{R}$ a bounded linear functional, and*

$j : V \to \overline{\mathbb{R}}$ a proper, convex, and l.s.c. functional on V. Then the EVI of the second kind (10.4.1) has a unique solution.

Proof. The uniqueness part is easy to prove. Since j is proper, $j(v_0) < \infty$ for some $v_0 \in V$. Thus a solution u of (10.4.1) satisfies

$$j(u) \le a(u, v_0 - u) + j(v_0) - \ell(v_0 - u) < \infty,$$

that is, $j(u)$ is a real number. Now let u_1 and u_2 denote two solutions of the problem (10.4.1); then

$$a(u_1, u_2 - u_1) + j(u_2) - j(u_1) \ge \ell(u_2 - u_1),$$
$$a(u_2, u_1 - u_2) + j(u_1) - j(u_2) \ge \ell(u_1 - u_2).$$

Adding the two inequalities, we get

$$-a(u_1 - u_2, u_1 - u_2) \ge 0$$

which implies, by the V-ellipticity of a, that $u_1 = u_2$.

The part on the existence is more involved. First consider the case in which $a(\cdot, \cdot)$ is symmetric; under this additional assumption, the variational inequality (10.4.1) is equivalent to the minimization problem

$$u \in V, \quad E(u) = \inf\{E(v) \mid v \in V\}, \tag{10.4.2}$$

where

$$E(v) = \frac{1}{2}a(v, v) + j(v) - \ell(v).$$

Since $j(\cdot)$ is proper, convex, and l.s.c., it is bounded below by a bounded affine functional,

$$j(v) \ge \ell_j(v) + c_0 \qquad \forall v \in V$$

where ℓ_j is a continuous linear form on V and $c_0 \in \mathbb{R}$ (see Lemma 10.4.2 below). Thus by the stated assumptions on a, j and ℓ, we see that J is proper, convex, and l.s.c., and has the property that

$$E(v) \to \infty \quad \text{as } \|v\| \to \infty.$$

We see that the problem (10.4.2), and hence the problem (10.4.1), has a solution.

Consider next the general case without the symmetry assumption. Again we will convert the problem into an equivalent fixed-point problem. For any $\theta > 0$, the problem (10.4.1) is equivalent to

$$u \in V, \ (u, v - u) + \theta\, j(v) - \theta\, j(u)$$
$$\ge (u, v - u) - \theta\, a(u, v - u) + \theta\, \ell(v - u) \ \forall v \in V.$$

Now for any $u \in V$, consider the problem

$$w \in V, \ (w, v - w) + \theta\, j(v) - \theta\, j(w)$$
$$\ge (u, v - w) - \theta\, a(u, v - w) + \theta\, \ell(v - w) \ \forall v \in V. \tag{10.4.3}$$

From the previous discussion this problem has a unique solution, which is denoted by $w = P_\theta u$. Obviously a fixed point of the mapping P_θ is a solution of the problem (10.4.1). We will see that for sufficiently small $\theta > 0$, P_θ is a contraction and hence has a unique fixed-point by the Banach fixed-point theorem (Theorem 4.1.3).

For any $u_1, u_2 \in V$, let $w_1 = P_\theta u_1$ and $w_2 = P_\theta u_2$. Then we have

$$(w_1, w_2 - w_1) + \theta\, j(w_2) - \theta\, j(w_1)$$
$$\geq (u_1, w_2 - w_1) - \theta\, a(u_1, w_2 - w_1) + \theta\, \ell(w_2 - w_1),$$
$$(w_2, w_1 - w_2) + \theta\, j(w_1) - \theta\, j(w_2)$$
$$\geq (u_2, w_1 - w_2) - \theta\, a(u_2, w_1 - w_2) + \theta\, \ell(w_1 - w_2).$$

Adding the two inequalities and simplifying, we get

$$\|w_1 - w_2\|^2 \leq (u_1 - u_2, w_1 - w_2) - \theta\, a(u_1 - u_2, w_1 - w_2)$$
$$= ((I - \theta\, A)(u_1 - u_2), w_1 - w_2),$$

where the operator A is defined by the relation $a(u, v) = (Au, v)$ for any $u, v \in V$. Hence

$$\|w_1 - w_2\| \leq \|(I - \theta\, A)(u_1 - u_2)\|.$$

Now for any $u \in V$,

$$\|(I - \theta\, A)u\|^2 = \|u - \theta\, Au\|^2$$
$$= \|u\|^2 - 2\,\theta\, a(u, u) + \theta^2 \|Au\|^2$$
$$\leq (1 - 2\,\theta\, \alpha + \theta^2 M^2)\, \|u\|^2.$$

Here M and α are the continuity and V-ellipticity constants of the bilinear form $a(\cdot, \cdot)$. Therefore, again, for $\theta \in (0, 2\,\alpha/M^2)$, the mapping P_θ is a contraction on the Hilbert space V. ∎

Lemma 10.4.2 Let V be a normed space. Assume $j : V \to \mathbb{R}$ is proper, convex and l.s.c. Then there exists a continuous linear functional $\ell_j \in V'$ and $c_0 \in \mathbb{R}$ such that

$$j(v) \geq \ell_j(v) + c_0 \quad \forall\, v \in V.$$

Proof. Since $j : V \to \mathbb{R}$ is proper, there exists $v_0 \in V$ with $j(v_0) \in \mathbb{R}$. Pick up a real number $a_0 < j(v_0)$. Consider the set

$$J = \{(v, a) \in V \times \mathbb{R} \mid j(v) \leq a\}.$$

This set is called the *epigraph* of j in the literature on convex analysis. From assumption that j is convex, it is easy to verify that J is closed in $V \times \mathbb{R}$. Since j is l.s.c., it is readily seen that J is closed in $V \times \mathbb{R}$. Thus the sets $\{(v_0, a_0)\}$ and J are disjoint, $\{(v_0, a_0)\}$ is a convex compact set, and J is a closed convex set. By Theorem 3.2.7, we can strictly separate the sets $\{(v_0, a_0)\}$ and J: There exists a non-zero continuous linear functional ℓ on

V and $\alpha \in \mathbb{R}$ such that

$$\ell(v_0) + \alpha\, a_0 < \ell(v) + \alpha\, a \quad \forall\, (v, a) \in J. \tag{10.4.4}$$

Taking $v = v_0$ and $a = j(v_0)$ in (10.4.4), we obtain

$$\alpha\,(j(v_0) - a_0) > 0.$$

Thus $\alpha > 0$. Divide the inequality (10.4.4) by α and let $a = j(v)$ to obtain

$$j(v) > \frac{-1}{\alpha}\,\ell(v) + a_0 + \frac{1}{\alpha}\,\ell(v_0).$$

Hence the result follows with $\ell_j(v) = -\ell(v)/\alpha$ and $c_0 = a_0 + \ell(v_0)/\alpha$. ∎

Example 10.4.3 *With the identification*

$$V = \{\mathbf{v} \in (H^1(\Omega))^d \mid \mathbf{v} = \mathbf{0} \text{ a.e. on } \Gamma_u\},$$

$$a(\mathbf{u}, \mathbf{v}) = \int_\Omega C_{ijkl} u_{i,j} v_{k,l}\, dx,$$

$$j(\mathbf{v}) = \int_{\Gamma_C} \mu_F G\, |\mathbf{v}_t|\, ds,$$

$$\ell(\mathbf{v}) = \int_\Omega \mathbf{f} \cdot \mathbf{v}\, dx + \int_{\Gamma_g} \mathbf{g} \cdot \mathbf{v}\, ds - \int_{\Gamma_C} G\, v_\nu\, ds,$$

we see that the frictional contact problem (10.1.16) is an elliptic variational inequality of the second kind. It is easy to verify that the conditions stated in Theorem 10.4.1 are satisfied, and hence the problem has a unique solution.

Exercise 10.4.1 *In addition to the conditions stated in Theorem 10.4.1, assume $j : V \to \mathbb{R}$. Show that u is the solution of the variational inequality (10.4.1) if and only if it satisfies the two relations:*

$$a(u, v) + j(v) \geq \ell(v) \quad \forall\, v \in V,$$
$$a(u, u) + j(u) = \ell(u).$$

Exercise 10.4.2 *The problem of the flow of a viscous plastic fluid in a pipe can be formulated as an EVI of the second kind. Let $\Omega \subseteq \mathbb{R}^2$ be a bounded domain with a Lipschitz boundary $\partial\Omega$. Define*

$$V = H_0^1(\Omega),$$

$$a(u, v) = \mu \int_\Omega \nabla u \cdot \nabla v\, dx,$$

$$\ell(v) = \int_\Omega f\, v\, dx,$$

$$j(v) = g \int_\Omega |\nabla v|\, dx,$$

where $\mu > 0$ and $g > 0$ are two parameters. Then the problem is

$$u \in V, \quad a(u, v - u) + j(v) - j(u) \geq \ell(v - u) \quad \forall\, v \in V.$$

Show that the problem has a unique solution.

   Like for the elasto-plastic torsion problem presented in Exercise 10.2.4, it is possible to find the exact solution with special data ([58, Chap. 2]). Assume Ω is the circle centered at the origin with radius R and $f = c_0 > 0$. Then if $g \geq c_0 R/2$, the solution is $u(\mathbf{x}) = 0$; while if $g \leq c_0 R/2$, the solution is given by

$$u(\mathbf{x}) = \begin{cases} \dfrac{R - R'}{2\,\mu} \left[\dfrac{c_0}{2} (R + R') - 2\,g \right], & \text{if } 0 \leq r \leq R', \\ \dfrac{R - r}{2\,\mu} \left[\dfrac{c_0}{2} (R + r) - 2\,g \right], & \text{if } R' \leq r \leq R, \end{cases}$$

where $R' = 2\,g/c_0$. Verify this result.

Exercise 10.4.3 *In this exercise, we consider a simplified version of the friction problem in linear elasticity. Let Ω be a bounded planar domain with a Lipschitz boundary Γ. The simplified friction problem is an EVI of the second kind with the data*

$$V = H^1(\Omega),$$

$$a(u, v) = \int_\Omega (\nabla u \cdot \nabla v + u\,v)\,dx,$$

$$\ell(v) = \int_\Omega f\,v\,dx,$$

$$j(v) = g \int_\Gamma |v|\,ds, \quad g > 0.$$

For simplicity, we assume $f \in L^2(\Omega)$. Show that the problem has a unique solution, which is also the unique minimizer of the functional

$$E(v) = \frac{1}{2}\,a(v, v) + j(v) - \ell(v)$$

over the space V. Also show that the variational inequality is formally equivalent to the boundary value problem:

$$\begin{cases} -\Delta u + u = f & \text{in } \Omega, \\ \left| \dfrac{\partial u}{\partial \nu} \right| \leq g, \; u\,\dfrac{\partial u}{\partial \nu} + g\,|u| = 0 & \text{on } \Gamma. \end{cases}$$

The boundary conditions can be expressed as

$$\left| \frac{\partial u}{\partial \nu} \right| \leq g$$

and

$$\left| \frac{\partial u}{\partial \nu} \right| < g \Longrightarrow u = 0,$$

$$\frac{\partial u}{\partial \nu} = g \Longrightarrow u \leq 0,$$

$$\frac{\partial u}{\partial \nu} = -g \Longrightarrow u \geq 0.$$

10.5 Approximation of EVIs of the second kind

As in the preceding section, let V be a real Hilbert space, $a(\cdot, \cdot)$ a V-elliptic, bounded bilinear form, ℓ a continuous linear functional on V. Also let $j(\cdot)$ be a proper, convex, and l.s.c. functional on V. Under these assumptions, by Theorem 10.4.1, there exists a unique solution of the elliptic variational inequality

$$u \in V, \quad a(u, v - u) + j(v) - j(u) \geq \ell(v - u) \quad \forall \, v \in V. \tag{10.5.1}$$

Let $V_h \subseteq V$ be a finite element space. Then the finite element approximation of the problem (10.5.1) is

$$u_h \in V_h, \quad a(u_h, v_h - u_h) + j(v_h) - j(u_h) \geq \ell(v_h - u_h) \quad \forall \, v_h \in V_h. \tag{10.5.2}$$

Assuming additionally that $j(\cdot)$ is proper also on V_h, as is always the case in applications, we can use Theorem 10.4.1 to conclude that the discrete problem (10.5.2) has a unique solution u_h and $j(u_h) \in \mathbb{R}$. We will now derive an abstract error estimate for $u - u_h$.

Theorem 10.5.1 *There is a constant $c > 0$ independent of h and u, such that*

$$\|u - u_h\| \leq c \inf_{v_h \in V_h} \Big\{ \|u - v_h\|$$
$$+ |a(u, v_h - u) + j(v_h) - j(u) - \ell(v_h - u)|^{1/2} \Big\}. \tag{10.5.3}$$

Proof. We let $v = u_h$ in (10.5.1) and add the resulting inequality to the inequality (10.3.1) to obtain an error relation

$$a(u, u_h - u) + a(u_h, v_h - u_h) + j(v_h) - j(u) \geq \ell(v_h - u) \quad \forall \, v_h \in V_h.$$

Using this error relation, together with the V-ellipticity and boundedness of the bilinear form, we have for any $v_h \in V_h$,

$$\begin{aligned}
\alpha \, \|u &- u_h\|^2 \\
&\leq a(u - u_h, u - u_h) \\
&= -a(u, u_h - u) - a(u_h, v_h - u_h) + a(u_h - u, v_h - u) + a(u, v_h - u) \\
&\leq a(u - u_h, u - v_h) + a(u, v_h - u) + j(v_h) - j(u) - \ell(v_h - u) \\
&\leq M \, \|u - u_h\| \, \|u - v_h\| + a(u, v_h - u) + j(v_h) - j(u) - \ell(v_h - u) \\
&\leq \tfrac{1}{2} \alpha \, \|u - u_h\|^2 + c \, \|u - v_h\|^2 + a(u, v_h - u) \\
&\quad + j(v_h) - j(u) - \ell(v_h - u),
\end{aligned}$$

from which it is easy to see that (10.5.3) holds. ∎

We observe that Theorem 10.5.1 is a generalization of Céa's lemma to the finite element approximation of elliptic variational inequalities of the second kind. The inequality (10.5.3) is the basis for order error estimates of finite element solutions of various application problems.

Let us apply the inequality (10.5.3) to derive an error estimate for some finite element solution of a model problem. Let $\Omega \subseteq \mathbb{R}^2$ be an open bounded set, with a Lipschitz domain $\partial\Omega$. We take

$$V = H^1(\Omega),$$

$$a(u, v) = \int_\Omega (\nabla u \, \nabla v + u \, v) \, dx,$$

$$\ell(v) = \int_\Omega f \, v \, dx,$$

$$j(v) = g \int_{\partial\Omega} |v| \, ds.$$

Here $f \in L^2(\Omega)$ and $g > 0$ are given. This problem is a simplified version of the friction problem in elasticity (cf. [45]). We choose this model problem for its simplicity while at the same time it contains the main feature of an elliptic variational inequality of the second kind. Applying Theorem 10.4.1, we see that the corresponding variational inequality problem

$$u \in V, \quad a(u, v - u) + j(v) - j(u) \geq \ell(v - u) \quad \forall v \in V \qquad (10.5.4)$$

has a unique solution. Given a finite element space V_h, let u_h denote the corresponding finite element solution defined in (10.5.2). To simplify the exposition, we will assume below that Ω is a polygonal domain, and write $\partial\Omega = \cup_{i=1}^{i_0} \Gamma_i$, where each Γ_i is a line segment. For an error estimation, we have the following result.

Theorem 10.5.2 *Assume, for the model problem, $u \in H^2(\Omega)$, and for each i, $u|_{\Gamma_i} \in H^2(\Gamma_i)$. Let V_h be a piecewise linear finite element space constructed from a regular partition of the domain Ω. Let $u_h \in V_h$ be the finite element solution defined by (10.5.2). Then we have the optimal order error estimate*

$$\|u - u_h\|_{H^1(\Omega)} \leq c(u) \, h. \qquad (10.5.5)$$

Proof. We apply the result of Theorem 10.5.1.

$$a(u, v_h - u) + j(v_h) - j(u) - \ell(v_h - u)$$
$$= \int_{\partial\Omega} \left[\frac{\partial u}{\partial \nu} (v_h - u) + g \left(|v_h| - |u| \right) \right] ds$$
$$+ \int_\Omega (-\Delta u + u - f) (v_h - u) \, dx$$
$$\leq \left(\left\| \frac{\partial u}{\partial \nu} \right\|_{L^2(\partial\Omega)} + g \sqrt{\text{meas}(\partial\Omega)} \right) \|v_h - u\|_{L^2(\partial\Omega)}$$
$$+ \| - \Delta u + u - f \|_{L^2(\Omega)} \|v_h - u\|_{L^2(\Omega)}.$$

Using (10.5.3), we get

$$\|u - u_h\|_{H^1(\Omega)}$$
$$\leq c(u) \inf_{v_h \in V_h} \left\{ \|v_h - u\|_{H^1(\Omega)} + \|v_h - u\|_{L^2(\partial\Omega)}^{1/2} + \|v_h - u\|_{L^2(\Omega)}^{1/2} \right\}.$$

Then the error estimate (10.5.5) follows from an application of the theory of finite element interpolation error estimates discussed in Chapter 9. ∎

Let us return to the general case. A major issue in solving the discrete system (10.5.2) is the treatment of the non-differentiable term. In practice, several approaches can be used, e.g., regularization technique, method of Lagrangian multipliers, method of numerical integration. We will briefly describe the regularization technique and the method of Lagrangian multipliers, and provides a detailed discussion of error analysis for the method of numerical integration.

10.5.1   Regularization technique

The basic idea of the regularization method is to approximate the non-differentiable term $j(\cdot)$ by a family of differentiable ones $j_\varepsilon(\cdot)$, where $\varepsilon > 0$ is a small regularization parameter. Convergence of the method is obtained when $\varepsilon \to 0$. Our presentation of the method is given on the continuous level; the extension of the method to the discrete level is straightforward. For the approximate solution of the variational inequality (10.5.1), we introduce the regularized problem

$$u_\varepsilon \in V: \quad a(u_\varepsilon, v - u_\varepsilon) + j_\varepsilon(v) - j_\varepsilon(u_\varepsilon) \geq \ell(v - u_\varepsilon) \quad \forall v \in V. \quad (10.5.6)$$

Since $j_\varepsilon(\cdot)$ is differentiable, the variational inequality (10.5.6) is actually a nonlinear equation:

$$u_\varepsilon \in V: \quad a(u_\varepsilon, v) + \langle j_\varepsilon'(u_\varepsilon), v \rangle = \ell(v) \quad \forall v \in V. \quad (10.5.7)$$

Many possible regularization functions can be used for this purpose. For example, in the case of the simplified friction problem mentioned earlier in this section,

$$j(v) = g \int_{\partial\Omega} |v| \, ds.$$

We approximate this functional by

$$j_\varepsilon(v) = g \int_{\partial\Omega} \phi_\varepsilon(v) \, ds;$$

here, $\phi_\varepsilon(t)$ is differentiable with respect to t and approximates $|t|$ as $\varepsilon \to 0$. We may choose

$$\phi_\varepsilon(t) = \begin{cases} t - \varepsilon/2 & \text{if } t \geq \varepsilon, \\ t^2/(2\varepsilon) & \text{if } |t| \leq \varepsilon, \\ -t - \varepsilon/2 & \text{if } t \leq -\varepsilon, \end{cases}$$

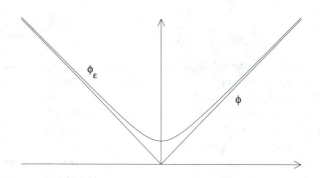

Figure 10.2. Regularization function

or

$$\phi_\varepsilon(t) = \begin{cases} t & \text{if } t \geq \varepsilon, \\ (t^2/\varepsilon + \varepsilon)/2 & \text{if } |t| \leq \varepsilon, \\ -t & \text{if } t \leq -\varepsilon, \end{cases}$$

or

$$\phi_\varepsilon(t) = \sqrt{t^2 + \varepsilon^2},$$

or

$$\phi_\varepsilon(t) = \frac{\varepsilon}{\varepsilon + 1} \left(\frac{|t|}{\varepsilon} \right)^{\varepsilon+1},$$

or

$$\phi_\varepsilon(t) = \frac{t^{\varepsilon+1}}{\varepsilon + 1}$$

and the list can expand without limit. Figure 10.2 shows graphs of the functions $\phi(t) = |t|$ and $\phi_\varepsilon(t) = \sqrt{t^2 + \varepsilon^2}$.

A general convergence result for the regularization method can be found in [59, 58]. The regularization method has been widely used in solving variational inequalities involving non-differentiable terms, see, e.g., [95, 134].

It is not difficult to derive a priori error estimates of the form

$$\|u - u_\varepsilon\|_V \leq c\varepsilon^\beta$$

for some exponent $\beta > 0$ (cf. the references mentioned above). The major problem associated with the regularization method is that the conditioning of a regularized problem deteriorates as $\varepsilon \to 0$. Thus, there is a tradeoff in the selection of the regularization parameter. Theoretically, to get more accurate approximations, we need to use smaller ε. Yet, if ε is too small, the numerical solution of the regularized problem cannot be computed accurately. It is highly desirable to have a posteriori error estimates that can give us computable error bounds once we have solutions of regularized problems. We can use the a posteriori error estimates in devising a stopping criterion in actual computations: If the estimated error is within the given error tolerance, we accept the solution of the regularized problem as the exact solution; and if the estimated error is large, then we need to use a smaller value for the regularization parameter ε. An adaptive algorithm can be developed based on the a posteriori error analysis. A posteriori error estimates of the form

$$\|u - u_\varepsilon\|_V \le F(u_\varepsilon),$$

where the error bound can be easily computed once the regularization solution u_ε is known, have been derived in several papers, see, e.g., [69, 71, 73, 80].

10.5.2   Method of Lagrangian multipliers

Again, here our presentation of the method is given on the continuous level. We take the simplified friction problem as an example. Let

$$\Lambda = \{\mu \in L^\infty(\partial\Omega) \mid |\mu(\mathbf{x})| \le 1 \text{ a.e. on } \partial\Omega\}.$$

Following [58], we have the following result.

Theorem 10.5.3 *The simplified friction problem* (10.5.4) *is equivalent to the problem of finding* $u \in V$ *and* $\lambda \in \Lambda$ *such that*

$$\int_\Omega (\nabla u \nabla v + u\,v)\,dx + g \int_{\partial\Omega} \lambda v\,ds = \int_\Omega f\,v\,dx \quad \forall\,v \in V, \qquad (10.5.8)$$

$$\lambda\,u = |u| \quad \text{a.e. on } \partial\Omega. \qquad (10.5.9)$$

λ *is called a Lagrangian multiplier.*

Proof. Let u be the solution of the variational inequality (10.5.4). Then from Exercise 10.4.1 we have

$$a(u, u) + j(u) = \ell(u) \qquad (10.5.10)$$

and

$$a(u, v) + j(v) \ge \ell(v) \quad \forall\,v \in V.$$

The latter relation implies

$$|\ell(v) - a(u, v)| \le j(v) \quad \forall\,v \in V. \qquad (10.5.11)$$

Denote $L(v) = \ell(v) - a(u, v)$. Then the value of $L(v)$ depends on the trace $v|_\Gamma$ only and $L(\cdot)$ is a linear continuous functional on $H^{1/2}(\Gamma)$. Moreover, we obtain from (10.5.11) the estimate

$$|L(v)| \leq g \, \|v\|_{L^1(\Gamma)} \quad \forall v \in H^{1/2}(\Gamma).$$

Since $H^{1/2}(\Gamma)$ is a subspace of $L^1(\Gamma)$, applying the Hahn-Banach theorem (Theorem 2.5.2) we can extend the functional L to $\tilde{L} \in (L^1(\Gamma))'$ such that

$$\|\tilde{L}\| = \|L\| \leq g.$$

Since $(L^1(\Gamma))' = L^\infty(\Gamma)$, we have the existence of a $\lambda \in \Lambda$ such that

$$\tilde{L}(v) = g \int_\Gamma \lambda v \, ds \quad \forall v \in L^1(\Gamma).$$

Therefore,

$$\ell(v) - a(u, v) = L(v) = \tilde{L}(v) = g \int_\Gamma \lambda v \, ds \quad \forall v \in V;$$

i.e., (10.5.8) holds.

Taking $v = u$ in (10.5.8) we obtain

$$a(u, u) + g \int_\Gamma \lambda u \, ds = \ell(u).$$

This relation and (10.5.10) together imply

$$\int_\Gamma (|u| - \lambda u) \, ds = 0.$$

Since $|\lambda| \leq 1$ a.e. on Γ, we must have (10.5.9).

Conversely, suppose we have $u \in V$ and $\lambda \in \Lambda$ satisfying (10.5.8) and (10.5.9). Then using (10.5.8) with v replaced by $v - u$, we obtain

$$a(u, v - u) + g \int_\Gamma \lambda v \, ds - g \int_\Gamma \lambda u \, ds = \ell(v - u).$$

Noticing that

$$g \int_\Gamma \lambda u \, ds = g \int_\Gamma |u| \, ds = j(u),$$

$$g \int_\Gamma \lambda v \, ds \leq g \int_\Gamma |v| \, ds = j(v),$$

we see that u solves the inequality (10.5.4).   ∎

It is then possible to develop an iterative solution procedure for the inequality problem. Let $\rho > 0$ be a parameter.

INITIALIZATION. Choose $\lambda_0 \in \Lambda$ (e.g. $\lambda_0 = 0$).

ITERATION. For $n = 0, 1, \ldots$, find $u_n \in V$ as the solution of the boundary value problem

$$a(u_n, v) = \ell(v) - g \int_\Gamma \lambda_n v \, ds \quad \forall v \in V,$$

and update the Lagrangian multiplier

$$\lambda_{n+1} = \mathcal{P}_\Lambda(\lambda_n + \rho g\, u_n).$$

Here $\mathcal{P}_\Lambda$ is a projection operator to Λ defined as

$$\mathcal{P}_\Lambda(\mu) = \sup(-1, \inf(1, \mu)) \quad \forall \mu \in L^\infty(\Gamma).$$

It can be shown that there exists a $\rho_0 > 0$ such that if $\rho \in (0, \rho_0)$, then the iterative method converges:

$$u_n \to u \text{ in } V, \ \lambda_n \to \lambda \text{ in } \Lambda.$$

An interested reader can consult [58, 69] for detailed discussion of the method of Lagrangian multipliers and convergence argument of the iterative method in the context of solving certain other variational inequalities.

10.5.3 Method of numerical integration

We follow [70] to analyze an approach by approximating $j(v_h)$ with $j_h(v_h)$, obtained through numerical integrations. Then the numerical method is

$$u_h \in V_h, \quad a(u_h, v_h - u_h) + j_h(v_h) - j_h(u_h) \geq \ell(v_h - u_h) \quad \forall v_h \in V_h. \tag{10.5.12}$$

For convergence analysis, there is a rather general result, proved in [58, 59].

Theorem 10.5.4 *Assume $\{V_h\}_h \subseteq V$ is a family of finite-dimensional subspaces, such that for a dense subset U of V, one can define mappings $r_h : U \to V_h$ with $\lim_{h\to 0} r_h v = v$ in V, for any $v \in U$. Assume j_h is convex, l.s.c., and uniformly proper in h, and if $v_h \rightharpoonup v$ in V, then $\liminf_{h\to 0} j_h(v_h) \geq j(v)$. Finally, assume $\lim_{h\to 0} j_h(r_h v) = j(v)$ for any $v \in U$. Then for the solution of (10.5.12), we have the convergence*

$$\lim_{h\to 0} \|u - u_h\| = 0.$$

In the above theorem, the functional family $\{j_h\}_h$ is said to be uniformly proper in h, if there exist $\ell_0 \in V^*$ and $c_0 \in \mathbb{R}$ such that

$$j_h(v_h) \geq \ell_0(v_h) + c_0 \quad \forall v_h \in V_h, \ \forall h.$$

In our application, $j(\cdot)$ is non-negative, as is $j_h(\cdot)$ (to be introduced below), so the family $\{j_h\}_h$ is trivially uniformly proper. Notice that Theorem 10.5.4 gives some general assumptions under which one can assert the convergence of the finite element solutions. However, Theorem 10.5.4 does not provide information on the convergence order of the approximations. To derive error estimates we need an inequality of the form (10.5.3).

Theorem 10.5.5 *Assume*

$$j(v_h) \leq j_h(v_h) \quad \forall\, v_h \in V_h. \tag{10.5.13}$$

Let u_h be defined by (10.5.12). Then

$$\|u - u_h\| \leq c \inf_{v_h \in V_h} \Big\{ \|u - v_h\| $$

$$+ |a(u, v_h - u) + j_h(v_h) - j(u) - \ell(v_h - u)|^{1/2} \Big\}. \tag{10.5.14}$$

Proof. Choosing $v = u_h$ in (10.5.1) and adding the resulting inequality to (10.5.12), we obtain

$$a(u, u_h - u) + a(u_h, v_h - u_h) + j(u_h) - j_h(u_h) + j_h(v_h) - j(u)$$
$$\geq \ell(v_h - u) \quad \forall\, v_h \in V_h.$$

Using the assumption (10.5.13) for $v_h = u_h$, we then have

$$a(u, u_h - u) + a(u_h, v_h - u_h) + j_h(v_h) - j(u) \geq \ell(v_h - u) \quad \forall\, v_h \in V_h.$$

The rest of the argument is similar to that in the proof of Theorem 10.5.4 and is hence omitted. ∎

Let us now comment on the assumption (10.5.13). In some applications, the functional $j(\cdot)$ is of the form $j(v) = I(g\,|v|)$ with I an integration operator, integrating over part or the whole domain or the boundary, $g \geq 0$ is a given non-negative function. One method to construct practically useful approximate functionals j_h is through numerical integrations, $j_h(v_h) = I_h(g\,|v_h|)$. Let $\{\phi_i\}_i$ be the set of functions chosen from a basis of the space V_h, which defines the functions v_h over the integration region. Assume the basis functions $\{\phi_i\}_i$ are non-negative. Writing

$$v_h = \sum_i v_i \phi_i \quad \text{on the integration region,}$$

we define

$$j_h(v_h) = \sum_i |v_i|\, I(g\,\phi_i). \tag{10.5.15}$$

Obviously the functional $j_h(\cdot)$ constructed in this way enjoys the property (10.5.13). We will see next in the analysis for solving the model problem that certain polynomial invariance property is preserved through a construction of the form (10.5.15). A polynomial invariance property is useful in deriving error estimates.

Let us again consider the model problem (10.5.4). Assume we use linear elements to construct the finite element space V_h. Denote $\{P_i\}$ the set of the nodes of the triangulation which lie on the boundary, numbered consecutively. Let $\{\phi_i\}$ be the canonical basis functions of the space V_h, corresponding to the nodes $\{P_i\}$. Obviously we have the non-negativity property for the basis functions, $\phi_i \geq 0$. Thus according to the formula

(10.5.15), we define

$$j_h(v_h) = g \sum_i |\overline{P_i P_{i+1}}| \frac{1}{2} \left(|v_h(P_i)| + |v_h(P_{i+1})| \right). \tag{10.5.16}$$

Here we use $\overline{P_i P_{i+1}}$ to denote the line segment between P_i and P_{i+1}, and $\overline{P_i P_{i+1}}$ for its length.

Assume $u \in H^2(\Omega)$. Applying Theorem 10.5.5, we have the following bound for the finite element solution error:

$$\|u - u_h\|_{H^1(\Omega)} \le c \left\{ \|u - \Pi_h u\|_{H^1(\Omega)} \right.$$
$$\left. + |a(u, \Pi_h u - u) + j_h(\Pi_h u) - j(u) - \ell(\Pi_h u - u)|^{1/2} \right\} \tag{10.5.17}$$

where $\Pi_h u \in V_h$ is the piecewise linear interpolant of the solution u. Let us first estimate the difference $j_h(\Pi_h u) - j(u)$. We have

$$j_h(\Pi_h u) - j(u) = g \sum_i \left\{ \frac{1}{2} |\overline{P_i P_{i+1}}| \left(|u(P_i)| + |u(P_{i+1})| \right) - \int_{P_i P_{i+1}} |u| \, ds \right\}. \tag{10.5.18}$$

Now if $u|_{\overline{P_i P_{i+1}}}$ keeps the same sign, then

$$\left| \frac{1}{2} |\overline{P_i P_{i+1}}| \left(|u(P_i)| + |u(P_{i+1})| \right) - \int_{P_i P_{i+1}} |u| \, ds \right|$$
$$= \left| \frac{1}{2} |\overline{P_i P_{i+1}}| \left(u(P_i) + u(P_{i+1}) \right) - \int_{P_i P_{i+1}} u \, ds \right|$$
$$= \left| \int_{P_i P_{i+1}} (u - \Pi_h u) \, ds \right|$$
$$\le \int_{P_i P_{i+1}} |u - \Pi_h u| \, ds.$$

Assume $u|_{\overline{P_i P_{i+1}}}$ changes its sign. It is easy to see that

$$\sup_{\overline{P_i P_{i+1}}} |u| \le h \, \|u\|_{W^{1,\infty}(P_i P_{i+1})}$$

if $u|_{\overline{P_i P_{i+1}}} \in W^{1,\infty}(P_i P_{i+1})$, which is guaranteed by $u|_{\Gamma_i} \in H^2(\Gamma_i)$, $i = 1, \dots, i_0$, an assumption made in Theorem 9.3.2. Thus,

$$\left| \frac{1}{2} |\overline{P_i P_{i+1}}| \left(|u(P_i)| + |u(P_{i+1})| \right) - \int_{P_i P_{i+1}} |u| \, ds \right| \le c \, h^2 \|u\|_{W^{1,\infty}(P_i P_{i+1})}.$$

Therefore, if the exact solution u changes its sign only finitely many times on $\partial\Omega$, then from (10.5.18) we find that

$$|j_h(\Pi_h u) - j(u)| \leq c h^2 \sum_{i=1}^{i_0} \|u\|_{W^{1,\infty}(\Gamma_i)} + c \|u - \Pi_h u\|_{L^1(\partial\Omega)}.$$

Using (10.5.17), we then get

$$\|u - u_h\|_{H^1(\Omega)} \leq c \Big\{ \|u - \Pi_h u\|_{H^1(\Omega)} + \left\|\frac{\partial u}{\partial\nu}\right\|_{L^2(\partial\Omega)} \|u - \Pi_h u\|_{L^2(\partial\Omega)}$$

$$+ h \Big(\sum_{i=1}^{i_0} \|u\|_{W^{1,\infty}(\Gamma_i)} \Big)^{1/2} + \|u - \Pi_h u\|_{L^1(\partial\Omega)}^{1/2}$$

$$+ \| -\Delta u + u - f\|_{L^2(\Omega)} \|u - \Pi_h u\|_{L^2(\Omega)} \Big\}.$$

In conclusion, if $u \in H^2(\Omega)$, $u|_{\Gamma_i} \in W^{1,\infty}(\Gamma_i)$ for $i = 1, \ldots, i_0$, and if $u|_{\partial\Omega}$ changes its sign only finitely many times, then we have the error estimate

$$\|u - u_h\|_{H^1(\Omega)} \leq c(u)\, h;$$

i.e., the approximation of j by j_h does not cause a degradation in the convergence order of the finite element method.

If quadratic elements are used, one can construct basis functions by using nodal shape functions and side modes (cf. [155]). Then the basis functions are non-negative, and an error analysis similar to the above one can be done.

Exercise 10.5.1 *Extend some of the discussions in this section for the numerical analysis of the variational inequalities studied in Example 10.4.3 and Exercise 10.4.2.*

Suggestion for Further Readings

Interest in variational inequalities originates in mechanical problems. An early reference is FICHERA [51]. The first rigorous comprehensive mathematical treatment seems to be LIONS AND STAMPACCHIA [110]. DUVAUT AND LIONS [45] formulated and studied many problems in mechanics and physics in the framework of variational inequalities. More recent references include FRIEDMAN [54] (mathematical analysis of various variational inequalities in mechanics); GLOWINSKI [58] and GLOWINSKI, LIONS, AND TRÉMOLIÈRES [59] (numerical analysis and solution algorithms), HAN AND REDDY [72] (mathematical and numerical analysis of variational inequalities arising in hardening plasticity); HASLINGER, HLAVÁČEK, AND NEČAS [75] and HLAVÁČEK, HASLINGER, NEČAS AND LOVÍŠEK [78] (numerical solution of variational inequalities in mechanics); KIKUCHI AND ODEN [95] (numerical analysis of various contact problems in elasticity); KINDERLEHRER AND STAMPACCHIA [96] (a mathematical introduction to the

theory of variational inequalities); and PANAGIOTOPOULOS [128] (theory
and numerical approximations of variational inequalities in mechanics). In
numerically solving higher-order variational inequalities, non-conforming
finite element methods offer a great advantage. Rigorous error analysis for
some non-conforming finite element methods in solving an EVI of the first
kind arising in unilateral problem can be found in [162]. This is a research
topic worth thorough investigation.

11

Numerical Solution of Fredholm Integral Equations of the Second Kind

Linear integral equations of the second kind,

$$\lambda u(x) - \int_D k(x, y)\, u(y)\, dy = f(x), \qquad x \in D \qquad (11.0.1)$$

were introduced in Chapter 2, and we note that they occur in a wide variety of physical applications. An important class of such equations are the *boundary integral equations*, about which more is said in Chapter 12. In the integral of 11.0.1, D is a *closed*, and often bounded, integration region. The integral operator is often a compact operator on $C(D)$ or $L^2(D)$, although not always. For the case that the integral operator is compact, a general solvability theory is given in Subsection 2.8.4 of Chapter 2. A more general introduction to the theory of such equations is given in Kress [100].

In this chapter, we look at the two most important classes of numerical methods for these equations: *projection methods* and *Nyström methods*. In Section 11.1, we introduce collocation and Galerkin methods, beginning with explicit definitions and followed by an abstract framework for the analysis of all projection methods. Illustrative examples are given in Section 11.2, and the *iterated projection method* is defined, analyzed, and illustrated in Section 11.3. The Nyström method is introduced and discussed in Section 11.4, and it is extended to the use of product integration in Section 11.5. We conclude the chapter in 11.6 by introducing and analyzing projection methods for solving some fixed point problems for nonlinear operators.

In this chapter, we use notation that is popular in the literature on the numerical solution of integral equations. For example, the spatial variable is denoted by x, not $\mathbf{x}$, in the multi-dimensional case.

11.1   Projection methods: General theory

With all projection methods, we consider solving (11.0.1) within the framework of some complete function space V, usually $C(D)$ or $L^2(D)$. We choose a sequence of finite-dimensional approximating subspaces $V_n \subseteq V$, $n \geq 1$, with V_n having dimension κ_n. Let V_n have a basis $\{\phi_1, \dots, \phi_\kappa\}$, with $\kappa \equiv \kappa_n$ for notational simplicity (which is done at various points throughout the chapter). We seek a function $u_n \in V_n$, which can be written as

$$u_n(x) = \sum_{j=1}^{\kappa_n} c_j \phi_j(x), \quad x \in D. \tag{11.1.1}$$

This is substituted into (11.0.1), and the coefficients $\{c_1, \dots, c_\kappa\}$ are determined by forcing the equation to be almost exact in some sense. For later use, introduce

$$r_n(x) = \lambda u_n(x) - \int_D k(x, y) u_n(y) \, dy - f(x)$$

$$= \sum_{j=1}^{\kappa} c_j \left\{ \lambda \phi_j(x) - \int_D k(x, y) \phi_j(y) \, dy \right\} - f(x), \tag{11.1.2}$$

for $x \in D$. This is called the *residual* in the approximation of the equation when using $u \approx u_n$. As usual, we write (11.0.1) in operator notation as

$$(\lambda - K) u = f. \tag{11.1.3}$$

Then the residual can be written as

$$r_n = (\lambda - K) u_n - f.$$

The coefficients $\{c_1, \dots, c_\kappa\}$ are chosen by forcing $r_n(x)$ to be approximately zero in some sense. The hope, and expectation, is that the resulting function $u_n(x)$ will be a good approximation of the true solution $u(x)$.

11.1.1   Collocation methods

Pick distinct node points $x_1, \dots, x_\kappa \in D$, and require

$$r_n(x_i) = 0, \quad i = 1, \dots, \kappa_n. \tag{11.1.4}$$

This leads to determining $\{c_1, \dots, c_\kappa\}$ as the solution of the linear system

$$\sum_{j=1}^{\kappa} c_j \left\{ \lambda \phi_j(x_i) - \int_D k(x_i, y) \phi_j(y) \, dy \right\} = f(x_i), \quad i = 1, \dots, \kappa. \tag{11.1.5}$$

An immediate question is whether this system has a solution and whether it is unique. If so, does u_n converge to u? Note also that the linear system contains integrals that must usually be evaluated numerically, a point we return to later. We should have written the node points as $\{x_{1,n}, \dots, x_{\kappa,n}\}$; but for notational simplicity, the explicit dependence on n has been suppressed, to be understood only implicitly.

The function space framework for collocation methods is often $C(D)$, which is what we use here. It is possible to use extensions of $C(D)$. For example, we can use $L^\infty(D)$, making use of the ideas of Example 2.5.3 from Section 2.5 to extend the idea of point evaluation of a continuous function to elements of $L^\infty(D)$. Such extensions of $C(D)$ are needed when the approximating functions u_n are not required to be continuous.

As a part of writing (11.1.5) in a more abstract form, we introduce a projection operator P_n that maps $V = C(D)$ onto V_n. Given $u \in C(D)$, define $P_n u$ to be that element of V_n that interpolates u at the nodes $\{x_1, \dots, x_\kappa\}$. This means writing

$$P_n u(x) = \sum_{j=1}^{\kappa_n} \alpha_j \phi_j(x)$$

with the coefficients $\{\alpha_j\}$ determined by solving the linear system

$$\sum_{j=1}^{\kappa_n} \alpha_j \phi_j(x_i) = u(x_i), \quad i = 1, \dots, \kappa_n.$$

This linear system has a unique solution if

$$\det\left[\phi_j(x_i)\right] \neq 0. \tag{11.1.6}$$

Henceforth in this chapter, we assume this is true whenever the collocation method is being discussed. By a simple argument, this condition also implies that the functions $\{\phi_1 \dots, \phi_\kappa\}$ are a linearly independent set over D. In the case of polynomial interpolation for functions of one variable and monomials $\{1, x, \dots, x^n\}$ as the basis functions, the determinant in (11.1.6) is referred to as the *Vandermonde determinant*.

To see more clearly that P_n is linear, and to give a more explicit formula, we introduce a new set of basis functions. For each i, $1 \leq i \leq \kappa_n$, let $\ell_i \in V_n$ be that element that satisfies the interpolation conditions

$$\ell_i(x_j) = \delta_{ij}, \quad j = 1, \dots, \kappa_n. \tag{11.1.7}$$

By (11.1.6), there is a unique such ℓ_i; and the set $\{\ell_1, \dots, \ell_\kappa\}$ is a new basis for V_n. With polynomial interpolation, such functions ℓ_i are called *Lagrange basis functions*; and we use this name with all types of approximating subspaces V_n. With this new basis, we can write

$$P_n u(x) = \sum_{j=1}^{\kappa_n} u(x_j) \ell_j(x), \quad x \in D. \tag{11.1.8}$$

Recall (3.1.1)–(3.1.3) in Chapter 3. Clearly, P_n is linear and finite rank. In addition, as an operator on $C(D)$ to $C(D)$,

$$\|P_n\| = \max_{x \in D} \sum_{j=1}^{\kappa_n} |\ell_j(x)|. \tag{11.1.9}$$

Example 11.1.1 *Let $V_n = \text{span}\{1, x, \ldots, x^n\}$. Recall the Lagrange interpolatory projection operator of Example 3.5.5 in Section 3.5 of Chapter 3:*

$$P_n g(x) \equiv \sum_{i=0}^{n} g(x_i) \ell_i(x) \tag{11.1.10}$$

with the Lagrange basis functions

$$\ell_i(x) = \prod_{\substack{j=0 \\ j \neq i}}^{n} \left(\frac{x - x_j}{x_i - x_j} \right), \qquad i = 0, 1, \ldots, n.$$

This is Lagrange's form of the interpolation polynomial. In Section 3.6.2 of Chapter 3, we denoted this projection operator by $\mathcal{I}_n$.

Returning to (11.1.8), we note that

$$P_n g = 0 \quad \text{if and only if} \quad g(x_j) = 0, \quad j = 1, \ldots, \kappa_n. \tag{11.1.11}$$

The condition (11.1.5) can now be rewritten as

$$P_n r_n = 0$$

or equivalently,

$$P_n(\lambda - K)u_n = P_n f, \quad u_n \in V_n. \tag{11.1.12}$$

We return to this below.

11.1.2 Galerkin methods

Let $V = L^2(D)$ or some other Hilbert function space, and let $(\cdot, \cdot)$ denote the inner product for V. Require r_n to satisfy

$$(r_n, \phi_i) = 0, \quad i = 1, \ldots, \kappa_n. \tag{11.1.13}$$

The left side is the Fourier coefficient of r_n associated with ϕ_i. If $\{\phi_1, \ldots, \phi_\kappa\}$ consists of the leading members of an orthonormal family $\Phi \equiv \{\phi_i\}_{i \geq 1}$ which spans V, then (11.1.13) requires the leading terms to be zero in the Fourier expansion of r_n with respect to Φ.

To find u_n, apply (11.1.13) to (11.0.1) written as $(\lambda - K)u = f$. This yields the linear system

$$\sum_{j=1}^{\kappa_n} c_j \{\lambda(\phi_j, \phi_i) - (K\phi_j, \phi_i)\} = (f, \phi_i), \quad i = 1, \ldots, \kappa_n. \tag{11.1.14}$$

This is Galerkin's method for obtaining an approximate solution to (11.0.1) or (11.1.3). Does the system have a solution? If so, is it unique? Does the resulting sequence of approximate solutions u_n converge to u in V? Does the sequence converge in $C(D)$; i.e., does u_n converge uniformly to u? Note also that the above formulation contains double integrals $(K\phi_j, \phi_i)$. These must often be computed numerically; and later, we return to a consideration of this.

As a part of writing (11.1.14) in a more abstract form, we recall the orthogonal projection operator P_n of Proposition 3.5.9 of Section 3.5 in Chapter 3, which maps V onto V_n. Recall that

$$P_n g = 0 \quad \text{if and only if} \quad (g, \phi_i) = 0, \qquad i = 1, \ldots, \kappa_n. \qquad (11.1.15)$$

With P_n, we can rewrite (11.1.13) as

$$P_n r_n = 0,$$

or equivalently,

$$P_n(\lambda - K)u_n = P_n f, \qquad u_n \in V_n. \qquad (11.1.16)$$

Note the similarity to (11.1.12).

There is a variant on Galerkin's method, known as the *Petrov-Galerkin method* (cf. Section 8.2). With it, we still choose $u_n \in V_n$; but now we require

$$(r_n, w) = 0 \qquad \forall\, w \in W_n$$

with W_n another finite-dimensional subspace, also of dimension κ_n. This method is not considered further in this chapter; but it is an important method when looking at the numerical solution of boundary integral equations. Another theoretical approach to Galerkin's method is to set it within a "variational framework," which is done in Chapter 9 and leads to finite element methods.

11.1.3   A general theoretical framework

Let V be a Banach space, and let $\{V_n \mid n \geq 1\}$ be a sequence of finite dimensional subspaces, say, of dimensions κ_n, with $\kappa_n \to \infty$ as $n \to \infty$. Let $P_n : V \to V_n$ be a bounded projection operator. This means that P_n is a bounded linear operator with

$$P_n u = u, \qquad u \in V_n.$$

Note that this implies $P_n^2 = P_n$, and thus

$$\|P_n\| = \|P_n^2\| \leq \|P_n\|^2,$$

$$\|P_n\| \geq 1. \qquad (11.1.17)$$

Recall the earlier discussion of projection operators from Section 3.5 in Chapter 3. We already have examples of P_n in the interpolatory projection operator and the orthogonal projection operator introduced above in defining the collocation and Galerkin methods, respectively.

Motivated by (11.1.12) and (11.1.16), we approximate the equation (11.1.3), $(\lambda - K)u = f$, by attempting to solve the problem

$$P_n(\lambda - K)u_n = P_n f, \quad u_n \in V_n. \tag{11.1.18}$$

This is the form in which the method is implemented, as it leads directly to equivalent finite linear systems such as (11.1.5) and (11.1.14). For the error analysis, however, we write (11.1.18) in an equivalent but more convenient form.

If u_n is a solution of (11.1.18), then by using $P_n u_n = u_n$, the equation can be written as

$$(\lambda - P_n K)u_n = P_n f, \quad u_n \in V. \tag{11.1.19}$$

To see that a solution of this is also a solution of (11.1.18), note that if (11.1.19) has a solution $u_n \in V$, then

$$u_n = \frac{1}{\lambda} [P_n f + P_n K u_n] \in V_n.$$

Thus $P_n u_n = u_n$,

$$(\lambda - P_n K)u_n = P_n(\lambda - K)u_n,$$

and this shows that (11.1.19) implies (11.1.18).

For the error analysis, we compare (11.1.19) with the original equation $(\lambda - K)u = f$ of (11.1.3), since both equations are defined on the original space V. The theoretical analysis is based on the approximation of $\lambda - P_n K$ by $\lambda - K$:

$$\lambda - P_n K = (\lambda - K) + (K - P_n K)$$
$$= (\lambda - K)[I + (\lambda - K)^{-1}(K - P_n K)]. \tag{11.1.20}$$

We use this in the following theorem.

Theorem 11.1.2 *Assume $K : V \to V$ is bounded, with V a Banach space; and assume $\lambda - K : V \overset{1-1}{\underset{onto}{\to}} V$. Further assume*

$$\|K - P_n K\| \to 0 \quad as \quad n \to \infty. \tag{11.1.21}$$

Then for all sufficiently large n, say, $n \geq N$, the operator $(\lambda - P_n K)^{-1}$ exists as a bounded operator from V to V. Moreover, it is uniformly bounded:

$$\sup_{n \geq N} \|(\lambda - P_n K)^{-1}\| < \infty. \tag{11.1.22}$$

For the solutions u_n (n sufficiently large) and u of (11.1.19) and (11.1.3), respectively, we have

$$u - u_n = \lambda(\lambda - P_nK)^{-1}(u - P_nu) \tag{11.1.23}$$

and the two-sided error estimate

$$\frac{|\lambda|}{\|\lambda - P_nK\|}\|u - P_nu\| \le \|u - u_n\| \le |\lambda|\,\|(\lambda - P_nK)^{-1}\|\,\|u - P_nu\|.$$

$$\tag{11.1.24}$$

This leads to a conclusion that $\|u - u_n\|$ converges to zero at exactly the same speed as $\|u - P_nu\|$.

Proof. (a) Pick N such that

$$\epsilon_N \equiv \sup_{n \ge N}\ \|K - P_nK\| < \frac{1}{\|(\lambda - K)^{-1}\|}. \tag{11.1.25}$$

Then the inverse $[I + (\lambda - K)^{-1}(K - P_nK)]^{-1}$ exists and is uniformly bounded by the geometric series theorem (cf. Theorem 2.3.1 in Chapter 2), and

$$\left\|[I + (\lambda - K)^{-1}(K - P_nK)]^{-1}\right\| \le \frac{1}{1 - \epsilon_N\,\|(\lambda - K)^{-1}\|}.$$

Using (11.1.20), $(\lambda - P_nK)^{-1}$ exists,

$$(\lambda - P_nK)^{-1} = [I + (\lambda - K)^{-1}(K - P_nK)]^{-1}(\lambda - K)^{-1},$$

$$\left\|(\lambda - P_nK)^{-1}\right\| \le \frac{\|(\lambda - K)^{-1}\|}{1 - \epsilon_N\,\|(\lambda - K)^{-1}\|} \equiv M. \tag{11.1.26}$$

This shows (11.1.22).

(b) For the error formula (11.1.23), apply P_n to the equation $(\lambda - K)u = f$, and then rearrange to obtain

$$(\lambda - P_nK)u = P_nf + \lambda(u - P_nu).$$

Subtract $(\lambda - P_nK)u_n = P_nf$ to get

$$(\lambda - P_nK)(u - u_n) = \lambda(u - P_nu). \tag{11.1.27}$$

Then

$$u - u_n = \lambda(\lambda - P_nK)^{-1}(u - P_nu),$$

which is (11.1.23). Taking norms and using (11.1.26),

$$\|u - u_n\| \le |\lambda|\,M\,\|u - P_nu\|. \tag{11.1.28}$$

Thus if $P_nu \to u$, then $u_n \to u$ as $n \to \infty$.

(c) The upper bound in (11.1.24) follows directly from (11.1.23), as we have just seen. The lower bound follows by taking bounds in (11.1.27), obtaining

$$|\lambda|\,\|u - P_nu\| \le \|\lambda - P_nK\|\,\|u - u_n\|.$$

This is equivalent to the lower bound in (11.1.24).

To obtain a lower bound that is uniform in n, note that for $n \geq N$,

$$\|\lambda - P_n K\| \leq \|\lambda - K\| + \|K - P_n K\|$$
$$\leq \|\lambda - K\| + \epsilon_N.$$

The lower bound in (11.1.24) can now be replaced by

$$\frac{|\lambda|}{\|\lambda - K\| + \epsilon_N} \|u - P_n u\| \leq \|u - u_n\|.$$

Combining this and (11.1.28), we have

$$\frac{|\lambda|}{\|\lambda - K\| + \epsilon_N} \|u - P_n u\| \leq \|u - u_n\| \leq |\lambda| M \|u - P_n u\|. \qquad (11.1.29)$$

This shows that u_n converges to u if and only if $P_n u$ converges to u. Moreover, if convergence does occur, then $\|u - P_n u\|$ and $\|u - u_n\|$ tend to zero with exactly the same speed. ∎

We note that in order for the theorem to be true, it is necessary only that (11.1.25) be valid, not the stronger assumption of (11.1.21). Nonetheless, the theorem is applied usually by proving (11.1.21). Therefore, to apply the above theorem we need to know whether $\|K - P_n K\| \to 0$ as $n \to \infty$. The following two lemmas address this question.

Lemma 11.1.3 *Let V, W be Banach spaces, and let $A_n : V \to W$, $n \geq 1$, be a sequence of bounded linear operators. Assume $\{A_n u\}$ converges for all $u \in V$. Then the convergence is uniform on compact subsets of V.*

Proof. By the principle of uniform boundedness (cf. Theorem 2.4.4 in Chapter 2), the operators A_n are uniformly bounded:

$$M \equiv \sup_{n \geq 1} \|A_n\| < \infty.$$

The functions A_n are also equicontinuous:

$$\|A_n u - A_n f\| \leq M \|u - f\|.$$

Let S be a compact subset of V. Then $\{A_n\}$ is a uniformly bounded and equicontinuous family of functions on the compact set S; and it is then a standard result of analysis (a straightforward generalization of Ascoli's Theorem 1.6.3 in the setting of Banach spaces) that $\{A_n u\}$ is uniformly convergent for $u \in S$. ∎

Lemma 11.1.4 *Let V be a Banach space, and let $\{P_n\}$ be a family of bounded projections on V with*

$$P_n u \to u \quad as \quad n \to \infty, \qquad u \in V. \qquad (11.1.30)$$

If $K : V \to V$ is compact, then

$$\|K - P_n K\| \to 0 \quad as \quad n \to \infty.$$

Proof. From the definition of operator norm,

$$\|K - P_n K\| = \sup_{\|u\| \leq 1} \|Ku - P_n Ku\| = \sup_{z \in K(U)} \|z - P_n z\|,$$

with $K(U) = \{Ku \mid \|u\| \leq 1\}$. The set $\overline{K(U)}$ is compact. Therefore, by the preceding Lemma 11.1.3 and the assumption (11.1.30),

$$\sup_{z \in K(U)} \|z - P_n z\| \to 0 \quad \text{as} \quad n \to \infty.$$

This proves the lemma. ∎

This last lemma includes most cases of interest, but not all. There are situations where $P_n u \to u$ for most $u \in V$, but not all u. In such cases, it is necessary to show directly that $\|K - P_n K\| \to 0$, if it is true. In such cases, of course, we see from (11.1.24) that $u_n \to u$ if and only if $P_n u \to u$; and thus the method is not convergent for some solutions u. This would occur, for example, if V_n is the set of polynomials of degree $\leq n$ and $V = C[a, b]$.

Exercise 11.1.1 *Prove the result stated in the last sentence of the proof of Lemma 11.1.3.*

Exercise 11.1.2 *Prove that the upper bound in (11.1.24) can be replaced by*

$$\|u - u_n\| \leq |\lambda| \, (1 + \gamma_n) \, \|(\lambda - K)^{-1}\| \, \|u - P_n u\|$$

with $\gamma_n \to 0$ as $n \to \infty$.

Exercise 11.1.3 *In Theorem 11.1.2, write*

$$u - u_n = e_n^{(1)} + e_n^{(2)}, \tag{11.1.31}$$

with

$$e_n^{(1)} = \lambda(\lambda - K)^{-1}(u - P_n u)$$

and $e_n^{(2)}$ defined implicitly by this and (11.1.31). Show that under the assumptions of Theorem 11.1.2,

$$\|e_n^{(2)}\| \leq \delta_n \|e_n^{(1)}\|$$

with $\delta_n \to 0$ as $n \to \infty$.

Exercise 11.1.4 *Let V and W be Banach spaces, let $A : V \xrightarrow[\text{onto}]{1-1} W$ be bounded, and let $B : V \to W$ be compact. Consider solving the equation $(A + B)u = f$ with the assumption that*

$$(A + B) v = 0 \quad \Rightarrow \quad v = 0.$$

Let V_n be an approximating finite-dimensional subspace of V, and further assume V_n is a subspace of W. Let $P_n : V \to V_n$ be a bounded projection for which

$$P_n v \to v \quad \text{as} \quad n \to \infty,$$

for all $v \in V$. Assume further that

$$P_n A = A P_n.$$

Consider the numerical approximation

$$P_n (A + B) u_n = P_n f, \qquad u_n \in V_n.$$

Develop a stability and convergence analysis for this numerical method. Hint: Consider the equation $(I + A^{-1}B) u = A^{-1}f$.

11.2 Examples

Most projection methods are based on the ways in which we approximate functions, and there are two main approaches.

- Decompose the approximation region D into elements $\Delta_1, \ldots, \Delta_m$; and then approximate a function $u \in C(D)$ by a low-degree polynomial over each of the elements Δ_i. These projection methods are often referred to as *piecewise polynomial methods* or *finite element methods*; and when D is the boundary of a region, such methods are often called *boundary element methods*.

- Approximate $u \in C(D)$ by using a family of functions which are defined globally over all of D; for example, use polynomials, trigonometric polynomials, or spherical polynomials. Often, these approximating functions are also infinitely differentiable. Sometimes these types of projection methods are referred to as *spectral methods*, especially when trigonometric polynomials are used.

We illustrate each of these, relying on approximation results introduced earlier in Chapter 3.

11.2.1 Piecewise linear collocation

We consider the numerical solution of the integral equation

$$\lambda u(x) - \int_a^b k(x, y) u(y) \, dy = f(x), \qquad a \le x \le b, \tag{11.2.1}$$

using piecewise linear approximating functions. Recall the definition of piecewise linear interpolation given in Subsection 3.1.3, including the piecewise linear interpolatory projection operator of (3.1.7). For convenience, we repeat those results here. Let $D = [a, b]$ and $n \ge 1$, and define $h = (b-a)/n$,

$$x_j = a + jh, \qquad j = 0, 1, \ldots, n.$$

The subspace V_n is the set of all functions that are continuous and piecewise linear on $[a, b]$, with breakpoints $\{x_0, \ldots, x_n\}$. Its dimension is $n + 1$.

Introduce the Lagrange basis functions for continuous piecewise linear interpolation:

$$\ell_i(x) = \begin{cases} 1 - \dfrac{|x - x_i|}{h}, & x_{i-1} \le x \le x_{i+1}, \\ 0, & \text{otherwise}. \end{cases} \qquad (11.2.2)$$

with the obvious adjustment of the definition for $\ell_0(x)$ and $\ell_n(x)$. The projection operator is defined by

$$P_n u(x) = \sum_{i=0}^{n} u(x_i)\ell_i(x). \qquad (11.2.3)$$

For convergence of $P_n u$, recall from (3.1.8) and (3.1.9) that

$$\|u - P_n u\|_\infty \le \begin{cases} \omega(u, h), & u \in C[a, b], \\ \dfrac{h^2}{8}\|u''\|_\infty, & u \in C^2[a, b]. \end{cases} \qquad (11.2.4)$$

This shows that $P_n u \to u$ for all $u \in C[a, b]$; and for $u \in C^2[a, b]$, the convergence order is 2. For any compact operator $K : C[a, b] \to C[a, b]$, Lemma 11.1.4 implies $\|K - P_n K\| \to 0$ as $n \to \infty$. Therefore the results of Theorem 11.1.2 can be applied directly to the numerical solution of the integral equation $(\lambda - K)u = f$. For sufficiently large n, say, $n \ge N$, the equation $(\lambda - P_n K)u_n = P_n f$ has a unique solution u_n for each $f \in C[a, b]$. Assuming $u \in C^2[a, b]$, (11.1.24) implies

$$\|u - u_n\|_\infty \le |\lambda| \, M \, \frac{h^2}{8} \|u''\|_\infty, \qquad (11.2.5)$$

with M a uniform bound on $(\lambda - P_n K)^{-1}$ for $n \ge N$.

The linear system (11.1.5) takes the simpler form

$$\lambda u_n(x_i) - \sum_{j=0}^{n} u_n(x_j) \int_a^b k(x_i, y)\ell_j(y) \, dy = f(x_i), \qquad i = 0, \dots, n.$$

$$(11.2.6)$$

The integrals can be simplified. For $j = 1, \dots, n - 1$,

$$\int_a^b k(x_i, y)\ell_j(y) \, dy = \frac{1}{h} \int_{x_{j-1}}^{x_j} k(x_i, y)(y - x_{j-1}) \, dy$$

$$+ \frac{1}{h} \int_{x_j}^{x_{j+1}} k(x_i, y)(x_j - y) \, dy. \qquad (11.2.7)$$

The integrals for $j = 0$ and $j = n$ are modified straightforwardly. These integrals must usually be calculated numerically; and we want to use a quadrature method that retains the order of convergence in (11.2.5) at a minimum cost in calculation time.

n	$E_n^{(1)}$	Ratio	$E_n^{(2)}$	Ratio
	$5.25E-3$		$2.32E-2$	
24	$1.31E-3$	4.01	$7.91E-3$	2.93
8	$3.27E-4$	4.01	$2.75E-3$	2.88
16	$8.18E-5$	4.00	$9.65E-4$	2.85
32	$2.04E-5$	4.00	$3.40E-4$	2.84
64	$5.11E-6$	4.00	$1.20E-4$	2.83
128	$1.28E-6$	4.00	$4.24E-5$	2.83

Table 11.1. Example of piecewise linear collocation for solving (11.2.8)

Example 11.2.1 *Consider the integral equation*

$$\lambda u(x) - \int_0^b e^{xy} u(y)\, dy = f(x), \qquad 0 \le x \le b. \tag{11.2.8}$$

The equation parameters are $b = 1$, $\lambda = 5$. We use the two unknowns

$$u^{(1)}(x) = e^{-x}\cos(x), \quad u^{(2)}(x) = \sqrt{x}, \quad 0 \le x \le b \tag{11.2.9}$$

and define $f(x)$ accordingly. The results of the use of piecewise linear collocation are given in Table 11.1. The errors given in the table are the maximum errors on the collocation node points,

$$E_n^{(k)} = \max_{0 \le i \le n} \left| u^{(k)}(x_i) - u_n^{(k)}(x_i) \right|.$$

The column labeled Ratio is the ratio of the successive values of $E_n^{(k)}$ as n is doubled.

The function $u^{(2)}(x)$ is not continuously differentiable on $[0, b]$, and we have no reason to expect a rate of convergence of $O(h^2)$. Empirically, the errors $E_n^{(2)}$ appear to be $O(h^{1.5})$. From (11.1.24), Theorem 11.1.2, we know that $\|u^{(2)} - u_n^{(2)}\|_\infty$ converges to zero at exactly the same speed as $\|u^{(2)} - P_n u^{(2)}\|_\infty$, and it can be shown that the latter is only $O(h^{0.5})$. This apparent contradiction between the empirical and theoretical rates is due to $u_n(t)$ being superconvergent at the collocation node points: for the numerical solution $u_n^{(2)}$,

$$\lim_{n \to \infty} \frac{E_n^{(2)}}{\|u^{(2)} - u_n^{(2)}\|_\infty} = 0.$$

This is examined in much greater detail in the following Section 11.3.

11.2.2 Trigonometric polynomial collocation

We solve the integral equation

$$\lambda u(x) - \int_0^{2\pi} k(x,y)u(y)\,dy = f(x), \qquad 0 \le x \le 2\pi, \qquad (11.2.10)$$

in which the kernel function is assumed to be continuous and 2π-periodic in both y and x:

$$k(x+2\pi,y) \equiv k(x,y+2\pi) \equiv k(x,y).$$

Let $V = C_p(2\pi)$, space of all 2π-periodic and continuous functions on $\mathbb{R}$. We consider the solution of (11.2.10) for $f \in C_p(2\pi)$, which then implies $u \in C_p(2\pi)$.

Since the solution $u(x)$ is 2π-periodic, we approximate it with *trigonometric polynomials;* and we use the general framework for trigonometric polynomial interpolation of Section 3.6.2 from Chapter 3. Let V_n denote the trigonometric polynomials of degree at most n; and recall V_n has dimension $\kappa_n = 2n+1$. Let $\{\phi_1(x),\ldots,\phi_\kappa(x)\}$ denote a basis for V_n, either $\{e^{ikx} \mid k = 0,\pm1,\ldots,\pm n\}$ or

$$\{1, \sin x, \cos x, \ldots, \sin nx, \cos nx\}. \qquad (11.2.11)$$

The interpolatory projection of $C_p(2\pi)$ onto V_n is given by

$$P_n u(x) = \sum_{j=1}^{\kappa_n} u(x_j)\ell_j(x), \qquad (11.2.12)$$

where the Lagrange basis functions $\ell_j(x)$ are given implicitly in the Lagrange formula (3.6.12) of Section 3.6.2. Note that P_n was denoted by $\mathcal{I}_n$ in that formula.

From (3.6.13), $\|P_n\| = O(\log n)$. Since $\|P_n\| \to \infty$ as $n \to \infty$, it follows from the principle of uniform boundedness that there exists $u \in C_p(2\pi)$ for which $P_n u$ does not converge to u in $C_p(2\pi)$ (cf. Theorem 2.4.4 in Chapter 2).

Consider the use of the above trigonometric interpolation in solving (11.2.10) by collocation. The linear system (11.1.5) becomes

$$\sum_{j=1}^{\kappa_n} c_j \left\{ \lambda\phi_j(x_i) - \int_0^{2\pi} k(x_i,y)\phi_j(y)\,dy \right\} = f(x_i), \qquad i = 1,\ldots,\kappa_n,$$

$$(11.2.13)$$

and the solution is

$$u_n(x) = \sum_{j=1}^{\kappa_n} c_j \phi_j(x).$$

The integrals in (11.2.13) are usually evaluated numerically, and for that, we recommend using the trapezoidal rule. With periodic integrands, the trapezoidal rule is very effective, as was noted earlier in Proposition 6.5.6 of Chapter 6.

To prove the convergence of this collocation method, we must show

$$\|K - P_n K\| \to 0 \quad \text{as } n \to \infty.$$

Since Lemma 11.1.4 cannot be used, we must examine $\|K - P_n K\|$ directly. The operator $P_n K$ is an integral operator, with

$$P_n K u(x) = \int_0^{2\pi} k_n(x, y) u(y) \, dy, \tag{11.2.14}$$

$$k_n(x, y) \equiv (P_n k_y)(x), \qquad k_y(x) \equiv k(x, y).$$

To show convergence of $\|K - P_n K\|$ to zero, we must prove directly that

$$\|K - P_n K\| = \max_{0 \le x \le 2\pi} \int_0^{2\pi} |k(x, y) - k_n(x, y)| \, dy \tag{11.2.15}$$

converges to zero. To do so, we use the result (3.6.14) on the convergence of trigonometric polynomial interpolation.

Assume that $k(x, y)$ satisfies, for some $\alpha > 0$,

$$|k(x, y) - k(\xi, y)| \le c(k) |x - \xi|^\alpha, \tag{11.2.16}$$

for all y, x, ξ. Then we leave it as Exercise 11.2.2 to prove that

$$\|K - P_n K\| \le \frac{c \log n}{n^\alpha}. \tag{11.2.17}$$

Since this converges to zero, we can apply Theorem 11.1.2 to the error analysis of the collocation method with trigonometric interpolation.

Assuming (11.2.10) is uniquely solvable, the collocation equation

$$(\lambda - P_n K) u_n = P_n f$$

has a unique solution u_n for all sufficiently large n; and $\|u - u_n\|_\infty \to 0$ if and only if $\|u - P_n u\|_\infty \to 0$. We know there are cases for which the latter is not true; but from (3.6.14) of Section 3.6 in Chapter 3, $\|u - u_n\|_\infty \nrightarrow 0$ only for functions u with very little smoothness (cf. Theorem 3.6.2 with $k = 0$). For functions u that are infinitely differentiable, the bound (3.6.14) shows the rate of convergence is very rapid, faster than $O(n^{-k})$ for any k.

There are kernel functions $k(x, y)$ that do not satisfy (11.2.16), but to which the above collocation method can still be applied. Their error analysis requires a more detailed knowledge of the smoothing properties of the operator K. Such cases occur when solving boundary integral equations with singular kernel functions, such as that defined in (6.5.16) of Section 6.5.

11.2.3 A piecewise linear Galerkin method

The error analysis of Galerkin methods is usually carried out in a Hilbert space, generally $L^2(D)$ or some Sobolev space $H^r(D)$. Following this, an analysis within $C(D)$ is often also given, to obtain results on uniform convergence of the numerical solutions.

We again consider the numerical solution of (11.2.1). Let $V = L^2(a, b)$, and let its norm and inner product be denoted by simply $\| \cdot \|$ and $(\cdot, \cdot)$, respectively. Let V_n be the subspace of continuous piecewise linear functions as described earlier in Subsection 11.2.1. The dimension of V_n is $n + 1$, and the Lagrange functions of (11.2.2) are a basis for V_n. However, now P_n denotes the orthogonal projection of $L^2(a, b)$ onto V_n. We begin by showing that $P_n u \to u$ for all $u \in L^2(a, b)$.

Begin by assuming $u(x)$ is continuous on $[a, b]$. Let $\mathcal{I}_n u(x)$ denote the piecewise linear function in V_n that interpolates $u(x)$ at $x = x_0, \ldots, x_n$; see (11.2.3). Recall that $P_n u$ minimizes $\|u - z\|$ as z ranges over V_n, a fact expressed in the identity in Proposition 3.5.9(c). Therefore,

$$\begin{aligned}
\|u - P_n u\| &\leq \|u - \mathcal{I}_n u\| \\
&\leq \sqrt{b - a}\, \|u - \mathcal{I}_n u\|_\infty \\
&\leq \sqrt{b - a}\, \omega(u; h). \quad\quad\quad (11.2.18)
\end{aligned}$$

The last inequality uses the error bound (11.2.4). This shows $P_n u \to u$ for all continuous functions u on $[a, b]$.

It is well known that the set of all continuous functions on $[a, b]$ is dense in $L^2(a, b)$ (see the comment following Theorem 1.5.7). Also, the orthogonal projection P_n satisfies $\|P_n\| = 1$; cf. (3.3.4) of Section 3.3. For a given $u \in L^2(a, b)$, let $\{u_m\}$ be a sequence of continuous functions that converge to u in $L^2(a, b)$. Then

$$\begin{aligned}
\|u - P_n u\| &\leq \|u - u_m\| + \|u_m - P_n u_m\| + \|P_n(u - u_m)\| \\
&\leq 2\,\|u - u_m\| + \|u_m - P_n u_m\|.
\end{aligned}$$

Given an $\epsilon > 0$, pick m such that $\|u - u_m\| < \epsilon/4$; and fix m. This then implies that for all n,

$$\|u - P_n u\| \leq \frac{\epsilon}{2} + \|u_m - P_n u_m\|.$$

We have

$$\|u - P_n u\| \leq \epsilon$$

for all sufficiently large values of n. Since ϵ was arbitrary, this shows that $P_n u \to u$ for general $u \in L^2(a, b)$.

For the integral equation $(\lambda - K)u = f$, we can use Lemma 11.1.4 to obtain $\|K - P_n K\| \to 0$. This justifies the use of Theorem 11.1.2 to carry out the error analysis for the Galerkin equation $(\lambda - P_n K)u_n = P_n f$. As before, $\|u - u_n\|$ converges to zero with the same speed as $\|u - P_n u\|$. For

$u \in C^2[a, b]$, we combine (11.1.28), (11.2.18), and (11.2.4), to obtain

$$\|u - u_n\| \leq |\lambda| \, M \, \|u - P_n u\|$$
$$\leq |\lambda| \, M \sqrt{b-a} \, \|u - \mathcal{I}_n u\|_\infty$$
$$\leq |\lambda| \, M \sqrt{b-a} \, \frac{h^2}{8} \, \|u''\|_\infty . \qquad (11.2.19)$$

For the linear system, we use the Lagrange basis functions of (11.2.2). These are not orthogonal, but they are still a very convenient basis with which to work. Moreover, producing an orthogonal basis for V_n is a nontrivial task. The solution u_n of $(\lambda - P_n K)u_n = P_n f$ is given by

$$u_n(x) = \sum_{j=0}^{n} c_j \ell_j(x).$$

The coefficients $\{c_j\}$ are obtained by solving the linear system

$$\sum_{j=0}^{n} c_j \left\{ \lambda(\ell_i, \ell_j) - \int_a^b \int_a^b k(x, y)\ell_i(x)\ell_j(y) \, dy \, dx \right\}$$
$$= \int_a^b f(x)\ell_i(x) \, dx, \quad i = 0, \dots, n. \qquad (11.2.20)$$

For the coefficients (ℓ_i, ℓ_j),

$$(\ell_i, \ell_j) = \begin{cases} 0, & |i - j| > 1, \\ \dfrac{2h}{3}, & 0 < i = j < n, \\ \dfrac{h}{3}, & i = j = 0 \text{ or } n, \\ \dfrac{h}{6}, & |i - j| = 1. \end{cases} \qquad (11.2.21)$$

The double integrals in (11.2.20) reduce to integrals over much smaller subintervals, because the basis functions $\ell_i(x)$ are zero over most of $[a, b]$. If these integrals are evaluated numerically, it is important to evaluate them with an accuracy consistent with the error bound in (11.2.19). Lesser accuracy degrades the accuracy of the Galerkin solution u_n; and greater accuracy is an unnecessary expenditure of effort.

Just as was true with collocation methods, we can easily generalize the above presentation to include the use of piecewise polynomial functions of any fixed degree. Since the theory is entirely analogous to that presented above, we omit it here and leave it as an exercise for the reader.

We defer to the following case a consideration of the uniform convergence of $u_n(x)$ to $u(x)$.

11.2.4 A Galerkin method with trigonometric polynomials

We consider again the use of trigonometric polynomials as approximations in solving the integral equation (11.2.10), with $k(x, y)$ and $f(x)$ being 2π-periodic functions as before. Initially, we use the space $V = L^2(0, 2\pi)$, the space of all complex-valued and square integrable Lebesgue measurable functions on $(0, 2\pi)$. The inner product is defined by

$$(u, v) = \int_0^{2\pi} u(x)\overline{v(x)}\, dx.$$

Later we consider the space $C_p(2\pi)$, the set of all complex-valued 2π-periodic continuous functions, with the uniform norm.

The approximating subspace V_n is again the set of all trigonometric polynomials of degree $\leq n$. As a basis, we use the complex exponentials,

$$\phi_j(x) = e^{ijx}, \quad j = 0, \pm 1, \ldots, \pm n.$$

Earlier, in Example 3.3.9 of Section 3.3 and in Subsection 3.6.1, we introduced and discussed the Fourier series

$$u(x) = \frac{1}{2\pi} \sum_{j=-\infty}^{\infty} (u, \phi_j)\phi_j(x), \quad u \in L^2(0, 2\pi), \tag{11.2.22}$$

with respect to the basis

$$\{1, \sin x, \cos x, \ldots, \sin nx, \cos nx, \ldots\}.$$

The basis

$$\{e^{ijx}, j = 0, \pm 1, \pm 2, \ldots\}$$

was used in defining the periodic Sobolev spaces $H^r(0, 2\pi)$ in Section 6.5 of Chapter 6. These two bases are equivalent, and it is straightforward to convert between them. It is well known that for $u \in L^2(0, 2\pi)$, the Fourier series converges in the norm of $L^2(0, 2\pi)$.

The orthogonal projection of $L^2(0, 2\pi)$ onto V_n is just the nth partial sum of this series,

$$P_n u(x) = \frac{1}{2\pi} \sum_{j=-n}^{n} (u, \phi_j)\phi_j(x), \tag{11.2.23}$$

which was denoted earlier by $\mathcal{F}_n u$ in Example 3.5.8 of Section 3.5. From the convergence of (11.2.22), it follows that $P_n u \to u$ for all $u \in L^2(0, 2\pi)$. Its rate of uniform convergence was considered in Subsection 3.6.1. For its rate of convergence in $H^r(0, 2\pi)$, it is straightforward to use the framework of Section 6.5 in Chapter 6 to prove the following:

$$\|u - P_n u\|_{L^2} \leq \frac{c}{n^r} \left[\frac{1}{2\pi} \sum_{|j|>n} |j|^{2r} |(u, \phi_j)|^2 \right]^{\frac{1}{2}} \leq \frac{c}{n^r} \|u\|_{H^r} \tag{11.2.24}$$

for $u \in H^r(0, 2\pi)$.

Using Lemma 11.1.4, we have that $\|K - P_n K\| \to 0$ as $n \to \infty$. Thus Theorem 11.1.2 can be applied to the error analysis of the approximating equation $(\lambda - P_n K)u_n = P_n f$. For all sufficiently large n, say, $n \geq N$, the inverses $(\lambda - P_n K)^{-1}$ are uniformly bounded; and $\|u - u_n\|$ can be bounded proportional to $\|u - P_n u\|$, and thus u_n converges to u. One result on the rate of convergence is obtained by applying (11.2.24) to (11.1.28):

$$\|u - u_n\| \leq \frac{c\,|\lambda|\,M}{n^r}\,\|u\|_{H^r}, \qquad n \geq N, \quad u \in H^r(0, 2\pi) \qquad (11.2.25)$$

with c the same as in (11.2.24).

With respect to the basis $\{e^{ijx}, j = 0, \pm 1, \pm 2, \dots\}$, the linear system (11.1.14) for

$$(\lambda - P_n K)u_n = P_n f$$

is given by

$$2\pi \lambda c_k - \sum_{j=-n}^{n} c_j \int_0^{2\pi} \int_0^{2\pi} e^{i(jy - kx)} k(x, y)\, dy\, dx$$

$$= \int_0^{2\pi} e^{-ikx} f(x)\, dx, \quad k = -n, \dots, n \qquad (11.2.26)$$

with the solution u_n given by

$$u_n(x) = \sum_{j=-n}^{n} c_j e^{ijx}.$$

The integrals in this system are usually evaluated numerically; and this is examined in some detail in [13, pp. 148–150]. Again, the trapezoidal rule is the standard form of quadrature used in evaluating these integrals; the fast Fourier transform can also be used to improve the efficiency of the quadrature process (cf., e.g., [11, p. 181]; [74, Chap. 13]).

Another important example of the use of globally defined and smooth approximations is the use of spherical polynomials (cf. (6.5.5) in Chapter 6) as approximations to functions defined on the unit sphere in $\mathbb{R}^3$.

Uniform convergence

We often are interested in obtaining uniform convergence of u_n to u. For this, we regard the operator P_n of (11.2.23) as an operator on $C_p(2\pi)$ to V_n, and we take $V = C_p(2\pi)$. Unfortunately, it is no longer true that $P_n u$ converges to u for all $u \in V$, and consequently, Lemma 11.1.4 cannot be applied. In fact, from (3.6.8)–(3.6.9) of Subsection 3.6.1,

$$\|P_n\| = O(\log n), \qquad (11.2.27)$$

which implies the sequence $\{P_n\}$ is not uniformly bounded and therefore we do not expect pointwise convergence of the sequence $\{P_n u\}$ to u for all $u \in V$.

We use the framework of (11.2.14)–(11.2.15) to examine whether the quantity $\|K - P_n K\|$ converges to zero or not. In the present case, the projection used in (11.2.14) is the orthogonal Fourier projection of (11.2.23), but otherwise the results are the same.

Assume $k(x, y)$ satisfies the Hölder condition

$$|k(x, y) - k(\xi, y)| \le c(K) |x - \xi|^\alpha ,$$

for all y, x, ξ, for some $0 < \alpha \le 1$. Then apply (3.6.11) to obtain

$$\|K - P_n K\| \le \frac{c \log n}{n^\alpha}$$

for a suitable constant c. With this, we can apply Theorem 11.1.2 and obtain a complete convergence analysis within $C_p(2\pi)$, thus obtaining results on the uniform convergence of u_n to u.

Another way of obtaining such uniform convergence results can be based on the ideas of the following section on the iterated projection method.

Conditioning of the linear system

We have omitted any discussion of the conditioning of the linear systems associated with either the Galerkin or collocation methods. This is important when implementing these methods, and the basis for V_n should be chosen with some care. The linear system is as well-conditioned as can be expected, based on the given equation $(\lambda - K) u = f$, if $[\phi_j(x_i)] = I$ for collocation or $[(\phi_j, \phi_i)] = I$ for the Galerkin method. It can still be well-conditioned without such a restriction, but the choice of basis must be examined more carefully. See [13, Section 3.6] for an extended discussion.

Exercise 11.2.1 *For the piecewise linear interpolatory projection operator of Subsection 11.2.1, calculate an explicit formula for the operator $P_n K$, showing it is a degenerate kernel integral operator. Be as explicit as possible in defining the degenerate kernel. Assuming $k(x, y)$ is twice continuously differentiable with respect to x, uniformly for $a \le y \le b$, show*

$$\|K - P_n K\| \le \frac{h^2}{8} \max_{a \le x \le b} \int_a^b \left| \frac{\partial^2 k(x, y)}{\partial x^2} \right| dy.$$

Exercise 11.2.2 *Prove (11.2.17).*
Hint: Apply Theorem 3.6.2.

Exercise 11.2.3 *Generalize the ideas of Subsection 11.2.1 to continuous piecewise polynomial collocation of degree $\kappa > 0$.*

Exercise 11.2.4 *Generalize the ideas of Subsection 11.2.3 to V_n the set of piecewise linear functions in which there is no restriction that the approximations be continuous.*

Exercise 11.2.5 *Generalize the ideas of Subsection 11.2.3 to continuous piecewise polynomial collocation of degree $\kappa > 0$.*

Exercise 11.2.6 *Prove* (11.2.24).

Exercise 11.2.7 *Give conditions on the data λ, k, and f so that a solution of the equation (11.2.1) has the regularity $u \in C^2[a,b]$. Note that this regularity is required in the error estimate (11.2.5).*

Exercise 11.2.8 *Let P_n be an interpolatory projection operator, and let*

$$Kv(x) = \int_a^b k(x,y)v(y)\,dy, \qquad a \le x \le b, \quad v \in C[a,b]$$

have a continuous kernel function $k(x,y)$. Show that $P_n K$ is a degenerate kernel integral operator. For the case of P_n the piecewise linear interpolatory operator of (11.2.3), write out an explicit formula for the degenerate kernel $k_n(x,y)$ and analyze the error $k(x,y) - k_n(x,y)$.

Exercise 11.2.9 *It is known that if $u \in C[-1,1]$, then the partial sums of the Chebyshev expansion*

$$u(x) = \frac{c_0}{2} + \sum_{i=1}^{\infty} c_i T_i(x), \qquad c_i = \frac{2}{\pi} \int_{-1}^1 \frac{u(y)T_i(y)}{\sqrt{1-y^2}}\,dy$$

are good uniform approximations of $u(x)$ when u is sufficiently smooth. This is an orthogonal polynomial expansion of u. The weight function is $w(y) = 1/\sqrt{1-y^2}$, and the associated orthogonal family is the Chebyshev polynomials $\{T_i(x)\}_{i \ge 0}$. We want to investigate the solution of

$$\lambda u(x) - \int_{-1}^1 k(x,y)u(y)\,dy = f(x), \qquad -1 \le x \le 1$$

using Galerkin's method with polynomial subspaces and the orthogonal projections

$$P_n v(x) = \frac{c_0}{2} + \sum_{i=1}^{\infty} c_i T_i(x), \qquad n \ge 1.$$

The space being used is $L_w^2(-1,1)$ with the $w(y)$ given above.
(a) Give the Galerkin method for solving the above integral equation.
(b) Give the coefficients of the linear system, and suggest a way for dealing with the singularity in the integrand (owing to the presence of the weight function w).
(c) If the true solution u is r-times continuously differentiable on $[-1,1]$, discuss the rate of convergence to zero of the error $\|u - u_n\|_{L_w^2}$. For an introductory account of Chebyshev polynomials and Chebyshev expansions, see [11, Sections 4.5–4.7].

Exercise 11.2.10 *Recall the linear system (11.1.5) for the collocation method, and consider it with the Lagrange basis $\{\ell_i(x)\}$ satisfying (11.1.7), with the associated projection operator P_n of (11.1.8). Denote this linear* *system by* $A_n \mathbf{u}_n = \mathbf{f}_n$, *with*

$$\mathbf{f}_n = [f(x_1), \dots, f(x_\kappa)]^T$$

and $\mathbf{u}_n$ *defined analogously. For* A_n,

$$(A_n)_{i,j} = \lambda \delta_{i,j} - \int_D k(x_i, y) \ell_j(y) \, dy.$$

Consider $A_n : \mathbb{R}^\kappa \to \mathbb{R}^\kappa$ *with the infinity norm, and find a bound for* $\|A_n^{-1}\|$, *using the row norm as the matrix norm. Find the bound in terms of* $\|(\lambda - P_n K)^{-1}\|$.

Hint: For arbitrary $\gamma \in \mathbb{R}^\kappa$, *let* $\mathbf{v} = A_n^{-1} \gamma$, *or equivalently,* $A_n \mathbf{v} = \gamma$. *You need to bound* $\mathbf{v}$ *in terms of* γ. *To do this, begin by showing you can construct* $g \in C(D)$ *with*

$$\gamma = [g(x_1), \dots, g(x_\kappa)]^T$$

and $\|g\|_\infty = \|\gamma\|_\infty$. *Define the function* $v \in C(D)$ *as the solution of*

$$(\lambda - P_n K) v = P_n g.$$

Then bound $\mathbf{v}$ *in terms of* $\|v\|_\infty$, *and bound the latter in terms of* $\|\gamma\|_\infty$.

11.3   Iterated projection methods

For the integral equation $(\lambda - K)u = f$, consider the following fixed point iteration which was considered earlier in (4.2.8) of Section 4.2, Chapter 4:

$$u^{(k+1)} = \frac{1}{\lambda} \left[f + K u^{(k)} \right], \qquad k = 0, 1, \dots$$

As earlier, this iteration can be shown to converge to the solution u if $\|K\| < |\lambda|$; and in that case

$$\left\| u - u^{(k+1)} \right\| \le \frac{\|K\|}{|\lambda|} \left\| u - u^{(k)} \right\|.$$

In [147], Sloan showed that one such iteration is always a good idea if K is a compact operator and if the initial guess is the solution u_n obtained by the Galerkin method, regardless of the size of $\|K\|$. We examine this idea and its consequences for projection methods.

Let u_n be the solution of the projection equation $(\lambda - P_n K)u_n = P_n f$. Define the *iterated projection solution* by

$$\widehat{u}_n = \frac{1}{\lambda} [f + K u_n]. \tag{11.3.1}$$

This new approximation $\widehat{u}_n$ is often an improvement on u_n. Moreover, it can often be used to better understand the behavior of the original projection solution u_n.

Applying P_n to both sides of (11.3.1), we have

$$P_n \widehat{u}_n = \frac{1}{\lambda} \left[P_n f + P_n K u_n \right];$$

i.e.,

$$P_n \widehat{u}_n = u_n. \qquad (11.3.2)$$

Thus, u_n is the projection of $\widehat{u}_n$ into V_n. Substituting into (11.3.1) and rearranging terms, we have $\widehat{u}_n$, which satisfies the equation

$$(\lambda - KP_n)\widehat{u}_n = f. \qquad (11.3.3)$$

Often we can directly analyze this equation; and then information can be obtained on u_n by applying (11.3.2).

Also, since

$$u - \widehat{u}_n = \frac{1}{\lambda} [f + Ku] - \frac{1}{\lambda} [f + Ku_n] = \frac{1}{\lambda} K(u - u_n), \qquad (11.3.4)$$

we have the error bound

$$\|u - \widehat{u}_n\| \le \frac{1}{|\lambda|} \|K\| \|u - u_n\|. \qquad (11.3.5)$$

This proves the convergence of $\widehat{u}_n$ to u is at least as rapid as that of u_n to u. Often it is more rapid, because operating on $u - u_n$ with K, as in (11.3.4), sometimes causes cancellation owing to the smoothing behavior of integration.

From the above, we see that if $(\lambda - P_n K)^{-1}$ exists, then so does $(\lambda - KP_n)^{-1}$. Moreover, from the definition of the solution u_n and (11.3.1), we have

$$\widehat{u}_n = \frac{1}{\lambda} [f + Ku_n] = \frac{1}{\lambda} \left[f + K(\lambda - P_n K)^{-1} P_n f \right];$$

and when combined with (11.3.3),

$$(\lambda - KP_n)^{-1} = \frac{1}{\lambda} \left[I + K(\lambda - P_n K)^{-1} P_n \right]. \qquad (11.3.6)$$

Conversely, if $(\lambda - KP_n)^{-1}$ exists, then so does $(\lambda - P_n K)^{-1}$. This follows from the general lemma given below, which also shows that

$$(\lambda - P_n K)^{-1} = \frac{1}{\lambda} \left[I + P_n(\lambda - KP_n)^{-1} K \right]. \qquad (11.3.7)$$

By combining (11.3.6) and (11.3.7), or by returning to the definitions of u_n and $\widehat{u}_n$, we also have

$$(\lambda - P_n K)^{-1} P_n = P_n(\lambda - KP_n)^{-1}. \qquad (11.3.8)$$

We can choose to show the existence of either $(\lambda - P_n K)^{-1}$ or $(\lambda - KP_n)^{-1}$, whichever is the more convenient; and the existence of the other inverse follows immediately. Bounds on one inverse in terms of the other can also be derived by using (11.3.6) and (11.3.7).

Lemma 11.3.1 *Let V be a Banach space, and let A, B be bounded linear operators on V to V. Assume $(\lambda - AB)^{-1}$ exists from V onto V. Then $(\lambda - BA)^{-1}$ also exists, and*

$$(\lambda - BA)^{-1} = \frac{1}{\lambda}\left[I + B(\lambda - AB)^{-1}A\right]. \tag{11.3.9}$$

Proof. Calculate

$$(\lambda - BA)\frac{1}{\lambda}\left[I + B(\lambda - AB)^{-1}A\right]$$

$$= \frac{1}{\lambda}\left\{\lambda - BA + (\lambda - BA)B(\lambda - AB)^{-1}A\right\}$$

$$= \frac{1}{\lambda}\left\{\lambda - BA + B(\lambda - AB)(\lambda - AB)^{-1}A\right\}$$

$$= \frac{1}{\lambda}\left\{\lambda - BA + BA\right\}$$

$$= I.$$

A similar proof works to show

$$\frac{1}{\lambda}\left[I + B(\lambda - AB)^{-1}A\right](\lambda - BA) = I. \tag{11.3.10}$$

This proves (11.3.9).   ∎

For the error in $\hat{u}_n$, first rewrite $(\lambda - K)u = f$ as

$$(\lambda - KP_n)u = f + Ku - KP_nu.$$

Subtract (11.3.3) to obtain

$$(\lambda - KP_n)(u - \hat{u}_n) = K(I - P_n)u. \tag{11.3.11}$$

Below we examine this apparently simple equation in much greater detail.

11.3.1   The iterated Galerkin method

Assume that V is a Hilbert space and that u_n is the Galerkin solution of the equation $(\lambda - K)u = f$ over a finite-dimensional subspace $V_n \subseteq V$. Then

$$(I - P_n)^2 = I - P_n.$$

and

$$\|K(I - P_n)u\| = \|K(I - P_n)(I - P_n)u\|$$
$$\leq \|K(I - P_n)\|\,\|(I - P_n)u\|. \tag{11.3.12}$$

Using the fact that we are in a Hilbert space and that P_n is a self-adjoint projection (cf. Theorem 3.3.7 in Section 3.3), we have

$$\|K(I - P_n)\| = \|[K(I - P_n)]^*\|$$
$$= \|(I - P_n)K^*\|. \tag{11.3.13}$$

The first line follows from the general principle that the norm of an operator is equal to the norm of its adjoint operator. The second line follows from Theorem 3.3.7 and properties of the adjoint operation.

With Galerkin methods, it is generally the case that when P_n is regarded as an operator on the Hilbert space V, then $P_n v \to v$ for all $v \in V$. This follows if we have that the sequence of spaces $\{V_n \mid n \geq 1\}$ has the *approximating property* on V: For each $v \in V$, there is a sequence $\{v_n\}$ with $v_n \in V_n$ and

$$\lim_{n \to \infty} \|v - v_n\| = 0. \tag{11.3.14}$$

When this is combined with the optimal approximation property of Proposition 3.5.9(c), we have $P_n v \to v$ for all $v \in V$.

Recall from Lemma 2.8.13 of Chapter 2 that if K is a compact operator, then so is its adjoint K^*. Combining this with Lemma 11.1.4 and the above assumption of the pointwise convergence of P_n to I on V, we have that

$$\lim_{n \to \infty} \|(I - P_n)K^*\| = 0. \tag{11.3.15}$$

We can also apply Theorem 11.1.2 to obtain the existence and uniform boundedness of $(\lambda - P_n K)^{-1}$ for all sufficiently large n, say, $n \geq N$. From (11.3.6), we also have that $(\lambda - K P_n)^{-1}$ exists and is uniformly bounded for $n \geq N$. Apply this and (11.3.12) to (11.3.11), to obtain

$$[c]ccl \, \|u - \widehat{u}_n\| \leq \left\|(\lambda - K P_n)^{-1}\right\| \, \|K(I - P_n)u\|$$
$$\leq c \, \|(I - P_n)K^*\| \, \|(I - P_n)u\|. \tag{11.3.16}$$

Combining this with (11.3.15), we see that $\|u - \widehat{u}_n\|$ converges to zero more rapidly than does $\|(I - P_n)u\|$, or equivalently, $\|u - u_n\|$. Thus

$$\lim_{n \to \infty} \frac{\|u - \widehat{u}_n\|}{\|u - u_n\|} = 0.$$

The quantity $\|(I - P_n)K^*\|$ can generally be estimated, in the same manner as is done for $\|(I - P_n)K\|$. Taking K to be an integral operator on $L^2(D)$, the operator K^* is an integral operator (cf. Example 2.6.1), with

$$K^* u(x) = \int_D k(y, x) u(y) \, dy, \quad u \in L^2(D). \tag{11.3.17}$$

Example 11.3.2 *Consider the integral equation*

$$\lambda u(x) - \int_0^1 e^{xy} u(y) \, dy = f(x), \quad 0 \leq x \leq 1 \tag{11.3.18}$$

with $\lambda = 50$ *and* $u(x) = e^x$. *For* $n \geq 1$, *define the meshsize* $h = 1/n$ *and the mesh* $x_j = jh$, $j = 0, 1, \ldots, n$. *Let* V_n *be the set of functions which are piecewise linear on* $[0, 1]$ *with breakpoints* $x_1, \ldots, x_{n-1}$, *without the continuity restriction of Section 11.2.3. The dimension of* V_n *is* $d_n = 2n$, *and this is also the order of the linear system associated with solving* $(\lambda - P_n K)u_n = P_n f$.

n	$\|u - u_n\|_\infty$	Ratio	$\|u - \widehat{u}_n\|_\infty$	Ratio
2	$4.66E - 2$		$5.45E - 6$	
4	$1.28E - 2$	3.6	$3.48E - 7$	15.7
8	$3.37E - 3$	3.8	$2.19E - 8$	15.9

Table 11.2. Piecewise linear Galerkin and iterated Galerkin method for solving (11.3.18)

It is straightforward to show $\|(I - P_n)K^*\| = O(h^2)$ in this case. Also, if $f \in C^2[0,1]$, then the solution u of (11.3.18) also belongs to $C^2[0,1]$; and consequently, we have $\|u - P_n u\| = O(h^2)$. These results lead to

$$\|u - u_n\| = O(h^2), \tag{11.3.19}$$

$$\|u - \widehat{u}_n\| = O(h^4). \tag{11.3.20}$$

This is confirmed empirically in the numerical calculations given in Table 11.2. The error columns give the maximum error rather than the norm of the error in $L^2(0,1)$. But it can be shown that (11.3.19)–(11.3.20) generalize to $C[0,1]$ with the uniform norm.

11.3.2   The iterated collocation solution

With collocation, the iterated solution $\widehat{u}_n$ is not always an improvement on the original collocation solution u_n, but it is for many cases of interest. The abstract theory is still applicable, and the error equation (11.3.11) is still the focus for the error analysis:

$$u - \widehat{u}_n = (\lambda - KP_n)^{-1}K(I - P_n)u. \tag{11.3.21}$$

Recall that the projection P_n is now an interpolatory operator, as in (11.1.8). In contrast to the iterated Galerkin method, we do not have that $\|K - KP_n\|$ converges to zero. In fact, it can be shown that

$$\|K(I - P_n)\| \geq \|K\|. \tag{11.3.22}$$

To show the possibly faster convergence of $\widehat{u}_n$, we must examine collocation methods on a case-by-case basis. With some, there is an improvement. We begin with a simple example to show one of the main tools used in proving higher orders of convergence.

Consider using collocation with piecewise quadratic interpolation to solve the integral equation

$$\lambda u(x) - \int_a^b k(x,y)u(y)\,dy = f(x), \qquad a \leq x \leq b. \tag{11.3.23}$$

Let $n \geq 2$ be an even integer. Define $h = (b - a)/n$ and $x_j = a + jh$, $j = 0, 1, \ldots, n$. Let V_n be the set of all continuous functions that are a quadratic polynomial when restricted to each of the subintervals $[x_0, x_2], \ldots, [x_{n-2}, x_n]$. Easily, the dimension of V_n is $\kappa_n = n + 1$, based on each element of V_n being completely determined by its values at the $n + 1$ nodes $\{x_0, \ldots, x_n\}$. Let P_n be the interpolatory projection operator from $V = C[a, b]$ to V_n.

We can write $P_n u$ in its Lagrange form:

$$P_n u(x) = \sum_{j=0}^{n} u(x_j) \ell_j(x). \tag{11.3.24}$$

For the Lagrange basis functions $\ell_j(x)$, we must distinguish the cases of even and odd indices j. For j odd,

$$\ell_j(x) = \begin{cases} -\dfrac{1}{h^2}(x - x_{j-1})(x - x_{j+1}), & x_{j-1} \leq x \leq x_{j+1}, \\ 0, & \text{otherwise.} \end{cases}$$

For j even, $2 \leq j \leq n - 2$,

$$\ell_j(x) = \begin{cases} \dfrac{1}{2h^2}(x - x_{j-1})(x - x_{j-2}), & x_{j-2} \leq x \leq x_j, \\ \dfrac{1}{2h^2}(x - x_{j+1})(x - x_{j+2}), & x_j \leq x \leq x_{j+2}, \\ 0, & \text{otherwise.} \end{cases}$$

The functions $\ell_0(x)$ and $\ell_n(x)$ are appropriate modifications of this last case.

For the interpolation error on $[x_{j-2}, x_j]$, for j even, we have two formulas:

$$u(x) - P_n u(x) = (x - x_{j-2})(x - x_{j-1})(x - x_j) u[x_{j-2}, x_{j-1}, x_j, x] \tag{11.3.25}$$

and

$$u(x) - P_n u(x) = \frac{(x - x_{j-2})(x - x_{j-1})(x - x_j)}{6} u'''(c_x), \quad x_{j-2} \leq x \leq x_j \tag{11.3.26}$$

for some $c_x \in [x_{j-2}, x_j]$, with $u \in C^3[a, b]$. The quantity $u[x_{j-2}, x_{j-1}, x_j, x]$ is a *Newton divided difference* of order three for the function $u(x)$. From the above formulas,

$$\|u - P_n u\|_\infty \leq \frac{\sqrt{3}}{27} h^3 \|u'''\|_\infty, \quad u \in C^3[a, b]. \tag{11.3.27}$$

See [11, pp. 143, 156] for details on this and more generally on divided differences.

In using piecewise quadratic functions to define the collocation method to solve (11.3.23), the result (11.3.27) implies

$$\|u - u_n\|_\infty = O(h^3) \tag{11.3.28}$$

if $u \in C^3[a, b]$. To examine the error in $\widehat{u}_n$, we make a detailed examination of $K(I - P_n)u$.

Using (11.3.24),

$$K(I - P_n)u(x) = \int_a^b k(x, y) \left\{ u(y) - \sum_{j=0}^n u(x_j)\ell_j(y) \right\} dy.$$

From (11.3.25),

$$K(I - P_n)u(x) = \sum_{k=1}^{n/2} \int_{x_{2k-2}}^{x_{2k}} k(x, y)(y - x_{2k-2})(y - x_{2k-1})(y - x_{2k})$$

$$\cdot u[x_{2k-2}, x_{2k-1}, x_{2k}, y] \, dy. \qquad (11.3.29)$$

To examine the integral in more detail, we write it as

$$\int_{x_{2k-2}}^{x_{2k}} g_x(y)\omega(y) \, dy \qquad (11.3.30)$$

with

$$\omega(y) = (y - x_{2k-2})(y - x_{2k-1})(y - x_{2k})$$

and

$$g_x(y) = k(x, y) \, u[x_{2k-2}, x_{2k-1}, x_{2k}, y].$$

Introduce

$$\nu(y) = \int_{x_{2k-2}}^y \omega(\xi) \, d\xi, \quad x_{2k-2} \le y \le x_{2k}.$$

Then $\nu'(y) = \omega(y)$, $\nu(y) \ge 0$ on $[x_{2k-2}, x_{2k}]$, and $\nu(x_{2k-2}) = \nu(x_{2k}) = 0$. The integral (11.3.30) becomes

$$\int_{x_{2k-2}}^{x_{2k}} g_x(y)\nu'(y) \, dy = \underbrace{\nu(y)g_x(y)|_{x_{2k-2}}^{x_{2k}}}_{=0} - \int_{x_{2k-2}}^{x_{2k}} g_x'(y)\nu(y) \, dy,$$

and so

$$\left| \int_{x_{2k-2}}^{x_{2k}} g_x'(y)\nu(y) \, dy \right| \le \|g_x'\|_\infty \int_{x_{2k-2}}^{x_{2k}} \nu(y) \, dy = \frac{4h^5}{15} \|g_x'\|_\infty.$$

In this,

$$g_x'(y) = \frac{\partial}{\partial y} \left\{ k(x, y) \, u[x_{2k-2}, x_{2k-1}, x_{2k}, y] \right\}$$

$$= \frac{\partial k(x, y)}{\partial y} \, u[x_{2k-2}, x_{2k-1}, x_{2k}, y]$$

$$+ k(x, y) \, u[x_{2k-2}, x_{2k-1}, x_{2k}, y, y].$$

The last formula uses a standard result for the differentiation of Newton divided differences (see [11, p. 147]). To have this derivation be valid, we must have $g \in C^1[a,b]$, and this is true if $u \in C^4[a,b]$ and $k_x \in C^1[a,b]$.

Combining these results, we have

$$K(I - P_n)u(x) = O(h^4). \tag{11.3.31}$$

With this, we have the following theorem.

Theorem 11.3.3 *Assume that the integral equation (11.3.23) is uniquely solvable for all $f \in C[a,b]$. Further assume that the solution $u \in C^4[a,b]$ and that the kernel function $k(x,y)$ is continuously differentiable with respect to y. Let P_n be the interpolatory projection (11.3.24) defined by piecewise quadratic interpolation. Then the collocation equation $(\lambda - P_n K)u_n = P_n f$ is uniquely solvable for all sufficiently large n, say, $n \geq N$; and the inverses $(\lambda - P_n K)^{-1}$ are uniformly bounded, say, by $M > 0$. Moreover,*

$$\|u - u_n\|_\infty \leq |\lambda|\, M\, \|u - P_n u\|_\infty \leq \frac{\sqrt{3}\,|\lambda|\, M}{27} h^3\, \|u'''\|_\infty, \quad n \geq N. \tag{11.3.32}$$

For the iterated collocation method,

$$\|u - \widehat{u}_n\|_\infty \leq ch^4 \tag{11.3.33}$$

for a suitable constant $c > 0$. Consequently,

$$\max_{j=0,\ldots,n} |u(x_j) - u_n(x_j)| = O(h^4). \tag{11.3.34}$$

Proof. Formula (11.3.32) and the remarks preceding it are just a restatement of results from Theorem 11.1.2, applied to the particular P_n being considered here. The final bound in (11.3.32) comes from (11.3.27). The bound (11.3.33) comes from (11.3.21) and (11.3.31). The final result (11.3.34) comes from noting first that the property $P_n \widehat{u}_n = u_n$ (cf. (11.3.2)) implies

$$u_n(x_j) = \widehat{u}_n(x_j), \qquad j = 0, \ldots, n$$

and second from applying (11.3.33). ∎

This theorem, and (11.3.33) in particular, shows that the iterated collocation method converges more rapidly when using the piecewise quadratic collocation method described preceding the theorem. However, when using piecewise linear interpolation to define P_n, the iterated collocation solution $\widehat{u}_n$ does not converge any more rapidly than the original solution u_n. In general, let V_n be the set of continuous piecewise polynomial functions of degree r with r an even integer, and let the collocation nodes be the breakpoints used in defining the piecewise quadratic functions. Then the iterated solution gains one extra power of h in its error bound. But this is not true if r is an odd integer.

The result (11.3.34) is an example of *superconvergence*. The rate of convergence of $u_n(x)$ at the node points $\{x_0, \ldots, x_n\}$ is greater than it is over the interval $[a, b]$ as a whole. There are a number of situations in the numerical solution of both differential and integral equations in which superconvergence occurs at special points in the domain over which the problem is defined. See Exercise 9.4.1 for a superconvergence result in the context of the finite element method. Also recall Example 11.2.1. For it, one can show that

$$K(I - P_n)u(x) = O(h^{1.5}) \qquad (11.3.35)$$

for the solution function $u(x) = \sqrt{x}$ of (11.2.9), thus proving superconvergence at the node points as was observed in Table 11.1.

The linear system for the iterated collocation solution

Let the interpolatory projection operator be written as

$$P_n u(x) = \sum_{j=1}^{\kappa_n} u(x_j)\ell_j(x), \qquad u \in C(D). \qquad (11.3.36)$$

When written out, the approximating equation $\lambda \widehat{u}_n - K P_n \widehat{u}_n = f$ becomes

$$\lambda \widehat{u}_n(x) - \sum_{j=1}^{\kappa_n} \widehat{u}_n(x_j) \int_D k(x, y)\ell_j(y)\, dy = f(x), \qquad x \in D. \qquad (11.3.37)$$

Evaluating at each node point x_i, we obtain the linear system

$$\lambda \widehat{u}_n(x_i) - \sum_{j=1}^{\kappa_n} \widehat{u}_n(x_j) \int_D k(x_i, y)\ell_j(y)\, dy = f(x_i), \qquad i = 1, \ldots, \kappa_n. \qquad (11.3.38)$$

This is also the linear system for the collocation solution at the node points, as given, for example, in (11.1.5) with $\phi_j = \ell_j$, or in (11.2.6). This is not surprising since u_n and $\widehat{u}_n$ agree at the node points.

The two solutions differ, however, at the remaining points in D. For general $x \in D$, $u_n(x)$ is based on the interpolation formula (11.3.36). However, the iterated collocation solution $\widehat{u}_n(x)$ is given by using (11.3.37) in the form

$$\widehat{u}_n(x) = \frac{1}{\lambda} \left\{ f(x) + \sum_{j=1}^{\kappa_n} \widehat{u}_n(x_j) \int_D k(x, y)\ell_j(y)\, dy \right\}, \qquad x \in D. \qquad (11.3.39)$$

In Section 11.5, we see that (11.3.37)–(11.3.39) is a special case of the Nyström method for solving $(\lambda - K)u = f$, and (11.3.39) is called the Nyström interpolation function. Generally, it is more accurate than ordinary polynomial interpolation.

Exercise 11.3.1 *Prove* (11.3.10).

Exercise 11.3.2 *Consider the piecewise linear Galerkin scheme of Subsection 11.2.3. Assume that $k(y, x)$ is twice continuously differentiable with respect to y, for $a \leq y, x \leq b$. Analyze the convergence of the iterated Galerkin method for this case. What is the rate of convergence of the iterated Galerkin solutions $\{\widehat{u}_n\}$?*

Exercise 11.3.3 *Derive the identity*

$$\lambda(u - \widehat{u}_n) = K(I - P_n)u + KP_n(u - \widehat{u}_n)$$

for the solution u of $(\lambda - K)u = f$ and the iterated projection solutions $\{\widehat{u}_n\}$. Using this, obtain results on the uniform convergence of $\{\widehat{u}_n\}$ and $\{u_n\}$ to u for the integral operator K of (11.2.1).
Hint: Write $K(I - P_n)u$ and $KP_n(u - \widehat{u}_n)$ as inner products to which the Cauchy-Schwartz inequality can be applied.

Exercise 11.3.4 *Prove* (11.3.22).
Hint: Look at functions v for which $\|v\| = 1$ and $\|K\| \approx \|Kv\|$. Then modify v to w with $P_n w$ and $\|Kv\| \approx \|Kw\|$.

Exercise 11.3.5 *Let $n > 0$, $h = (b - a)/n$, and $\tau_j = a + jh$ for $j = 0, 1, \ldots, n$. Let V_n be the set of piecewise linear functions, linear over each subinterval $[\tau_{j-1}, \tau_j]$, with no restriction that the functions be continuous. The dimension of V_n is $2n$. To define the collocation nodes, introduce*

$$\mu_1 = \frac{3 - \sqrt{3}}{6}, \qquad \mu_2 = \frac{3 + \sqrt{3}}{6}.$$

On each subinterval $[\tau_{j-1}, \tau_j]$, define two collocation nodes by

$$x_{2j-1} = \tau_{j-1} + h\mu_1, \qquad x_{2j} = \tau_{j-1} + h\mu_2.$$

Define a collocation method for solving (11.3.23) *using the approximating subspace V_n and the collocation nodes $\{x_1, \ldots, x_{2n}\}$. Assume $u \in C^4[a, b]$ and assume $k(x, y)$ is twice continuously differentiable with respect to y, for $a \leq y, x \leq b$. It can be shown from the methods of Section 11.5 that $(\lambda - P_n K)^{-1}$ exists and is uniformly bounded for all sufficiently large n, say $n \geq N$. Assuming this, show that*

$$\|u - u_n\|_\infty = O\left(h^2\right),$$
$$\|u - \widehat{u}_n\|_\infty = O\left(h^4\right),$$
$$\max_{j=1,\ldots,2n} |u(x_j) - u_n(x_j)| = O(h^4).$$

Hint: The polynomial $(\mu - \mu_1)(\mu - \mu_2)$ is the Legendre polynomial of degree 2 on $[0, 1]$, and therefore it is orthogonal over $[0, 1]$ to all polynomials of lesser degree.

Exercise 11.3.6 *Consider the integral equation* $(\lambda - K)\,u = f$, *with*

$$Kv(x) = \int_a^b k(x,y)v(y)\,dy, \quad y \in [a,b], \quad v \in C[a,b].$$

Assume that u is continuous, but that its first derivative is discontinuous. Moreover, assume the kernel function $k(x,y)$ is several times continuously differentiable with respect to x. Write

$$u = \frac{1}{\lambda}\,(f + Ku) \equiv \frac{1}{\lambda}\,(f + w).$$

Prove that w satisfies the equation $(\lambda - K)\,w = Kf$. Show that w is smoother than u. Be as precise as possible in stating your results.

Exercise 11.3.7 *(continuation of Exercise 11.3.6). Apply a projection method to the solution of the modified equation $(\lambda - K)\,w = Kf$, denoting the approximate solution by w_n. Then define $u_n = \frac{1}{\lambda}\,(f + w_n)$. Analyze the convergence of $\{u_n\}$. Compare the method to the original projection method applied directly to $(\lambda - K)\,u = f$.*

Exercise 11.3.8 *Derive (11.3.35) in the case the integral operator K has the kernel function $k \equiv 1$.*

Exercise 11.3.9 *Generalize Exercise 11.3.8 to the case of a general kernel function $k(x,y)$ that is once continuously differentiable with respect to the variable y on the interval of integration.*

11.4   The Nyström method

The Nyström method was originally introduced to handle approximations based on numerical integration of the integral operator in the equation

$$\lambda u(x) - \int_D k(x,y)u(y)\,dy = f(x), \qquad x \in D. \tag{11.4.1}$$

The resulting solution is found first at the set of quadrature node points, and then it is extended to all points in D by means of a special, and generally quite accurate, interpolation formula. The numerical method is much simpler to implement on a computer, but the error analysis is more sophisticated than for projection methods. The resulting theory has taken an abstract form that also includes an error analysis of projection methods, although the latter are probably still best understood as distinct methods of interest in their own right.

11.4.1   The Nyström method for continuous kernel functions

Let a numerical integration scheme be given:

$$\int_D g(y)\,dy \approx \sum_{j=1}^{q_n} w_{n,j} g(x_{n,j}), \qquad g \in C(D), \tag{11.4.2}$$

with an increasing sequence of values of n. We assume that for every $g \in C(D)$, the numerical integrals converge to the true integral as $n \to \infty$. As in Subsection 2.4.4, this implies

$$c_I \equiv \sup_{n \geq 1} \sum_{j=1}^{q_n} |w_{n,j}| < \infty. \tag{11.4.3}$$

To simplify the notation, we omit the subscript n, so that $w_{n,j} \equiv w_j$, $x_{n,j} \equiv x_j$; but the presence of n is to be understood implicitly. On occasion, we also use $q \equiv q_n$.

Let $k(x, y)$ be continuous for all $x, y \in D$, where D is a closed and bounded set in $\mathbb{R}^d$ for some $d \geq 1$. Usually, in fact, we want $k(x, y)$ to be several times continuously differentiable. Using the above quadrature scheme, approximate the integral in (11.4.1), obtaining a new equation:

$$\lambda u_n(x) - \sum_{j=1}^{q_n} w_j k(x, x_j) u_n(x_j) = f(x), \qquad x \in D. \tag{11.4.4}$$

We write this as an exact equation with a new unknown function $u_n(x)$. To find the solution at the node points, let x run through the quadrature node points x_i. This yields

$$\lambda u_n(x_i) - \sum_{j=1}^{q_n} w_j k(x_i, x_j) u_n(x_j) = f(x_i), \qquad i = 1, \ldots, q_n \tag{11.4.5}$$

which is a linear system of order q_n. The unknown is a vector

$$\underline{u}_n \equiv [u_n(x_1), \ldots, u_n(x_q)]^{\mathrm{T}}.$$

Each solution $u_n(x)$ of (11.4.4) furnishes a solution to (11.4.5): merely evaluate $u_n(x)$ at the node points. The converse is also true. To each solution $\underline{z} \equiv [z_1, \ldots, z_q]^{\mathrm{T}}$ of (11.4.5), there is a unique solution of (11.4.4) which agrees with $\underline{z}$ at the node points. If one solves for $u_n(x)$ in (11.4.4), then $u_n(x)$ is determined by its values at the node points $\{x_j\}$. Therefore, when given a solution $\underline{z}$ to (11.4.5), define

$$z(x) = \frac{1}{\lambda} \left[f(x) + \sum_{j=1}^{q_n} w_j k(x, x_j) z_j \right], \qquad x \in D. \tag{11.4.6}$$

This is an interpolation formula. In fact,

$$z(x_i) = \frac{1}{\lambda} \left[f(x_i) + \sum_{j=1}^{q_n} w_j k(x_i, x_j) z_j \right] = z_i$$

for $i = 1, \ldots, q_n$. The last step follows from $\underline{z}$ being a solution to (11.4.5). Using this interpolation result in (11.4.6), we have that $z(x)$ solves (11.4.4). The uniqueness of the relationship between $\underline{z}$ and $z(x)$ follows from the solutions $u_n(x)$ of (11.4.4) being completely determined by their values at the nodes $\{x_i\}$.

The formula (11.4.6) is called the *Nyström interpolation formula*. In the original paper of Nyström [125], the author uses a highly accurate Gaussian quadrature formula with a very small number of quadrature nodes (e.g., $q = 3$). He then uses (11.4.6) in order to extend the solution to all other $x \in D$ while retaining the accuracy found in the solution at the node points. The formula (11.4.6) is usually a very good interpolation formula.

Example 11.4.1 *Consider the integral equation*

$$\lambda u(x) - \int_0^1 e^{yx} u(y) \, dy = f(x), \quad 0 \le x \le 1, \tag{11.4.7}$$

with $\lambda = 2$ and $u(x) = e^x$. Since $\|K\| = e - 1 \doteq 1.72$, the geometric series theorem (cf. Theorem 2.3.1 in Chapter 2) implies the integral equation is uniquely solvable for any given $f \in C[0, 1]$.

Consider first using the three-point Simpson rule to approximate (11.4.7), with nodes $\{0, 0.5, 1\}$. Then the errors at the nodes are collectively

$$\begin{bmatrix} u(0) \\ u(.5) \\ u(1) \end{bmatrix} - \begin{bmatrix} u_3(0) \\ u_3(.5) \\ u_3(1) \end{bmatrix} \doteq \begin{bmatrix} -0.0047 \\ -0.0080 \\ -0.0164 \end{bmatrix}, \tag{11.4.8}$$

which are reasonably small errors. For comparison, use Gauss-Legendre quadrature with three nodes,

$$\int_0^1 g(x) \, dx \approx \frac{1}{18} \left[5g(x_1) + 8g(x_2) + 5g(x_3) \right],$$

where

$$x_1 = \frac{1 - \sqrt{0.6}}{2} \doteq 0.11270167, \quad x_2 = 0.5, \quad x_3 = \frac{1 + \sqrt{0.6}}{2} \doteq 0.88729833.$$

The errors at the nodes in solving (11.4.7) with the Nyström method are now collectively

$$\begin{bmatrix} u(x_1) \\ u(x_2) \\ u(x_3) \end{bmatrix} - \begin{bmatrix} u_3(x_1) \\ u_3(x_2) \\ u_3(x_3) \end{bmatrix} \doteq \begin{bmatrix} 2.10 \times 10^{-5} \\ 3.20 \times 10^{-5} \\ 6.32 \times 10^{-5} \end{bmatrix}, \tag{11.4.9}$$

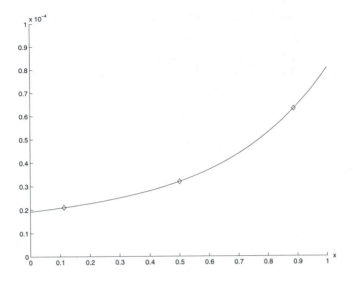

Figure 11.1. Error in Nyström interpolation with three point Gauss-Legendre quadrature

and these errors are much smaller than with Simpson's rule when using an equal number of node points. Generally, Gaussian quadrature is much superior to Simpson's rule; but it results in the answers' being given at the Gauss-Legendre nodes, which is usually not a convenient choice for subsequent uses of the answers.

Quadratic interpolation can be used to extend the numerical solution to all other $x \in [0,1]$, but it generally results in much larger errors. For example,

$$u(1.0) - P_2 u_3(1.0) \doteq 0.0158,$$

where $P_2 u_3(x)$ denotes the quadratic polynomial interpolating the Nyström solution at the Gaussian quadrature node points given above. In contrast, the Nyström formula (11.4.6) gives errors that are consistent in size with those in (11.4.9). For example,

$$u(1.0) - u_3(1.0) \doteq 8.08 \times 10^{-5}.$$

A graph of the error in $u_3(x)$ over $[0,1]$ is given in Figure 11.1, with the errors at the node points indicated by $\diamond$.

11.4.2   Properties and error analysis of the Nyström method

The Nyström method is implemented with the finite linear system (11.4.5), but the formal error analysis is done using the functional equation (11.4.4). As before, we write the integral equation (11.4.1) in abstract form as

$$(\lambda - K)u = f;$$

and we write the numerical integral equation (11.4.4) as

$$(\lambda - K_n)u_n = f.$$

The Banach space for our initial error analysis is $V = C(D)$. The numerical integral operator

$$K_n u(x) \equiv \sum_{j=1}^{q_n} w_j k(x, x_j) u(x_j), \qquad x \in D, \quad u \in C(D), \qquad (11.4.10)$$

is a bounded, finite-rank linear operator on $C(D)$ to $C(D)$, with

$$\|K_n\| = \max_{x \in D} \sum_{j=1}^{q_n} |w_j k(x, x_j)|. \qquad (11.4.11)$$

The error analyses of projection methods depended on showing $\|K - K_n\|$ converges to zero as n increases, with $K_n = P_n K$ the approximation to the integral operator K. This cannot be done here; and in fact,

$$\|K - K_n\| \geq \|K\|. \qquad (11.4.12)$$

We leave the proof of this as an exercise for the reader. Because of this result, the standard type of perturbation analysis that was used earlier needs to be modified. We begin by looking at quantities that do converge to zero as $n \to \infty$.

Lemma 11.4.2 *Let D be a closed, bounded set in $\mathbb{R}^d$; and let $k(x, y)$ be continuous for $x, y \in D$. Let the quadrature scheme (11.4.2) be convergent for all continuous functions on D. Define*

$$e_n(x, y) = \int_D k(x, v)k(v, y)\, dv - \sum_{j=1}^{q_n} w_j k(x, x_j)k(x_j, y)\,, \quad x, y \in D, \ n \geq 1,$$

$$(11.4.13)$$

the numerical integration error for the integrand $k(x, \cdot)k(\cdot, y)$. Then for $z \in C(D)$,

$$(K - K_n)Kz(x) = \int_D e_n(x, y)z(y)\, dy, \qquad (11.4.14)$$

$$(K - K_n)K_n z(x) = \sum_{j=1}^{q_n} w_j e_n(x, x_j)z(x_j). \qquad (11.4.15)$$

In addition,

$$\|(K - K_n)K\| = \max_{x \in D} \int_D |e_n(x,y)| \, dy, \tag{11.4.16}$$

$$\|(K - K_n)K_n\| = \max_{x \in D} \sum_{j=1}^{q_n} |w_j e_n(x, x_j)|. \tag{11.4.17}$$

Finally, the numerical integration error E_n converges to zero uniformly on D,

$$c_E \equiv \lim_{n \to \infty} \max_{x,y \in D} |e_n(x,y)| = 0 \tag{11.4.18}$$

and thus

$$\|(K - K_n)K\|, \ \|(K - K_n)K_n\| \to 0 \quad as \quad n \to \infty. \tag{11.4.19}$$

Proof. The proofs of (11.4.14) and (11.4.15) are straightforward manipulations and we omit them. The quantity $(K - K_n)K$ is an integral operator on $C(D)$, by (11.4.14); and therefore, we have (11.4.16) for its bound. The proof of (11.4.17) is also straightforward and we omit it.

To prove (11.4.18), we begin by showing that $\{e_n(x,y) \mid n \geq 1\}$ is a uniformly bounded and equicontinuous family that is pointwise convergent to 0 on the closed bounded set D; and then $e_n(x,y) \to 0$ uniformly on D by the Ascoli theorem. By the assumption that the quadrature rule of (11.4.2) converges for all continuous functions g on D, we have that for each $x, y \in D$, $e_n(x,y) \to 0$ as $n \to \infty$.

To prove boundedness,

$$|e_n(x,y)| \leq (c_D + c_I) c_K^2$$

with

$$c_D = \int_D dy, \qquad c_K = \max_{x,y \in D} |k(x,y)|$$

and c_I the constant from (11.4.3). For equicontinuity,

$$|e_n(x,y) - e_n(\xi,\eta)| \leq |e_n(x,y) - e_n(\xi,y)| + |e_n(\xi,y) - e_n(\xi,\eta)|,$$

$$|e_n(x,y) - e_n(\xi,y)| \leq c_K(c_D + c_I) \max_{y \in D} |k(x,y) - k(\xi,y)|,$$

$$|e_n(\xi,y) - e_n(\xi,\eta)| \leq c_K(c_D + c_I) \max_{x \in D} |k(x,y) - k(x,\eta)|.$$

By the uniform continuity of $k(x,y)$ on the closed bounded set D, this shows the equicontinuity of $\{e_n(x,y)\}$. This also completes the proof of (11.4.18).

For (11.4.19) we notice that

$$\|(K - K_n)K\| \leq c_D \max_{x,y \in D} |e_n(x,y)|, \tag{11.4.20}$$

$$\|(K - K_n)K_n\| \leq c_I \max_{x,y \in D} |e_n(x,y)|. \tag{11.4.21}$$

This completes the proof. ∎

To carry out an error analysis for the Nyström method (11.4.4)–(11.4.6), we need the following perturbation theorem. It furnishes an alternative to the perturbation arguments based on the geometric series theorem (e.g., Theorem 2.3.5 in Section 2.3).

Theorem 11.4.3 *Let V be a Banach space, let S, T be bounded operators on V to V, and let S be compact. For given $\lambda \neq 0$, assume $\lambda - T : V \xrightarrow[\text{onto}]{1-1} V$, which implies $(\lambda - T)^{-1}$ exists as a bounded operator on V to V. Finally, assume*

$$\|(T - S)S\| < \frac{|\lambda|}{\|(\lambda - T)^{-1}\|}. \tag{11.4.22}$$

Then $(\lambda - S)^{-1}$ exists and is bounded on V to V, with

$$\|(\lambda - S)^{-1}\| \leq \frac{1 + \|(\lambda - T)^{-1}\| \, \|S\|}{|\lambda| - \|(\lambda - T)^{-1}\| \, \|(T - S)S\|}. \tag{11.4.23}$$

If $(\lambda - T)u = f$ and $(\lambda - S)z = f$, then

$$\|u - z\| \leq \|(\lambda - S)^{-1}\| \, \|Tu - Su\|. \tag{11.4.24}$$

Proof. Consider that if $(\lambda - S)^{-1}$ were to exist, then it would satisfy the identity

$$(\lambda - S)^{-1} = \frac{1}{\lambda} \left\{ I + (\lambda - S)^{-1}S \right\}. \tag{11.4.25}$$

Without any motivation at this point, consider the approximation

$$(\lambda - S)^{-1} \approx \frac{1}{\lambda} \left\{ I + (\lambda - T)^{-1}S \right\}. \tag{11.4.26}$$

To check this approximation, compute

$$\frac{1}{\lambda} \left\{ I + (\lambda - T)^{-1}S \right\} (\lambda - S) = \left\{ I + \frac{1}{\lambda}(\lambda - T)^{-1}(T - S)S \right\}. \tag{11.4.27}$$

The right side is invertible by the geometric series theorem, because (11.4.22) implies

$$\frac{1}{|\lambda|} \|(\lambda - T)^{-1}\| \, \|(T - S)S\| < 1.$$

In addition, the geometric series theorem implies, after simplification, that

$$\left\| \left[\lambda + (\lambda - T)^{-1}(T - S)S \right]^{-1} \right\| \leq \frac{1}{|\lambda| - \|(\lambda - T)^{-1}\| \, \|(T - S)S\|}. \tag{11.4.28}$$

Since the right side of (11.4.27) is invertible, the left side is also invertible. This implies that $\lambda - S$ is one-to-one, as otherwise the left side would not be invertible. Since S is compact, the Fredholm alternative theorem (cf.

Theorem 2.8.10 of Section 2.8.4) implies $(\lambda - S)^{-1}$ exists and is bounded on V to V. In particular,

$$(\lambda - S)^{-1} = \left[\lambda + (\lambda - T)^{-1}(T - S)S\right]^{-1}\left\{I + (\lambda - T)^{-1}S\right\}. \quad (11.4.29)$$

The bound (11.4.23) follows directly from this and (11.4.28).

For the error $u - z$, rewrite $(\lambda - T)u = f$ as

$$(\lambda - S)u = f + (T - S)u.$$

Subtract $(\lambda - S)z = f$ to get

$$(\lambda - S)(u - z) = (T - S)u, \quad (11.4.30)$$

$$u - z = (\lambda - S)^{-1}(T - S)u, \quad (11.4.31)$$

$$\|u - z\| \le \left\|(\lambda - S)^{-1}\right\| \|(T - S)u\|,$$

which proves (11.4.24). ∎

Using this theorem, we can give a complete convergence analysis for the Nyström method (11.4.4)–(11.4.6).

Theorem 11.4.4 Let D be a closed, bounded set in $\mathbb{R}^d$; and let $k(x, y)$ be continuous for $x, y \in D$. Assume the quadrature scheme (11.4.2) is convergent for all continuous functions on D. Further, assume that the integral equation (11.4.1) is uniquely solvable for given $f \in C(D)$, with $\lambda \ne 0$. Then for all sufficiently large n, say, $n \ge N$, the approximate inverses $(\lambda - K_n)^{-1}$ exist and are uniformly bounded,

$$\left\|(\lambda - K_n)^{-1}\right\| \le \frac{1 + \left\|(\lambda - K)^{-1}\right\| \|K_n\|}{|\lambda| - \left\|(\lambda - K)^{-1}\right\| \|(K - K_n)K_n\|} \le c_y, \qquad n \ge N, \quad (11.4.32)$$

with a suitable constant $c_y < \infty$. For the equations $(\lambda - K)u = f$ and $(\lambda - K_n)u_n = f$, we have

$$\|u - u_n\|_\infty \le \left\|(\lambda - K_n)^{-1}\right\| \|(K - K_n)u\|_\infty$$
$$\le c_y \|(K - K_n)u\|_\infty, \qquad n \ge N. \quad (11.4.33)$$

Proof. The proof is a simple application of the preceding theorem, with $S = K_n$ and $T = K$. From Lemma 11.4.2, we have $\|(K - K_n)K_n\| \to 0$, and therefore (11.4.22) is satisfied for all sufficiently large n, say, $n \ge N$. From (11.4.11), the boundedness of $k(x, y)$ over D, and (11.4.3),

$$\|K_n\| \le c_I c_K, \quad n \ge 1.$$

Then

$$c_y \equiv \sup_{n \ge N} \frac{1 + \left\|(\lambda - K)^{-1}\right\| \|K_n\|}{|\lambda| - \left\|(\lambda - K)^{-1}\right\| \|(K - K_n)K_n\|} < \infty. \quad (11.4.34)$$

This completes the proof. ∎

This last theorem gives complete information for analyzing the convergence of the Nyström method (11.4.4)–(11.4.6). The term $\|(K - K_n)K_n\|$ can be analyzed from (11.4.17) by analyzing the numerical integration error $e_n(x, y)$ of (11.4.13). From the error bound (11.4.33), the speed with which $\|u - u_n\|_\infty$ converges to zero is bounded by that of the numerical integration error

$$\|(K - K_n)u\|_\infty = \max_{x \in D} \left| \int_D k(x, y)u(y) \, dy - \sum_{j=1}^{q_n} w_j k(x, x_j)u(x_j) \right|.$$

$$(11.4.35)$$

In fact, the error $\|u - u_n\|_\infty$ converges to zero with exactly this speed. Recall from applying (11.4.30) that

$$(\lambda - K_n)(u - u_n) = (K - K_n)u. \qquad (11.4.36)$$

From bounding this,

$$\|(K - K_n)u\|_\infty \le \|\lambda - K_n\| \, \|u - u_n\|_\infty \, .$$

When combined with (11.4.33), this shows the assertion that $\|u - u_n\|_\infty$ and $\|(K - K_n)u\|_\infty$ converge to zero with the same speed.

There is a very large literature on bounding and estimating the errors for the common numerical integration rules. Thus the speed of convergence with which $\|u - u_n\|_\infty$ converges to zero can be determined by using results on the speed of convergence of the integration rule (11.4.2) when it is applied to the integral

$$\int_D k(x, y)u(y) \, dy.$$

Example 11.4.5 *Consider the trapezoidal numerical integration rule*

$$\int_a^b g(y) \, dy \approx h \sum_{j=0}^{n} {}'' g(x_j) \qquad (11.4.37)$$

with $h = (b-a)/n$ and $x_j = a + jh$ for $j = 0, \ldots, n$. The notation Σ'' means the first and last terms are to be halved before summing. For the error,

$$\int_a^b g(y) \, dy - h \sum_{j=0}^{n} {}'' g(x_j) = -\frac{h^2(b - a)}{12} g''(\xi_n), \quad g \in C^2[a, b], \quad n \ge 1,$$

$$(11.4.38)$$

with ξ_n some point in $[a, b]$. There is also the asymptotic error formula

$$\int_a^b g(y) \, dy - h \sum_{j=0}^{n} {}'' g(x_j) = -\frac{h^2}{12} [g'(b) - g'(a)] + O(h^4), \quad g \in C^4[a, b],$$

$$(11.4.39)$$

and we make use of it in a later example. For a derivation of these formulas, see [11, p. 285].

When this is applied to the integral equation

$$\lambda u(x) - \int_a^b k(x,y)u(y)\,dy = f(x), \quad a \le x \le b, \tag{11.4.40}$$

we obtain the approximating linear system

$$\lambda u_n(x_i) - h\sum_{j=0}^{n}{}''k(x_i,x_j)u_n(x_j) = f(x_i), \quad i = 0,1,\ldots,n, \tag{11.4.41}$$

which is of order $q_n = n+1$. The Nyström interpolation formula is given by

$$u_n(x) = \frac{1}{\lambda}\left[f(x) + h\sum_{j=0}^{n}{}''k(x,x_j)u_n(x_j)\right], \quad a \le x \le b. \tag{11.4.42}$$

The speed of convergence is based on the numerical integration error

$$(K - K_n)u(y) = -\frac{h^2(b-a)}{12}\left.\frac{\partial^2 k(x,y)u(y)}{\partial y^2}\right|_{y=\xi_n(x)} \tag{11.4.43}$$

with $\xi_n(x) \in [a,b]$. From (11.4.39), the asymptotic integration error is

$$(K - K_n)u(y) = -\frac{h^2}{12}\left.\frac{\partial k(x,y)u(y)}{\partial y}\right|_{y=a}^{y=b} + O(h^4). \tag{11.4.44}$$

From (11.4.43), we see the Nyström method converges with an order of $O(h^2)$, provided $k(x,y)u(y)$ is twice continuously differentiable with respect to y, uniformly in x.

An asymptotic error estimate

In those cases for which the quadrature formula has an asymptotic error formula, as in (11.4.39), we can give an asymptotic estimate of the error in solving the integral equation using the Nyström method. Returning to (11.4.36), we can write

$$u - u_n = (\lambda - K_n)^{-1}(K - K_n)u = \epsilon_n + r_n \tag{11.4.45}$$

with

$$\epsilon_n = (\lambda - K)^{-1}(K - K_n)u$$

and

$$\begin{aligned} r_n &= \left[(\lambda - K_n)^{-1} - (\lambda - K)^{-1}\right](K - K_n)u \\ &= (\lambda - K_n)^{-1}(K_n - K)(\lambda - K)^{-1}(K - K_n)u. \end{aligned} \tag{11.4.46}$$

The term r_n generally converges to zero more rapidly than the term ϵ_n, although showing this is dependent on the quadrature rule being used.

Assuming the latter to be true, we have

$$u - u_n \approx \epsilon_n \qquad (11.4.47)$$

with ϵ_n satisfying the original integral equation with the integration error $(K - K_n)u$ as the right-hand side,

$$(\lambda - K)\epsilon_n = (K - K_n)u. \qquad (11.4.48)$$

At this point, one needs to consider the quadrature rule in more detail.

Example 11.4.6 *Consider again the earlier example (11.4.37)–(11.4.44) of the Nyström method with the trapezoidal rule. Assume further that $k(x,y)$ is four times continuously differentiable with respect to both y and x, and assume $u \in C^4[a,b]$. Then from the asymptotic error formula (11.4.44), we can decompose the right side $(K - K_n)u$ of (11.4.48) into two terms, of sizes $O(h^2)$ and $O(h^4)$. Introduce the function $\gamma(y)$ satisfying the integral equation*

$$\lambda\gamma(x) - \int_a^b k(x,y)\gamma(y)\,dy = -\frac{1}{12}\left.\frac{\partial k(x,y)u(y)}{\partial y}\right|_{y=a}^{y=b}, \quad a \le x \le b.$$

Then the error term ϵ_n in (11.4.47)–(11.4.48) is dominated by $\gamma(x)h^2$. By a similar argument, it can also be shown that the term $r_n = O(h^4)$. Thus we have the asymptotic error estimate

$$u - u_n \approx \gamma(x)h^2 \qquad (11.4.49)$$

for the Nyström method with the trapezoidal rule.

Conditioning of the linear system

Let A_n denote the matrix of coefficients for the linear system (11.4.5):

$$(A_n)_{i,j} = \lambda\delta_{i,j} - w_j k(x_i, x_j).$$

We want to bound $\mathrm{cond}(A_n) = \|A_n\|\|A_n^{-1}\|$.
    For general $z \in C(D)$,

$$\max_{i=1,\dots,q_n} \left|\lambda z(x_i) - \sum_{j=1}^{q_n} w_j k(x_i, x_j)z(x_j)\right|$$

$$\le \sup_{x \in D} \left|\lambda z(x) - \sum_{j=1}^{q_n} w_j k(x, x_j)z(x_j)\right|.$$

This shows

$$\|A_n\| \le \|\lambda - K_n\|. \qquad (11.4.50)$$

    For A_n^{-1},

$$\|A_n^{-1}\| = \sup_{\substack{\gamma \in \mathbb{R}^{q_n} \\ \|\gamma\|_\infty = 1}} \|A_n^{-1}\gamma\|_\infty.$$

For such γ, let $z = A_n^{-1}\gamma$ or $\gamma = A_n z$. Pick $f \in C(D)$ such that

$$f(x_i) = \gamma_i, \quad i = 1, \ldots, q_n$$

and $\|f\|_\infty = \|\gamma\|_\infty$. Let $u_n = (\lambda - K_n)^{-1}f$, or equivalently, $(\lambda - K_n)u_n = f$. Then from the earlier discussion of the Nyström method,

$$u_n(x_i) = z_i, \quad i = 1, \ldots, q_n.$$

Then

$$\begin{aligned}
\left\|A_n^{-1}\gamma\right\|_\infty &= \|z\|_\infty \\
&\leq \|u_n\|_\infty \\
&\leq \left\|(\lambda - K_n)^{-1}\right\| \, \|f\|_\infty \\
&= \left\|(\lambda - K_n)^{-1}\right\| \, \|\gamma\|_\infty.
\end{aligned}$$

This proves

$$\left\|A_n^{-1}\right\|_\infty \leq \left\|(\lambda - K_n)^{-1}\right\|. \tag{11.4.51}$$

Combining these results,

$$\mathrm{cond}(A_n) \leq \|\lambda - K_n\| \, \left\|(\lambda - K_n)^{-1}\right\| \equiv \mathrm{cond}(\lambda - K_n). \tag{11.4.52}$$

Thus if the operator equation $(\lambda - K_n)u_n = f$ is well-conditioned, then so is the linear system associated with it. We leave as an exercise the development of the relationship between $\mathrm{cond}(\lambda - K_n)$ and $\mathrm{cond}(\lambda - K)$.

11.4.3 Collectively compact operator approximations

The error analysis of the Nyström method was developed mainly during the period 1940 to 1970, and a number of researchers were involved. Initially, the only goal was to show that the method was stable and convergent, and perhaps, to obtain computable error bounds. As this was accomplished, a second goal emerged of creating an abstract framework for the method and its error analysis, a framework in the language of functional analysis that referred only to mapping properties of the approximate operators and not to properties of the particular integral operator, function space, or quadrature scheme being used. The final framework developed is due primarily to P. Anselone, and he gave to it the name of the *theory of collectively compact operator approximations*. A complete presentation of it is given in his book [2], and we present only a portion of it here. With this framework, it has been possible to analyze a number of important extensions of the Nyström method, including those discussed in the following Section 11.5.

Within a functional analysis framework, how does one characterize the numerical integral operators $\{K_n \mid n \geq 1\}$? We want to know the characteristic properties of these operators that imply that $\|(K - K_n)K_n\| \to 0$ as $n \to \infty$. Then the earlier Theorem 11.4.3 remains valid, and the Nyström method and its error analysis can be extended to other situations, some of which are discussed in later sections.

We assume that $\{K_n \mid n \geq 1\}$ satisfies the following properties.

A1. V is a Banach space; and K and K_n, $n \geq 1$, are linear operators on V into V.

A2. $K_n u \to K u$ as $n \to \infty$, for all $u \in V$.

A3. The set $\{K_n \mid n \geq 1\}$ is *collectively compact*, which means that the set

$$S = \{K_n u \mid n \geq 1 \text{ and } \|u\| \leq 1\} \qquad (11.4.53)$$

has compact closure in V.

These assumptions are an excellent abstract characterization of the numerical integral operators introduced earlier in (11.4.10) of this chapter. We refer to a family $\{K_n\}$ that satisfies **A1–A3** as a *collectively compact family of pointwise convergent operators*.

Lemma 11.4.7 *Assume the above properties **A1–A3**. Then*

1. K *is compact;*

2. $\{K_n \mid n \geq 1\}$ *is uniformly bounded;*

3. *For any compact operator* $M : V \to V$,

$$\|(K - K_n)M\| \to 0 \quad as \quad n \to \infty;$$

4. $\|(K - K_n)K_n\| \to 0$ *as* $n \to \infty$.

Proof. (1) To show K is compact, it is sufficient to show that the set

$$\{Ku \mid \|u\| \leq 1\}$$

has compact closure in V. By **A2**, this last set is contained in $\overline{S}$, and it is compact by **A3**.

(2) This follows from the definition of operator norm and the boundedness of the set $\overline{S}$.

(3) Using the definition of operator norm,

$$
\begin{aligned}
\|(K - K_n)M\| &= \sup_{\|u\| \leq 1} \|(K - K_n)Mu\| \\
&= \sup_{z \in M(B)} \|(K - K_n)z\| \qquad (11.4.54)
\end{aligned}
$$

with $B = \{u \mid \|u\| \leq 1\}$. From the compactness of M, the set $M(B)$ has compact closure. Using Lemma 11.4.2, we then have that the last quantity in (11.4.54) goes to zero as $n \to \infty$.

(4) Again, using the definition of operator norm,

$$\|(K - K_n)K_n\| = \sup_{\|u\| \leq 1} \|(K - K_n)K_n\| = \sup_{z \in S} \|(K - K_n)z\|. \qquad (11.4.55)$$

Using **A3**, S has compact closure; and then using Lemma 11.4.2, we have that the last quantity in (11.4.55) goes to zero as $n \to \infty$. ∎

As a consequence of this lemma, we can apply Theorem 11.4.3 to any set of approximating equations $(\lambda - K_n)u_n = f$ where the set $\{K_n\}$ satisfies **A1–A3**. This extends the idea of the Nyström method, and the product integration methods of the following section is analyzed using this more abstract framework.

Returning to the proof of Theorem 11.4.3, we can better motivate an argument used there. With $S = K_n$ and $T = K$, the statements (11.4.25) and (11.4.26) become

$$(\lambda - K_n)^{-1} = \frac{1}{\lambda}\left[I + (\lambda - K_n)^{-1}K_n \right], \tag{11.4.56}$$

$$(\lambda - K_n)^{-1} \approx \frac{1}{\lambda}\left[I + (\lambda - K)^{-1}K_n \right]. \tag{11.4.57}$$

Since K_n is not norm convergent to K, we cannot expect $(\lambda - K)^{-1} \approx (\lambda - K_n)^{-1}$ to be a good approximation. However, it becomes a much better approximation when the operators are restricted to act on a compact subset of V. Since the family $\{K_n\}$ is collectively compact, (11.4.57) is a good approximation of (11.4.56).

Exercise 11.4.1 *Prove* (11.4.12).
Hint: Recall the discussion in Exercise 11.3.4.

Exercise 11.4.2 *Derive* (11.4.14)–(11.4.17).

Exercise 11.4.3 *Obtain a bound for* cond$(\lambda - K_n)$ *in terms of* cond$(\lambda - K)$. *More generally, explore the relationship between these two condition numbers.*
Hint: Use Theorem 11.4.4.

Exercise 11.4.4 *Generalize Example 11.4.5 to Simpson's rule.*

Exercise 11.4.5 *Generalize Example 11.4.6 to Simpson's rule.*

11.5 Product integration

We now consider the numerical solution of integral equations of the second kind in which the kernel function $k(x, y)$ is not continuous, but for which the associated integral operator K is still compact on $C(D)$ into $C(D)$. The main ideas we present will extend to functions in any finite number of variables; but it is more intuitive to first present these ideas for integral equations for functions of a single variable, and in particular,

$$\lambda u(x) - \int_a^b k(x, y)u(y)\, dy = f(x), \qquad a \le x \le b. \tag{11.5.1}$$

In this setting, most such discontinuous kernel functions $k(x, y)$ have an infinite singularity; and the most important examples are $\log |x - y|$, $|x - y|^{\gamma-1}$ for some $\gamma > 0$ (although it is only singular for $0 < \gamma < 1$), and variants of them.

We introduce the idea of product integration by considering the special case of

$$\lambda u(x) - \int_a^b l(x, y) \log |y - x| \, u(y) \, dy = f(x), \qquad a \leq x \leq b, \qquad (11.5.2)$$

with the kernel

$$k(x, y) = l(x, y) \log |y - x| . \qquad (11.5.3)$$

We assume that $l(x, y)$ is a well-behaved function (i.e., it is several times continuously differentiable), and initially, we assume the unknown solution $u(x)$ is also well-behaved. To solve (11.5.2), we define a method called the *product trapezoidal rule*.

Let $n \geq 1$ be an integer, $h = (b - a)/n$, and $x_j = a + jh$, $j = 0, 1, \dots, n$. For general $u \in C[a, b]$, define

$$[l(x, y)u(y)]_n = \frac{1}{h} [(x_j - y)l(x, x_{j-1})u(x_{j-1}) + (y - x_{j-1})l(x, x_j)u(x_j)], \qquad (11.5.4)$$

for $x_{j-1} \leq y \leq x_j$, $j = 1, \dots, n$ and $a \leq x \leq b$. This is piecewise linear in y, and it interpolates $l(x, y)u(y)$ at $y = x_0, \dots, x_n$, for all $x \in [a, b]$. Define a numerical approximation to the integral operator in (11.5.2) by

$$K_n u(x) \equiv \int_a^b [l(x, y)u(y)]_n \log |y - x| \, dy, \qquad a \leq x \leq b. \qquad (11.5.5)$$

This can also be written as

$$K_n u(x) = \sum_{j=0}^n w_j(x) l(x, x_j) u(x_j), \qquad u \in C[a, b], \qquad (11.5.6)$$

with weights

$$w_0(x) = \frac{1}{h} \int_{x_0}^{x_1} (x_1 - y) \log |x - y| \, dy,$$
$$w_n(x) = \frac{1}{h} \int_{x_{n-1}}^{x_n} (y - x_{n-1}) \log |x - y| \, dy, \qquad (11.5.7)$$

$$w_j(x) = \frac{1}{h} \int_{x_{j-1}}^{x_j} (y - x_{j-1}) \log |x - y| \, dy$$
$$+ \frac{1}{h} \int_{x_j}^{x_{j+1}} (x_{j+1} - y) \log |x - y| \, dy, \qquad j = 1, \dots, n - 1. \qquad (11.5.8)$$

To approximate the integral equation (11.5.2), we use

$$\lambda u_n(x) - \sum_{j=0}^{n} w_j(x)l(x, x_j)u_n(x_j) = f(x), \qquad a \leq x \leq b. \qquad (11.5.9)$$

As with the Nyström method (11.4.4)–(11.4.6), this is equivalent to first solving the linear system

$$\lambda u_n(x_i) - \sum_{j=0}^{n} w_j(x_i)l(x_i, x_j)u_n(x_j) = f(x_i), \qquad i = 0, \ldots, n, \qquad (11.5.10)$$

and then using the Nyström interpolation formula

$$u_n(x) = \frac{1}{\lambda} \left[f(x) + \sum_{j=0}^{n} w_j(x)l(x, x_j)u_n(x_j) \right], \qquad a \leq x \leq b. \qquad (11.5.11)$$

We leave it to the reader to check these assertions, since it is quite similar to what was done for the original Nyström method. With this method, we approximate those parts of the integrand in (11.5.2) that can be well-approximated by piecewise linear interpolation, and we integrate exactly the remaining more singular parts of the integrand.

Rather than using piecewise linear interpolation, other more accurate interpolation schemes could have been used to obtain a more rapidly convergent numerical method. Later in the section, we consider and illustrate the use of piecewise quadratic interpolation. We have also used evenly spaced node points $\{x_i\}$, but this is not necessary. The use of such evenly spaced nodes is an important case; but we will see later in the section that special choices of nonuniformly spaced node points are often needed for solving an integral equation such as (11.5.2).

Other singular kernel functions can be handled in a manner analogous to what has been done for (11.5.2). Consider the equation

$$\lambda u(x) - \int_a^b l(x, y)g(x, y)u(y)\, dy = f(x), \qquad a \leq x \leq b, \qquad (11.5.12)$$

in which $g(x, y)$ is singular, with $l(x, y)$ and $u(x)$ as before. An important case is to take

$$g(x, y) = \frac{1}{|x - y|^{1-\gamma}}$$

for some $\gamma > 0$. To approximate (11.5.12), use the earlier approximation (11.5.4). Then

$$K_n u(x) = \int_a^b [l(x, y)u(y)]_n\, g(x, y)\, dy, \qquad a \leq x \leq b. \qquad (11.5.13)$$

All arguments proceed exactly as before. To evaluate $K_n u(x)$, we need to evaluate the analogues of the weights in (11.5.7)–(11.5.8), where $\log |x - y|$

is replaced by $g(x, y)$. We assume these weights can be calculated in some practical manner, perhaps analytically. We consider further generalizations later in the section.

11.5.1   Error analysis

We consider the equation (11.5.12), with $l(x, y)$ assumed to be continuous. Further, we assume the following for $g(x, y)$:

$$c_g \equiv \sup_{a \leq x \leq b} \int_a^b |g(x, y)| \, dy < \infty, \tag{11.5.14}$$

$$\lim_{h \searrow 0} \omega_g(h) = 0, \tag{11.5.15}$$

where

$$\omega_g(h) \equiv \sup_{\substack{|x - \tau| \leq h \\ a \leq x, \tau \leq b}} \int_a^b |g(x, y) - g(\tau, y)| \, dy.$$

These two properties can be shown to be true for both $\log |x - y|$ and $|x - y|^{\gamma - 1}$, $\gamma > 0$. Such assumptions were used earlier in Subsection 2.8.1 in showing compactness of integral operators on $C[a, b]$, and we refer to that earlier material.

Theorem 11.5.1 *Assume the function $g(x, y)$ satisfies (11.5.14)–(11.5.15), and assume $l(x, y)$ is continuous for $a \leq x, y \leq b$. For a given $f \in C[a, b]$, assume the integral equation*

$$\lambda u(x) - \int_a^b l(x, y) g(x, y) u(y) \, dy = f(x), \qquad a \leq x \leq b,$$

is uniquely solvable. Consider the numerical approximation (11.5.13), with $[l(x, y) u(y)]_n$ defined with piecewise linear interpolation, as in (11.5.4). Then for all sufficiently large n, say $n \geq N$, the equation (11.5.13) is uniquely solvable, and the inverse operators are uniformly bounded for such n. Moreover,

$$\|u - u_n\|_\infty \leq c \|Ku - K_n u\|_\infty, \qquad n \geq N, \tag{11.5.16}$$

for suitable $c > 0$.

Proof. We can show that the operators $\{K_n\}$ of (11.5.13) are a collectively compact and pointwise convergent family on $C[a, b]$ to $C[a, b]$. This will prove the abstract assumptions **A1**–**A3** in the Subsection 11.4.3; and by using Lemma 11.4.7, we can then apply Theorem 11.4.4. We note that **A1** is obvious from the definitions of K and K_n.

Let $\mathcal{S} = \{K_n u \mid n \geq 1 \text{ and } \|u\|_\infty \leq 1\}$. For bounds on $\|K_n u\|_\infty$, first note that the piecewise linear interpolant z_n of a function $z \in C[a, b]$ satisfies

$$\|z_n\|_\infty \leq \|z\|_\infty.$$

With this, it is straightforward to show

$$\|K_n u\|_\infty \le c_l c_g, \quad u \in C[a,b], \qquad \|u\|_\infty \le 1,$$

with

$$c_l \equiv \max_{a \le x, y \le b} |l(x,y)|.$$

This also shows the uniform boundedness of $\{K_n\}$, with

$$\|K_n\| \le c_l c_g, \qquad n \ge 1.$$

For equicontinuity of $\mathcal{S}$, write

$$K_n u(x) - K_n u(\xi) = \int_a^b [l(x,y)u(y)]_n \, g(x,y) \, dy$$

$$- \int_a^b [l(\xi,y)u(y)]_n \, g(\tau,y) \, dy$$

$$= \int_a^b [\{l(x,y) - l(\xi,y)\} \, u(y)]_n \, g(x,y) \, dy$$

$$+ \int_a^b [l(\xi,y)u(y)]_n \, \{g(x,y) - g(\xi,y)\} \, dy.$$

This uses the linearity in z of the piecewise linear interpolation being used in defining $[z(y)]_n$. The assumptions on $g(x,y)$ and $l(x,y)$, together with $\|u\|_\infty \le 1$, now imply

$$\left| \int_a^b [\{l(x,y) - l(\xi,y)\} \, u(y)]_n \, g(x,y) dy \right| \le c_g \|u\|_\infty \max_{a \le y \le b} |l(x,y) - l(\xi,y)|.$$

Also,

$$\left| \int_a^b [l(\xi,y)u(y)]_n \, \{g(x,y) - g(\xi,y)\} \, dy \right| \le c_l \|u\|_\infty \, \omega_g(|x - \xi|).$$

Combining these results shows the desired equicontinuity of $\mathcal{S}$, and it completes the proof of the abstract property **A3** needed in applying the collectively compact operator framework.

We leave the proof of **A2** as an exercise for the reader. To complete the proof of the theorem, we apply Lemma 11.4.7 and Theorem 11.4.4. The constant c is the uniform bound on $\|(\lambda - K_n)^{-1}\|$ for $n \ge N$. ∎

Example 11.5.2 *Let $z_n(y)$ denote the piecewise linear interpolant of $z(y)$, as used above in defining the product trapezoidal rule. It is a well-known standard result that*

$$|z(y) - z_n(y)| \le \frac{h^2}{8} \|z''\|_\infty, \qquad z \in C^2[a,b].$$

Thus if $l(x, \cdot) \in C^2[a, b]$, $a \leq x \leq b$, and if $u \in C^2[a, b]$, then (11.5.16) *implies*

$$\|u - u_n\|_\infty \leq \frac{ch^2}{8} \max_{a \leq x, y \leq b} \left| \frac{\partial^2 l(x, y) u(y)}{\partial y^2} \right|, \qquad n \geq N. \qquad (11.5.17)$$

The above ideas for solving (11.5.12) *will generalize easily to higher degrees of piecewise polynomial interpolation. All elements of the above proof also generalize, and we obtain a theorem analogous to Theorem 11.5.1. In particular, suppose $[l(\tau, y) u(y)]_n$ is defined using piecewise polynomial interpolation of degree $m \geq 0$. Assume $l(x, \cdot) \in C^{m+1}[a, b]$, $a \leq x \leq b$, and $u \in C^{m+1}[a, b]$. Then*

$$\|u - u_n\|_\infty \leq ch^{m+1} \max_{a \leq x, y \leq b} \left| \frac{\partial^{m+1} l(x, y) u(y)}{\partial y^{m+1}} \right|, \qquad n \geq N, \qquad (11.5.18)$$

for a suitable constant $c > 0$. When using piecewise quadratic interpolation, the method (11.5.13) *is called the product Simpson rule; and according to* (11.5.18), *its rate of convergence is at least $O(h^3)$.*

11.5.2   Generalizations to other kernel functions

Many singular integral equations are not easily written in the form (11.5.12) with a function $l(x, y)$ that is smooth and a function $g(x, y)$ for which weights such as those in (11.5.7)–(11.5.8) can be easily calculated. For such equations, we assume instead that the singular kernel function $k(x, y)$ can be written in the form

$$k(x, y) = \sum_{j=1}^{r} l_j(x, y) g_j(x, y) \qquad (11.5.19)$$

with each $l_j(x, y)$ and $g_j(x, y)$ satisfying the properties listed above for $l(x, y)$ and $g(x, y)$. We now have an integral operator written as a sum of integral operators of the form used in (11.5.12):

$$\mathcal{K}u(x) = \sum_{j=1}^{r} \mathcal{K}_j u(x) = \sum_{j=1}^{r} \int_a^b l_j(x, y) g_j(x, y) u(y) \, dy, \qquad u \in C[a, b].$$

Example 11.5.3 *Consider the integral equation*

$$u(x) - \int_0^\pi u(y) \log |\cos x - \cos y| \, dy = 1, \qquad 0 \leq x \leq \pi. \qquad (11.5.20)$$

One possibility for the kernel function $k(x, y) = \log |\cos x - \cos y|$ is to write

$$k(x, y) = \underbrace{|x - y|^{\frac{1}{2}} \log |\cos x - \cos y|}_{=l(x,y)} \underbrace{|x - y|^{-\frac{1}{2}}}_{=g(x,y)}.$$

	Product trapezoidal		Product Simpson	
n	$\|u - u_n\|_\infty$	Ratio	$\|u - u_n\|_\infty$	Ratio
2	$9.50E-3$		$2.14E-4$	
4	$2.49E-3$	3.8	$1.65E-5$	13.0
8	$6.32E-4$	3.9	$1.13E-6$	14.6
16	$1.59E-4$	4.0	$7.25E-8$	15.6
32	$3.98E-5$	4.0	$4.56E-9$	15.9

Table 11.3. Product trapezoidal and product Simpson examples for (11.5.20)

Unfortunately, this choice of $l(x,y)$ is continuous without being differentiable; and the function $l(x,y)$ needs to be differentiable in order to have the numerical method converge with sufficient speed. A better choice is to use

$$
k(x,y) = \log \left| 2 \sin \frac{1}{2}(x-y) \sin \frac{1}{2}(x+y) \right|
$$

$$
= \log \left\{ \frac{2 \sin \frac{1}{2}(x-y) \sin \frac{1}{2}(x+y)}{(x-y)(x+y)(2\pi - x - y)} \right\} + \log |x-y|
$$

$$
+ \log(x+y) + \log(2\pi - x - y). \tag{11.5.21}
$$

This is of the form (11.5.19) with $g_1 = l_2 = l_3 = l_4 \equiv 1$ and

$$
l_1(x,y) = \log \left\{ \frac{2 \sin \frac{1}{2}(x-y) \sin \frac{1}{2}(x+y)}{(x-y)(x+y)(2\pi - x - y)} \right\},
$$

$$
g_2(x,y) = \log |x-y|,
$$

$$
g_3(x,y) = \log(x+y),
$$

$$
g_4(x,y) = \log(2\pi - x - y).
$$

This is of the form (11.5.19). The function $l_1(x,y)$ is infinitely differentiable on $[0, 2\pi]$; and the functions g_2, g_3, and g_4 are singular functions for which the needed integration weights are easily calculated.

We solve (11.5.20) with both the product trapezoidal rule and the product Simpson rule, and error results are given in Table 11.3. The decomposition (11.5.21) is used to define the approximating operators. With the operator with kernel $l_1(x,y)g_1(x,y)$, we use the regular Simpson rule. The true solution of the equation is

$$
u(x) \equiv \frac{1}{1 + \pi \log 2} \doteq 0.31470429802.
$$

Note that the error for the product trapezoidal rule is consistent with (11.5.17). But for the product Simpson rule, we appear to have an error behavior of $O(h^4)$, whereas that predicted by (11.5.18) is only $O(h^3)$. This is discussed further below.

11.5.3   Improved error results for special kernels

If we consider again the error formula (11.5.16), the error result (11.5.18) was based on applying standard error bounds for polynomial interpolation to bounding the numerical integration error $\|Ku - K_n u\|_\infty$. We know that for many ordinary integration rules (e.g., Simpson's rule), there is an improvement in the speed of convergence over that predicted by the polynomial interpolation error, and this improvement is made possible by fortuitous cancellation of errors when integrating. Thus it is not surprising that the same type of cancellation occurs with the error $\|Ku - K_n u\|_\infty$ in product Simpson integration, as is illustrated in Table 11.3.

For the special cases of $g(x, y)$ equal to $\log |x - y|$ and $|x - y|^{\gamma-1}$, de-Hoog and Weiss [79] improved on the bound (11.5.18). In [79], they first extended known asymptotic error formulas for ordinary composite integration rules to product integration formulas; and then these results were further extended to estimate $Ku - K_n u$ for product integration methods of solving singular integral equations. For the case of the product Simpson's rule, their results state that if $u \in C^4[a, b]$, then

$$\|Ku - K_n u\|_\infty \leq \begin{cases} ch^4 \left|\log h\right|, & g(x, y) = \log |x - y|, \\ ch^{3+\gamma}, & g(x, y) = |x - y|^{\gamma-1}. \end{cases} \tag{11.5.22}$$

This is in agreement with the results in Table 11.3.

11.5.4   Product integration with graded meshes

The rate of convergence results (11.5.18) and (11.5.22) both assume that the unknown solution $u(x)$ possesses several continuous derivatives. In fact, $u(x)$ seldom is smoothly differentiable, but rather has somewhat singular behavior in the neighborhood of the endpoints of the interval $[a, b]$ on which the integral equation is being solved. In the following, this is made more precise; and we also give a numerical method that restores the speed of convergence seen above with smoothly differentiable unknown solution functions.

To examine the differentiability of the solution $u(x)$ of a general integral equation $(\lambda - K)u = f$, the differentiability of the kernel function $k(x, y)$ allows the smoothness of $f(x)$ to be carried over to that of $u(x)$: Use

$$\frac{d^j u(x)}{dx^j} = \frac{1}{\lambda} \left[\frac{d^j f(x)}{dx^j} + \int_a^b \frac{\partial^j k(x, y)}{\partial x^j} u(y) dy \right].$$

But if the kernel function is not differentiable, then the integral operator need not be smoothing. To see that the integral operator K with kernel $\log |x - y|$ is not smoothing in the manner that is true with differentiable kernel functions, let $u_0(x) \equiv 1$ on the interval $[0, 1]$, and calculate $Ku_0(x)$:

$$Ku_0(x) = \int_0^1 \log|x - y| \, dy = x \log x + (1 - x) \log(1 - x) - 1, \quad 0 \leq x \leq 1.$$
$$(11.5.23)$$

The function $Ku_0(x)$ is not continuously differentiable on $[0, 1]$, whereas the function $u_0(x)$ is a C^∞ function. This formula also contains the typical type of singular behavior that appears in the solution when solving a second-kind integral equation with a kernel function $k(x, y) = l(x, y) \log|x - y|$.

We give the main result of Schneider [144] on the regularity behavior of solutions of $(\lambda - K)u = f$ for special weakly singular kernel functions. As notation, introduce the following spaces:

$$C^{(0,\beta)}[a, b] = \left\{ g \in C[a, b] \,\bigg|\, d_\beta(g) \equiv \sup_{a \leq x, \xi \leq b} \frac{|g(x) - g(\xi)|}{|x - \xi|^\beta} < \infty \right\}$$
$$(11.5.24)$$

for $0 < \beta < 1$, and

$$C^{(0,1)}[a, b] = \left\{ g \in C[a, b] \,\bigg|\, \sup_{a \leq x, \xi \leq b} \frac{|g(x) - g(\xi)|}{|x - \xi| \log|B/(x - \xi)|} < \infty \right\},$$

for some $B > b - a$. For $0 < \beta < 1$, $C^{(0,\beta)}[a, b]$ are the standard Hölder spaces introduced in Subsection 1.4.1 of Chapter 1.

Theorem 11.5.4 *Let $k \geq 0$ be an integer, and let $0 < \gamma \leq 1$. Assume $f \in C^{(0,\gamma)}[a, b]$, $f \in C^k(a, b)$, and*

$$(x - a)^i (b - x)^i f^{(i)}(x) \in C^{(0,\gamma)}[a, b], \quad i = 1, \ldots, k.$$

Also assume $L \in C^{k+1}(D)$ with $D = [a, b] \times [a, b]$. Finally, assume the integral equation

$$\lambda u(x) - \int_a^b l(x, y) g_\gamma(x - y) u(y) \, dy = f(x), \quad a \leq x \leq b \qquad (11.5.25)$$

with

$$g_\gamma(u) \equiv \begin{cases} u^{\gamma - 1}, & 0 < \gamma < 1, \\ \log|u|, & \gamma = 1 \end{cases}$$

is uniquely solvable. Then

    a. *The solution $u(x)$ satisfies $u \in C^{(0,\gamma)}[a, b]$, $u \in C^k(a, b)$, and*

$$u_i(x) \equiv (x - a)^i (b - x)^i u^{(i)}(x) \in C^{(0,\gamma)}[a, b], \quad i = 1, \ldots, k.$$
$$(11.5.26)$$

    Further, $u_i(a) = u_i(b) = 0$, $i = 1, \ldots, k$.

b. *For* $0 < \gamma < 1$,

$$\left|u^{(i)}(x)\right| \le c_i(x-a)^{\gamma-i}, \quad a < x \le \frac{1}{2}(a+b), \quad i = 1, \ldots, k.$$
(11.5.27)

With $\gamma = 1$, *for any* $\epsilon \in (0,1)$,

$$\left|u^{(i)}(x)\right| \le c_i(x-a)^{1-\epsilon-i}, \quad a < x \le \frac{1}{2}(a+b), \quad i = 1, \ldots, k,$$
(11.5.28)

with c_i *dependent on* ϵ. *Analogous results are true for* x *in a neighborhood of* b, *with* $x - a$ *replaced by* $b - x$.

A proof of this theorem is given in [144, p. 63]. In addition, more detail on the asymptotic behavior of $u(x)$ for x near to either a or b is given in the same reference and in Graham [63], bringing in functions of the type seen on the right side of (11.5.23) for the case of logarithmic kernel functions.

This theorem says we should expect endpoint singularities in $u(x)$ of the form $(x - a)^\gamma$ and $(b - x)^\gamma$ for the case $g(x, y) = |x - y|^{\gamma-1}$, $0 < \gamma < 1$, with corresponding results for the logarithmic kernel. Thus the approximation of the unknown $u(x)$ should be based on such behavior. We do so by introducing the concept of a *graded mesh*, an idea developed in Rice [137] for the types of singular functions considered here.

We first develop the idea of a graded mesh for functions on $[0,1]$ with the singular behavior in the function occurring at 0; and then the construction is extended to other situations by a simple change of variables. The singular behavior in which we are interested is $u(x) = x^\gamma$, $\gamma > 0$. For a given integer $n \ge 1$, define

$$x_j = \left(\frac{j}{n}\right)^q, \quad j = 0, 1, \ldots, n,$$
(11.5.29)

with the real number $q \ge 1$ to be specified later. For $q > 1$, this is an example of a *graded mesh*, and it is the one introduced and studied in Rice [137]. For a given integer $m \ge 0$, let a partition of $[0,1]$ be given:

$$0 \le \mu_0 < \cdots < \mu_m \le 1.$$
(11.5.30)

Define interpolation nodes on each subinterval $[x_{j-1}, x_j]$ by

$$x_{ji} = x_{j-1} + \mu_i h_j, \quad i = 0, 1, \ldots, m, \quad h_j \equiv x_j - x_{j-1}.$$

Let $P_n u(x)$ be the piecewise polynomial function that is of degree $\le m$ on each subinterval $[x_{j-1}, x_j]$ and that interpolates $u(x)$ at the nodes $\{x_{j0}, \ldots, x_{jm}\}$ on that subinterval. To be more explicit, let

$$L_i(\mu) = \prod_{\substack{k=0 \\ k \ne i}}^{m} \frac{\mu - \mu_k}{\mu_i - \mu_k}, \quad i = 0, 1, \ldots, m,$$

which are the basis functions associated with interpolation at the nodes of (11.5.30). Then

$$P_n u(x) = \sum_{i=0}^{m} L_i \left(\frac{x - x_{j-1}}{h_j} \right) u(x_{ji}), \quad x_{j-1} \le x \le x_j, \quad j = 1, \ldots, n.$$

$$(11.5.31)$$

If $\mu_0 > 0$ or $\mu_m < 1$, then $P_n u(x)$ is likely to be discontinuous at the interior breakpoints $x_1, \ldots, x_{n-1}$. We now present the main result from Rice [137].

Lemma 11.5.5 *Let n, m, $\{x_j\}$, $\{x_{ji}\}$, and P_n be as given in the preceding paragraph. For $0 < \gamma < 1$, assume $u \in C^{(0,\gamma)}[0,1] \cap C^{m+1}(0,1]$, with*

$$\left| u^{(m+1)}(x) \right| \le c_{\gamma,m}(u) x^{\gamma - (m+1)}, \quad 0 < x \le 1. \tag{11.5.32}$$

Then for

$$q \ge \frac{m+1}{\gamma}, \tag{11.5.33}$$

we have

$$\|u - P_n u\|_\infty \le \frac{c}{n^{m+1}}, \tag{11.5.34}$$

with c a constant independent of n. For $1 \le p < \infty$, let

$$q > \frac{p(m+1)}{1 + p\gamma}. \tag{11.5.35}$$

Then

$$\|u - P_n u\|_p \le \frac{c}{n^{m+1}}, \tag{11.5.36}$$

with $\|\cdot\|_p$ denoting the standard p-norm for $L^p(0,1)$. (In the language of Rice [137], the function $u(x)$ is said to be of Type($\gamma, m+1$)).

A proof of the result can be found in [13, p. 128].

The earlier product integration methods were based on using interpolation on a uniform subdivision of the interval $[a, b]$. Now we use the same form of interpolation, but base it on a graded mesh for $[a, b]$. Given an even $n \ge 2$, define

$$x_j = a + \left(\frac{2j}{n} \right)^q \left(\frac{b - a}{2} \right), \quad x_{n-j} = b + a - x_j, \quad j = 0, 1, \ldots, \frac{n}{2}.$$

Use the partition (11.5.30) as the basis for polynomial interpolation of degree m on each of the intervals $[x_{j-1}, x_j]$, for $j = 1, \ldots, \frac{1}{2}n$, just as was done in (11.5.31); and use the partition

$$0 \le 1 - \mu_m < \cdots < 1 - \mu_0 \le 1$$

when defining the interpolation on the subintervals $[x_{j-1}, x_j]$ of the remaining half $[\frac{1}{2}(a+b), b]$. In the integral equation (11.5.25), replace $[l(x, y) u(y)]$

with $[l(x,y)u(y)]_n$ using the interpolation just described. For the resulting approximation

$$\lambda u_n(x) - \int_a^b [l(x,y)u_n(y)]_n \, g_\gamma(x-y) \, dy = f(x), \quad a \le x \le b, \quad (11.5.37)$$

we have the following convergence result.

Theorem 11.5.6 *Consider again the integral equation (11.5.25), and assume the same assumptions as for Theorem 11.5.4, but with the integer k replaced by $m+1$, where m is the integer used in defining the interpolation of the preceding paragraph. Then the approximating equation (11.5.37) is uniquely solvable for all sufficiently large n, say, $n \ge N$, and the inverse operator for the equation is uniformly bounded for $n \ge N$. If $0 < \gamma < 1$, then choose the grading exponent q to satisfy*

$$q \ge \frac{m+1}{\gamma}. \quad (11.5.38)$$

If $\gamma = 1$, then choose

$$q > m+1. \quad (11.5.39)$$

With such choices, the approximate solution u_n satisfies

$$\|u - u_n\|_\infty \le \frac{c}{n^{m+1}}. \quad (11.5.40)$$

Proof. The proof is a straightforward generalization of the method of proof used in Theorem 11.5.1, resulting in the error bound

$$\|u - u_n\|_\infty \le c \|Ku - K_n u\|_\infty, \quad n \ge N.$$

Combine Theorem 11.5.4 and Lemma 11.5.5 to complete the proof.

This theorem is from Schneider [145, Theorem 2], and he also allows for greater generality in the singularity in $u(x)$ than has been assumed here. In addition, he extends results of deHoog and Weiss [79], such as (11.5.22), to the use of graded meshes.   ∎

Graded meshes are used with other problems in which there is some kind of singular behavior in the functions being considered. For example, they are used in solving boundary integral equations for the planar Laplace's equation for regions whose boundaries have corners.

11.5.5   The relationship of product integration and collocation methods

Recall the earlier discussion of the collocation method in Section 11.1 and Section 11.3. It turns out that collocation methods can be regarded as product integration methods, and occasionally there is an advantage to doing so.

Recalling this earlier discussion, let P_n be the interpolatory projection operator from $C(D)$ onto the interpolatory approximating space V_n. Then the collocation solution of $(\lambda - K)u = f$ can be regarded abstractly as $(\lambda - P_n K)u_n = P_n f$, and the iterated collocation solution

$$\widehat{u}_n = \frac{1}{\lambda}[f + Ku_n]$$

is the solution of the equation

$$(\lambda - KP_n)\widehat{u}_n = f. \tag{11.5.41}$$

Define a numerical integral operator by

$$K_n u(x) = KP_n u(x) = \int_D k(x, y)(P_n u)(y)\, dy. \tag{11.5.42}$$

This is product integration with $l(x, y) \equiv 1$ and $g(x, y) = k(x, y)$. Thus the iterated collocation solution $\widehat{u}_n$ of (11.5.41) is simply the Nyström solution when defining K_n using the simple product integration formula (11.5.42). Since the collocation solution $u_n = P_n \widehat{u}_n$, we can use results from the error analysis of product integration methods to analyze collocation methods.

Exercise 11.5.1 *Prove the property $A2$ in the proof of Theorem 11.5.1.*

Exercise 11.5.2 *Develop a practical product trapezoidal rule (with even spacing) for the numerical solution of*

$$\lambda u(x) - \int_0^\pi u(y) \log|\sin(x - y)|\, dy = f(x), \qquad 0 \le x \le \pi,$$

assuming λ is so chosen that the integral equation is uniquely solvable. Program your procedure. Solve the equation approximately with $f(x) = 1$ and $f(x) = e^{\sin x}$. Do numerical examples with varying values of n, as in Example 11.5.3.

Exercise 11.5.3 *Develop a product integration Nyström method for solving*

$$\lambda u(x) - \int_0^1 \frac{c\, \ell(x, y)u(y)}{c^2 + (x - y)^2}\, dy = f(x), \qquad 0 \le x \le 1$$

where c is a very small positive number. Assume $\ell(x, y)$ and its low-order derivatives are "well-behaved" functions; and note that the above kernel function is very peaked for small values of c. Define the numerical integration operators, and discuss the error in them as approximations to the original operator. Discuss convergence of your Nyström method.

Exercise 11.5.4 *Consider numerically approximating*

$$I = \int_0^1 x^\alpha dx,$$

$0 < \alpha < 1$, *using the trapezoidal rule with a graded mesh of the form* (11.5.29). *How should the grading parameter q be chosen so as to insure that the rate of convergence is $O(n^{-2})$?*
Hint: Consider the error on each subinterval $[x_{i-1}, x_i]$, and consider separately the cases of $i = 1$ and $i > 1$. Choose q to make the error on $[x_0, x_1]$ of size $O(n^{-2})$. Then examine the error on the remaining subintervals and the total on $[x_1, 1]$. Recall the use of integrals to approximate summations.

11.6   Projection methods for nonlinear equations

Recall the material of Sections 4.3–4.5 of Chapter 4 on nonlinear fixed point problems. We will define and analyze projection methods for the discretization of fixed-point problems

$$u = T(u) \tag{11.6.1}$$

with $T : H \subseteq V \to V$ a completely continuous nonlinear operator. The space V is a Banach space, and H is an open subset of V. The prototype example of T is the Urysohn integral equation of Example 4.3.10:

$$T(u)(t) = g(t) + \int_a^b k(t, s, u(s)) \, ds. \tag{11.6.2}$$

The function $k(t, s, u)$ is to possess such properties as to ensure it is a completely continuous operator on some open set $H \subseteq C[a, b]$ (cf. Section 4.3.10).

Recall the theoretical framework of Subsection 11.1.3. We define the projection method for solving (11.6.1) as follows. For a given discretization parameter n, find $u_n \in V_n$ satisfying the equation

$$u_n = P_n T(u_n). \tag{11.6.3}$$

We can illustrate the method in analogy with Section 11.2, but defer this to later in this section.

There are two major approaches to the error analysis of (11.6.3): (1) Linearize the problem and apply Theorem 4.1.3, the Banach fixed-point theorem; (2) Apply the theory associated with the rotation of a completely continuous vector field (cf. Section 4.5).

11.6.1   *Linearization*

We begin the linearization process by discussing the error in the linearization of $T(v)$ about a point v_0:

$$R(v; v_0) \equiv T(v) - [T(v_0) + T'(v_0)(v - v_0)]. \tag{11.6.4}$$

Lemma 11.6.1 *Let V be a Banach space, and let H be an open subset of V. Let $T : H \subseteq V \to V$ be twice continuously differentiable with $T''(v)$ bounded over any bounded subset of H. Let $B \subseteq H$ be a closed, bounded, and convex set with a non-empty interior. Let v_0 belong to the interior of B, and define $R(v; v_0)$, as above. Then for all $v_1, v_2 \in B$,*

$$\|R(v_2; v_1)\| \leq \frac{1}{2} M \|v_1 - v_2\|^2 \tag{11.6.5}$$

with $M = \sup_{v \in B} \|T''(v)\|$. Moreover,

$$\|T'(v_2) - T'(v_1)\| \leq M \|v_2 - v_1\|, \tag{11.6.6}$$

implying $T'(v)$ is Lipschitz continuous; and

$$\|R(v_1; v_0) - R(v_2; v_0)\| \leq M \left[\|v_1 - v_0\| + \tfrac{1}{2} \|v_1 - v_2\| \right] \|v_1 - v_2\|. \tag{11.6.7}$$

Proof. The result (11.6.5) is immediate from Proposition 4.3.12 of Section 4.3; and the proof of (11.6.6) can be based on Proposition 4.3.11 when applied to $T'(v)$. The proof of (11.6.7) is let as an exercise. ∎

As earlier, assume $T : H \subseteq V \to V$ is a completely continuous nonlinear operator. Assume (11.6.1) has an isolated solution $u^* \in H$, and assume it is unique within the ball

$$B(u^*, \epsilon) = \{ v \mid \|v - u^*\| \leq \epsilon \}$$

for some $\epsilon > 0$ and with $B(u^*, \epsilon) \subseteq H$. We assume T is twice continuously differentiable over H, with $T''(v)$ uniformly bounded over all bounded neighborhoods, such as $B(u^*, \epsilon)$:

$$M(u^*, \epsilon) \equiv \sup_{v \in B(u^*, \epsilon)} \|T''(v)\| < \infty.$$

Assume that 1 is not an eigenvalue of $T'(u^*)$. This then implies that $I - T'(u^*)$ is a bijective mapping from V to V and that it has a bounded inverse. For a proof, invoke Proposition 4.5.5 to show $T'(u^*)$ is a compact linear operator, and then apply Theorem 2.8.10, the Fredholm alternative theorem. Henceforth, we let $L = T'(u^*)$.

Assume that the projections $\{P_n\}$ are pointwise convergent to the identity on V,

$$P_n v \to v \quad \text{as} \quad n \to \infty \qquad \forall v \in V. \tag{11.6.8}$$

Then from Proposition 4.5.5 and Lemma 11.1.4,

$$\|(I - P_n) L\| \to 0 \quad \text{as} \quad n \to \infty.$$

From Theorem 11.1.2, $(I - P_n L)^{-1}$ exists for all sufficiently large n and is uniformly bounded with respect to all such n.

We want to show that for all sufficiently large n, (11.6.3) has a unique solution within $B(u^*, \epsilon_1)$ for some $0 < \epsilon_1 \leq \epsilon$. We also would like to obtain bounds on the rate of convergence of u_n to u^*. In (11.6.3), expand $T(u_n)$ about u^*, obtaining

$$T(u_n) = T(u^*) + L(u_n - u^*) + R(u_n; u^*).$$

Equation (11.6.3) can be rewritten as the equivalent equation

$$(I - P_n L)(u_n - u^*) = P_n u^* - u^* + P_n R(u_n; u^*) \qquad (11.6.9)$$

Introduce a new unknown $\delta_n = u_n - u^*$, and then write

$$\begin{aligned} \delta_n &= (I - P_n L)^{-1}(P_n u^* - u^*) + (I - P_n L)^{-1} R(\delta_n + u^*; u^*) \\ &\equiv F_n(\delta_n) \end{aligned} \qquad (11.6.10)$$

We are interested in showing that on some ball about the origin in V, of radius $\epsilon_1 \leq \epsilon$, this fixed-point equation has a unique solution δ_n, provided only that n is chosen sufficiently large. This can be done by showing that F_n is a contractive mapping on a ball $B(0, \epsilon_1)$ provided that $\epsilon_1 > 0$ is chosen sufficiently small. To do this requires showing the two main hypotheses of Theorem 4.1.3, the Banach contractive mapping theorem. Namely, show that if n is sufficiently large, there exists ϵ_1 for which

1.

$$F_n : B(0, \epsilon_1) \to B(0, \epsilon_1), \qquad (11.6.11)$$

2.

$$\|F_n(\delta_{n,1}) - F_n(\delta_{n,2})\| \leq \alpha \|\delta_{n,1} - \delta_{n,2}\|, \qquad \delta_{n,1}, \delta_{n,2} \in B(0, \epsilon_1), \qquad (11.6.12)$$

with $\alpha < 1$ and independent of n, provided n is chosen to be sufficiently large.

The ϵ_1 can be made independent of n, provided n is sufficiently large. These two properties can be proven using the various results and assumptions we have made regarding T and $\{P_n\}$, and we leave their demonstration as an exercise for the reader. This proves that for all sufficiently large n, the approximating equation (11.6.3) has a unique solution u_n in some ball of fixed radius about u^*.

There are a number of results on the rate of convergence of u_n to u^*, and we quote only one of them. With the same hypotheses on T and $\{P_n\}$ as above,

$$\|u^* - u_n\|_V \leq \|(I - T'(u^*))^{-1}\| (1 + \gamma_n) \|u^* - P_n u^*\|_V \qquad (11.6.13)$$

with $\gamma_n \to 0$ as $n \to \infty$. A proof of this result is given in [16, Theorem 2.2]. This error bound is somewhat comparable to the bound (11.1.24) given earlier for linear projection methods; also see Exercise 11.1.2.

11.6.2   A homotopy argument

This mode of analysis of projection methods for the discretization of fixed-point problems (11.6.1) requires fewer assumptions on the nonlinear operator T, and there is no assumption on the differentiability of T. As before, we assume $T : H \subseteq V \to V$ is a completely continuous operator. Let u^* be an isolated fixed point of T, and assume u^* is isolated within the ball $B(u^*, \epsilon)$ for some $\epsilon > 0$. Further, assume that u^* has a non-zero index (recall the discussion of *index* as discussed in **P3** of Subsection 4.5.1 in Chapter 4). The discussion in **P4** of Subsection 4.5.1 assures us that the index of u^* is non-zero if $I - T'(u^*)$ is a bijective linear operator; but the index can be non-zero under weaker assumptions on u^*; for example, see **P5** of Subsection 4.5.1.

Let S denote the boundary of $B(u^*, \epsilon)$. Recalling Subsection 4.5.1, we have the concept of the quantity $\mathrm{Rot}(\Phi)$, the *rotation* of the completely continuous vector field

$$\Phi(v) = v - T(v), \qquad v \in B(u^*, \epsilon).$$

Also, introduce the approximating vector field

$$\Phi_n(v) = v - P_n T(v), \qquad v \in B(u^*, \epsilon).$$

By our assumptions on u^*, $\Phi(v) \neq 0$ for all $v \in S$, and consequently $\mathrm{Rot}(\Phi) \neq 0$ (and in fact equals the index of the fixed-point u^*). We introduce the homotopy

$$X(v,t) = v - (1-t)T(v) - tP_n T(v), \qquad v \in B(u^*, \epsilon) \qquad (11.6.14)$$

for $0 \leq t \leq 1$. We show that for all sufficiently large values of n, say $n \geq N(\epsilon)$, this homotopy satisfies the hypotheses of **P2** of Subsection 4.5.1; and consequently, the index of Φ_n will be the same as that of Φ, namely, non-zero. In turn, this implies that Φ_n contains zeros within the ball $B(u^*, \epsilon)$, or equivalently, the approximating equation (11.6.3) has solutions within this ϵ-neighborhood of u^*.

Recalling the four hypotheses of **P2** of Subsection 4.5.1, only the fourth one is difficult to show, namely, that

$$X(v,t) \neq 0, \qquad \forall v \in S, \quad 0 \leq t \leq 1, \qquad (11.6.15)$$

for all sufficiently large values of n. To examine this, rewrite (11.6.14) as

$$X(v,t) = [v - T(v)] + t[T(v) - P_n T(v)]. \qquad (11.6.16)$$

We note as a preliminary lemma that

$$\alpha \equiv \inf_{v \in S} \|v - T(v)\| > 0. \qquad (11.6.17)$$

To prove this, assume the contrary. Then there exists a sequence $\{v_m\} \subseteq S$ for which

$$v_m - T(v_m) \to 0 \quad \text{as} \quad m \to \infty. \qquad (11.6.18)$$

Since S is bounded and T is completely continuous, the sequence $\{T(v_m)\}$ has a convergent subsequence, say,

$$T(v_{m_j}) \to w \quad \text{as} \quad m_j \to \infty.$$

When combined with (11.6.18), this implies $v_{m_j} \to w$; and the closedness of S then implies $w \in S$. The continuity of T implies $v = T(v)$, contradicting the assumption that S contains no fixed points of S. This proves (11.6.17).

Returning to (11.6.16), we have,

$$\|X(v,t)\| \geq \alpha - t\,\|T(v) - P_n T(v)\|\,, \qquad v \in S, \quad 0 \leq t \leq 1. \qquad (11.6.19)$$

We assert that

$$\sup_{v \in S} \|T(v) - P_n T(v)\| \to 0 \quad \text{as} \quad n \to \infty.$$

This follows by writing this in the equivalent form

$$\sup_{w \in T(S)} \|w - P_n w\| \to 0 \quad \text{as} \quad n \to \infty.$$

This results follows from Lemma 11.1.3, (11.6.8), and the precompactness of $T(S)$.

When combined with (11.6.19), we have that for all sufficiently large n, say, $n \geq N(\epsilon)$,

$$\|X(v,t)\| \geq \frac{\alpha}{2}, \qquad v \in S, \quad 0 \leq t \leq 1.$$

This completes the proof of (11.6.15), the fourth hypothesis of **P2** of Subsection 4.5.1. As discussed earlier, this implies that (11.6.3) has solutions within $B(u^*, \epsilon)$. As we make $\epsilon \to 0$, this construction also implies the existence of a sequence of approximating solutions u_n that converges to u^* as $n \to \infty$. The analysis of the preceding few paragraphs is essentially the argument given in Krasnoselskii [98, Section 3.3] for the convergence of Galerkin's method for solving (11.6.1).

This is a powerful means of argument for the existence and convergence of approximation solutions of a completely continuous fixed-point problem. But it does not imply that the equations (11.6.3) are uniquely solvable, and indeed they may not be. For an example in which $v = T(v)$ has a isolated fixed-point u^*, but one for which the approximating equations are not *uniquely* solvable in any neighborhood of u^*, see [10, p. 590].

11.6.3 The approximating finite-dimensional problem

Consider solving the Urysohn nonlinear equation

$$u(x) = f(x) + \int_D k(x, y, u(y))\, dy \equiv T(u)(x), \qquad x \in D$$

for an integration region $D \subseteq \mathbb{R}^d$. We denote by V the function space for the consideration of this equation, and we let V_n denote the finite-dimensional subspace from which our approximation will be chosen,

$$u_n(x) = \sum_{j=1}^{\kappa_n} c_j \phi_j(x), \quad x \in D. \tag{11.6.20}$$

In this discussion, recall the general framework of Section 11.1 and the specific examples of Section 11.2.

To be more specific, let V be a space of continuous functions, and let P_n be an interpolatory projection operator from V to V_n, based on node points $\{x_j \mid 1 \leq j \leq \kappa_n\}$. Then the approximating equation (11.6.3) is equivalent to choosing u_n as in (11.6.20) with $\{c_j\}$ satisfying the nonlinear system

$$\sum_{j=1}^{\kappa_n} c_j \phi_j(x_i) = f(x_i) + \int_D k\left(x_i, y, \sum_{j=1}^{\kappa_n} c_j \phi_j(y)\right) dy, \quad i = 1, \ldots, \kappa_n.$$
$$\tag{11.6.21}$$

This is a nontrivial system to solve, and usually some variant of Newton's method is used to find an approximating solution. From a practical perspective, a major difficulty is that the integral will need to be numerically evaluated repeatedly with varying x_i and varying iterates $\{c_j^{(k)}\}$, where k is an index for the iterative solution of the system.

An important variant is possible for the Hammerstein equation

$$u(x) = f(x) + \int_D k(x, y) g(y, u(y)) \, dy \equiv T(u)(x), \quad x \in D. \tag{11.6.22}$$

We can convert this problem to one for which the number of needed numerical integrations is reduced greatly. Introduce a new unknown

$$w(x) = g(x, u(x)), \quad x \in D.$$

Then u can be recovered from w using

$$u(x) = f(x) + \int_D k(x, y) w(y) \, dy, \quad x \in D. \tag{11.6.23}$$

To solve for w, use the equation

$$w(x) = g\left(x, f(x) + \int_D k(x, y) w(y) \, dy\right), \quad x \in D. \tag{11.6.24}$$

If we examine the nonlinear system (11.6.21) for this equation, we can greatly minimize the needed integrations, needing only the evaluations of the integrals

$$\int_D k(x_i, y) \phi_j(y) \, dy, \quad i, j = 1, \ldots, \kappa_n$$

The integrations need not be recomputed for each iteration of the solution of the nonlinear system. We leave the further analysis of this to Exercise 11.6.4. For further literature on this method, see [104].

Exercise 11.6.1 *Prove* (11.6.7) *of Lemma 11.6.1.*
Hint: Use the definition (11.6.4) *to write out both* $R(v_1; v_0)$ *and* $R(v_2; v_0)$.
Simplify; then apply (11.6.5) *and* (11.6.6).

Exercise 11.6.2 *Prove* (11.6.11) *and* (11.6.12), *provided that* $\epsilon_1 > 0$ *is chosen sufficiently small and* n *is chosen sufficiently large.*

Exercise 11.6.3 *Using* (11.6.9), *prove a weaker form of* (11.6.13), *namely,*

$$\|u^* - u_n\|_V \le c \|u^* - P_n u^*\|_V, \qquad n \ge N$$

for some $N \ge 1$, *with* c *a constant (dependent on* N).

Exercise 11.6.4 *Fill in the details of the solution of the nonlinear system* (11.6.21) *for the equation* (11.6.24).

Exercise 11.6.5 *Do a detailed presentation and analysis of the solution of*

$$u(t) = g(t) + \int_a^b k(t, s, u(s)) \, ds, \qquad a \le t \le b$$

using piecewise linear collocation (as in Subsection 11.2.1). Include a discussion of the nonlinear system that you must setup and solve.

Exercise 11.6.6 *Repeat Exercise 11.6.5 for the equations* (11.6.23)–(11.6.24).

Exercise 11.6.7 *Recall the material of Section 11.3 on iterated projection methods. Define the iterated projection solution for 11.6.3 as*

$$\widehat{u}_n = T(u_n).$$

Show $P_n \widehat{u}_n = u_n$ *and* $\widehat{u}_n = T(P_n \widehat{u}_n)$. *This can be used as a basis for a direct analysis of the convergence of* $\{\widehat{u}_n\}$.

Suggestion for Further Readings

Parts of this chapter are a modification of portions of the presentation in ATKINSON [13, Chaps. 3, 4]. Another introduction to the numerical solution of integral equations is given in KRESS [100]. The first general treatment of projection methods appears to have been due to L.V. Kantorovich in 1948, and those arguments appear in an updated form in KANTOROVICH AND AKILOV [88]. The general theory of collectively compact operator approximations was created by P. ANSELONE, and the best introduction to it is his book [2]. For a survey of numerical methods for solving nonlinear integral equations, see ATKINSON [12]. Extensions of the ideas of Section 11.6 to Nyström's method for nonlinear equations are given in ATKINSON [8] and ATKINSON AND POTRA [16].

12
Boundary Integral Equations

In Chapter 9, we examined finite element methods for the numerical solution of Laplace's equation. In this chapter, we propose an alternative approach. We introduce the idea of reformulating Laplace's equation as a *boundary integral equation* (BIE), and then we consider the numerical solution of Laplace's equation by numerically solving its reformulation as a BIE. Some of the most important boundary value problems for elliptic partial differential equations have been studied and solved numerically by this means; and depending on the requirements of the problem, the use of BIE reformulations may be the most efficient means of solving these problems. Examples of other equations solved by use of BIE reformulations are the Helmholtz equation ($\Delta u + \lambda u = 0$) and the biharmonic equation ($\Delta^2 u = 0$). We consider here the use of boundary integral equations in solving only planar problems for Laplace's equation. For the domain D for the equation, we restrict it or its complement to be a simply connected set with a smooth boundary S. Most of the results and methods given here will generalize to other equations (e.g., Helmholtz's equation).

In this chapter, Section 12.1 contains a theoretical framework for BIE reformulations of Laplace's equation in $\mathbb{R}^2$, giving the most popular of such boundary integral equations. For much of the history of BIE, those of the second kind have been the most popular; this includes the work of Ivar Fredholm, Carl Neumann, David Hilbert, and others in the late 1800s and early 1900s. In Section 12.2, we discuss the numerical solution of such BIE of the second kind. In Section 12.3, we introduce briefly the study of BIE of the first kind, and we discuss the use of Fourier series as a means of studying these equations and numerical methods for their solution.

As in the preceding Chapter 11, here we use notation that is popular in the literature on boundary integral equations.

12.1   Boundary integral equations

Let D be a bounded, open, simply connected region in the plane, and let its boundary be denoted by S. At a point $P \in S$, let $\mathbf{n}_P$ denote the inner unit normal to S. We restate the principal boundary value problems of interest when solving Laplace's equation on D.

The interior Dirichlet problem:
    Find $u \in C(\overline{D}) \cap C^2(D)$ that satisfies

$$\begin{aligned} \Delta u(P) &= 0, & P &\in D \\ u(P) &= f(P), & P &\in S \end{aligned} \qquad (12.1.1)$$

with $f \in C(S)$ a given boundary function.

The interior Neumann problem:
    Find $u \in C^1(\overline{D}) \cap C^2(D)$ that satisfies

$$\begin{aligned} \Delta u(P) &= 0, & P &\in D \\ \frac{\partial u(P)}{\partial \mathbf{n}_P} &= f(P), & P &\in S \end{aligned} \qquad (12.1.2)$$

with $f \in C(S)$ a given boundary function.

Another important boundary value problem is that with a mixture of Neumann and Dirichlet boundary conditions on different sections of the boundary, or perhaps some combination of them. The techniques introduced here can also be used to study and solve such mixed boundary value problems, but we omit any such discussion here. Corresponding to the interior Dirichlet and Neumann problems given above, there are corresponding exterior problems. These are discussed later in the section. Functions satisfying Laplace's equation are often called "harmonic functions." The study of Laplace's equation is often referred to as "potential theory," since many applications involve finding a potential function u in order to construct a conservative vector field ∇u.

The above boundary value problems have been discussed earlier, in Chapter 7. Here we give a theorem summarizing the main results on their solvability, in the form needed here.

Theorem 12.1.1 *Let the function $f \in C(S)$; and assume S can be parameterized by a twice continuously differentiable function. Then*

1. The Dirichlet problem (12.1.1) has a unique solution.

2. *The Neumann problem (12.1.2) has a unique solution, up to the addition of an arbitrary constant, provided*

$$\int_S f(Q)\,dS = 0. \tag{12.1.3}$$

12.1.1   Green's identities and representation formula

A very important tool for studying elliptic partial differential equations is the *divergence theorem* or *Gauss's theorem*. This was given earlier in Section 6.6 of Chapter 6 (cf. Proposition 6.6.1); but we re-state it in the form needed for the planar Laplace equation, a form usually called "Green's theorem." We state the result for regions Ω that are not simply connected and whose boundaries need not be smooth. This form is needed when proving Green's representation formula (12.1.23).

Let Ω denote an open planar region. Let its boundary Γ consist of $m+1$ distinct simple closed curves, $m \geq 0$,

$$\Gamma = \Gamma_0 \cup \cdots \cup \Gamma_m.$$

Assume $\Gamma_1, \ldots, \Gamma_m$ are contained in the interior of Γ_0. For each $i = 1, \ldots, m$, let Γ_i be exterior to the remaining curves $\Gamma_1, \ldots, \Gamma_{i-1}, \Gamma_{i+1}, \ldots, \Gamma_m$. Further, assume each curve Γ_i is a piecewise smooth curve. We say a curve γ is *piecewise smooth* if—

1. It can be broken into a finite set of curves $\gamma_1, \ldots, \gamma_k$ with each γ_j having a parametrization that is at least twice continuously differentiable.

2. The curve γ does not contain any cusps, meaning that each pair of adjacent curves γ_i and γ_{i+1} join at an interior angle in the interval $(0, 2\pi)$.

The region Ω is interior to Γ_0, but it is exterior to each of the curves $\Gamma_1, \ldots, \Gamma_m$. The orientation of Γ_0 is to be counterclockwise, while the curves $\Gamma_1, \ldots, \Gamma_m$ are to be clockwise.

Theorem 12.1.2 (The divergence theorem) *Assume* $\mathbf{F} : \overline{\Omega} \to \mathbb{R}^2$ *with each component of F contained in* $C^1(\overline{\Omega})$. *Then—*

$$\int_\Omega \nabla \cdot \mathbf{F}(Q)\,d\Omega = -\int_\Gamma \mathbf{F}(Q) \cdot \mathbf{n}(Q)\,d\Gamma. \tag{12.1.4}$$

This important result, which generalizes the fundamental theorem of the calculus, is proven in most standard textbooks on "advanced calculus." It is also a special case of Proposition 6.6.1 from Chapter 6.

Using the divergence theorem, one can obtain *Green's identities* and *Green's representation formula*. Assuming $u \in C^1(\overline{\Omega})$ and $w \in C^2(\overline{\Omega})$, one can prove *Green's first identity* by letting $\mathbf{F} = u\nabla w$ in (12.1.4):

$$\int_\Omega u\Delta w\,d\Omega + \int_\Omega \nabla u \cdot \nabla w\,d\Omega = -\int_\Gamma u\frac{\partial w}{\partial \mathbf{n}}\,d\Gamma. \tag{12.1.5}$$

(This was given earlier in (6.6.4) of Chapter 6.)

Next, assume $u, w \in C^2(\overline{\Omega})$. Interchanging the roles of u and w in (12.1.5), and then subtracting the two identities, one obtains *Green's second identity*:

$$\int_\Omega [u\Delta w - w\Delta u]\,d\Omega = \int_\Gamma \left[w\frac{\partial u}{\partial \mathbf{n}} - u\frac{\partial w}{\partial \mathbf{n}}\right]\,d\Gamma. \tag{12.1.6}$$

The identity (12.1.5) can be used to prove (i) if the Neumann problem (12.1.2) has a solution, then it is unique up to the addition of an arbitrary constant; and (ii) if the Neumann problem is to have a solution, then the condition (12.1.3) is necessary. The identity (12.1.5) also leads to a proof of the uniqueness of possible solutions of the Dirichlet problem.

Return to the original domain D on which the problems (12.1.1) and (12.1.2) are posed, and assume $u \in C^2(\overline{D})$. Let $u(Q)$ be a solution of Laplace's equation, and let $w(Q) = \log|A - Q|$, with $A \in D$. Here $|A - Q|$ denotes the ordinary Euclidean length of the vector $A - Q$. Define Ω to be D after removing the small disk $B(A, \epsilon) \equiv \{Q \mid |A - Q| \le \epsilon\}$, with $\epsilon > 0$ so chosen that $B(A, 2\epsilon) \subset D$. Note that for the boundary Γ of Ω,

$$\Gamma = S \cup \{Q \mid |A - Q| = \epsilon\}.$$

Apply (12.1.6) with this choice of Ω, and then let $\epsilon \to 0$. Doing so, and then carefully computing the various limits, we obtain *Green's representation formula*:

$$u(A) = \frac{1}{2\pi}\int_S \left[\frac{\partial u(Q)}{\partial n_Q}\log|A - Q| - u(Q)\frac{\partial}{\partial n_Q}\left[\log|A - Q|\right]\right]dS_Q, \quad A \in D. \tag{12.1.7}$$

This expresses u over D in terms of the boundary values of u and its normal derivative on S.

From hereon in this chapter, we assume S has a parametrization $\mathbf{r}(t)$ *that is in* C^2. Some of the results given here are still true if S is only piecewise smooth; but we refer to [13, Chaps. 7–9] for a more complete treatment.

We can take limits in (12.1.7) as A approaches a point on the boundary S. Let $P \in S$. Then after a careful calculation,

$$\lim_{A \to P} \int_S \frac{\partial u(Q)}{\partial \mathbf{n}_Q} \log |A - Q| \, dS_Q = \int_S \frac{\partial u(Q)}{\partial \mathbf{n}_Q} \log |P - Q| \, dS_Q,$$

$$\lim_{\substack{A \to P \\ A \in D}} \int_S u(Q) \frac{\partial}{\partial \mathbf{n}_Q} \left[\log |A - Q| \right] \, dS_Q \qquad (12.1.8)$$

$$= -\pi u(P) + \int_S u(Q) \frac{\partial}{\partial \mathbf{n}_Q} \left[\log |P - Q| \right] \, dS_Q.$$

A proof of (12.1.8), and of the associated limit in (12.1.26), can be found in [39, pp. 197–202] or in many other texts on Laplace's equation.

Using these limits in (12.1.7) yields the relation

$$u(P) = \frac{1}{\pi} \int_S \left[\frac{\partial u(Q)}{\partial \mathbf{n}_Q} \log |P - Q| - u(Q) \frac{\partial}{\partial \mathbf{n}_Q} \left[\log |P - Q| \right] \right] dS_Q, \quad P \in S$$

$$(12.1.9)$$

which gives a relationship between the values of u and its normal derivative on S.

The formula (12.1.9) is an example of a boundary integral equation; and it can be used to create other such boundary integral equations. First, however, we need to look at solving Laplace's equation on exterior regions $D_e = \mathbb{R}^2 \backslash \overline{D}$ and to obtain formulas that correspond to (12.1.7)–(12.1.9) for such exterior regions. We also use the notation $D_i = D$ in some places, to indicate clearly that an interior region is being used.

12.1.2   The Kelvin transformation and exterior problems

Define a transformation $\mathcal{T} : \mathbb{R}^2 \backslash \{\mathbf{0}\} \to \mathbb{R}^2 \backslash \{\mathbf{0}\}$,

$$\mathcal{T}(x, y) = (\xi, \eta) \equiv \frac{1}{r^2}(x, y), \qquad r = \sqrt{x^2 + y^2}. \qquad (12.1.10)$$

In polar coordinates,

$$\mathcal{T}(r \cos \theta, r \sin \theta) = \frac{1}{r}(\cos \theta, \sin \theta).$$

Thus a point (x, y) is mapped onto another point (ξ, η) on the same ray emanating from the origin, and we call (ξ, η) the inverse of (x, y) with respect to the unit circle. Note that $\mathcal{T}(\mathcal{T}(x, y)) = (x, y)$, so that $\mathcal{T}^{-1} = \mathcal{T}$. The Jacobian matrix for $\mathcal{T}$ is

$$J(\mathcal{T}) = \begin{pmatrix} \dfrac{\partial \xi}{\partial x} & \dfrac{\partial \xi}{\partial y} \\ \dfrac{\partial \eta}{\partial x} & \dfrac{\partial \eta}{\partial y} \end{pmatrix} = \frac{1}{r^2} H \qquad (12.1.11)$$

with

$$H = \begin{pmatrix} \dfrac{y^2 - x^2}{r^2} & \dfrac{-2xy}{r^2} \\[2ex] \dfrac{-2xy}{r^2} & \dfrac{x^2 - y^2}{r^2} \end{pmatrix}.$$

The matrix H is orthogonal with determinant -1, and

$$\det[J(\mathcal{T}(x,y))] = -\frac{1}{r^2}.$$

Assume the bounded open region $D \equiv D_i$ contains the origin $\mathbf{0}$. For a function $u \in C(\overline{D}_e)$, define

$$\widehat{u}(\xi, \eta) = u(x, y), \qquad (\xi, \eta) = \mathcal{T}(x, y), \qquad (x, y) \in \overline{D}_e. \tag{12.1.12}$$

This is called the *Kelvin transformation* of u. Introduce the interior region $\widehat{D} = \mathcal{T}(D_e)$, and let $\widehat{S}$ denote the boundary of $\widehat{D}$. The boundaries S and $\widehat{S}$ have the same degree of smoothness. In addition, the condition $(\xi, \eta) \to \mathbf{0}$ in $\widehat{D}$ corresponds to $r \to \infty$ for points $(x, y) \in D_e$. For a function u satisfying Laplace's equation on D, it is a straightforward calculation to show

$$\Delta \widehat{u}(\xi, \eta) = r^4 \Delta u(x, y) = 0, \qquad (\xi, \eta) = \mathcal{T}(x, y), \qquad (x, y) \in D_e, \tag{12.1.13}$$

thus showing $\widehat{u}$ to be harmonic on $\widehat{D}$. We can pass from the solution of Laplace's equation on the unbounded region D_e to the bounded open region $\widehat{D}$.

If we were to impose the Dirichlet condition $u = f$ on the boundary S, this is equivalent to the Dirichlet condition

$$\widehat{u}(\xi, \eta) = f(\mathcal{T}^{-1}(\xi, \eta)), \qquad (\xi, \eta) \in \widehat{S}.$$

From the existence and uniqueness result of Theorem 12.1.1, the interior Dirichlet problem on $\widehat{D}$ will have a unique solution. This leads us to considering the following problem.

The exterior Dirichlet problem:
    Find $u \in C(\overline{D}_e) \cap C^2(D_e)$ that satisfies

$$\begin{aligned} \Delta u(P) &= 0, & P &\in D_e \\ u(P) &= f(P), & P &\in S \\ \lim_{r \to \infty} \sup_{|P| \geq r} |u(P)| &< \infty, \end{aligned} \tag{12.1.14}$$

with $f \in C(S)$ a given boundary function.

Using the above discussion on the Kelvin transform, this converts to the interior Dirichlet problem

$$\begin{aligned} \Delta \widehat{u}(\xi, \eta) &= 0, & (\xi, \eta) &\in \widehat{D} \\ u(\xi, \eta) &= f(\mathcal{T}^{-1}(\xi, \eta)), & (\xi, \eta) &\in \widehat{S}, \end{aligned} \tag{12.1.15}$$

and Theorem 12.1.1 guarantees the unique solvability of this problem. The condition on $u(x, y)$ as $r \to \infty$ can be used to show that $\widehat{u}(\xi, \eta)$ has a removable singularity at the origin; and $\widehat{u}(0, 0)$ will be the value of $u(x, y)$ as $r \to \infty$. Thus the above exterior Dirichlet problem has a unique solution.

For functions $u \in C^1(\overline{D}_e)$,

$$\frac{\partial u(x, y)}{\partial \mathbf{n}(x, y)} = -\rho^2 \frac{\partial \widehat{u}(\xi, \eta)}{\partial \widehat{\mathbf{n}}(\xi, \eta)}, \qquad \rho = \frac{1}{r} = \sqrt{\xi^2 + \eta^2} \qquad (12.1.16)$$

with $\widehat{\mathbf{n}}(\xi, \eta)$ the unit interior normal to $\widehat{S}$ at (ξ, η). Thus the Neumann condition

$$\frac{\partial u(x, y)}{\partial \mathbf{n}(x, y)} = f(x, y), \qquad (x, y) \in S$$

is equivalent to

$$\frac{\partial \widehat{u}(\xi, \eta)}{\partial \widehat{\mathbf{n}}(\xi, \eta)} = -\frac{1}{\rho^2} f(\mathcal{T}^{-1}(\xi, \eta)) \equiv \widehat{f}(\xi, \eta), \qquad (\xi, \eta) \in \widehat{S}. \qquad (12.1.17)$$

Also,

$$\int_S \frac{\partial u}{\partial \mathbf{n}} \, dS = -\int_{\widehat{S}} \frac{\partial \widehat{u}}{\partial \widehat{\mathbf{n}}} \, d\widehat{S}. \qquad (12.1.18)$$

Using this information, consider the following problem.

The exterior Neumann problem:
Find $u \in C^1(\overline{D}_e) \cap C^2(D_e)$ that satisfies

$$\Delta u(P) = 0, \qquad P \in D_e$$
$$\frac{\partial u(P)}{\partial \mathbf{n}_P} = f(P), \qquad P \in S \qquad (12.1.19)$$

$$u(r \cos\theta, r \sin\theta) = O\left(\frac{1}{r}\right), \qquad \frac{\partial u(r \cos\theta, r \sin\theta)}{\partial r} = O\left(\frac{1}{r^2}\right) \qquad (12.1.20)$$

as $r \to \infty$, uniformly in θ. The function $f \in C(S)$ is assumed to satisfy

$$\int_S f(Q) \, dS = 0 \qquad (12.1.21)$$

just as in (12.1.3) for the interior Neumann problem.

Combining (12.1.18) with (12.1.21) yields

$$\int_{\widehat{S}} \widehat{f}(\xi, \eta) \, d\widehat{S} = 0. \qquad (12.1.22)$$

The problem (12.1.19) converts to the equivalent interior problem of finding $\widehat{u}$ satisfying

$$\Delta\widehat{u}(\xi,\eta) = 0, \qquad\qquad (\xi,\eta) \in \widehat{D},$$

$$\frac{\partial\widehat{u}(\xi,\eta)}{\partial\widehat{\mathbf{n}}(\xi,\eta)} = \widehat{f}(\xi,\eta), \qquad (\xi,\eta) \in \widehat{S}, \qquad\qquad (12.1.23)$$

$$\widehat{u}(0,0) = 0.$$

By Theorem 12.1.31 and (12.1.22), this has a unique solution $\widehat{u}$. This gives a complete solvability theory for the exterior Neumann problem.

The converted problems (12.1.15) and (12.1.23) can also be used for numerical purposes, and later we will return to these reformulations of exterior problems for Laplace's equation.

Green's representation formula on exterior regions

From the form of solutions to the interior Dirichlet problem, and using the Kelvin transform, we can assume the following form for potential functions u defined on D_e:

$$u(r\cos\theta, r\sin\theta) = u(\infty) + \frac{c(\theta)}{r} + O\left(\frac{1}{r^2}\right) \qquad (12.1.24)$$

as $r \to \infty$ and with $c(\theta) = A\cos\theta + B\sin\theta$ for suitable constants A, B. The notation $u(\infty)$ denotes the limiting value of $u(r\cos\theta, r\sin\theta)$ as $r \to \infty$. From this, we can use the Green's representation formulas (12.1.7)–(12.1.9) for interior regions to obtain the following Green's representation formula for potential functions on exterior regions.

$$u(A) = u(\infty) - \frac{1}{2\pi}\int_S \frac{\partial u(Q)}{\partial\mathbf{n}_Q} \log|A - Q|\, dS_Q$$

$$+ \frac{1}{2\pi}\int_S u(Q)\frac{\partial}{\partial\mathbf{n}_Q}\left[\log|A - Q|\right] dS_Q, \qquad A \in D_e. \qquad (12.1.25)$$

To obtain a limiting value as $A \to P \in S$, we need the limit

$$\lim_{\substack{A \to P \\ A \in D_e}} \int_S u(Q)\frac{\partial}{\partial\mathbf{n}_Q}\left[\log|A - Q|\right] dS_Q$$

$$= \pi u(P) + \int_S u(Q)\frac{\partial}{\partial\mathbf{n}_Q}\left[\log|P - Q|\right] dS_Q. \qquad (12.1.26)$$

Note the change of sign of $u(P)$ when compared to (12.1.8). Using this in (12.1.25), we obtain

$$u(P) = 2u(\infty) - \frac{1}{\pi}\int_S \frac{\partial u(Q)}{\partial\mathbf{n}_Q} \log|P - Q|\, dS_Q$$

$$+ \frac{1}{\pi}\int_S u(Q)\frac{\partial}{\partial\mathbf{n}_Q}\left[\log|P - Q|\right] dS_Q, \qquad P \in S. \qquad (12.1.27)$$

12.1.3 Boundary integral equations of direct type

The equations (12.1.7) and (12.1.25) give representations for functions harmonic in D_i and D_e, respectively, in terms of u and $\partial u/\partial n$ on the boundary S of these regions. When given one of these boundary functions, the equations (12.1.9) and (12.1.27) can often be used to obtain the remaining boundary function. Numerical methods based on (12.1.9) and (12.1.27) are said to be of "direct type," as they find u or $\partial u/\partial n$ on the boundary and these are quantities that are often of immediate physical interest. We will illustrate some of the possible BIE of direct type, leaving others as problems for the reader.

The interior Dirichlet problem (12.1.1)

The boundary condition is $u(P) = f(P)$ on S; and using it, (12.1.9) can be written as

$$\frac{1}{\pi} \int_S \rho(Q) \log |P - Q| \, dS_Q = g(P), \qquad P \in S. \tag{12.1.28}$$

To emphasize the form of the equation, we have introduced

$$\rho(Q) \equiv \frac{\partial u(Q)}{\partial n_Q}, \qquad g(P) \equiv f(P) + \frac{1}{\pi} \int_S f(Q) \frac{\partial}{\partial n_Q} \left[\log |P - Q| \right] dS_Q.$$

The equation (12.1.28) is of the first kind, and it is often used as the prototype for studying boundary integral equations of the first kind. In Section 12.3, we discuss the solution of (12.1.28) in greater detail.

The interior Neumann problem (12.1.2)

The boundary condition is $\partial u/\partial n = f$ on S; and using it, we write (12.1.9) as

$$u(P) + \frac{1}{\pi} \int_S u(Q) \frac{\partial}{\partial n_Q} \left[\log |P - Q| \right] dS_Q$$
$$= \frac{1}{\pi} \int_S f(Q) \log |P - Q| \, dS_Q, \qquad P \in S. \tag{12.1.29}$$

This is an integral equation of the second kind. Unfortunately, it is not uniquely solvable; and this should not be surprising when given the lack of unique solvability for the Neumann problem itself. The homogeneous equation has $u \equiv 1$ as a solution, as can be seen by substituting the harmonic function $u \equiv 1$ into (12.1.9). The equation (12.1.29) is solvable if and only if the boundary function f satisfies the condition (12.1.3). The simplest way to deal with the lack of uniqueness in solving (12.1.29) is to introduce an additional condition such as

$$u(P^*) = 0$$

for some fixed point $P^* \in S$. This will lead to a unique solution for (12.1.29). Combine this with the discretization of the integral equation, to obtain a suitable numerical approximation for u. There are other ways of converting (12.1.29) to a uniquely solvable equation, and some of these are explored in [6]. However, there are preferable alternative ways to solve the interior Neumann problem. One of the simplest is simply to convert it to an equivalent exterior Neumann problem, using the Kelvin transform given earlier; and then use techniques for the exterior problem, such as the BIE given in (12.1.30) below.

The exterior Neumann problem (12.1.19)

The boundary condition is $\partial u/\partial \mathbf{n} = f$ on S, and u also satisfies $u(\infty) = 0$. Using this, (12.1.27) becomes

$$
\begin{aligned}
u(P) - \frac{1}{\pi} \int_S u(Q) \frac{\partial}{\partial \mathbf{n}_Q} \left[\log |P - Q| \right] dS_Q \\
= -\frac{1}{\pi} \int_S f(Q) \log |P - Q| \, dS_Q, \qquad P \in S.
\end{aligned}
\tag{12.1.30}
$$

This equation is uniquely solvable, as will be discussed in greater detail below, following (12.2.3) in Section 12.2. This is considered a practical approach to solving the exterior Neumann problem, especially when one wants to find only the boundary data $u(P)$, $P \in S$. The numerical solution of the exterior Neumann problem using this approach is given following (12.2.24) in Section 12.2.

As above, we assume the boundary S is a smooth simple closed curve with a twice continuously differentiable parametrization. More precisely, let S be parameterized by

$$
\mathbf{r}(t) = (\xi(t), \eta(t)), \qquad 0 \le t \le L
\tag{12.1.31}
$$

with $\mathbf{r} \in C^2[0, L]$ and $|\mathbf{r}'(t)| \ne 0$ for $0 \le t \le L$. We assume the parametrization traverses S in a counter-clockwise direction. We usually consider $\mathbf{r}(t)$ as being extended periodically from $[0, L]$ to $(-\infty, \infty)$; and we write $\mathbf{r} \in C_p^2(L)$, generalizing from the definition of $C_p^2(2\pi)$ given in Chapter 1. Introduce the interior unit normal $\mathbf{n}(t)$ that is orthogonal to the curve S at $\mathbf{r}(t)$:

$$
\mathbf{n}(t) = \frac{(-\eta'(t), \xi'(t))}{\sqrt{\xi'(t)^2 + \eta'(t)^2}}.
$$

Using this representation $\mathbf{r}(t)$ for S, and multiplying in (12.1.30) by $-\pi$, we can rewrite (12.1.30) as

$$
-\pi u(t) + \int_0^L k(t, s) u(s) \, ds = g(t), \qquad 0 \le t \le L,
\tag{12.1.32}
$$

where

$$k(t,s) = \frac{\eta'(s)[\xi(t) - \xi(s)] - \xi'(s)[\eta(t) - \eta(s)]}{[\xi(t) - \xi(s)]^2 + [\eta(t) - \eta(s)]^2}$$

$$= \frac{\eta'(s)\xi[s,s,t] - \xi'(s)\eta[s,s,t]}{|\mathbf{r}[s,t]|^2}, \quad s \neq t, \qquad (12.1.33)$$

$$k(t,t) = \frac{\eta'(t)\xi''(t) - \xi'(t)\eta''(t)}{2\{\xi'(t)^2 + \eta'(t)^2\}} \qquad (12.1.34)$$

and

$$g(t) = \int_0^L f(\mathbf{r}(s))\sqrt{\xi'(s)^2 + \eta'(s)^2} \log|\mathbf{r}(t) - \mathbf{r}(s)| \, ds. \qquad (12.1.35)$$

In (12.1.32), we have used $u(t) \equiv u(\mathbf{r}(t))$, for simplicity in notation. The second fraction in (12.1.33) uses first- and second-order Newton divided differences, to obtain the limiting value $k(t,t)$ of (12.1.34) more easily. The value of $k(t,t)$ is one-half the curvature of S at $\mathbf{r}(t)$.

As in earlier chapters, we write (12.1.32) symbolically as

$$(-\pi + K)u = g. \qquad (12.1.36)$$

By examining the formulas for $k(t,s)$, we have

$$\mathbf{r} \in C^\kappa[0,L] \implies k \in C^{\kappa-2}([0,L] \times [0,L]). \qquad (12.1.37)$$

The kernel function k is periodic in both variables, with period L, as are also the functions u and g.

Recall from Example 1.2.22(a) of Chapter 1 the space $C_p^\ell(2\pi)$ of all ℓ-times continuously differentiable and periodic functions on $(-\infty, \infty)$. Since the parameterization $\mathbf{r}(t)$ is on $[0,L]$, we generalize $C_p^\ell(2\pi)$ to $C_p^\ell(L)$, with functions having period L on $(-\infty, \infty)$. The norm is

$$\|h\|_\ell = \max\left\{\|h\|_\infty, \|h'\|_\infty, \ldots, \|h^{(\ell)}\|_\infty\right\}$$

with the maximum norm taken over the interval $[0,L]$. We always assume for the parametrization that $\mathbf{r} \in C_p^\kappa(L)$, with $\kappa \geq 2$; and therefore the integral operator K is a compact operator from $C_p(L)$ to $C_p(L)$. Moreover, from (12.1.37), K maps $C_p(L)$ to $C_p^{\kappa-2}(L)$. The numerical solution of (12.1.32) is examined in detail in Section 12.2, along with related integral equations.

Using the Kelvin transform, the interior Neumann problem (12.1.1) can be converted to an equivalent exterior Neumann problem, as was done in passing between (12.1.19) and (12.1.23). Solving the exterior problem will correspond to finding that solution to the interior Neumann problem that is zero at the origin [where we assume $\mathbf{0} \in D_i$].

The exterior Dirichlet problem (12.1.14)

The Kelvin transform can also be used to convert the exterior Dirichlet problem (12.1.14) to an equivalent interior Dirichlet problem. After doing so, there are many options for solving the interior problem, including using the first kind boundary integral equation (12.1.28). The value of $u(\infty)$ can be obtained as the value at $\mathbf{0}$ of the transformed problem.

Boundary integral equations of indirect type

Indirect BIE methods are based on representing the unknown harmonic function u as either a *single layer potential*,

$$u(A) = \int_S \rho(Q) \log |A - Q| \, dS_Q, \qquad A \in \mathbb{R}^2, \tag{12.1.38}$$

or a *double layer potential*,

$$u(A) = \int_S \rho(Q) \frac{\partial}{\partial \mathbf{n}_Q} \left[\log |A - Q| \right] dS_Q, \qquad A \in \mathbb{R}^2. \tag{12.1.39}$$

These have physical interpretations, for example, letting ρ denote a given *charge density* on S or a *dipole charge density* on S. For a classical interpretation of such potentials, see Kellogg [93]. Both of these formulas satisfy Laplace's equation for $A \in \mathbb{R}^2 \backslash S$. The density ρ is to be chosen such that u satisfies given boundary conditions on S.

Double layer potentials

Suppose the function u is the solution of the interior Dirichlet problem with $u \equiv f$ on S. Then use (12.1.8) to take limits in (12.1.39) as $A \to P \in S$. This yields the boundary integral equation

$$-\pi \rho(P) + \int_S \rho(Q) \frac{\partial}{\partial \mathbf{n}_Q} \left[\log |P - Q| \right] dS_Q = f(P), \qquad P \in S. \tag{12.1.40}$$

Note that the form of the left side of this equation is exactly that of (12.1.32) for the exterior Neumann problem. We discuss in detail the numerical solution of this and related equations in Section 12.2. Ivar Fredholm used (12.1.39) to show the solvability of the interior Dirichlet problem for Laplace's equation, and he did so by showing (12.1.40) is uniquely solvable for all $f \in C(S)$.

The use of (12.1.40) gives a BIE of "indirect type," as the solution ρ is usually of only indirect interest, it being a means of obtaining u using (12.1.39). Usually, ρ has no immediate physical significance.

Single layer potentials

The single layer potentials are also used to solve interior and exterior problems, for both Dirichlet and Neumann problems. The single layer potential (12.1.38) satisfies Laplace's equation in $D_i \cup D_e$, and it is continuous in $\mathbb{R}^2$, provided $\rho \in L^1(S)$. For example, to solve the interior Dirichlet problem with boundary data $u = f$ on S, we must solve the first kind integral equation

$$\int_S \rho(Q) \log |P - Q| \, dS_Q = f(P), \qquad P \in S. \tag{12.1.41}$$

Some additional properties of the single layer potential are examined in the exercises at the end of this section.

If we seek the solution of the interior Neumann problem (12.1.2) as a single layer potential (12.1.38), with boundary data f on S, then the density ρ must satisfy

$$\pi\rho(P) + \int_S \rho(Q)\frac{\partial}{\partial \mathbf{n}_P}\left[\log|P - Q|\right]dS_Q = f(P), \qquad P \in S. \quad (12.1.42)$$

To obtain this, begin by forming the normal derivative of (12.1.38),

$$\frac{\partial u(A)}{\partial \mathbf{n}_P} = \mathbf{n}_P \cdot \nabla_A\left[\int_S \rho(Q)\log|A - Q|\,dS_Q\right], \qquad A \in D_i, \quad P \in S.$$

Take the limit as $A \to P \in S$. Using an argument similar to that used in obtaining (12.1.8), and applying the boundary condition $\partial u/\partial \mathbf{n}_P = f$, we obtain (12.1.42). The integral operator in (12.1.42) is the adjoint to that in (12.1.40); and the left side of the integral equation is the adjoint of the left side of (12.1.29).

The adjoint equation (12.1.29) is not uniquely solvable, as $\rho \equiv 1$ is a solution of the homogeneous equation. To see this, let $u \equiv 1$ (and $f \equiv 0$) in (12.1.29), thus showing that (12.1.29) is not a uniquely solvable equation. Since this is the adjoint equation to the homogeneous form of (12.1.42), we have that the latter is also not uniquely solvable (cf. Theorem 2.8.14 in Subsection 2.8.5). An examination of how to obtain uniquely solvable variants of (12.1.42) is given in [6].

The single and double layer potentials of (12.1.38)–(12.1.39) can be given additional meaning by using the Green's representation formulas of this section. The density ρ can be related to the difference on the boundary S of solutions or their normal derivatives for Laplace's equation on the regions that are interior and exterior to S; see [13, pp. 317–320].

There are also additional representation formulas and boundary integral equations that can be obtained by other means. For example, representation formulas can be obtained from the Cauchy integral formula for functions of a complex variable. All analytic functions $f(z)$ can be written in the form

$$f(z) = u(x, y) + i\,v(x, y).$$

Using the Cauchy-Riemann equations for u and v, it follows that both u and v are harmonic functions in the domain of analyticity for f. For results obtained from this approach, see Mikhlin [118]. Most of the representation formulas and BIE given in this section can also be obtained by using Cauchy's integral formula.

Exercise 12.1.1 *Derive* (12.1.5)–(12.1.6).

Exercise 12.1.2 *Using* (12.1.5), *show that if the interior Dirichlet problem* (12.1.1) *has a solution, then it is unique.*

Exercise 12.1.3 *Derive (12.1.7), using the ideas sketched preceding the formula.*

Exercise 12.1.4 *Derive (12.1.11) and (12.1.13).*

Exercise 12.1.5 *Assume S is a smooth simple closed curve $(\mathbf{r} \in C_p^2(L))$. Prove*

$$\int_S \log |A - Q| \, dS_Q = 2\pi, \qquad A \in D.$$

What is the value of this integral if $A \in S$? If $A \in D_e$?

Exercise 12.1.6 *Assume S is a smooth simple closed curve $(\mathbf{r} \in C_p^2(L))$. What are the values of*

$$\int_S \frac{\partial}{\partial \mathbf{n}_Q} \left[\log |A - Q| \right] dS_Q$$

for the three cases of $A \in D$, $A \in S$, and $A \in D_e$?

Exercise 12.1.7 *Derive the formulas given in (12.1.33)–(12.1.34), and then show (12.1.37).*

Exercise 12.1.8 *Consider the single layer potential u of (12.1.38). Show that*

$$u(A) \approx c \log |A| \quad as \quad |A| \to \infty.$$

What is c? Suppose you are solving the exterior Dirichlet problem by representing it as the single layer potential in (12.1.38), say, with boundary data f on S. Then the density function ρ must satisfy the integral equation

$$\int_S \rho(Q) \log |P - Q| \, dS_Q = f(P), \quad P \in S.$$

In order to assure that this single layer potential u represents a function bounded at ∞, what additional condition must be imposed on the density function ρ?

Exercise 12.1.9 *Derive the analogue of (12.1.33)–(12.1.34) for the integral operator in (12.1.42).*

Exercise 12.1.10 *Let the boundary parameterization for S be*

$$\mathbf{r}(t) = \gamma(t) \, (\cos t, \sin t), \qquad 0 \le t \le 2\pi,$$

with $\gamma(t)$ a twice continuously and positive 2π-periodic function on $[0, 2\pi]$. Find the kernel function $k(t, s)$ of (12.1.32)–(12.1.34) for this boundary, simplifying as much as possible. What happens when $s - t \to 0$?

Exercise 12.1.11 *Generalize the preceding Exercise 12.1.10 to the boundary parameterization*

$$\mathbf{r}(t) = \gamma(t) \, (a \cos t, b \sin t), \qquad 0 \le t \le 2\pi,$$

with $a, b > 0$, and $\gamma(t)$ a twice continuously and positive 2π-periodic function on $[0, 2\pi]$. Find the kernel function $k(t, s)$ of (12.1.32)–(12.1.34) for this boundary, simplifying as much as possible.

12.2   Boundary integral equations of the second kind

The original theory developed by Ivar Fredholm for the solvability of integral equations was for the the boundary integral equations of the second kind introduced in the preceding section; and these equations have also long been used as a means to solve boundary value problems for Laplace's equation. In this section, we consider the numerical solution of these boundary integral equations of the second kind. We begin with a classic indirect method for solving the interior Dirichlet problem for Laplace's equation; and then the results for this method are extended to integral equations for the interior and exterior Neumann problems.

Recall the double layer representation (12.1.39) for a function u harmonic on the interior region D_i:

$$u(A) = \int_S \rho(Q) \frac{\partial}{\partial \mathbf{n}_Q} \left[\log |A - Q| \right] dS_Q, \qquad A \in D_i. \tag{12.2.1}$$

To solve the interior Dirichlet problem (12.1.1), the density ρ is obtained by solving the boundary integral equation given in (12.1.40), namely,

$$-\pi \rho(P) + \int_S \rho(Q) \frac{\partial}{\partial \mathbf{n}_Q} \left[\log |P - Q| \right] dS_Q = f(P), \qquad P \in S, \tag{12.2.2}$$

with f the given value of u on S. This is basically the same form of integral equation as in (12.1.30) for the exterior Neumann problem, with a different right-hand function. When the representation $\mathbf{r}(t) = (\xi(t), \eta(t))$ of (12.1.31) for S is applied, this integral equation becomes

$$-\pi \rho(t) + \int_0^L k(t, s) \rho(s) \, ds = f(t), \qquad 0 \le t \le L, \tag{12.2.3}$$

with $k(t, s)$ given in (12.1.33)–(12.1.34) and $f(t) \equiv f(\mathbf{r}(t))$. The smoothness and periodicity of k is discussed in and following (12.1.37); and the natural function space setting for studying (12.2.3) is $C_p(L)$ with the uniform norm. Symbolically, we write (12.2.3) as $(-\pi + K)\rho = f$.

The equation (12.2.2) has been very well studied, for over a century; for example, see the references and discussion of this equation in Colton [39, p. 216], Kress [100, p. 71], and Mikhlin [118, Chap. 4]. From this work, $(-\pi + K)^{-1}$ exists as a bounded operator from $C_p(L)$ to $C_p(L)$.

The functions $f, \rho \in C_p(L)$, and the kernel k is periodic in both variables, with period L, over $(-\infty, \infty)$; and in addition, both k and ρ are usually smooth functions. Thus the most efficient numerical method for solving the equation (12.2.3) is generally the Nyström method with the trapezoidal rule as the numerical integration rule. Recall the extensive discussion of the trapezoidal rule in Proposition 6.5.6 of Chapter 6.

Because of the periodicity, the trapezoidal rule simplifies further, and the approximating equation takes the form

$$-\pi \rho_n(t) + h \sum_{j=1}^{n} k(t, t_j) \rho_n(t_j) = f(t), \qquad 0 \le t \le L, \qquad (12.2.4)$$

with $h = L/n$, $t_j = jh$ for $j = 1, 2, \ldots, n$. Symbolically, we write this as $(-\pi + K_n)\rho_n = f$, with the numerical integration operator K_n defined implicitly by (12.2.4). Collocating at the node points, we obtain the linear system

$$-\pi \rho_n(t_i) + h \sum_{j=1}^{n} k(t_i, t_j) \rho_n(t_j) = f(t_i), \qquad i = 1, \ldots, n, \qquad (12.2.5)$$

whose solution is $[\rho_n(t_1), \ldots, \rho_n(t_n)]^{\mathrm{T}}$. Then the Nyström interpolation formula can be used to obtain $\rho_n(t)$:

$$\rho_n(t) = \frac{1}{\pi} \left[-f(t) + h \sum_{j=1}^{n} k(t, t_j) \rho_n(t_j) \right], \qquad 0 \le t \le L. \qquad (12.2.6)$$

This is a simple method to program; and usually the value of n is not too large, so that the linear system (12.2.5) can be solved directly, without iteration.

The error analysis for the above is straightforward from Theorem 11.4.4 of Chapter 11. This theorem shows that (12.2.4) is uniquely solvable for all sufficiently large values of n, say $n \ge N$; and moreover,

$$\|\rho - \rho_n\|_\infty \le \|(-\pi + K_n)^{-1}\| \, \|K\rho - K_n\rho\|_\infty, \qquad n \ge N. \qquad (12.2.7)$$

It is well known that the trapezoidal rule is very rapidly convergent when the integrand is periodic and smooth; and consequently, $\rho_n \to \rho$ with a similarly rapid rate of convergence.

Example 12.2.1 *Let the boundary S be the ellipse*

$$\mathbf{r}(t) = (a \cos t, b \sin t), \qquad 0 \le t \le 2\pi. \qquad (12.2.8)$$

In this case, the kernel k of (12.1.33) can be reduced to

$$k(t, s) = \kappa \left(\frac{s + t}{2} \right), \qquad \kappa(\theta) = \frac{-ab}{2 \left[a^2 \sin^2 \theta + b^2 \cos^2 \theta \right]} \qquad (12.2.9)$$

n	$(a, b) = (1, 2)$	$(a, b) = (1, 5)$	$(a, b) = (1, 8)$
8	$3.67E - 3$	$4.42E - 1$	$3.67E + 0$
16	$5.75E - 5$	$1.13E - 2$	$1.47E - 1$
32	$1.34E - 14$	$1.74E - 5$	$1.84E - 3$
64		$3.96E - 11$	$6.66E - 7$
128			$7.23E - 14$

Table 12.1. Errors in density function ρ_n for (12.2.10)

and the integral equation (12.2.3) becomes

$$-\pi\rho(t) + \int_0^{2\pi} \kappa\left(\frac{s+t}{2}\right)\rho(s)\,ds = f(t), \qquad 0 \le t \le 2\pi. \qquad (12.2.10)$$

In Table 12.1, we give results for solving this equation with

$$f(x, y) = e^x \cos y, \qquad (x, y) \in S. \qquad (12.2.11)$$

The true solution ρ is not known explicitly; but we obtain a highly accurate solution by using a large value of n, and then this solution is used to calculate the errors shown in the table.

Results are given for $(a, b) = (1, 2)$ and $(1, 5)$. The latter ellipse is somewhat elongated, and this causes the kernel k to be more peaked. In particular, introduce the peaking factor

$$p(a, b) \equiv \frac{\max |k(t, s)|}{\min |k(t, s)|} = \left[\frac{\max\{a, b\}}{\min\{a, b\}}\right]^2.$$

Then $p(1, 2) = 4$, $p(1, 5) = 25$, $p(1, 8) = 64$. As the peaking factor becomes larger, it is necessary to increase n in order to retain comparable accuracy in approximating the integral $K\rho$, and the consequences of this can be seen in the table.

A graph of ρ is given in Figure 12.1 for $(a, b) = (1, 5)$, and it shows a somewhat rapid change in the function around $t = 0$ or $(x, y) = (a, 0)$ on S. For the same curve S, a graph of the error $\rho(t) - \rho_n(t)$, $0 \le t \le 2\pi$, is given in Figure 12.2 for the case $n = 32$. Perhaps surprisingly in light of Figure 12.1, the error is largest around $t = \pi$ or $(x, y) = (-a, 0)$ on S, where ρ is better behaved.

12.2.1 Evaluation of the double layer potential

When using the representation $\mathbf{r}(s) = (\xi(s), \eta(s))$ of (12.1.31) for S, the double layer integral formula (12.2.1) takes the form

$$u(x, y) = \int_0^L M(x, y, s)\rho(s)\,ds, \qquad (x, y) \in D_i \qquad (12.2.12)$$

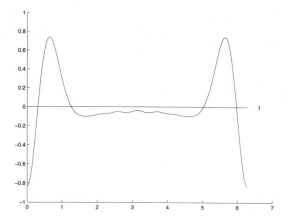

Figure 12.1. The density ρ for (12.2.10) with $(a, b) = (1, 5)$

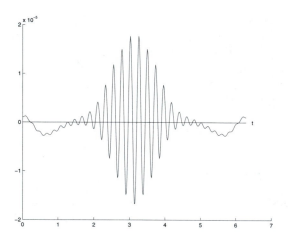

Figure 12.2. The error $\rho - \rho_{32}$ for (12.2.10) with $(a, b) = (1, 5)$

where

$$M(x, y, s) = \frac{-\eta'(s)[\xi(s) - x] + \xi'(s)[\eta(s) - y]}{[\xi(s) - x]^2 + [\eta(s) - y]^2}. \qquad (12.2.13)$$

This kernel is increasingly peaked as (x, y) approaches S. To see this more clearly, let S be the unit circle given by $\mathbf{r}(s) = (\cos s, \sin s)$, $0 \le s \le 2\pi$. Then

$$M(x, y, s) = \frac{-\cos s[\cos s - x] - \sin s[\sin s - y]}{[\cos s - x]^2 + [\sin s - y]^2}.$$

To see the near-singular behavior more clearly, let (x, y) approach the point $(\cos s, \sin s)$ along the line

$$(x, y) = q(\cos s, \sin s), \qquad 0 \le q < 1.$$

Then after simplifying,

$$M(q \cos s, q \sin s, s) = \frac{1}{1 - q}.$$

The integrand of (12.2.12) is increasingly peaked as $q \nearrow 1$.

We use numerical integration to approximate (12.2.12); and since the integrand is periodic in s, the trapezoidal rule is an optimal choice when choosing among regular quadrature rules with uniformly distributed quadrature nodes. As (x, y) approaches S, the needed number of integration nodes will need to be increased in order to retain equivalent accuracy in the approximate values of $u(x, y)$. For (x, y) very close to S, other means should be used to approximate the integral (12.2.12), since the trapezoidal rule will be very expensive.

To solve the original Dirichlet problem (12.1.1), we first approximate the density ρ, obtaining ρ_n; and then we numerically integrate the double layer integral based on ρ_n. To aid in studying the resulting approximation of $u(x, y)$, introduce the following notation. Let $u_n(x, y)$ be the double layer potential using the approximate density ρ_n obtained by the Nyström method of (12.2.4):

$$u_n(x, y) = \int_0^L M(x, y, s)\rho_n(s)\, ds, \qquad (x, y) \in D_i. \qquad (12.2.14)$$

Let $u_{n,m}(x, y)$ denote the result of approximating $u_n(x, y)$ using the trapezoidal rule:

$$u_{n,m}(x, y) = h \sum_{i=1}^{m} M(x, y, t_i)\rho_n(t_i), \qquad (x, y) \in D_i. \qquad (12.2.15)$$

For the error in u_n, note that $u - u_n$ is a harmonic function; and therefore, by the maximum principle for such functions,

$$\max_{(x,y) \in \overline{D}_i} |u(x, y) - u_n(x, y)| = \max_{(x,y) \in S} |u(x, y) - u_n(x, y)|. \qquad (12.2.16)$$

Since $u - u_n$ is also a double layer potential, the argument that led to the original integral equation (12.2.1) also implies

$$u(P) - u_n(P) = -\pi[\rho(P) - \rho_n(P)]$$
$$+ \int_S [\rho(Q) - \rho_n(Q)] \frac{\partial}{\partial \mathbf{n}_Q} [\log |P - Q|]\, dS_Q, \qquad P \in S. \qquad (12.2.17)$$

Taking bounds,

$$|u(P) - u_n(P)| \le [\pi + \|K\|]\, \|\rho - \rho_n\|_\infty, \qquad P \in S. \qquad (12.2.18)$$

Combined with (12.2.16),

$$\max_{(x,y)\in D_i} |u(x,y) - u_n(x,y)| \leq [\pi + \|K\|] \, \|\rho - \rho_n\|_\infty. \qquad (12.2.19)$$

If the region D_i is convex, then the double layer kernel is strictly negative; and it can then be shown that

$$\|K\| = \pi. \qquad (12.2.20)$$

For convex regions, therefore,

$$\max_{(x,y)\in \overline{D}_i} |u(x,y) - u_n(x,y)| \leq 2\pi \, \|\rho - \rho_n\|_\infty. \qquad (12.2.21)$$

An algorithm for solving the interior Dirichlet problem (12.1.1) can be based on first solving for ρ_n to a prescribed accuracy. Then (12.2.19) says u_n has comparable accuracy uniformly on D_i. To complete the task of evaluating $u_n(x,y)$ for given values of (x,y), one can use the trapezoidal rule (12.2.15), varying m to obtain desired accuracy in $u_{n,m}(x,y)$. The total error is then given by

$$u(x,y) - u_{n,m}(x,y) = [u(x,y) - u_n(x,y)] + [u_n(x,y) - u_{n,m}(x,y)]. \qquad (12.2.22)$$

Ideally, the two errors on the right side should be made comparable in size, to make the algorithm as efficient as possible. A Fortran program implementing these ideas is given in [15], and it also uses a slight improvement on (12.2.15) when (x,y) is near to S.

r	$m = 32$	$m = 64$	$m = 128$	$m = 256$
0	$-1.34E-2$	$-2.20E-5$	$-1.68E-6$	$-1.68E-6$
.20	$1.60E-2$	$6.89E-5$	$-1.82E-6$	$-1.82E-6$
.40	$1.14E-3$	$1.94E-5$	$-1.49E-7$	$-1.58E-7$
.60	$-7.88E-2$	$-3.5E-3$	$4.63E-6$	$2.22E-6$
.80	$5.28E-1$	$2.33E-2$	$-1.31E-3$	$4.71E-6$
.90	$-1.13E+0$	$4.82E-1$	$3.12E-2$	$-2.64E-4$
.94	$-1.08E+0$	$-8.44E-1$	$2.05E-1$	$1.85E-3$

Table 12.2. Errors $u(\mathbf{c}(q)) - u_{n,m}(\mathbf{c}(q))$ with $n = 32$

Example 12.2.2 *We continue with the preceding example (12.2.8)– (12.2.11), noting that the true solution is also given by (12.2.11). For the case $(a,b) = (1,5)$ and $n = 32$, we examine the error in the numerical solutions u_n and $u_{n,m}$ along the line*

$$\mathbf{c}(q) = q(a\cos\frac{\pi}{4}, b\sin\frac{\pi}{4}), \qquad 0 \leq q < 1. \qquad (12.2.23)$$

A graph of the error $u(\mathbf{c}(q)) - u_n(\mathbf{c}(q))$, $0 \leq q \leq .94$, is shown in Figure 12.3. Note that the size of the error is around 20 times smaller than is

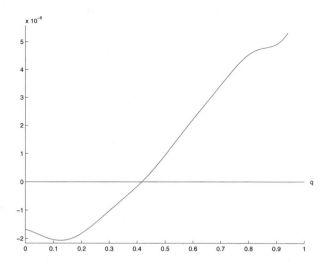

Figure 12.3. The errors $u(\mathbf{c}(q)) - u_n(\mathbf{c}(q))$ with $n = 32$

predicted from the error of $\|\rho - \rho_{32}\|_\infty = 1.74 \times 10^{-5}$ of Table 12.1 and the bound (12.2.21). Table 12.2 contains the errors $u(\mathbf{c}(q)) - u_{n,m}(\mathbf{c}(q))$ for selected values of q and m, with $n = 32$. Graphs of these errors are given in Figure 12.4. Compare these graphs with that of Figure 12.3, noting the quite different vertical scales. It is clear that increasing m decreases the error, up to the point that the dominant error is that of $u(x,y) - u_n(x,y)$ in (12.2.22).

12.2.2   The exterior Neumann problem

Recall the solving of the exterior Neumann problem (12.1.2) by means of the integral representation formula (12.1.25) and the boundary integral equation of (12.1.30). We rewrite the latter as

$$-\pi u(P) + \int_S u(Q) \frac{\partial}{\partial \mathbf{n}_Q} \left[\log |P - Q|\right] dS_Q \qquad (12.2.24)$$
$$= \int_S f(Q) \log |P - Q| \, dS_Q, \qquad P \in S.$$

The left side of this equation is the same as that of (12.2.2) for the interior Dirichlet problem; and it is therefore only the evaluation of the right side which concerns us here. Recalling (12.1.35), the right side is

$$g(t) = \int_0^L f(\mathbf{r}(s)) \sqrt{\xi'(s)^2 + \eta'(s)^2} \log |\mathbf{r}(t) - \mathbf{r}(s)| \, ds. \qquad (12.2.25)$$

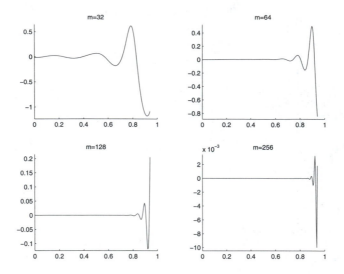

Figure 12.4. The errors $u(\mathbf{c}(q)) - u_{n,m}(\mathbf{c}(q))$ with $n = 32$

This could be approximated using the product integration techniques of Section 11.5 in Chapter 11; but we consider a more efficient method.

To simplify the notation, the parametrization $\mathbf{r}(t)$ of (12.2.1) is assumed to be defined on the standard interval $[0, 2\pi]$. Also, introduce

$$\varphi(s) = f(\mathbf{r}(s))\sqrt{\xi'(s)^2 + \eta'(s)^2}. \tag{12.2.26}$$

The integral (12.2.25) becomes

$$g(t) = \int_0^{2\pi} \varphi(s) \log |\mathbf{r}(t) - \mathbf{r}(s)| \, ds, \qquad 0 \le t \le 2\pi. \tag{12.2.27}$$

We write the kernel of this integral in the form

$$\log |\mathbf{r}(t) - \mathbf{r}(s)| = \log \left| 2e^{-\frac{1}{2}} \sin\left(\frac{t-s}{2}\right) \right| - \pi b(t, s) \tag{12.2.28}$$

with

$$b(t, s) = \begin{cases} -\dfrac{1}{\pi} \log \dfrac{\left| e^{\frac{1}{2}}[\mathbf{r}(t) - \mathbf{r}(s)] \right|}{\left| 2 \sin\left(\dfrac{t-s}{2}\right) \right|}, & t - s \ne 2m\pi, \\[6mm] -\dfrac{1}{\pi} \log \left| e^{\frac{1}{2}} \mathbf{r}'(t) \right|, & t - s = 2m\pi. \end{cases} \tag{12.2.29}$$

The integral (12.2.27) becomes

$$g(t) = -\pi \left[-\frac{1}{\pi} \int_0^{2\pi} \varphi(s) \log \left| 2e^{-\frac{1}{2}} \sin \left(\frac{t-s}{2} \right) \right| ds + \int_0^{2\pi} b(t,s)\varphi(s)\, ds \right]$$

$$\equiv -\pi[\mathcal{A}\varphi(t) + \mathcal{B}\varphi(t)].$$

$$(12.2.30)$$

Assuming $\mathbf{r} \in C_p^\kappa(2\pi)$, the kernel function $b \in C^{\kappa-1}([0,2\pi] \times [0,2\pi])$; and b is periodic in both variables t and s. Consequently, the second integral $\mathcal{B}\varphi(t)$ in (12.2.30) can be accurately and efficiently approximated using the trapezoidal rule.

The first integral in (12.2.30) is a minor modification of the integral operator associated with the kernel $\log |P - Q|$ for S equal to the unit circle about the origin, where we have

$$\mathcal{A}\varphi(t) = -\frac{1}{\pi} \int_0^{2\pi} \varphi(s) \log \left| 2e^{-\frac{1}{2}} \sin \left(\frac{t-s}{2} \right) \right| ds, \qquad 0 \le t \le 2\pi.$$

$$(12.2.31)$$

This operator was introduced in Section 6.5.4 and some properties of it were given there.

In particular,

$$\mathcal{A}\varphi(t) = \frac{1}{\sqrt{2\pi}} \left[\widehat{\varphi}(0) + \sum_{|m|>0} \frac{\widehat{\varphi}(m)}{|m|} e^{imt} \right],$$

$$(12.2.32)$$

based on the Fourier series

$$\varphi(s) = \frac{1}{\sqrt{2\pi}} \sum_{m=-\infty}^{\infty} \widehat{\varphi}(m)e^{ims}$$

for an arbitrary $\varphi \in L^2(0, 2\pi)$. This is an expansion of $\mathcal{A}\varphi$ using the eigenfunctions $\psi_m(t) \equiv e^{imt}$ and the corresponding eigenvalues of $\mathcal{A}$. For a proof of this result, and for a much more extensive discussion of the properties of $\mathcal{A}$, see Yan and Sloan [168].

As noted in Section 6.5.4, (12.2.32) can be used to show that $\mathcal{A}$ is a bijective bounded linear operator from $H^0(2\pi) \equiv L^2(0, 2\pi)$ to $H^1(2\pi)$, with $\|\mathcal{A}\| = 1$ for this mapping. The Sobolev space $H^1(2\pi)$ was introduced in Definition 6.5.1 of Chapter 6. When $\mathcal{A}$ is considered as an operator from $C_p(2\pi)$ to $C_p(2\pi)$, we can show

$$\|\mathcal{A}\| \le \sqrt{1 + \frac{\pi^2}{3}} \doteq 2.07.$$

$$(12.2.33)$$

For a derivation of this last bound, see [13, p. 330].

To approximate $\mathcal{A}\varphi$, we approximate φ using trigonometric interpolation; and then (12.2.32) is used to evaluate exactly the resulting approximation of $\mathcal{A}\varphi$. Let $n \geq 1$, $h = 2\pi/(2n+1)$, and

$$t_j = jh, \quad j = 0, \pm 1, \pm 2, \ldots \tag{12.2.34}$$

Let $\mathcal{Q}_n\varphi$ denote the trigonometric polynomial of degree $\leq n$ which interpolates $\varphi(t)$ at the nodes $\{t_0, t_1, \ldots, t_{2n}\}$, and by periodicity at all other nodes t_j (cf. Theorem 6.5.7 of Chapter 6). Also, let $T_k(\varphi)$ denote the trapezoidal rule on $[0, 2\pi]$ with k subdivisions:

$$T_k(\varphi) = \frac{2\pi}{k} \sum_{j=0}^{k-1} \varphi\left(\frac{2\pi j}{k}\right), \quad \varphi \in C_p(2\pi).$$

The interpolation polynomial can be written as

$$\mathcal{Q}_n\varphi(t) = \sum_{j=-n}^{n} \alpha_j e^{ijt}. \tag{12.2.35}$$

The coefficients $\{\alpha_j\}$ can be obtained as numerical quadratures of the standard Fourier coefficients of φ; see [13, p. 331].

For the error in $\mathcal{Q}_n\varphi$, recall the error bound (3.6.14) in Chapter 3. Then

$$\|\varphi - \mathcal{Q}_n\varphi\|_\infty = O\left(\frac{\log n}{n^{\ell+\alpha}}\right), \quad \varphi \in C_p^{\ell,\alpha}(2\pi). \tag{12.2.36}$$

In this, φ is assumed to be ℓ-times continuously differentiable, and $\varphi^{(\ell)}$ is assumed to satisfy the Hölder condition

$$\left|\varphi^{(\ell)}(s) - \varphi^{(\ell)}(t)\right| \leq c\,|s - t|^\alpha, \quad -\infty < s, t < \infty,$$

with c a finite constant.

We approximate $\mathcal{A}\varphi(t)$ using $\mathcal{A}\mathcal{Q}_n\varphi(t)$. From (12.2.32),

$$\mathcal{A}\psi_j = \begin{cases} 1, & j = 0, \\ \dfrac{1}{|j|}e^{ijt}, & |j| > 0. \end{cases} \tag{12.2.37}$$

Applying this with (12.2.35),

$$\mathcal{A}\varphi(t) \approx \mathcal{A}\mathcal{Q}_n\varphi(t) = \alpha_0 + \sum_{\substack{j=-n \\ j \neq 0}}^{n} \frac{\alpha_j}{|j|} e^{ijt}, \quad -\infty < t < \infty. \tag{12.2.38}$$

To bound the error in $\mathcal{A}\mathcal{Q}_n\varphi$, we apply (12.2.33), yielding

$$\|\mathcal{A}\varphi - \mathcal{A}\mathcal{Q}_n\varphi\|_\infty \leq \|\mathcal{A}\|\,\|\varphi - \mathcal{Q}_n\varphi\|_\infty.$$

Using (12.2.36), this bound implies

$$\|\mathcal{A}\varphi - \mathcal{A}\mathcal{Q}_n\varphi\|_\infty = O\left(\frac{\log n}{n^{\ell+\alpha}}\right), \quad \varphi \in C_p^{\ell,\alpha}(2\pi), \tag{12.2.39}$$

provided $\ell + \alpha > 0$. The approximation $\mathcal{A}\mathcal{Q}_n\varphi$ is rapidly convergent to $\mathcal{A}\varphi$.

To complete the approximation of the original integral (12.2.30), approximate $B\varphi(t)$ using the trapezoidal rule with the nodes $\{t_j\}$ of (12.2.34):

$$B\varphi(t) \approx T_{2n+1}(b(t,\cdot)\varphi)$$

$$= \frac{2\pi}{2n+1} \sum_{k=0}^{2n} b(t,t_k)\varphi(t_k) \qquad (12.2.40)$$

$$\equiv B_n\varphi(t).$$

To bound the error, we can use the standard Euler-MacLaurin error formula [13, p. 285] to show

$$|B\varphi(t) - B_n\varphi(t)| \le O(n^{-\ell}), \qquad \varphi \in C_p^\ell(2\pi). \qquad (12.2.41)$$

This assumes that $\mathbf{r} \in C_p^\kappa(2\pi)$ with $\kappa \ge \ell + 1$.

To solve the original integral equation (12.2.24), we use the Nyström method of (12.2.4)–(12.2.6) based on the trapezoidal numerical integration method with the $2n + 1$ nodes $\{t_0, \ldots, t_{2n}\}$ of (12.2.34). The right side g of (12.2.30) is approximated by using (12.2.38) and (12.2.40), yielding the approximation

$$(-\pi + K_n)u_n = -\pi\left[A\mathcal{Q}_n\varphi + B_n\varphi(t)\right]. \qquad (12.2.42)$$

Error bounds can be produced by combining (12.2.39) and (12.2.41) with the earlier error analysis based on (12.2.7). We leave it as an exercise to show that if $\varphi \in C_p^\ell(2\pi)$ for some $\ell \ge 1$, and if $\mathbf{r} \in C_p^\kappa(2\pi)$ with $\kappa \ge \ell + 1$, then the approximate Nyström solution u_n of (12.2.24) satisfies

$$\|u - u_n\|_\infty \le O\left(\frac{\log n}{n^\ell}\right). \qquad (12.2.43)$$

Example 12.2.3 *We solve the exterior Neumann problem on the region outside the ellipse*

$$\mathbf{r}(t) = (a\cos t, b\sin t), \qquad 0 \le t \le 2\pi.$$

For purposes of illustration, we use a known true solution,

$$u(x,y) = \frac{x}{x^2 + y^2}.$$

This function is harmonic; and $u(x,y) \to 0$ as $x^2 + y^2 \to \infty$. The Neumann boundary data is generated from u. Numerical results for $(a,b) = (1,2)$ are given in Table 12.3; and in it, $m = 2n+1$ is the order of the linear system being solved by the Nyström method. A graph of the error $u(\mathbf{r}(t)) - u_n(\mathbf{r}(t))$ is given in Figure 12.5 for the case $n = 16$.

Exercise 12.2.1 *Derive the integral equation (12.2.9)–(12.2.10) for solving the interior Dirichlet problem over an elliptical domain.*

n	m	$\|u - u_n\|_\infty$
8	17	$3.16E - 2$
16	33	$3.42E - 4$
32	65	$4.89E - 8$
64	129	$1.44E - 15$

Table 12.3. The error $\|u - u_n\|_\infty$ for (12.2.42)

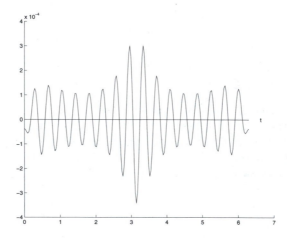

Figure 12.5. The error $u(\mathbf{r}(t)) - u_n(\mathbf{r}(t))$ for $n = 16$ and $(a, b) = (1, 2)$

Exercise 12.2.2 *Write a program to solve (12.2.10), implementing the Nyström method (12.2.4)–(12.2.6), as in Example 12.2.1. Experiment with varying values for n, (a, b), and boundary function f. For the latter, do experiments when f has a singular derivative (with respect to arc-length) on the boundary.*

Exercise 12.2.3 *Using the computation of Exercise 12.1.11, develop and program a numerical method for the parameterization*

$$\mathbf{r}(t) = (2 + \cos t)\,(a\cos t, b\sin t), \qquad 0 \le t \le 2\pi.$$

Do so for a variety of values of the positive constants a, b.

Exercise 12.2.4 *Fill in the details of the arguments for the results given in (12.2.17)–(12.2.19).*

Exercise 12.2.5 *Prove that for D a bounded convex region, $\|K\| = \pi$, thus proving (12.2.20).*

Exercise 12.2.6 *Confirm the formulas (12.2.27)–(12.2.29).*

Exercise 12.2.7 *Assume* $\mathbf{r} \in C_p^{\kappa}(2\pi)$. *Show that the kernel function* $b(t, s)$ *of* (12.2.29) *belongs to* $C^{\kappa-1}([0, 2\pi] \times [0, 2\pi])$ *and that it is periodic in both variables* t *and* s.

Exercise 12.2.8 *Derive the results in the paragraph preceding* (12.2.33). *In particular, show* $\mathcal{A}$ *is a bijective mapping of* $L^2(0, 2\pi)$ *to* $H^1(2\pi)$ *with* $\|\mathcal{A}\| = 1$ *for this mapping.*

12.3   A boundary integral equation of the first kind

Most of the original theoretical work with boundary integral equations was for integral equations of the second kind, and consequently, these types of boundary integral equations came to be the principal type used in applications. In addition, some integral equations of the first kind can be quite ill-conditioned, and this led some people to avoid such equations in general. Finally, numerical methods for integral equations of the first kind were difficult to analyze until somewhat recently.

Boundary integral equations of the first kind, however, are generally quite well behaved; and recently, they have been an increasingly popular approach to solving various boundary value problems. In this section, we look at a well-studied boundary integral equation of the first kind, and we introduce some general analytical tools by means of doing an error analysis of a numerical method for solving this integral equation.

Returning to Section 12.1, the BIE (12.1.28) is an integral equation of the first kind of direct type. Introducing a change of sign, we write this integral equation as

$$-\frac{1}{\pi} \int_S \rho(Q) \log |P - Q| \, dS_Q = g(P), \qquad P \in S. \tag{12.3.1}$$

In this case, the unknown density ρ is the value of the normal derivative on S of the unknown harmonic function u. This integral equation also arises as an indirect BIE for solving the interior Dirichlet problem for Laplace's equation (cf. (12.1.41) when $A \in S$). In this section, we consider various numerical methods for solving this integral equation, building on the ideas introduced in Section 12.2 following (12.2.25).

The solvability theory for (12.3.1) is well-developed, and an excellent presentation of it is given in Yan and Sloan [168]. In particular, if

$$\text{diameter}(D_i) < 1, \tag{12.3.2}$$

then the equation (12.3.1) is uniquely solvable for all $g \in H^1(S)$. [The space $H^1(S)$ is equivalent to the space $H^1(2\pi)$ that was introduced in Definition 6.5.1 of Chapter 6, provided S is a smooth simple closed curve, as is assumed for this chapter.] More generally, the integral equation is uniquely solvable if the equation

$$\int_S \psi(Q) \log |P - Q| \, dS_Q = 1, \qquad P \in S \qquad (12.3.3)$$

does not possess a solution. This is assured if (12.3.2) is satisfied; and since harmonic functions remain such under uniform scalar change of variables, we can assume (12.3.1) with no loss of generality. Curves S for which (12.3.3) has a solution are called "Γ-contours," and they are discussed at length in [168].

Write the first kind boundary integral equation (12.3.1) in the form

$$-\frac{1}{\pi} \int_0^{2\pi} \varphi(s) \log \left| 2e^{-\frac{1}{2}} \sin \left(\frac{t-s}{2} \right) \right| ds$$
$$-\int_0^{2\pi} b(t,s)\varphi(s)\,ds = g(t), \quad 0 \le t \le 2\pi \qquad (12.3.4)$$

with $\varphi(s) \equiv \rho(\mathbf{r}(s)) |\mathbf{r}'(s)|$. This decomposition of the integral operator of (12.3.1) was given earlier in (12.2.27)–(12.2.29). We write (12.3.4) in operator form as

$$\mathcal{A}\varphi + \mathcal{B}\varphi = g. \qquad (12.3.5)$$

Because of the continuity and smoothness properties of b, the operator $\mathcal{B}$ maps $H^q(2\pi)$ into $H^{q+2}(2\pi)$, at least. Using the embedding result that $H^{q+2}(2\pi)$ is compactly embedded in $H^{q+1}(2\pi)$ (cf. Theorem 6.3.11), it follows that $\mathcal{B}$ is a compact operator when considered as an operator from $H^q(2\pi)$ into $H^{q+1}(2\pi)$. Also, recall from (6.5.18) that

$$\mathcal{A} : H^q(2\pi) \overset{1-1}{\underset{onto}{\to}} H^{q+1}(2\pi), \qquad q \ge 0. \qquad (12.3.6)$$

On account of these mapping properties of $\mathcal{A}$ and $\mathcal{B}$, we consider the integral equation (12.3.5) with the assumption $g \in H^{q+1}(2\pi)$; and we seek a solution $\varphi \in H^q(2\pi)$ to the equation.

From (12.3.6), the equation (12.3.5) is equivalent to

$$\varphi + \mathcal{A}^{-1}\mathcal{B}\varphi = \mathcal{A}^{-1}g. \qquad (12.3.7)$$

This is an integral equation of the second kind on $H^q(2\pi)$; and $\mathcal{A}^{-1}\mathcal{B}$ is a compact integral operator when regarded as an operator on $H^q(2\pi)$ into itself. Consequently, the standard Fredholm alternative theorem applies; and if the homogeneous equation $\varphi + \mathcal{A}^{-1}\mathcal{B}\varphi = 0$ has only the zero solution, then the original nonhomogeneous equation has a unique solution for all right sides $\mathcal{A}^{-1}g$. From [168], if S is not a Γ-contour, then the homogeneous version of the original integral equation (12.3.5) has only the zero solution; and thus by means of the Fredholm alternative theorem applied to (12.3.7), the integral equation (12.3.5) is uniquely solvable for all $g \in H^{q+1}(2\pi)$.

12.3.1   A numerical method

We give a numerical method for solving the first kind single layer equation
(12.3.1) in the space $L^2(0, 2\pi)$. The method is a Galerkin method using
trigonometric polynomials as approximations. We assume that the integral
equation (12.3.1) is uniquely solvable for all $g \in H^1(2\pi)$.

For a given $n \geq 0$, introduce

$$V_n = \text{span}\{\psi_{-n}, \dots, \psi_0, \dots, \psi_n\}$$

with $\psi_j(t) = e^{ijt}/\sqrt{2\pi}$; and let $\mathcal{P}_n$ denote the orthogonal projection of
$L^2(0, 2\pi)$ onto V_n (cf. Section 3.6.1). For $\varphi = \sum a_m \psi_m$, it is straightforward
that

$$\mathcal{P}_n \varphi(s) = \sum_{m=-n}^{n} a_m \psi_m(s)$$

the truncation of the Fourier series for φ.

Recall the decomposition (12.3.4)–(12.3.5) of (12.3.1),

$$\mathcal{A}\varphi + \mathcal{B}\varphi = g \tag{12.3.8}$$

with $\mathcal{A}\varphi$ given in (12.2.32). It is immediate that

$$\mathcal{P}_n \mathcal{A} = \mathcal{A}\mathcal{P}_n, \qquad \mathcal{P}_n \mathcal{A}^{-1} = \mathcal{A}^{-1}\mathcal{P}_n. \tag{12.3.9}$$

Approximate (12.3.8) by the equation

$$\mathcal{P}_n(\mathcal{A}\varphi_n + \mathcal{B}\varphi_n) = \mathcal{P}_n g, \qquad \varphi_n \in V_n. \tag{12.3.10}$$

Letting

$$\varphi_n(s) = \sum_{m=-n}^{n} a_m^{(n)} \psi_m(s)$$

and recalling (12.2.34), the equation (12.3.10) implies that the coefficients
$\{a_m^{(n)}\}$ are determined from the linear system

$$\frac{a_k^{(n)}}{\max\{1, |k|\}} + \sum_{m=-n}^{n} a_m^{(n)} \int_0^{2\pi} \int_0^{2\pi} b(t, s)\psi_m(s)\overline{\psi_k(t)}\, ds\, dt$$

$$= \int_0^{2\pi} g(t)\overline{\psi_k(t)}\, dt, \qquad k = -n, \dots, n. \tag{12.3.11}$$

Generally these integrals must be evaluated numerically.

The equation (12.3.8) is equivalent to

$$\varphi + \mathcal{A}^{-1}\mathcal{B}\varphi = \mathcal{A}^{-1}g. \tag{12.3.12}$$

The right-side function $\mathcal{A}^{-1}g \in L^2(0, 2\pi)$, by (12.3.6) and by the earlier
assumption that $g \in H^1(2\pi)$. From the discussion following (12.3.7), $\mathcal{A}^{-1}\mathcal{B}$
is a compact mapping from $L^2(0, 2\pi)$ into $L^2(0, 2\pi)$, and thus (12.3.12) is a
Fredholm integral equation of the second kind. By the earlier assumption on

the unique solvability of (12.3.8), we have $(I+\mathcal{A}^{-1}B)^{-1}$ exists on $L^2(0,2\pi)$ to $L^2(0,2\pi)$.

Using (12.3.9), the approximating equation (12.3.10) is equivalent to

$$\varphi_n + \mathcal{P}_n\mathcal{A}^{-1}B\varphi_n = \mathcal{P}_n\mathcal{A}^{-1}g. \tag{12.3.13}$$

Equation (12.3.13) is simply a standard Galerkin method for solving the equation (12.3.12), and it is exactly of the type discussed in Subsection 11.2.4.

Since $\mathcal{P}_n\varphi \to \varphi$, for all $\varphi \in L^2(0,2\pi)$, and since $\mathcal{A}^{-1}B$ is a compact operator, we have

$$\|(I - \mathcal{P}_n)\mathcal{A}^{-1}B\| \to 0 \quad \text{as} \quad n \to \infty$$

from Lemma 11.1.4 in Subsection 11.1.3 of Chapter 11. Then by standard arguments, the existence of $(I+\mathcal{A}^{-1}B)^{-1}$ implies that of $(I+\mathcal{P}_n\mathcal{A}^{-1}B)^{-1}$, for all sufficiently large n. This is simply a repetition of the general argument given in Theorem 11.1.2, in Subsection 11.1.3. From (11.1.24) of that theorem,

$$\|\varphi - \varphi_n\|_0 \leq \left\|(I + \mathcal{P}_n\mathcal{A}^{-1}B)^{-1}\right\| \|\varphi - \mathcal{P}_n\varphi\|_0, \tag{12.3.14}$$

where $\|\cdot\|_0$ is the norm for $H^0(2\pi) \equiv L^2(0,2\pi)$. For more detailed bounds on the rate of convergence, apply Theorem 6.5.7 of Section 6.5, obtaining

$$\|\varphi - \varphi_n\|_0 \leq \frac{c}{n^q}\|\varphi\|_q, \qquad \varphi \in H^q(2\pi), \tag{12.3.15}$$

for any $q > 0$.

A fully discretized variant of (12.3.13) is given in [13, p. 351], including numerical examples.

Exercise 12.3.1 *Let k be a non-negative integer. Solve the integral equation*

$$-\frac{1}{\pi}\int_0^{2\pi} \varphi(s)\log\left|2e^{-\frac{1}{2}}\sin\left(\frac{t-s}{2}\right)\right| ds = \cos(kt), \qquad 0 \leq t \leq 2\pi.$$

Exercise 12.3.2 *Obtain an explicit formula for the function $b(t,s)$ when the boundary S is the ellipse of (12.2.36). Simplify it as much as possible.*

Suggestion for Further Readings

Parts of this chapter are modifications of portions of ATKINSON [13, Chap. 7]. Chapters 7–9 of the latter contain a more complete and extensive introduction to boundary integral equation reformulations and their numerical solution, again for only Laplace's equation; and a very large set of references are given there. More complete introductions to boundary integral equations and their analysis can be found in KRESS [100], MIKHLIN [119], and POGORZELSKI [130]. From the perspective of applications of BIE, see JASWON AND SYMM [83], MCLEAN [115], and POZRIKIDIS [131].

A comprehensive survey of numerical methods for planar BIE of both the first and second kinds is given by SLOAN [148]. An important approach to the study and solution of BIE, one that we have omitted here, is to regard BIEs as strongly elliptic pseudo-differential operator equations between suitably chosen Sobolev spaces. Doing such, we can apply Galerkin and finite-element methods to the BIE, in much the manner of Chapters 8 and 9. There is no other numerical method known for solving and analyzing some BIEs. As important examples of this work, see WENDLAND [163]–[165] and ARNOLD AND WENDLAND [5]. An introduction is given in [13, Section 7.4].

References

[1] R.A. Adams, *Sobolev Spaces*, Academic Press, New York, 1975.

[2] P. Anselone, *Collectively Compact Operator Approximation Theory and Applications to Integral Equations*, Prentice-Hall, Englewood Cliffs, NJ, 1971.

[3] H. Anton and C. Rorres, *Elementary Linear Algebra*, 7th ed., John Wiley, New York, 1994.

[4] T. Apostol, *Mathematical Analysis*, Addison-Wesley, Reading, MA, 1957.

[5] D. Arnold and W. Wendland, The convergence of spline collocation for strongly elliptic equations on curves, *Numer. Math.* **47** (1985), 317–341.

[6] K. Atkinson, The solution of non-unique linear integral equations, *Numer. Math.* **10** (1967), 117–124.

[7] K. Atkinson, The numerical solution of the eigenvalue problem for compact integral operators, *Trans. Amer. Math. Soc.* **129** (1967), 458–465.

[8] K. Atkinson, The numerical evaluation of fixed points for completely continuous operators, *SIAM J. Num. Anal.* **10** (1973), 799–807.

[9] K. Atkinson, Convergence rates for approximate eigenvalues of compact integral operators, *SIAM J. Num. Anal.* **12** (1975), 213–222.

[10] K. Atkinson, The numerical solution of a bifurcation problem, *SIAM J. Num. Anal.* **14** (1977), 584–599.

[11] K. Atkinson, *An Introduction to Numerical Analysis*, 2nd ed., John Wiley, New York, 1989.

[12] K. Atkinson, A survey of numerical methods for solving nonlinear integral equations, *J. Int. Eqns. & Applics* **4** (1992), 15–46.

[13] K. Atkinson, *The Numerical Solution of Integral Equations of the Second Kind*, Cambridge University Press, Cambridge, 1997.

[14] K. Atkinson, I. Graham, and I. Sloan, Piecewise continuous collocation for integral equations, *SIAM J. Numer. Anal.* **20** (1983), 172–186.

[15] K. Atkinson and Y.-M. Jeon, Algorithm 788: Automatic boundary integral equation programs for the planar Laplace equation, *ACM Trans. Math. Software* **24** (1998), 395–417.

[16] K. Atkinson and F. Potra, Projection and iterated projection methods for nonlinear integral equations, *SIAM J. Numer. Anal.* **24** (1987), 1352–1373.

[17] J.-P. Aubin, *Applied Functional Analysis,* John Wiley, 1979.

[18] O. Axelsson, *Iterative Solution Methods,* Cambridge University Press, Cambridge, 1996.

[19] I. Babuška and A.K. Aziz, Survey lectures on the mathematical foundations of the finite element method, in A.K. Aziz, ed., *The Mathematical Foundations of the Finite Element Method with Applications to Partial Differential Equations,* Academic Press, New York, 1972, 3–359.

[20] J.W. Barrett and W.B. Liu, Finite element approximation of the p-Laplacian, *Math. Comp.* **61** (1993), 523–537.

[21] M. Berger, *Nonlinearity and Functional Analysis,* Academic Press, New York, 1977.

[22] J. Bergh and J. Löfström, *Interpolation Spaces, An Introduction,* Springer-Verlag, Berlin, 1976.

[23] C. Bernardi and Y. Maday, Spectral methods, in P.G. Ciarlet and J.-L. Lions, eds., *Handbook of Numerical Analysis,* Vol. V, North-Holland, Amsterdam, 1997, 209–485.

[24] M. Bernkopf, The development of function spaces with particular reference to their origins in integral equation theory, *Archive for History of Exact Sciences* **3** (1966), 1–96.

[25] G. Birkhoff, M.H. Schultz and R.S. Varga, Piecewise Hermite interpolation in one and two variables with applications to partial differential equations, *Numer. Math.* **11** (1968), 232–256.

[26] C. de Boor, *A Practical Guide to Splines,* Springer-Verlag, New York, 1978.

[27] D. Braess, *Finite Elements: Theory, Fast Solvers, and Applications in Solid Mechanics,* Cambridge University Press, Cambridge, 1997.

[28] S.C. Brenner and L.R. Scott, *The Mathematical Theory of Finite Element Methods,* Springer-Verlag, New York, 1994.

[29] F. Brezzi and M. Fortin, *Mixed and Hybrid Finite Element Methods,* Springer-Verlag, Berlin, 1991.

[30] C. Canuto, M.Y. Hussaini, A. Quarteroni and T.A. Zang, *Spectral Methods in Fluid Dynamics,* Springer-Verlag, New York, 1988.

[31] C. Canuto and A. Quarteroni, Approximation results for orthogonal polynomials in Sobelov spaces, *Math. Comput.* **38** (1982), 67–86.

[32] J. Céa, Approximation variationnelle des problémes aux limites, *Ann. Inst. Fourier (Grenoble)* **14** (1964), 345–444.

[33] F. Chatelin, *Spectral Approximation of Linear Operators,* Academic Press, New York, 1983.

[34] C. Chui, *An Introduction to Wavelets*, Academic Press, New York, 1992.

[35] P.G. Ciarlet, *The Finite Element Method for Elliptic Problems*, North Holland, Amsterdam, 1978.

[36] P.G. Ciarlet, Basic error estimates for elliptic problems, in P.G. Ciarlet and J.-L. Lions, eds., *Handbook of Numerical Analysis*, Vol. II, North-Holland, Amsterdam, 1991, 17–351.

[37] P. Clément, Approximation by finite element functions using local regularization, *RAIRO Anal. Numer.* **9R2** (1975), 77–84.

[38] L. Collatz, *Functional Analysis and Numerical Mathematics*, Academic Press, New York, 1966.

[39] D. Colton, *Partial Differential Equations: An Introduction*, Random House, New York, 1988.

[40] J. Conway, *A Course in Functional Analysis*, 2nd ed., Springer-Verlag, New York, 1990.

[41] C. Cryer, *Numerical Functional Analysis*, Clarendon Press, Oxford, 1982.

[42] P. Davis, *Interpolation and Approximation*, Blaisdell, New York, 1963.

[43] J. Demmel, *Applied Numerical Linear Algebra*, SIAM Pub., Philadelphia, 1997.

[44] N. Dunford and J. Schwartz, *Linear Operators—Part I: General Theory*, Interscience Pub., New York, 1964.

[45] G. Duvaut and J.-L. Lions, *Inequalities in Mechanics and Physics*, Springer-Verlag, Berlin, 1976.

[46] R. Edwards, *Functional Analysis*, Holt, Rinehart and Winston, New York, 1965.

[47] I. Ekeland and R. Temam, *Convex Analysis and Variational Problems*, North-Holland, Amsterdam, 1976.

[48] L.C. Evans, *Partial Differential Equations*, Berkeley Mathematics Lecture Notes, Volume 3, 1994.

[49] R.S. Falk, Error estimates for the approximation of a class of variational inequalities, *Math. Comp.* **28** (1974), 963–971.

[50] I. Fenyö and H. Stolle, *Theorie und Praxis der linearen Integralgleichungen—2*, Birkhäuser-Verlag, Basel, 1983.

[51] G. Fichera, Problemi elastostatici con vincoli unilaterali: il problema di Signorini con ambigue condizioni al contorno, *Mem. Accad. Naz. Lincei* **8** (7) (1964), 91–140.

[52] J. Flores, The conjugate gradient method for solving Fredholm integral equations of the second kind, *Intern. J. Computer Math.* **48** (1993), 77–94.

[53] J. Franklin, *Methods of Mathematical Economics*, Springer-Verlag, New York, 1980.

[54] A. Friedman, *Variational Principles and Free-boundary Problems*, John Wiley, New York, 1982.

[55] W. Freeden, T. Gervens, and M. Schreiner, *Constructive Approximation on the Sphere, with Applications to Geomathematics*, Oxford Univ. Press, Oxford, 1998.

[56] R. Freund, F. Golub, and N. Nachtigal, Iterative solution of linear systems, *Acta Numerica—1992*, Cambridge University Press, Cambridge, pp. 57–100.

[57] V. Girault and P.-A. Raviart, *Finite Element Methods for Navier-Stokes Equations, Theory and Algorithms*, Springer-Verlag, Berlin, 1986.

[58] R. Glowinski, *Numerical Methods for Nonlinear Variational Problems*, Springer-Verlag, New York, 1984.

[59] R. Glowinski, J.-L. Lions and R. Trémolières, *Numerical Analysis of Variational Inequalities*, North-Holland, Amsterdam, 1981.

[60] E. Godlewski and P.-A. Raviart, *Numerical Approximation of Hyperbolic Systems of Conservation Laws*, Springer-Verlag, New York, 1996.

[61] G.H. Golub and C.F. Van Loan, *Matrix Computations*, 3rd edition, The Johns Hopkins University Press, Baltimore, 1996.

[62] D. Gottlieb and S.A. Orszag, *Numerical Analysis of Spectral Methods: Theory and Applications*, SIAM, Philadelphia, 1977.

[63] I. Graham, Singularity expansions for the solution of second kind Fredholm integral equations with weakly singular convolution kernels, *J. Integral Eqns* **4** (1982), 1–30.

[64] P. Grisvard, *Elliptic Problems in Nonsmooth Domains*, Pitman, Boston, 1985.

[65] C. Groetsch, *Inverse Problems in the Mathematical Sciences*, Vieweg Pub., Braunschweig/Wiesbaden, 1993.

[66] T. Gronwall, On the degree of convergence of Laplace's series, *Trans. Amer. Math. Soc.* **15**(1914), 1–30.

[67] C.A. Hall and T.A. Porsching, *Numerical Analysis of Partial Differential Equations*, Prentice Hall, Englewood Cliffs, NJ, 1990.

[68] W. Han, The best constant in a trace inequality, *Journal of Mathematical Analysis and Applications* **163** (1992), 512–520.

[69] W. Han, Finite element analysis of a holonomic elastic-plastic problem, *Numer. Math.* **60** (1992), 493–508.

[70] W. Han, S. Jensen, and B.D. Reddy, Numerical approximations of internal variable problems in plasticity: Error analysis and solution algorithms, *Numerical Linear Algebra with Applications* **4** (1997), 191–204.

[71] W. Han and B.D. Reddy, On the finite element method for mixed variational inequalities arising in elastoplasticity, *SIAM J. Numer. Anal.* **32** (1995), 1778–1807.

[72] W. Han and B.D. Reddy, *Plasticity: Mathematical Theory and Numerical Analysis*, Springer-Verlag, New York, 1999.

[73] W. Han, B.D. Reddy, and G.C. Schroeder, Qualitative and numerical analysis of quasistatic problems in elastoplasticity, *SIAM J. Numer. Anal.* **34** (1997), 143–177.

[74] P. Henrici, *Applied and Computational Complex Analysis*, Vol. 3, John Wiley, New York, 1986.

[75] J. Haslinger, I. Hlaváček and J. Nečas, Numerical methods for unilateral problems in solid mechanics, in P.G. Ciarlet and J.-L. Lions, eds., *Handbook of Numerical Analysis*, Vol. IV, North-Holland, Amsterdam, 1996, 313–485.

[76] N. Higham, *Accuracy and Stability of Numerical Algorithms*, SIAM, Philadelphia, 1996.

[77] E. Hille and J. Tamarkin, On the characteristic values of linear integral equations, *Acta Math.* **57** (1931), 1–76.

[78] I. Hlaváček, J. Haslinger, J. Nečas and J. Lovíšek, *Solution of Variational Inequalities in Mechanics*, Springer-Verlag, New York, 1988.

[79] F. de Hoog and R. Weiss, Asymptotic expansions for product integration, *Math. Comp.* **27** (1973), 295–306.

[80] H. Huang, W. Han, and J. Zhou, The regularization method for an obstacle problem, *Numer. Math.* **69** (1994), 155–166.

[81] V. Hutson and J.S. Pym, *Applications of Functional Analysis and Operator Theory*, Academic Press, London, 1980.

[82] E. Isaacson and H. Keller, *Analysis of Numerical Methods*, John Wiley, New York, 1966.

[83] M. Jaswon and G. Symm, *Integral Equation Methods in Potential Theory and Elastostatics*, Academic Press, 1977.

[84] C. Johnson, *Numerical Solutions of Partial Differential Equations by the Finite Element Method*, Cambridge University Press, Cambridge, 1987.

[85] G. Kaiser, *A Friendly Guide to Wavelets*, Birkhäuser, Boston, 1994.

[86] S. Kakutani, Topological properties of the unit sphere of Hilbert space, *Proc. Imp. Acad. Tokyo* **19** (1943), 269–271.

[87] L. Kantorovich, Functional analysis and applied mathematics, *Uspehi Mat. Nauk* **3** (1948), 89–185.

[88] L. Kantorovich and G. Akilov, *Functional Analysis in Normed Spaces*, 2nd ed., Pergamon Press, New York, 1982.

[89] L. Kantorovich and V. Krylov *Approximate Methods of Higher Analysis*, Noordhoff, Groningen, 1964.

[90] H. Kardestuncer and D.H. Norrie (eds.), *Finite Element Handbook*, McGraw-Hill Book Company, New York, 1987.

[91] C. T. Kelley, *Iterative Methods for Linear and Nonlinear Equations*, SIAM, Philadelphia, 1995.

[92] C. T. Kelley, *Iterative Methods for Optimization*, SIAM, Philadelphia, 1999.

[93] O. Kellogg, *Foundations of Potential Theory*, reprinted by Dover Pub. 1929.

[94] S. Kesavan, *Topics in Functional Analysis and Applications*, John Wiley, New Delhi, 1989.

[95] N. Kikuchi and J.T. Oden, *Contact Problems in Elasticity: A Study of Variational Inequalities and Finite Element Methods*, SIAM, Philadelphia, 1988.

[96] D. Kinderlehrer and G. Stampacchia, *An Introduction to Variational Inequalities and their Applications*, Academic Press, New York, 1980.

[97] V.A. Kozlov, V.G. Maz'ya and J. Rossmann, *Elliptic Boundary Value Problems in Domains with Point Singularities*, American Mathematical Society, 1997.

[98] M. A. Krasnosel'skii, *Topological Methods in the Theory of Nonlinear Integral Equations*, Pergamon Press, New York, 1964.

[99] M. Krasnosel'skii and P. Zabreyko, *Geometric Methods of Nonlinear Analysis*, Springer-Verlag, Berlin, 1984.

[100] R. Kress, *Linear Integral Equations*, Springer-Verlag, Berlin, 1989.

[101] R. Kress, *Numerical Analysis*, Springer, New York, 1998.

[102] R. Kress and I. Sloan, On the numerical solution of a logarithmic integral equation of the first kind for the Helmholtz equation, *Numerische Math.* **66** (1993), 199–214.

[103] E. Kreyszig, *Introductory Functional Analysis with Applications*, John Wiley, New York, 1978.

[104] S. Kumar and I. Sloan, A new collocation-type method for Hammerstein equations, *Math. Comp.* **48** (1987), 585–593.

[105] V.I. Lebedev, *An Introduction to Functional Analysis in Computational Mathematics*, Birkhäuser, Boston, 1997.

[106] D. Luenberger, *Linear and Nonlinear Programming*, 2nd ed., Addison-Wesley, Reading, MA., 1984.

[107] R.J. LeVeque, *Numerical Methods for Conservation Laws*, 2nd ed., Birkhäuser, Boston, 1992.

[108] P. Linz, *Theoretical Numerical Analysis, An Introduction to Advanced Techniques*, John Wiley, New York, 1979.

[109] J.L. Lions and E. Magenes, *Nonhomogeneous Boundary Value Problems and Applications*, three volumes, Springer-Verlag, New York, 1968.

[110] J.-L. Lions and G. Stampacchia, Variational inequalities, *Comm. Pure Appl. Math.* **20** (1967), 493–519.

[111] T. MacRobert, *Spherical Harmonics*, 3rd ed., Pergamon Press, New York, 1967.

[112] Y. Maday and A. Quarteroni, Legendre and Chebyshev spectral approximations of Burgers' equation, *Numer. Math.* **37** (1981), 321–332.

[113] G. I. Marchuk, Splitting and alternating direction methods, in P.G. Ciarlet and J.-L. Lions, eds., *Handbook of Numerical Analysis*, Vol. I, North-Holland, Amsterdam, 1990, 197–464.

[114] G.I. Marchuk and V.V. Shaidurov, *Difference Methods and Their Extrapolations*, Springer-Verlag, New York, 1983.

[115] W. McLean, *Strongly Elliptic Systems and Boundary Integral Equations*, Cambridge University Press, Cambridge, 2000.

[116] R. McOwen, *Partial Differential Equations, Methods and Applications*, Prentice Hall, NJ, 1996.

[117] G. Meinardus, *Approximation of Functions: Theory and Numerical Methods*, Springer-Verlag, New York, 1967.

[118] S. Mikhlin, *Integral Equations*, 2nd ed., Pergamon Press, New York, 1964.

[119] S. Mikhlin, *Mathematical Physics: An Advanced Course*, North-Holland Pub., Amsterdam, 1970.

[120] L. Milne-Thomson, *Theoretical Hydrodynamics*, 5th ed., Macmillan, New York, 1968.

[121] R.E. Moore, *Computational Functional Analysis*, Ellis Horwood Limited, Chichester, 1985.

[122] J. Nečas, Sur une méthode pour resoudre les equations aux derivées partielles du type elliptique, voisine de la variationnelle, *Ann. Scuola Norm. Sup. Pisa* **16** (1962), 305–326.

[123] J. Nečas and I. Hlavaček, *Mathematical Theory of Elastic and Elastoplastic Bodies: An Introduction*, Elsevier, Amsterdam, 1981.

[124] O. Nevanlinna, *Convergence of Iterations for Linear Equations*, Birkhäuser, Basel, 1993.

[125] E. Nyström, Über die praktische Auflösung von Integralgleichungen mit Anwendungen auf Randwertaufgaben, *Acta Math.* **54** (1930), 185–204.

[126] J.T. Oden and J.N. Reddy, *An Introduction to the Mathematical Theory of Finite Elements*, John Wiley, New York, 1976.

[127] J. Ortega and W. Rheinboldt, *Iterative Solution of Nonlinear Equations in Several Variables*, Academic Press, New York, 1970.

[128] P.D. Panagiotopoulos, *Inequality Problems in Mechanics and Applications*, Birkhäuser, Boston, 1985.

[129] W. Patterson, *Iterative Methods for the Solution of a Linear Operator Equation in Hilbert Space—A Survey*, Springer Lecture Notes in Mathematics, No. 394, Springer-Verlag, 1974.

[130] W. Pogorzelski, *Integral Equations and Their Applications*, Pergamon Press, New York, 1966.

[131] C. Pozrikidis, *Boundary Integral and Singularity Methods for Linearized Viscous Flow*, Cambridge Univ. Press, Cambridge, 1992.

[132] A. Quarteroni and A. Valli, *Numerical Approximation of Partial Differential Equations*, Springer-Verlag, New York, 1994.

[133] B.D. Reddy, *Introductory Functional Analysis with Applications to Boundary Value Problems and Finite Elements*, Springer, New York, 1998.

[134] B.D. Reddy and T.B. Griffin, Variational principles and convergence of finite element approximations of a holonomic elastic-plastic problem, *Numer. Math.* **52** (1988), 101–117.

[135] D. Ragozin, Constructive polynomial approximation on spheres and projective spaces, *Trans. Amer. Math. Soc.* **162** (1971), 157–170.

[136] D. Ragozin, Uniform convergence of spherical harmonic expansions, *Math. Annalen* **195** (1972), 87–94.

[137] J. Rice, On the degree of convergence of nonlinear spline approximation, in *Approximations with Special Emphasis on Spline Functions*, ed. by I. J. Schoenberg, Academic Press, New York, 1969, 349–365.

[138] T. Rivlin, *Chebyshev Polynomials*, 2nd ed., John Wiley, New York, 1990.

[139] J.E. Roberts and J.-M. Thomas, Mixed and hybrid methods, in P.G. Ciarlet and J.-L. Lions, eds., *Handbook of Numerical Analysis*, Vol. II, North-Holland, Amsterdam, 1991, 523–639.

[140] R.T. Rockafellar, *Convex Analysis*, Princeton University Press, Princeton, NJ, 1970.

[141] H. Royden, *Real Analysis*, 3rd ed., Collier-MacMillan, New York, 1988.

[142] W. Rudin, *Real and Complex Analysis*, 3rd ed., McGraw-Hill, New York, 1987.

[143] W. Rudin, *Functional Analysis*, 2nd ed., McGraw-Hill, New York, 1991.

[144] C. Schneider, Regularity of the solution to a class of weakly singular Fredholm integral equations of the second kind, *Integral Eqns & Operator Thy* **2** (1979), 62–68.

[145] C. Schneider, Product integration for weakly singular integral equations, *Math. Comp.* **36** (1981), 207–213.

[146] Ch. Schwab, *p- and hp-Finite Element Methods*, Oxford University Press, Oxford, 1998.

[147] I. Sloan, Improvement by iteration for compact operator equations, *Math. Comp.* **30** (1976), 758–764.

[148] I. Sloan, Error analysis of boundary integral methods, *Acta Numerica* **1** (1992), 287–339.

[149] I. Stakgold, *Green's Functions and Boundary Value Problems*, John Wiley, New York, 1979.

[150] E.M. Stein, *Singular Integrals and Differentiability Properties of Functions*, Princeton University Press, Princeton, NJ, 1970.

[151] F. Stenger, *Numerical Methods Based on Sinc and Analytic Functions*, Springer-Verlag, New York, 1993.

[152] G.W. Stewart, *Introduction to Matrix Computations*, Academic Press, New York, 1973.

[153] G. Strang and G. Fix, *An Analysis of the Finite Element Method*, Prentice-Hall, Englewood Cliffs, NJ, 1973.

[154] J.C. Strikwerda, *Finite Difference Schemes and Partial Differential Equations*, Wadsworth & Brooks/Cole Advanced Books & Software, 1989.

[155] B. Szabó and I. Babuška, *Finite Element Analysis*, John Wiley, Inc., New York, 1991.

[156] G. Szegö, *Orthogonal Polynomials*, revised edition, American Mathematical Society, New York, 1959.

[157] V. Thomée, *Galerkin Finite Element Methods for Parabolic Problems*, Springer Lecture Notes in Mathematics, No. 1054, 1984.

[158] V. Thomée, Finite difference methods for linear parabolic equations, in P.G. Ciarlet and J.-L. Lions, eds., *Handbook of Numerical Analysis*, Vol. I, North-Holland, Amsterdam, 1990, 5–196.

[159] V. Thomée, *Galerkin Finite Element Methods for Parabolic Problems*, Springer, New York, 1997.

[160] L. Trefethen and D. Bau, *Numerical Linear Algebra*, SIAM Pub., Philadelphia, 1997.

[161] H. Triebel, *Interpolation Theory, Function Spaces, Differential Operators*, North-Holland Publ. Comp., Amsterdam, 1978.

[162] Lie-heng Wang, Nonconforming finite element approximations to the unilateral problem, *Journal of Computational Mathematics* **17** (1999), 15–24.

[163] W. Wendland, Boundary element methods and their asymptotic convergence, in *Theoretical Acoustics and Numerical Techniques*, ed. by P. Filippi, Springer-Verlag. CISM Courses and Lectures No. 277, International Center for Mechanical Sciences, 1983.

[164] W. Wendland, Strongly elliptic boundary integral equations, in *The State of the Art in Numerical Analysis*, A. Iserles and M. Powell, eds., Clarendon Press, Oxford, 1987, 511–562.

[165] W. Wendland, Boundary element methods for elliptic problems, in *Mathematical Theory of Finite and Boundary Element Methods*, by A. Schatz, V. Thomée, and W. Wendland, Birkhäuser, Boston, 1990, pp. 219–276.

[166] R. Winther, Some superlinear convergence results for the conjugate gradient method, *SIAM J. Numer. Anal.* **17** (1980), 14–17.

[167] J. Wloka, *Partial Differential Equations*, Cambridge University Press, Cambridge, 1987.

[168] Y. Yan and I. Sloan, On integral equations of the first kind with logarithmic kernels, *J. Integral Eqns & Applics* **1** (1988), 517–548.

[169] N. N. Yanenko, *The Method of Fractional Steps*, Springer-Verlag, New York, 1971.

[170] E. Zeidler, *Nonlinear Functional Analysis and its Applications. I: Fixed-point Theorems*, Springer-Verlag, New York, 1985.

[171] E. Zeidler, *Nonlinear Functional Analysis and its Applications. II/A: Linear Monotone Operators*, Springer-Verlag, New York, 1990.

[172] E. Zeidler, *Nonlinear Functional Analysis and its Applications. II/B: Nonlinear Monotone Operators*, Springer-Verlag, New York, 1990.

[173] E. Zeidler, *Nonlinear Functional Analysis and its Applications. III: Variational Methods and Optimization*, Springer-Verlag, New York, 1986.

[174] E. Zeidler, *Applied Functional Analysis: Applications of Mathematical Physics*, Springer-Verlag, New York, 1995.

[175] E. Zeidler, *Applied Functional Analysis: Main Principles and Their Applications*, Springer-Verlag, New York, 1995.

[176] A. Zygmund, *Trigonometric Series*, Vols. I and II, Cambridge Univ. Press, Cambridge, 1959.

Index

Texts in Applied Mathematics

(continued from page ii)